TABOR HIGH SCHOOL,
BRAINTREE.

BIOLOGY STUDY GUIDE II

Revised Nuffield Advanced Science

General editor,
Revised Nuffield
Advanced Biology
Grace Monger

Editors of Part Three,
'Inheritance and development'
John A. Barker
Grace Monger
Ianto Stevens

Editors of Part Four,
'Ecology and evolution'
Dr T. J. King
Grace Monger

Contributors to this book
John A. Barker
S. L. Brooks
Dr Alun Brown
Dr A. F. Dyer
Peter L. Forey
Wilma George
Graham Goldsworthy
Professor Brian Goodwin
Professor A. Hallam
Carolyn Halliday
Dr Tim Halliday
Dr J. W. Hannay
Dr Jennifer Jones
Dr T. J. King
Grace Monger
Dr Alan Radford
Ianto Stevens
Dr J. A. Bird Stewart
Dr Peter Thorogood
Cheryll Tickle
Dr A. J. Trewavas
Professor L. Wolpert, F.R.S.

The General Editor would like to acknowledge with thanks the helpful advice of Professor John Harper on Chapters 26, 27, 28, and 29, and that of Dr Geoffrey Harper on Chapter 30.

The Nuffield–Chelsea Curriculum Trust is grateful to the authors and editors of the first edition:

Organizers, P. J. Kelly, W. H. Dowdeswell; **Editors**, John A. Barker, John H. Gray, P. J. Kelly, Margaret K. Sands, C. F. Stoneman; **Contributors**, John A. Barker, L. C. Comber, J. F. Eggleston, Dr P. Fleetwood-Walker, W. H. Freeman, Peter Fry, Dr R. Gliddon, John H. Gray, Stephen W. Hurry, P. J. Kelly, R. E. Lister, Dr R. Lowery, Diana E. Manuel, Brian Mowl, M. B. V. Roberts, Margaret K. Sands, C. F. Stoneman, K. O. Turner, Dr A. Upshall.

BIOLOGY STUDY GUIDE II

PART THREE INHERITANCE AND DEVELOPMENT

PART FOUR ECOLOGY AND EVOLUTION

Revised Nuffield Advanced Science
Published for the Nuffield–Chelsea Curriculum Trust
by Longman Group Limited

Longman Group Limited
Longman House, Burnt Mill, Harlow, Essex CM20 2JE, England and
Associated Companies throughout the World

First published 1970
Revised edition first published 1986
Copyright © 1970, 1986. The Nuffield–Chelsea Curriculum Trust

Design and art direction by Ivan Dodd
New diagrams by Oxford Illustrators Limited

Filmset in Times Roman and Univers
and made and printed in Great Britain
by Hazell Watson & Viney Limited, Aylesbury

ISBN 0 582 35432 3

All rights reserved. No part of this publication may be reproduced, stored in a retrieval system, or transmitted in any form or by any means – electronic, mechanical, photocopying, recording or otherwise – without the prior written permission of the Publishers.

The Nuffield–Chelsea Curriculum Trust acknowledges with thanks the permission granted by the Joint Matriculation Board for the reproduction of some material from past Nuffield Advanced Biology examination papers.

Cover photograph

Scale on a rainbow trout. *Salmo gairdneri* (× 4500).

Nick Taylor/London Scientific Films Ltd.

CONTENTS

Foreword page *vi*

To the students who use this book *viii*

Part Three **INHERITANCE AND DEVELOPMENT**

Chapter 15 Cell development and differentiation page *1*

Chapter 16 The cell nucleus and inheritance *31*

Chapter 17 Variation and its causes *60*

Chapter 18 The nature of genetic material *93*

Chapter 19 Gene action *125*

Chapter 20 Population genetics and selection *154*

Chapter 21 The principles and applications of biotechnology *192*

Chapter 22 Methods of reproduction *216*

Chapter 23 The nature of development *274*

Chapter 24 Control and integration through the internal environment *312*

Chapter 25 Development and the external environment *377*

Part Four **ECOLOGY AND EVOLUTION**

Chapter 26 The organism and its environment *415*

Chapter 27 Organisms and their biotic environments *451*

Chapter 28 Population dynamics *480*

Chapter 29 Communities and ecosystems *509*

Chapter 30 Evolution *547*

Index *579*

FOREWORD

When the Nuffield Advanced Science series first appeared on the market in 1970, they were rapidly accepted as a notable contribution to the choices for the sixth form science curriculum. Devised by experienced teachers working in consultation with the universities and examination boards, and subjected to extensive trials in schools before publication, they introduced a new element of intellectual excitement into the work of A-level students. Though the period since publication has seen many debates on the sixth form curriculum, it is now clear that the Advanced Level framework of education will be with us for some years in its established form. That period saw various proposals for change in structure which were not accepted but the debate to which we contributed encouraged us to start looking at the scope and aims of our A-level courses and at the ways they were being used in schools. Much of value was learned during those investigations and has been extremely useful in the planning of the present revision. The time since first publication has also seen a remarkable expansion in the number of candidates taking A-level biology and it is encouraging to us to know that we helped in this development.

The revision of the biology series under the general editorship of Grace Monger has been conducted with the help of a committee under the chairmanship of Arthur Lucas, Professor of Curriculum Studies, CSME, Chelsea College, University of London. We are grateful to him and to the committee. We also owe a considerable debt to the Joint Matriculation Board which for many years has been responsible for the special Nuffield examinations in biology, and to the representatives of the Board who sat on the advisory committee and who have given help in many other ways.

The Nuffield–Chelsea Curriculum Trust is also grateful for the advice and recommendations received from its Advisory Committee, a body containing representatives from the teaching profession, the Association for Science Education, Her Majesty's Inspectorate, universities, and local authority advisers; the committee is under the chairmanship of Professor P. J. Black, academic adviser to the Trust.

Our appreciation also goes to the editors and authors of the first edition of Nuffield Advanced Biological Science, who worked under the joint direction of W. H. Dowdeswell and P. J. Kelly, the project organizers. Their team of editors and writers included John A. Barker, John H. Gray, Margaret K. Sands, and C. F. Stoneman. The present revision has only been possible because of their original work.

I particularly wish to record our gratitude to Grace Monger, the General Editor of the revision. This is the second occasion on which we have asked her to undertake the revision of one of our biology series, as she was responsible for the highly successful O-level Biology revision. We are therefore doubly grateful to Miss C. M. Holland, Headmistress of the Holt School, Wokingham and the Berkshire Education Authority for agreeing to her secondment. Grace Monger has had a particularly onerous task because the many topics that biology covers have been subject to an exceptional number of changes and new discoveries in recent years. She and her team of editors have been fortunate in being able to draw on the help, as writers and consultants, of experts in their fields in universities, teaching hospitals, and other institutions of learning. To Grace Monger and her editors, John A. Barker, T. J. King, M. B. V. Roberts, Ianto Stevens, Tim Turvey, and Colin Wood-Robinson, and to the many contributors, we offer our most sincere thanks.

I would also like to acknowledge the work of William Anderson, publications manager to the Trust, his colleagues, and our publishers, the Longman Group, for their assistance in the publication of these books. The editorial and publishing skills they contribute are essential to effective curriculum development.

K. W. Keohane
Chairman, Nuffield–Chelsea Curriculum Trust

TO THE STUDENTS WHO USE THIS BOOK

This *Study guide* is one of the revised Nuffield Advanced Biology publications. It continues a special approach to biology which was developed in the original edition first published in 1970. Through this approach it was hoped that students would gain a broad knowledge of biological science, an understanding of the processes by which biologists acquire knowledge, and awareness of the significance of the subject to human society. It was also hoped that the publications would give students the opportunity to apply the knowledge they gained, in a creative manner.

In writing the new edition the editors have been able to draw on the experience of many teachers and students who have used the first edition. They found that the approach of the first edition was successful in developing an understanding of biological principles and they also found that there were several ways in which students could make better use of the materials. These suggestions and improvements have been incorporated in the new publications.

The *Study guide* now contains substantial passages of descriptive text on a range of biological subjects, in addition to a collection of biological data which are presented as Study items. The different kinds of investigation will provide you with experience in:

1 Critical reading, writing, and discussion on biological issues.
2 Making observations and asking relevant questions about them.
3 Analysing data and drawing conclusions from them.
4 Handling quantitative information and assessing error.
5 Working out hypotheses and assessing those proposed by others.
6 Designing investigations.
7 Evaluating the implications of biological knowledge for society.

Whatever method is used to establish a concept the intention is the same – that it should lead to an understanding of that concept. Questions are asked in the text as well as in Study items and we hope you will be stimulated to ask further questions and so learn more about the subject.

The *Study guide* inevitably contains data collected by other people; your own practical investigations will enable you to collect data for yourselves. Suggestions for relevant practical investigations are in a series of *Practical guides* which are cross-referenced to the *Study guide* so that knowledge gained from these two sources can be linked.

It is inevitable, in studying any subject, to view it initially as a series of topics. It is nevertheless important, from time to time, to stop and take a broader view and to see how topics are related and what underlying principles apply to them all. The following questions should help you make these connections.

1 To what extent are organisms similar and different?

2 How are the organisms adapted in their environment?
3 How are the matter and energy contained in organisms obtained, utilized, and replaced in relation to the environment of the organisms?
4 How does an organism function as a whole?
5 How are the structure of organisms related to their functions?
6 How do organisms maintain themselves in a balanced state both within themselves and with their environment?
7 How is one generation of organisms related to the next?
8 How do organisms develop?
9 What are the features of biological investigations?
10 How is biological research affected by the people who undertake it and the society in which they live?
11 In what way do the findings of research biologists influence society?
12 How can you link work in biological science with that in other subjects?
13 How has biological knowledge developed in the past and how is it developing now?

It is worth remembering that to arrive at a satisfactory answer one must ask the right question. When using this *Study guide* and when carrying out the relevant practical investigations we hope you will ask many questions and find their answers. Above all, we hope that as a result of using these materials you will find the study of biology a satisfying and exciting experience, and one that will continue to interest and fascinate you.

Grace Monger
General Editor

PART THREE INHERITANCE AND DEVELOPMENT

Note
Throughout this **Study guide** the end of a Study item is indicated by the symbol □.

CHAPTER 15 CELL DEVELOPMENT AND DIFFERENTIATION

15.1 Introduction

Before we can claim to understand living processes we must explain the continuity of life. A famous example of this continuity is the brachiopod *Lingula* which seems to have changed hardly at all if one compares fossil shells about 600 million years old with those from living specimens found in coastal waters of the Pacific today. There are many other examples in the fossil record of organisms which, judging from the parts which have been preserved, have changed little over millions of years and countless generations (see *figure 1*). Much more familiar examples of continuity are the 'thoroughbred' domestic animals and named varieties of plants which reliably breed 'true to type'. The resemblance we see daily between parents and children again demonstrates continuity.

Life does not only show continuity but incredible diversity. This diversity shows itself in two ways; first there is the enormous number of different types of organism, each type showing its own characteristic range of individual variation. Second, there is the range of cell types which differentiate in the development of a single individual.

This chapter will be concerned with the mechanism which ensures that characteristics of a cell pass unchanged to its descendants – the mechanism underlying continuity. It will also be concerned with cell differentiation. Both these themes will reappear and be explored further in later chapters.

We might start off by asking why cell diversity evolved in the first place. It seems highly probable that it was a consequence of achieving a multicellular state. A unicellular organism has to cope with all the

Figure 1
The tooth on the right is from a tiger shark, *Galeocerdo cuvieri*, from the Western Atlantic. The one in the centre is a Miocene fossil approximately 30 million years old from North Carolina, USA, and the one on the left is an Eocene fossil about 50 million years old from Gosport, Hampshire. The sizes are different but the structures clearly very similar. **Scale:** 1 square = 1 cm^2.
Photograph, D. J. Kemp Collection, Gosport Museum, Hampshire.

Figure 2
The green alga, *Pleodorina californica* (× 350).
Photograph, Biophoto Associates.

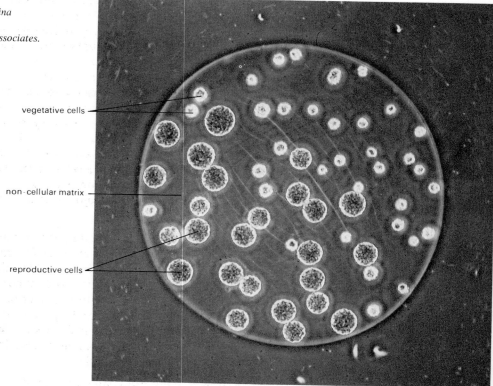

demands of independent life, such as digestion, osmotic balance, respiration, reproduction, and sometimes motility. The transition to a multicellular organization, however, opens up the possibility of division of labour between cell types. A limited range of cell diversity can be seen even among some of the simplest multicellular organisms. Green algae such as *Pandorina* and *Pleodorina*, which may consist of as few as 8 and 32 cells respectively, show a division of labour between vegetative flagellated cells and reproductive cells involved in asexual and sexual propagation. As the degree of multicellularity increases, so the range of cell diversity widens; in other words, the 'labour' of survival and effective propagation is divided up progressively among the various populations of cells which comprise a particular organism. Consequently, in the so-called 'higher' plants and animals, we find a multiplicity of different cell types, each one specialized for support, or regulation of water loss, or any one of a large number of functions. (*Homo sapiens* is estimated to have a repertoire of approximately 120 cell types.) Although such specialization permits greater efficiency, it is achieved at a price. Unicellular organisms are autonomous units which can often survive in a wide range of environmental conditions. In contrast, cells of a multicellular organism are not autonomous or even semi-autonomous, but heavily dependent upon each other for their well-being and survival. Thus single cells isolated from metazoans cannot normally survive outside the complex and carefully stabilized internal environment of the organism. (Exceptions to this are, first, the growth of cells in the rather specialized and unnatural conditions of tissue culture and, second, the release of gametes into the external environment by those organisms displaying external fertilization.) Even the vegetative cells of the algae mentioned earlier are unable to survive disruption of the colonial mode of life. We can conclude that cell diversity and interdependence between cells are essential features of the multicellular state.

If a single cell, the zygote, gives rise to the whole organism it follows that the zygote must contain information ('instructions') specifying the various materials which the organism will synthesize. There must also be information determining the types of cell which will differentiate and the way these cell types will become arranged into tissues and organs. The material carrying information in a cell is called the genetic material and we refer to the units of information as genes.

Theoretically there are two major types of strategy by which different types of cells might be generated.

1 During division, cells might lose parts of the genetic material and retain only those parts which are vital for basic functions such as respiration, and for one or a few specialized functions. For example, cells destined to form liver might retain only those genes specifying liver function. Cells differentiating into neurones might lose the liver-type genes but retain genes needed for the production of neurotransmitters and other nerve cell functions. In such a system, only those cells destined to give rise to the next generation, that is, reproductive cells, would have to retain the entire genetic material.

2 All cells of the organism could contain all the genetic material as it initially existed in the zygote, and genes would become active (switched on) or inactive (switched off) to produce the appropriate range of functions.

If all cells possess the entire genetic material, one might predict that differentiated cells could give rise to an embryo and develop into a complete and fertile organism. Testing this prediction is not easy, since a negative result could easily be a false negative. A differentiated cell might fail to develop into a whole organism, not because its genetic material was incomplete, but rather because it had become so specialized that it could not survive in an environment in which differentiation could be reversed.

STUDY ITEM

15.11 Plant tissue culture

Pieces of differentiated plant tissue, such as storage tissue from tubers of Jerusalem artichokes, phloem from carrot roots, or pith from tobacco plants, can be cut from the plant under sterile conditions and transferred to complex growth media. Absolute sterility is essential since the growth media used are ideal habitats for micro-organisms.

After a variable period of apparent dormancy, cells from the pieces of tissue start to divide and lose their characteristic differentiated state, that is, they become less and less like pith cells or phloem cells. If the medium

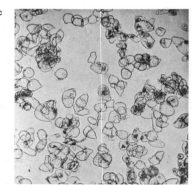

Figure 3
Tissue and cell cultures of belladonna (*Atropa belladonna*).
a Actively growing undifferentiated callus growing on a medium containing $2\,\mathrm{mg\,dm^{-3}}$ of the auxin naphthalene–ethanoic acid and $3.75\,\mathrm{mg\,dm^{-3}}$ kinetin.
b Cell suspension cultures growing in a defined culture medium similar to that used to grow the callus (**a**), but with the agar omitted. During incubation the culture is continuously swirled in the culture flasks by the action of the platform shaker shown. From an initial cell density of approximately $20\,000$ cells $\mathrm{cm^{-3}}$ the cell population rises to over a million cells $\mathrm{cm^{-3}}$ during 21 days' incubation.
c Isolated cells from a liquid suspension culture.
d Development of a leafy shoot from a callus cultured on a medium from which auxin has been omitted but which contains $0.45\,\mathrm{mg\,dm^{-3}}$ kinetin.
Photographs, the Botanical Laboratories, University of Leicester.

is a solid (agar-containing) one, the resulting cells form a mass called a callus (see *figure 3a*). If a liquid medium is used, it must be constantly aerated or shaken (see *figure 3b*); the cells are then released into a suspension of single cells and small clumps (see *figure 3c*).

Plant cells cultured in this way are called tissue cultures. Before any plant tissue can be cultured, a suitable medium must be found. Some successful media are very complex, containing materials such as coconut milk and yeast extract as well as simple sugars, salts, and plant hormones. Small changes in the concentration of sucrose or of a hormone often produce marked changes in the rate of cell division of the cultured cells and in their appearance and metabolism.

If the conditions of culture are appropriate, small masses of undifferentiated cells in a liquid culture, or large masses in a callus culture, will sprout shoots and roots, giving rise to small plantlets (see *figure 3d*). These can be further cultured until they are large enough to be transplanted into compost and grown normally. Thousands of plantlets may be formed in a single flask of tissue culture.

A plant such as a carrot grows if provided with carbon dioxide, water, and a variety of salts such as nitrates. (See page 307.)

a   *Why do carrot cells in culture require many organic compounds?*

Large numbers of normal, fertile, carrot plants may arise from differentiated phloem cells as a result of tissue culture.

b   *What does this show about the genetic material of the original differentiated cells?*

c   *Suggest a practical application for plant tissue culture.*

Differentiated cells rarely start to divide immediately they are transferred to a suitable culture medium.

d   *Why do you think that a period of adjustment is needed?*

STUDY ITEM
15.12 The influence of nucleus and cytoplasm

Acetabularia is a green alga about 2 to 3 cm in length, living in tropical seas. It is sometimes called the mermaid's wineglass, because it consists of a long stem with finger-like projections (called a rhizoid) at one end, and a saucer-like structure (called a 'hat' or 'cap') at the other end.

Acetabularia normally has only one nucleus situated in the rhizoid, which means that it consists of one very large cell. It has several biological features which make it suitable for use in experiments designed to study the roles of the nucleus and the cytoplasm in development. It is uninucleate, with a fairly simple structure, yet relatively large and easy to handle, and it is very hardy: it can recover from microsurgery and pieces of it regenerate well.

Scientists, among whom Joachim Hammerling may be particularly mentioned, have carried out investigations on the effects of isolating portions of *Acetabularia* cells or interchanging parts of two different cells. In an experiment illustrated in *figure 4*, a young plant (1) which had not developed a 'hat' was cut into three parts. The stem did not develop (2). The tip portion developed a fully formed 'hat' and a stem (3). From the rhizoid a fully formed plant developed (4).

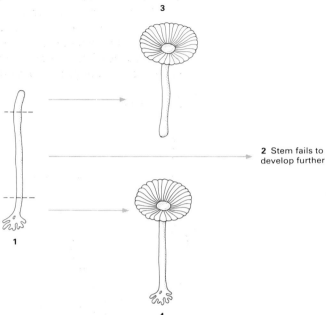

Figure 4
Separating tip, stem, and rhizoid.

a **From this experiment, what can be concluded about the location of the genetic material in the *Acetabularia* cell?**

In the experiment illustrated in *figure 5*, the tip portion was removed (1) as in the previous investigation, but the remainder of the cell was left for a few days (2). The rhizoid was then removed (3) and a 'hat' developed on the remaining stem (4).

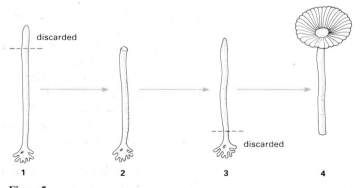

Figure 5
Introducing a time factor.

b How does the result of this investigation compare with that of the previous one? *aquires ability to develop 'hat' by being attached to Rhizoid + nuch.*

c Suggest a hypothesis to explain this result. *Some chemical passes from rhizoid up stem + results in it ability to make Hat.*

The nucleus of *Acetabularia* is extremely hardy. It can be taken from a cell with a fine pipette by the use of micro-dissection techniques, and washed in a suitable solution to remove most of the cytoplasm. Then it can be transplanted into another cell or part of one.

Figure 6 illustrates an *Acetabularia* cell which has been cut into three (1). The nucleus from the rhizoid is then transferred to the stem (2). A complete cell develops (3) from this nucleated stem.

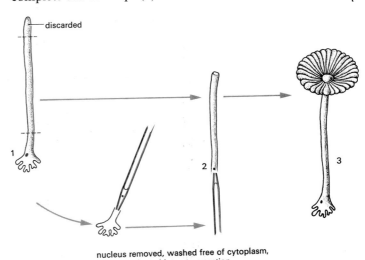

Figure 6
Transplanting a nucleus.

d In what ways do the results of the investigation illustrated in **figure 6** support or modify the hypothesis you put forward in answer to a? *Nucleus not Rhizoid is controlling factor*

During the life-cycle of *Acetabularia*, the 'hat' acts for some time as a photosynthetic structure and the nucleus remains in the rhizoid. When the plant has fully matured, the nucleus becomes smaller and more condensed and then migrates up the stem into the 'hat'. It then divides repeatedly to produce a large number of reproductive cells which are released in small cysts. The parent cell then dies.

Figure 7 shows an experiment where the stem and 'hat' of a fully mature *Acetabularia* were removed just before the nucleus was ready to condense and migrate to the 'hat'. In contrast to the first investigation which used a young cell, the rhizoid developed only a little and the nucleus became smaller.

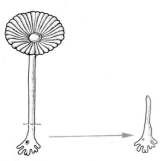

Figure 7
The effect on the rhizoid of removing 'hat' and stem.

e Compare the behaviour of an isolated rhizoid of a mature *Acetabularia* with an isolated, differentiated cell of a multicellular organism. *Both have lost ability to develop into a complete org.*

If a 'hat' from a fully mature cell is transplanted onto the rhizoid of a young, developing cell (*figure 8*), the nucleus of the young cell becomes smaller, migrates up the stem to the 'hat', and divides to produce reproductive cells.

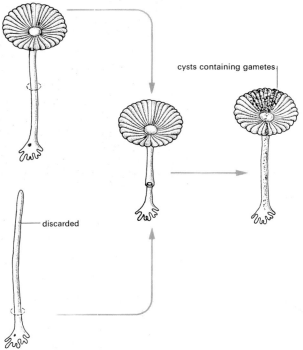

Figure 8
Transferring the 'hat'.

f Can differentiated cytoplasm control the activity of the nucleus?

The species *Acetabularia mediterranea* and *Acetabularia crenulata* differ in the appearance of their 'hats' (see *figure 9*). The stems were removed

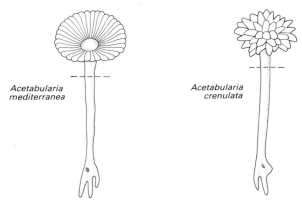

Figure 9
Two species of *Acetabularia*.

8   Inheritance and development

Figure 10
Acetabularia on a stone lifted from the sea at Corfu.
Photograph, David Hosking.

from the rhizoids of young specimens of each species and transplanted onto the rhizoids of the other species. That is,

developing stem of *A. mediterranea* ⟶ rhizoid of *A. crenulata*
developing stem of *A. crenulata* ⟶ rhizoid of *A. mediterranea*.

In each of these experiments, the cells developed 'hats' intermediate in form between those of the two species. If the intermediate 'hats', or the stem tips producing them, were removed and discarded, new 'hats' were formed which resembled the species which had provided the nucleus of each composite cell.

> **g** ***Explain how this final investigation validates and extends the conclusions already reached.***

STUDY ITEM

15.13 Transplantation of nuclei

Experiments such as those on *Acetabularia* reported above show that the nucleus controls the differentiation of the cytoplasm of a cell. They also show that differentiated cytoplasm can, in turn, regulate and limit the activity of the nucleus.

In order to test the developmental potential of the nucleus of a differentiated animal cell, that nucleus must be removed from its differentiated cytoplasm and placed into the cytoplasm of a fertilized egg. The zygote's own nucleus must be either inactivated by irradiation or physically removed, so that it can no longer play any role in subsequent events. The technique of 'nuclear transplantation' is difficult, and negative results may merely indicate that a cell has been damaged. In a significant number of cases, composite cells, comprising differentiated cell nucleus and zygote cytoplasm, will develop into embryos and subsequently into fertile adult organisms. The technique has been used in a range of different animal types, including mammals, but applied with

Figure 11
The procedure for nuclear transplantation.

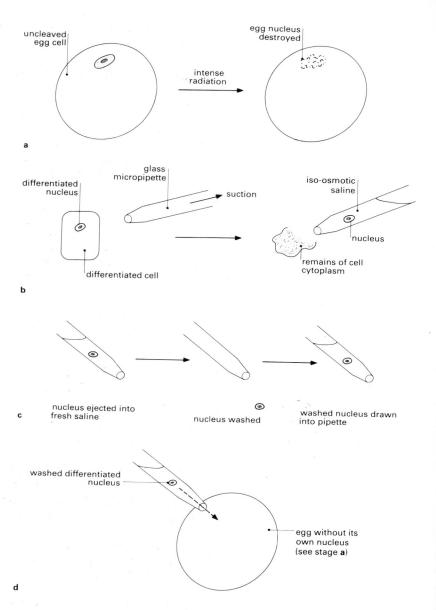

most success to insects and amphibia. It is noteworthy that these are animals which produce large durable eggs which will withstand the micro-injection technique whereby the donor nucleus is transferred into the recipient egg cytoplasm.

Table 1 shows the results of experiments in which nuclei of endoderm (gut) cells were injected into uncleaved eggs, with inactivated nuclei, of the clawed toad, *Xenopus laevis*. An outline of the experimental procedure is given in *figure 11*. Each operated host egg which had been implanted with a nucleus from an endoderm cell was allowed to develop under carefully controlled conditions and close observation. The embryonic stage to which each egg developed before dying was recorded. Control experiments established that differences in development were not due to operative damage, or to cytoplasm being transferred with the nucleus.

a **Indicate the nature of the control experiments.** — portions of different cytoplasm from donor cells / removal + immediate return

Table 1
Results of the transplants of nucleus into newly fertilized eggs.

Age and stage of development of embryos donating nuclei	Percentage of transplanted eggs surviving to the named stage of development				
	late blastula (per cent)	early gastrula (per cent)	late gastrula and neural folds (per cent)	swimming tadpole (per cent)	normal feeding tadpole (per cent)
1 6 hour	100	98	94	85	81
2 12 hour	100	100	98	79	77
3 25–30 hour	100	96	79	53	52
4 Muscular response: 39 hour	100	94	75	47	41
5 Heart beat and hatching tadpoles: 58 hour	100	76	59	36	27
6 Swimming tadpoles: 120 hour	100	80	54	19	15

b **Using the information in table 1, plot six graphs, on the same axes, of the percentage survival of the operated eggs against the developmental stage they had reached. There should be one line for each stage of embryo donating nuclei.**

One hypothesis to explain the data is that in endoderm nuclei of *Xenopus* the switching off of some of the genetic material is less easily reversed as differentiation proceeds.

c **Put forward an alternative hypothesis to this idea.**
More easily damaged as differentiation occurs

In experiments of this kind, the animal producing the eggs is usually chosen for its distinctive colour (e.g. albino), while the one from which the transplanted nuclei are obtained is of a contrasting colour (e.g. black).

d **Why is this colour contrast an advantage?**
indicates which nucleus had the influence

In the experiment reported in table 1, the nuclei of differentiated gut cells of swimming tadpoles were, in a significant number of cases, able to control the development of a normal individual. The egg cell cytoplasm presumably exercised a redirecting influence on the differentiated nucleus.

15.2 Cell division during growth and development

In the first part of this chapter, we have reviewed some evidence that the genetic information of a cell is in its nucleus and that all cells of a multicellular organism have a full complement of genetic material. What are the organization and structure of a nucleus, and how is the genetic information in it so faithfully transmitted to each new generation of cells?

The prokaryotes (see *Systematics and classification* and *Study guide I*, Chapter 5) have no nucleus. We shall be mostly concerned with events in eukaryote cells. There are a few examples of cells from eukaryotes which have no nucleus (red blood cells and phloem sieve tubes) but these all start their life with one and lose it as they develop.

Viewed with the light microscope, the nucleus of an animal or plant cell looks denser than the surrounding cytoplasm and can be made clearly visible, using stains.

a *There are dyes which stain all types of nuclei, from whatever type of cell, in a similar way. What can be deduced from this?*

In the nucleus of a non-dividing cell, there are usually one or more regions which stain differently from the other material. Such a differentially-staining region is called a nucleolus.

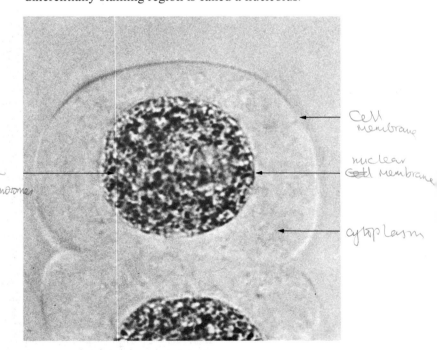

Figure 12
A young plant cell and some of its parts (organelles).
From McLeish, J. and Snoad, B., Looking at chromosomes, *revised edition*, Macmillan, 1966.

b *What are the parts to which the arrows point?*

The rest of the material in the non-dividing nucleus is called chromatin. It seems to be a rather featureless granular material, and it gets its name because of the ease with which early microscopists could stain it. There are two types of chromatin: heterochromatin and euchromatin. Heterochromatin stains readily in the non-dividing nucleus and is electron-dense in electron microscope preparations. Euchromatin only stains readily in dividing cells and is more easily penetrated by electrons in the electron microscope.

The nucleus is bounded by a double phospholipid membrane (see *Study guide I*, Chapter 5) which is perforated by a large number of pores (*figures 13* and *14*). This nuclear membrane becomes disorganized during division, a new one being reconstituted around each daughter nucleus.

At the time of cell division, chromatin becomes organized into thread-like chromosomes. The process whereby the chromosomes in a nucleus are duplicated and distributed to achieve two identical daughter nuclei is called *mitosis*. This occurs whenever cells increase in number during growth or asexual reproduction.

A nucleus in which the chromosomes are not visible and only the granules of heterochromatin and a nucleolus can be observed is said to be in interphase. (See *figure 12*). It is possible to measure the amount of genetic material in individual interphase nuclei of a population of cells dividing by mitosis. It is found that some nuclei have twice the amount of material existing in others, and that a few nuclei show an intermediate quantity.

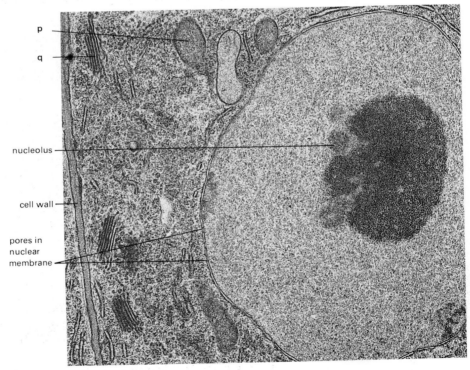

Figure 13
An electronmicrograph of a section through a nuclear membrane ($\times 16\,000$). Its double nature and the pores can clearly be seen.
What are the structures labelled p and q?
Photograph, Biophoto Associates.

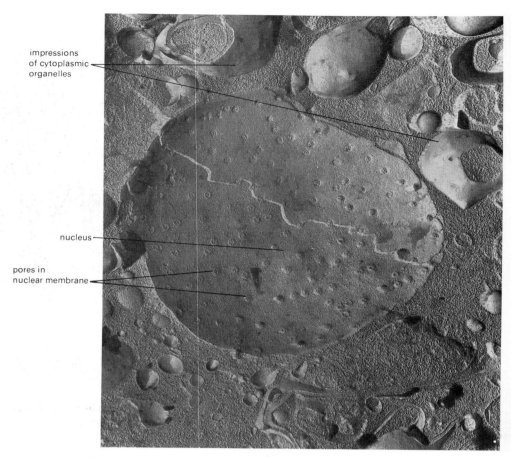

Figure 14
A scanning electronmicrograph of a freeze etched nuclear surface, showing the pores. ($\times 16\,900$).
Photograph, Biophoto Associates.

c *Suggest why there are different quantities of genetic material found in a population of interphase nuclei.*

Some have replicated genetic material

When the chromatin first becomes condensed into chromosomes, these are a highly complex 'tangle' of very thin threads. However, they become steadily shorter and thicker and thus easier to resolve individually. Even in interphase, the chromatin has a linear arrangement appearing granular (heterochromatic) because some small parts of each chromosome remain condensed while other regions completely uncoil (euchromatic). Good microscopic preparations often show chromosome threads coiling up as they condense (*figure 15*.) (A crude model of chromosome condensation is achieved by twisting a piece of elastic. The end in your left hand is held still and the end in the right hand twisted many times, rather as one would wring out a wet garment.)

As chromosome condensation proceeds, the nucleolus becomes smaller and stains less deeply until it becomes undetectable. The size of chromosomes varies greatly from species to species. Where they are

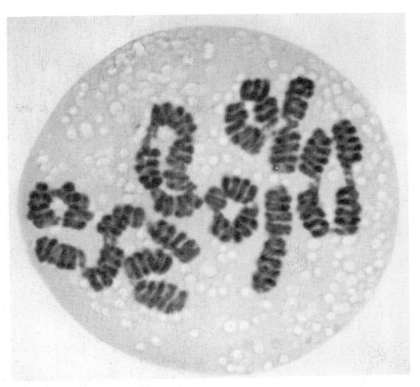

Figure 15
Chromosomes of *Tradescantia virginiana*, showing the spiral structure (×2470).
Photograph, Dr C. G. Vosa, The Botany School, Oxford.

fairly large, condensation shows that each thread is double. The double nature of each chromosome results from the replication of the genetic material in the previous interphase.

The events described above, that is, chromosome spiralization and the disorganization of the nucleolus, are described as prophase of mitosis.

The products of chromosome replication are called chromatids. They become more easily visible as prophase proceeds. They are identical, lying alongside each other and physically joined at one point where the original chromosome has not yet replicated. This point is the centromere.

The end of prophase is marked by the breakdown of the nuclear membrane. With this breakdown, the nuclear spindle is able to develop (*figure 16*). This is composed of microtubules – fine thread-like protein structures which are able to shorten. They radiate from two regions called the spindle poles. These are associated with centrioles. A centriole is found near the nuclear membrane in the cytoplasm of a cell during interphase. This is replicated at the beginning of prophase and the daughter centrioles migrate to opposite sides of the nucleus. When the spindle forms, its poles are closely associated with the centrioles; it has been suggested that the centrioles organize the formation of the spindle.

Echinoderms (starfish and sea urchins), and also all flowering plants,

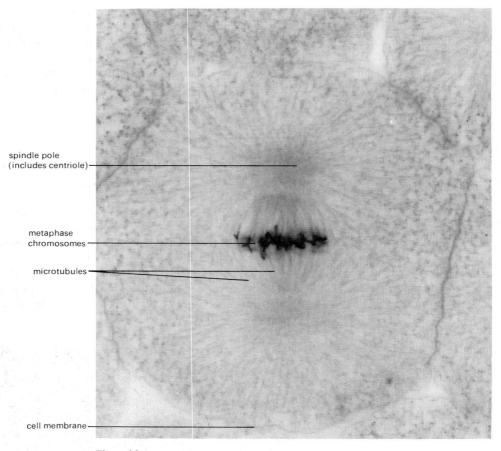

Figure 16
A photomicrograph showing the nuclear spindle (× 1400).
Photograph, Biophoto Associates.

have no centrioles in their cells. Spindles in these groups seem very similar in structure and behaviour to those of other eukaryotes.

d *What does this suggest about the role of centrioles in mitosis?*

Under the influence of the spindle microtubules, the centromeres of the chromosomes are pulled to the equatorial plane of the spindle. The processes of spindle formation and the alignment of centromeres on its equatorial plane is called metaphase. The spindle equatorial plane is therefore called the metaphase plate.

Only when the centromeres are all located on the metaphase plate do they become visibly duplicated. The spindle fibres linking centromeres to the poles shorten, pulling chromatids by their centromeres, one of each pair to each pole. The 'arms' of the chromatids trail behind the centromeres as they are pulled towards the poles. The processes of separation of sister chromatids by centromere duplication and the migration of the resulting daughter chromosomes to the polar regions is called anaphase.

With the separation of daughter chromosomes into two distinct parts of the cell, the spindle structure disintegrates and a nuclear membrane forms around each of the two groups of chromosomes, which can now be called daughter nuclei. The chromosomes begin to despiralize and eventually assume an interphase appearance with nucleoli and granules of heterochromatin. These final events in the process of mitosis are called telophase.

With certain exceptions, division of the cytoplasm follows the division of the nucleus.

e *Do you know of any of these exceptions, that is, structures having many nuclei in a common cytoplasm?*

Division of the cytoplasm in animals is normally by constriction of the equatorial region until two completely separate cells result. This happens during telophase. In plants, material accumulates across the equator of the cell and forms a cell plate (*figure 17*). This gives rise to a pair of membranes separated by a layer of polysaccharide which becomes the new cell wall (*figure 18*). The daughter cells are connected by threads of cytoplasmic material called plasmodesmata which pass through pores in the new cell wall.

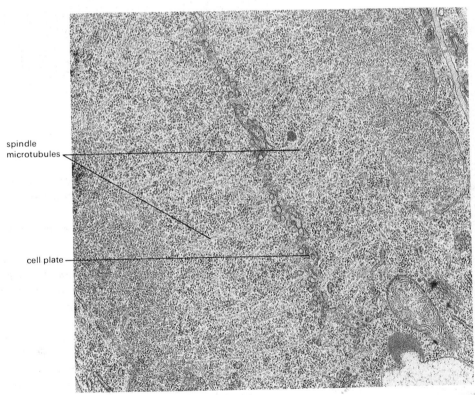

Figure 17
A photomicrograph of a meristem cell of a root tip of *Arabidopsis thaliana*, showing the cell plate forming and the remains of the spindle (× 23 000).
Photograph, Biophoto Associates.

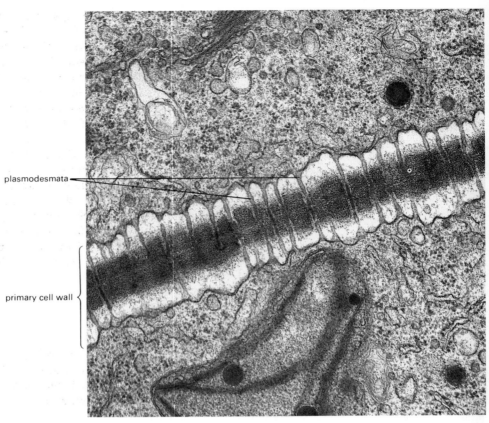

Figure 18
An electronmicrograph showing a young primary wall and plasmodesmata ($\times$ 30 750). *Photograph, Biophoto Associates.*

Since plant cells are usually immobile and physically joined to their neighbours, it is most important that the plane in which cell division occurs is determined so that an organized structure results.

Practical investigation. *Practical guide 5*, **investigation 15A, 'Mitosis in a plant meristem'.**

The process of nuclear division described above was first discovered by immobilizing cell contents, using a suitable protein coagulant. This is called fixation. It turns the protoplasm to a more or less solid gel in which biochemical processes cease. Fixation is followed by other histochemical techniques, for example, staining with toluidine (2-methylphenylamine) blue which makes the nuclei visible. (See the Foreword to *Practical guide 5*, and also investigation 15A.)

As a completely different example, the Roman city of Pompeii was overwhelmed by a mass of volcanic ash and dust in A.D. 79. Archaeologists have been able to infer a great deal about the Roman way of life by examining objects found under this volcanic deposit, which can be thought of as having 'fixed' a number of households at a particular moment. In a similar way, the microscopist can infer the process of mitosis as a hypothesis to account for the differences between nuclei in the same tissue. A description of the whole process can be arrived at by examining a large number of cells which have been killed when they have reached different points in the cycle of division. The cells are classified into groups which have reached a similar point in the cycle. Mitosis thus became regarded as a series of stages (interphase, prophase, metaphase, anaphase and telophase), as already described.

Tissues can be viewed after illuminating them in such a way that different transparent objects within the cells become visible. Three types of lighting system have been employed: polarized light, phase contrast, and dark ground. Phase contrast has been particularly useful in studying nuclear division, and you will probably have seen film taken using phase contrast of chromosomes actually condensing and moving on the spindle. Such film shows that the idea of stages (phases) is a bit artificial, because it is a continuous process. However, the names of the stages are useful for reference, rather as the day–night cycle is artificially divided into hours.

STUDY ITEM

15.21 Evidence for mitosis

Figures 19a to j (overleaf) are photographs, not all at the same magnification, of nuclei of the plant *Crocus candidus*. They are arranged in a random order. In some cases, a drawing is given which interprets the structure in a corresponding photograph.

a *List the photographs in the order in which they should appear as a cell goes through the mitotic cycle starting with interphase.*

b *If a photograph has no corresponding drawing, make one and label it.*

c *Explain the appearance of the pair of chromatids labelled 'r' in figure 19.*

d *How many chromosomes are there in the cells of this plant?*

e *What is the evidence that spindle fibres (microtubules) attach to chromosomes only at their centromeres?*

f *The spindle is a transparent structure and may not show up if a cell is stained to reveal chromosomes. Make two diagrams to show how the chromosomes and their centromeres would appear at metaphase if viewed:*
  1 from above one of the spindle poles
  2 from the side of the spindle in the same plane as its equator.

g *What would be the result of failure of a centromere to divide at metaphase in terms of chromosome content of the two daughter cells?*

h *Of all cell components, only the chromosomes have a mechanism to ensure regular and precisely equal division into the daughter cells. Why should this be so?*

It is essential to remember that, during the cycle of mitosis, each chromosome is duplicated exactly and one copy passes to each daughter nucleus. This is the basis for the continuity of life – of the similarity between fossil and modern brachiopods such as *Lingula* over a period of 600 million years.

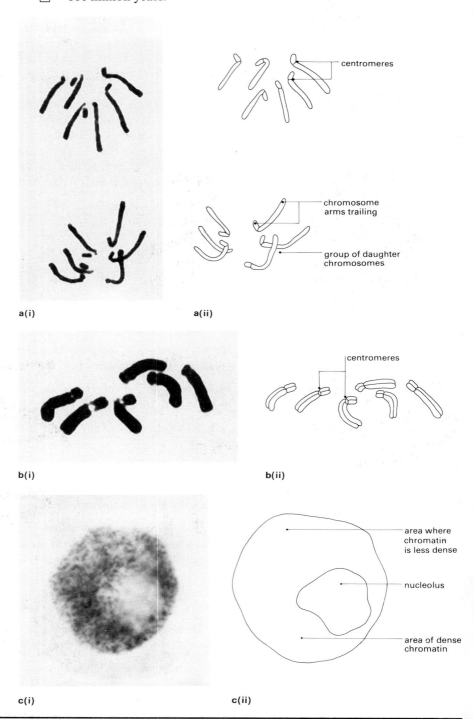

Figure 19
Phases of mitosis of *Crocus candidus*.
Photographs, Gareth Jones.

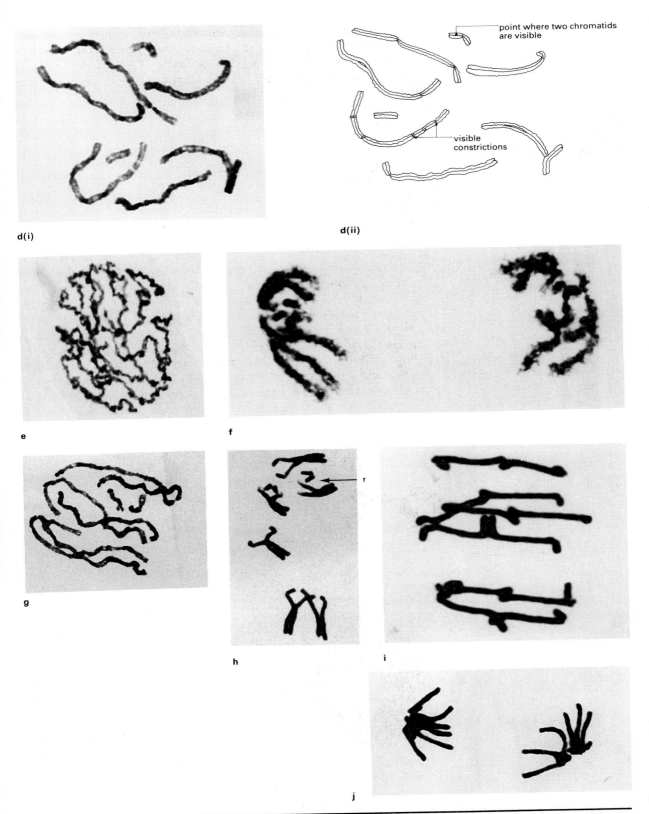

15.3 A review of cell diversity

This section does not attempt to describe the general ultrastructure of a cell – for that you should refer to *Study guide I*, Chapter 5; neither does it give a list of particular ultrastructural features characteristic of certain types of differentiation. Instead, we will try to define some general categories of ultrastructural change which might be predicted to accompany differentiation, and we will also attempt to explain why they occur.

Nuclear changes

Can we expect to be able to see changes in patterns of gene expression with time, or differences in gene expression between tissues, reflected in the morphology of the nucleus and its contents? Generally the answer is no, as these events occur at a molecular level, and would not be detectable even using an electron microscope. What is sometimes seen during differentiation is a gradual increase in the ratio of heterochromatin to euchromatin as, presumably, an increasing proportion of 'irrelevant' genes are switched off (that is, become inactive) while a smaller proportion of 'relevant' genes remain very active. A second aspect of nuclear morphology which can be an index of the differentiative state of the cell is the relative prominence of the nucleolus. The nucleolus is the site of ribosomal RNA synthesis (see page 115), and therefore a cell making a great number of ribosomes (and consequently synthetically active) is likely to possess a prominent nucleolus.

There are often changes in the volume of nuclei during differentiation. (In mice, the range of nuclear volume encountered in different tissues is between $24\,\mu m^3$ and $460\,\mu m^3$.) Volume changes are likely to be correlated with other changes already mentioned.

We have tried to show that differentiation can occur by alterations in the activity of genes rather than by changes in the genetic material itself. There are, however, many instances where the amount of genetic material in the nucleus increases during differentiation and others in which it is rearranged. Such changes will be discussed further in Chapter 19.

Cytoplasmic changes

This is a very broad category of change, and just about any cytoplasmic organelle could reflect a differentiative change in the cell. For instance, a cell with increased metabolism might display an increase in the number of mitochondria it possesses. This may be paralleled by an increased packing density of the cristae within each mitochondrion. Photosynthetically active cells may have increased numbers of chloroplasts. The increase in ribosomal numbers in actively synthesizing cells has already been mentioned, but their arrangement will vary according to whether the cell is synthesizing materials to be retained within the cell or destined for export from the cell. In the former case, individual ribosomes or polyribosome clusters will be seen 'free' in the cytoplasm. In the latter case, ribosomes will be found attached to the endoplasmic

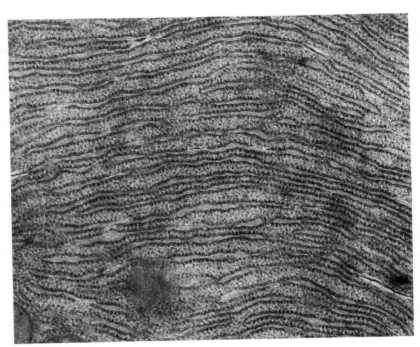

Figure 20
Electronmicrograph of pancreatic exocrine cell of a rat, showing rough endoplasmic reticulum (× 30 000).
Photograph, Biophoto Associates.

reticulum system, conferring upon it a 'studded' appearance – hence the term 'rough endoplasmic reticulum' (RER). (See *figure 20*.) The spaces or cisternae of the RER provide a means of transport to the exterior, but all newly synthesized molecules destined for export are usually processed in the Golgi apparatus before their ultimate release. So, not surprisingly, a cell synthesizing and exporting a lot of material will not only possess an extensive RER system but will also display a prominent dilated Golgi apparatus. When the material has been processed by the Golgi apparatus, it may be stored prior to release and a number of synthetic cell types are found to contain electron-dense storage granules. These are exemplified by the zymogen granules of the pancreatic exocrine cells. Some differentiating cells, while being synthetically active, will lack prominent RER and Golgi systems, simply because the products of their synthesis are to be retained within the cell. In such cell types we may see any one of a range of new cytoplasmic organelles or inclusions arising. An example, which also displays a very high order of organization, is the myofibrillar system in cardiac and skeletal muscle cells, where actin and myosin associate to form interdigitating arrays of characteristic morphology (see *Study Guide I*, Chapter 11).

Cell surface changes

There are a great variety of gross surface changes which take place during differentiation, such as the acquisition of cilia by tracheal epithelium or microvilli by cells of a nephron. At a finer level of organization, there are subtle molecular changes which may only be detectable using rather specialized techniques. As a cell differentiates, the array of integral membrane proteins in the plasma membrane will change too (see the discussion of membrane structure in *Study guide I*, section 5.6, 'The importance of cell membranes'). This is not simply a strategy whereby a cell can display to its neighbours what its differentiative state is, but almost always will have an important functional significance. That is to say, how a cell will respond to its environment will be determined largely by the composition of its plasma membrane and, in particular, the array of proteins in that membrane. As cells differentiate, so too does their response to environmental change; and remember that environment means not only predictable components such as hormones and neurotransmitters but also other cells. So it is not surprising that, along with the nuclear and cytoplasmic changes, we find changes in membrane structure. A good example is the acquisition of hormone-sensitivity by a differentiating population of cells.

Although there are many other examples of cell surface changes during differentiation in animal systems, examples from plant systems are more difficult to select. The reason for this is not that parallel changes do not occur – they almost certainly do – but that obtaining pure membrane preparations from plant cells is notoriously difficult. Consequently, rigorous demonstration of compositional changes in the plasma membranes of plant cells as they differentiate is a much greater problem.

a *Why is it difficult to obtain samples of plant cell membranes?*

Extra-cellular matrix changes

Virtually all cell types, whether animal or plant, appear to be surrounded by greater or lesser amounts of extracellular matrix. In plants this is in the form of a cell wall. In fact, the evolution of the ability to synthesize and accumulate matrix components was probably a fundamental feature of multicellular evolution. Even some of the simplest multicellular organisms, such as *Gonium* (see *figure 21*), consist of cells embedded in a mucilaginous matrix; and the sponges contain banded collagen in the matrix of their mesoglea. (Collagen is the single most abundant protein in the human body. Up to one-third of our total body protein is collagen.)

Often during differentiation there may be dramatic changes in the amount of matrix made and in its composition. This is usually the case where tissue function is dependent upon the presence of an extensive extracellular matrix. Two obvious examples are the secondary walls of xylem tracheids and vessel cells, and the mineralized matrix of bone. In both cases it is the extracellular matrix component of the tissue which confers upon it the appropriate functional properties, not only by virtue of its composition, but also in the way in which it has been deposited. In some cases, the composition and accumulation of matrix may even bring

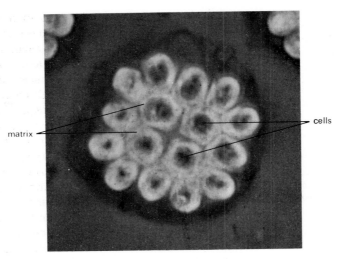

Figure 21
A photomicrograph of *Gonium*, showing cells embedded in mucilaginous matrix (× 130).
Photograph, Biophoto Associates.

about the death of the cells within it; for instance the deposition of lignin in the secondary plant cell wall renders the wall impermeable and leads to the death of the cell within.

After reading this section, you may find it useful to refer to your results for the following practical investigations.

> **Practical investigations.** *Practical guide 5*, investigation 15B, 'Differentiation in stems and roots', and 15C, 'Differentiation in cells'.

15.4 The 'one gene–one polypeptide' hypothesis

Changes in the structure and behaviour of differentiating cells are preceded by changes in the levels of substances synthesized by the cells. Nearly all biochemical changes in cells are brought about by enzymes, which are of course proteins. Proteins are also of great importance as structural components, and cell surface antigens are usually protein.

The properties of a particular protein depend largely on the sequence of amino acids joined by peptide bonds in the polypeptide chain or chains of which it is composed, that is, its primary structure. (See *Study guide I*, Chapter 6.) A polypeptide becomes folded and coiled into a characteristic three-dimensional shape, partly through the spatial orientation of its peptide bonds and partly as a result of interactions of the side-chains of the constituent amino acids. Non-protein substances may become loosely or firmly attached to the polypeptide and become an important part of the complete protein molecule. Such attachments depend on the chemical properties of particular amino acid side-chains and on the position of these amino acids in the primary structure.

Haemoglobin is an important protein, and you will already be familiar with some of its properties in relation to oxygen transport and storage (see *Study guide I*, Chapter 4). Red blood cells contain little besides haemoglobin, so large pure samples can be obtained fairly easily. The molecule contains four polypeptides of two different types, and the primary structures of these polypeptides are well known. Automated techniques using computers allow the sequence of amino acids in a peptide to be determined, and haemoglobin samples from many sources have been analysed.

There is a great deal of variation in amino acid sequence shown by different haemoglobin samples. As one might expect, different species, for example pigeon and human, have characteristic differences in sequence, and these are associated with different oxygen-dissociation curves. (See table 2.) Such differences are determined by the genetic material of the two species. Each type of haemoglobin polypeptide requires a gene to specify its primary structure.

Table 2
Sequence of amino acids in β-polypeptide of haemoglobin from three mammals. For the full names of amino acids see the genetic code on page 114. Variations from human sequence are shown in bold.

	Order of aa sequence	1	2	3	4	5	6	7	8	9	10
Haemoglobin chains	HUMAN – beta	VAL–	HIS–	LEU–	THR–	PRO–	GLU–	GLU–	LYS–	SER–	ALA–
	GORILLA – beta	VAL–	HIS–	LEU–	THR–	PRO–	GLU–	GLU–	LYS–	SER–	ALA–
	HORSE – beta	VAL–	**GLU–**	LEU–	**SER–**	**GLY–**	GLU–	GLU–	LYS–	**ALA–**	ALA–

	Order of aa sequence	11	12	13	14	15	16	17	18	19
Haemoglobin chains	HUMAN – beta	VAL–	THR–	ALA–	LEU–	TRY–	GLY–	LYS–	VAL–	ASP–
	GORILLA – beta	VAL–	THR–	ALA–	LEU–	TRY–	GLY–	LYS–	VAL–	ASP–
	HORSE – beta	VAL–	**LEU–**	ALA–	LEU–	TRY–	**ASP–**	LYS–	VAL–	ASP–

It seems logical that, for every polypeptide an organism is able to synthesize, there must be a gene which is transmitted from generation to generation, storing the information on which the polypeptide sequence depends. This is the 'one gene–one polypeptide' hypothesis.

STUDY ITEM
15.41 Sequential activation of haemoglobin genes in humans

During the development of a human, three different types of red blood cell are formed. The first or embryonic cells are produced during the earliest stages of foetal life. They are large and nucleated, like the red cells of lower vertebrates. Later generations of cells are intermediate in size and not nucleated. They are called foetal cells. Towards the end of intra-uterine life, typical small cells characteristic of the adult are formed. (There is a similar development sequence in other mammals, such as mice, though the time scale of development is different, of course.)

Figure 22 shows the proportions of the different types of red blood cell found in a typical human at different ages before and after birth.

Figure 22
The three generations of red blood cells (embryonic, foetal, and adult) in the human foetus and the newborn child. The horizontal scale is in months up to and beyond birth. The vertical scale is in percentage. The curves make it possible to read off the proportion of each of the three generations of red blood cells present in the blood at any age.
Based on Allison, A. C., New Biology No. 21, 'Human haemoglobin types', Penguin, September 1956.

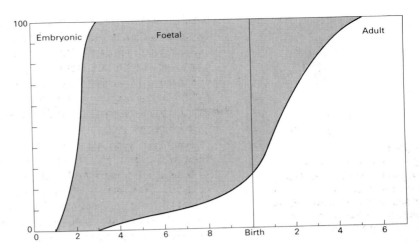

a From figure 22, what percentage of cells of each type would be expected in an individual:
   1 2 months after conception?
   2 6 months after conception?
   3 2 months after birth?

Parallel with these changes in the types of red cell produced are changes in the nature of the haemoglobin found in the blood. Embryonic haemoglobin has a very steep oxygen-dissociation curve (*Study guide I*, Chapter 4). Foetal haemoglobin has a curve of intermediate steepness, and adult haemoglobin has a relatively shallow curve.

b Why are three different haemoglobins, each with a different oxygen-dissociation curve, an advantage?

c It has been suggested that each type of red blood cell produces its own type of haemoglobin. How could this hypothesis be tested?

The three types of haemoglobin are produced by the activity of different genes. Genetic abnormalities, such as thalassaemia and sickle-cell anaemia, which result in an unusual type of adult haemoglobin have no influence on foetal haemoglobin, so that sufferers from these conditions show few symptoms until after birth. In mice, a gene is known which affects foetal haemoglobin but not adult haemoglobin.

d What is the implication of the appearance of three types of haemoglobin at different stages during development for the hypothesis that genetic material is selectively activated or inactivated as cells differentiate?

STUDY ITEM

15.42 Haemoglobins and inherited variation

Electrophoresis is a powerful technique used to detect different types of protein molecule. It can reveal very subtle differences, perhaps caused by

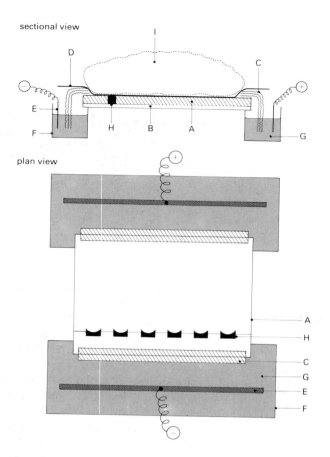

Figure 23
An assembly for starch gel electrophoresis.
A, gel; B, glass plate; C, wicks; D, plastic sheet; E, platinum electrodes; F, electrode vessel; G, bridge buffer solution; H, sample; I, ice pack.

the change of a single amino acid, in a molecule containing some hundreds of amino acid units.

An apparatus used to carry out electrophoresis is shown in *figure 23*. A stiff jelly (gel) is made using starch or polyacrylamide and a buffer solution so that the pH will remain constant. Precautions are taken to achieve a sheet of gel of uniform thickness and composition which is free of air bubbles.

Samples of the materials, the proteins of which are to be compared, are placed in small slots cut in the surface of the gel sheet. The slots are arranged in a line near the cathode.

An ice pack is necessary to keep the gel very cool.

a *Why is this necessary?*

A large potential difference is applied across the gel and charged particles move in the gel in the same way as simple ions during electrolysis. Most protein molecules carry a net negative charge and move towards the anode (+ electrode). The rate of movement of the molecules depends on the potential gradient, pH, temperature, and gel composition. As these factors are the same for all the samples being examined, differences in rates of movement depend on subtle differences between molecules. If a sample contains more than one protein, it will be

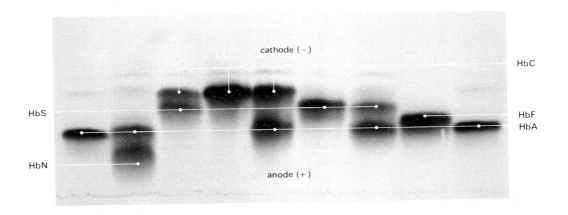

Figure 24
Hospital records showing results of electrophoresis of blood samples. HbA is the common type of haemoglobin, HbF is foetal haemoglobin and HbS is associated with sickle cell anaemia.
Photograph, Malcolm Lill, Department of Haematology, St Helier Hospital, Carshalton, Surrey.

separated into a series of bands, each corresponding to a different protein moving at its characteristic rate.

Enzyme proteins in the gel can be detected, using staining techniques which depend on a reaction catalysed by the enzyme under consideration. Coloured substances, such as haemoglobin, can be detected by simple inspection.

Electrophoresis has been used to screen thousands of blood samples from blood banks for variations from the normal types of human haemoglobin.

Figure 24 is a photograph of the results of electrophoresis of blood samples from a number of different people.

b *Discuss these results briefly.*

c *If blood from a person is shown by electrophoresis to contain an unusual type of haemoglobin, what investigations might follow up this discovery – assuming the cooperation of the individual concerned?*

Variations in adult human haemoglobin fall into two groups, those that involve amino acid changes in the α peptide chain and those in which the β chain is affected (*Study guide I*, Chapter 5). Each chain is synthesized as a result of activity by a different gene. Slightly different versions of the genes are in each case responsible for the unusual types of haemoglobin detected.

Most unusual haemoglobins seem to carry on their biological functions more or less normally. In two cases a single amino acid change results in a drastic alteration of function. Each of these changes is associated with an inherited disease – thalassaemia and sickle-cell anaemia. These conditions will be further discussed in Chapter 20.

Electrophoresis has detected variation in many different proteins apart from haemoglobin. In plants, animals, and micro-organisms these changes can be shown to be inherited, and they all provide support for the 'one gene–one polypeptide' hypothesis.

Summary

1. Organisms can reproduce themselves exactly for millions of generations (**15.1**).
2. Multicellular organisms show cell differentiation, and this involves loss of independence and specialization of function (**15.1**).
3. Genetic material contains instructions which specify the materials a cell can make (**15.1**).
4. Cells might differentiate by the loss of some of their genetic material or by the selective expression of only some of their genes (**15.1**).
5. Complete plants can develop from cultured cells, showing that the genetic material of these cells was intact (**15.1**).
6. Experiments with single-celled *Acetabularia* show that the genetic material is in the nucleus, that the nucleus determines the development of the rest of the cell, and that developed cytoplasm can in turn control the nucleus (**15.1**).
7. The nuclei of differentiated animal cells can be put into the cytoplasm of fertilized eggs and will often control the development of complete individuals, showing that they had lost none of their genetic material (**15.1**).
8. The nucleus has its genetic material organized into chromosomes which undergo a highly organized process of replication and distribution between the products of cell division. This is called mitosis (**15.2**).
9. During differentiation, cells show changes in the nucleus, the cytoplasm, the cell surface, and the type and quantity of external matrix they secrete (**15.3**).
10. The structure of the polypeptides a cell is able to synthesize is determined by particular genes (**15.4**).
11. The example of haemoglobin is used to show that, during the development of an individual, genes are activated in sequence (**15.4**).
12. Inherited differences between people are sometimes the result of changes in the amino acid sequence of the polypeptides they produce (**15.4**).

Suggestions for further reading

BLACK, M. and EDELMAN, J. *Plant growth.* Heinemann, 1970. (A clear and readable account of cell growth and differentiation in plants.)

GURDON, J. B. Carolina Biology Readers No. 25, *Gene expression during cell differentiation.* Carolina Biological Supply Company, distributed by Packard Publishing Ltd, 1978.

JOHN, B. and LEWIS, K. R. Carolina Biology Readers No. 26, *Somatic cell division.* 2nd revised edn. Carolina Biological Supply Company, distributed by Packard Publishing Ltd, 1981.

PRESCOTT, D. M. Carolina Biology Readers No. 96, *The reproduction of eukaryotic cells.* Carolina Biological Supply Company, distributed by Packard Publishing Ltd, 1978. (This assumes a knowledge of much material found in Chapter 18 of this volume.)

CHAPTER 16 THE CELL NUCLEUS AND INHERITANCE

16.1 Do male and female contribute equally to inheritance?

This chapter will be concerned with events in cell nuclei which must precede fertilization, and which impose a similar pattern on the life cycles of all eukaryotic organisms. Fertilization is the fusion of two cells called gametes. The cell resulting from fertilization is called a zygote. Each chromosome which was present in either of the gamete nuclei finds itself in the zygote nucleus, and it is replicated every time a cell descended from the zygote undergoes mitosis.

Gametes of multicellular organisms are very different in size, and when these differences were discovered by the early microscopists, they helped to fuel a controversy which had existed for centuries about the relative importance of males and females in reproduction. Some human societies have believed that our inherent qualities come largely from our mother. In these cultures, family name and sometimes property are inherited through women, and men achieve their position in society by reason of their relationship to women. Such a society is said to be matrilineal. (Examples are the Iroquois of North America and the Ashanti of West Africa.)

A more common type of social organization has been the patrilineal one, in which family name and social position are inherited through men. Western European societies since Roman times have been patrilineal, and have been based on the hypothesis that the life of a child comes from the father's semen (Latin *semen*: a seed). The mother's womb was believed to provide merely a suitable environment for the development of male 'seeds'. So convinced of the truth of this view were some early microscopists that they imagined little men (homunculi) curled up inside the heads of human sperm cells. (see *figure 25*).

The patrilineal view of inheritance has had a most profound effect on British history because it resulted in doubt about whether the 'royal blood' could be inherited through a woman.

During the twentieth century, women have increasingly looked for and have in some measure achieved equality with men, and it is interesting to speculate whether disproof of patrilineal and matrilineal hypotheses about inheritance was necessary for the achievement of this important change.

A true breeding variety (sometimes called a pure strain) is a group of related organisms which show no inheritable variation in some well-defined character or characters, and will always produce offspring with the same character if they are mated together. The results of a number of

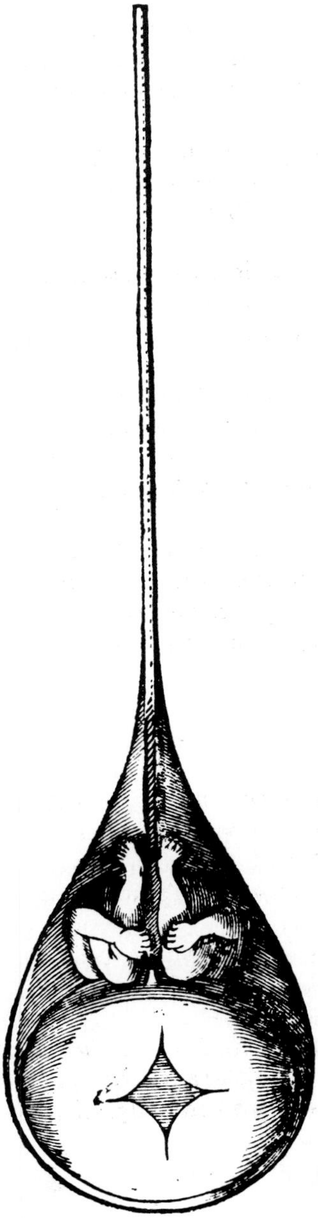

Figure 25
A late seventeenth century drawing by the Dutch scientist Niklaas Hartsoeker of a sperm with a miniature man – the homunculus – inside it.
From *Essai de Dioptrique* by Niklaas Hartsoeker, published in Paris, 1694. © The British Library, 537, k19.

crosses between pure strains are shown in table 3. Reciprocal crosses are crosses where each strain is used to provide both male and female parents in parallel experiments.

	Character shown by true breeding male parents	Character shown by true breeding female parents	Character shown by the first generation offspring (F_1)
Cross 1 (mice)	black coat	brown coat	black coat
reciprocal cross	brown coat	black coat	black coat
Cross 2 (Drosophila)	normal (long) wings	vestigial wings	normal wings
reciprocal cross	vestigial wings	normal (long) wings	normal wings

Table 3
The offspring of reciprocal crosses.

a *Do the results of these crosses support ideas of matrilineal or patrilineal inheritance?*

While most reciprocal crosses produce the same result, there are exceptions which are most instructive. One group of exceptions is that which shows sex-linked inheritance, discussed in section 16.5.

In plants, some types of variegation are inherited only from the female parent. Investigation has shown that this results from the presence of genetically abnormal chloroplasts. Chloroplasts and mitochondria are self-replicating structures in the cytoplasm. They are under the influence of the chromosomal genetic material, but have some genes of their own. They are transmitted by the female gamete with its larger cytoplasmic contribution to the zygote and not by the male. However, as we shall see, most genes show a pattern consistent with a chromosomal theory of inheritance.

Gametes are vastly different in size, but they have very similar sets of chromosomes. A maternal and a paternal set unite to form the zygote nucleus. This accounts for the similar results of the reciprocal crosses shown in table 3. Because gametes contribute similar sets of chromosomes to the zygote, cells which descend by mitosis from the zygote have chromosomes that can be arranged into pairs; one of a pair comes from the maternal chromosome set and the other from the paternal set.

A cell with two sets of chromosomes is *diploid*. Cells which have only one set, and in which there are no pairs of chromosomes, are *haploid*. Gametes are haploid and zygotes are diploid. In some individuals all the cells in the body are haploid. Examples are honey bees (drones) (see also page 228) and the gametophytes of moss plants. Such haploid individuals can produce gametes by mitosis. Diploid individuals must produce gametes by a special type of nuclear division in which the two sets of chromosomes are re-organized into haploid nuclei having only single sets. This special type of division in which the diploid state is reduced to the haploid is called *meiosis*. The relationship between haploid and diploid states in the life cycles of any type of organism is summarized in *figure 26*.

Figure 26
A generalized life cycle. 'n' represents one set of chromosomes.

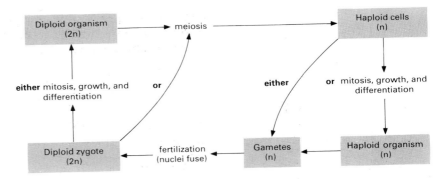

If you carry out either of the following two investigations you will be able to observe meiosis at first hand.

> **Practical investigations.** *Practical guide* 5, investigation 16A, 'Gamete production by a plant' and investigation 16B, 'Gamete production in an animal'.

16.2 Evidence that chromosome changes are associated with inherited characteristics

Before examining the process of meiosis, let us look at the structure of chromosomes and at some of the evidence that they are the site of most of the genetic material of the cell. A chromosome is composed of nucleic acid and proteins (see page 93). When it is first observable at the start of nuclear division, it is a very long thread; this changes to a flexible rod as a result of spiralization. (See *figure 28*, overleaf.) When it has reached its state of maximum spiralization, certain features become visible. Every chromosome has a short region, often seen as a constriction, which is the centromere, already mentioned in Chapter 15 on page 15. This is the region to which the microtubules of the spindle are attached. It also holds a pair of sister chromatids together so that their behaviour can be co-ordinated. Each chromosome in the haploid set of a species has its centromere in a characteristic position. It may be in or near the middle, towards one end, or actually at the end.

In addition to a centromere there may be other, secondary constrictions, the most noticeable being those associated with nucleoli. A region of chromosome arm distal to a secondary constriction (on the opposite side of the secondary constriction to the centromere) is called a satellite. The relative sizes of chromosomes, the position of the centromere, and the presence and position of secondary constrictions are all features which allow identification of individual chromosomes.

A nucleus is a three-dimensional structure; a two-dimensional microscope image of it may be confusing since some chromosomes overlap. A great deal of overlap is likely if there are many chromosomes. It is possible to obtain non-overlapping photographic images of all the chromosomes in a nucleus by treating dividing cells with a drug such as colchicine. This prevents the formation of the spindle, and when cells are in metaphase of mitosis or meiosis the nuclear membrane disrupts and the chromosomes become dispersed in the cell instead of attaching to spindle fibres. They can be photographed in this state and, if the nucleus is diploid, the images can be cut out and pasted on card next to the one they most resemble. This identifies pairs of chromosomes. The largest chromosome and its partner are called chromosome 1, the second largest 2, and so on until the whole chromosome content (karyotype) has been numbered as in *figure 27*.

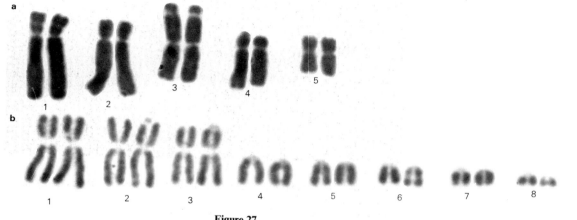

Figure 27
Karyotypes of cells in mitosis, treated so that pairs of chromosomes can be identified and numbered.
a *Puschkinia libanotica* ($\times 2500$ approximately) from a root tip cell.
b *Chorthippus parallelus* ($\times 2700$ approximately) from a cell of an embryo. As with all grasshoppers and locusts, the cells of the male have a single X chromosome. (See also *figure 38*.)
Photographs, Dr C. G. Vosa, The Botany School, Oxford.

STUDY ITEM

16.21 Chromosome banding

Figure 28 shows the karyotype of a tip cell, which is diploid.

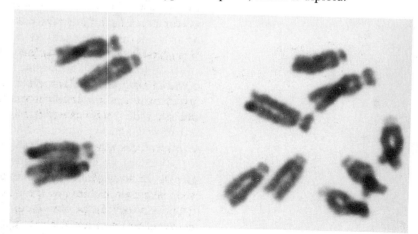

Figure 28
A cell nucleus from the root tip of star of Bethlehem (*Ornithogalum zeiheri*), showing chromosomes ($\times 2600$).
Photograph, Dr C. G. Vosa, The Botany School, Oxford.

a *How many chromosomes are there?*

b *How many sets of chromosomes are there?*

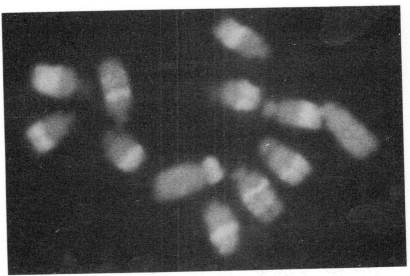

Figure 29
A root tip cell of *Ornithogalum zeiheri* similar to the one in *figure 28*, but stained with quinacrine (× 2475).
Photograph, Dr. C. G. Vosa, The Botany School, Oxford.

Figure 29 shows a cell similar to the one in *figure 28* but it is stained with quinacrine stain. This stain reveals a pattern of bands on the chromosome and their number, size, and position are found to be constant, though what these bands represent is uncertain. They can, however, be very useful as a means of identifying chromosomes.

A number of other stains besides quinacrine have been used to reveal bands on chromosomes. Such techniques not only help homologous pairs of chromosomes to be identified in the karyotype of an individual, but also allow chromosomes of different individuals to be compared in detail. Such comparisons have revealed individuals or populations in which the sequence of bands on one or more chromosomes is different from that of other individuals of that species.

c *Suggest a hypothesis to explain this.*

If the karyotypes of a chimpanzee and a man are compared, using a stain which produces bands on chromosomes, some regions of chromosomes seem to have a sequence of bands identical in the two species.

d *Suggest an explanation for this.*

The salivary gland cells of flies and midges are very large, and appear to have correspondingly large chromosomes which remain visible even when the cell is not dividing. They are called *polytene* or *giant chromosomes* and are the result of repeated replication of the genetic material without a corresponding replication of the chromosome itself. (See page 325.) The bands are called *chromomeres* and they are revealed by most types of stain.

In the fruit fly *Drosophila melanogaster* a well-known inherited variation, bar eye, reduces the number of facets in the compound eye of the adult fly. In a fly culture true breeding for bar eye, the salivary glands of the larvae always have a particular band duplicated on the X chromosome (see *figure 30*).

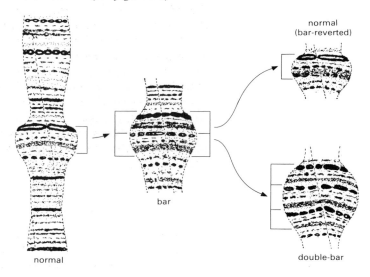

Figure 30
Portions of X chromosomes from the salivary glands of *Drosophila melanogaster*.
Based on Bridges, C. B. 'The bar "gene", a duplication', Science, **83**, 210, 1936.

☐ e *What does this observation suggest?*

16.3 Polyploidy and aneuploidy

Both meiosis and mitosis are processes whereby the products of chromosome replication are distributed in a regular manner between the daughter nuclei. Mitosis results in the maintenance of the chromosome content of the cell, meiosis in a reduction from the diploid to the haploid state.

Failures of meiosis and mitosis are possible, and such failure happens in a very small proportion of cell divisions. (The frequency of failure can be increased by a variety of environmental influences such as radiation and certain chemicals).

If daughter chromatids do not separate at anaphase of mitosis after the centromeres have divided, a new nuclear membrane arises and produces a nucleus in which the chromosome number has doubled. (This can be induced artificially by the drug colchicine.)

$n \longrightarrow 2n$ or $2n \longrightarrow 4n$

Cells which contain more than 2 sets of chromosomes are said to be polyploid. The commonest polyploids are triploids and tetraploids. Triploids contain 3 sets of chromosomes and arise by fusion of a normal haploid gamete with an abnormal 2n gamete. Tetraploids may arise by failure of anaphase of mitosis in a diploid and have 4 sets of chromosomes. Polyploidy is common in organisms which have a means of asexual reproduction during the diploid phase of the life cycle.

Mitosis in a polyploid occurs quite normally and perpetuates the polyploid state. Meiosis in some types of polyploid is abnormal for reasons which are discussed later on page 47.

Polyploids are frequently larger than their diploid relatives, possibly because their cells and nuclei are larger. Much more dramatic changes in the whole organism result from the gain or loss of particular chromosomes as opposed to whole sets of chromosomes.

If a centromere fails to divide at the beginning of anaphase of mitosis, a pair of sister chromatids may move together to one pole of the spindle so that nuclei result, one of which is deficient in a chromosome while the other has an extra chromosome. If chromosomes carry genetic material, one would expect the gain or loss of a particular chromosome to produce corresponding changes in the character of the organism, and this is indeed the case. Organisms composed of cells which have gained or lost particular chromosomes (as opposed to whole sets) are called aneuploids.

Down syndrome: an example of human aneuploidy

Dr John Down was an unusual man who won many distinctions in his medical training, and yet chose the unfashionable career of superintendent of an institution for the mentally retarded. He first described the syndrome which bears his name in 1866. (A syndrome is a set of symptoms which usually appear together and may be causally linked.) Down thought his patients had an oriental appearance and called the condition 'mongolism'. This was because of the skeletal abnormalities which produce upward and outward slanting eyes, flattened bridge of the nose, and other characters. The main features include mental retardation and inadequate muscle tone. There is an increased risk of contracting certain illnesses, including respiratory disorders and leukaemia. The life expectancy of people with the condition is less than usual. It is often claimed that they tend to have a particularly happy, loving, and carefree disposition – important qualities which are difficult to measure objectively. The degree of mental retardation is variable, but with extra stimulation approximately sixty per cent of Down syndrome children may be classified as only mildly mentally handicapped. An increasing number attend normal schools, but, where necessary, special schooling is provided in the U.K. either for day or boarding pupils. The financial and personal problems caused by having a totally dependent member of the family sometimes result in children with Down syndrome being placed in a residential home, although it has been shown that the intelligence and general abilities of these children are much better if they are brought up in a family.

The son of Dr John Down, Dr R. Down, pointed out in 1906 that the facial features of people with the syndrome do not resemble those of any particular oriental group. It is now realized that the condition can occur in people of any national origin and the term Down syndrome is preferred to mongolism. (*Figure 31.*)

The condition is relatively rare in babies born to women under twenty and becomes more likely as the mother gets older; at the age of

Figure 31
Unlike her brothers, the child in the centre was born with Down syndrome.
Photograph, John May.

forty, the mother has a risk of between 1 and 100 and 1 in 200 of having a baby with the disorder. The mean incidence for all mothers is between 1 and 2 in 1000 births. Because most babies are born to women under thirty years old, so are most Down syndrome babies – in spite of the increased risk to older mothers.

In 1959 Lejeune showed, by orcein staining of metaphase chromosomes, that the condition is associated with 3 copies of chromosome number 21 instead of the 2 normal for a diploid. Possession of an extra chromosome is called trisomy and Down syndrome is associated with trisomy 21. Trisomy results either because the paired chromosomes 21 fail to separate at meiosis or else because the chromatids do not separate, also at meiosis. This results in some gamates with no chromosome 21 and others with an extra chromosome 21. The cause of this failure of chromosome 21 to undergo a proper meiosis is unknown. It used to be thought to occur only during the production of eggs, but there is positive evidence that in about a quarter of the individuals with Down syndrome the extra chromosome originated with the father's sperm. (It must be remembered that while older mothers have an increased risk of having a Down syndrome baby, this does not itself prove they are the cause of the abnormality. Older mothers are usually married to older fathers!)

If a gamete which completely lacks a chromosome 21 achieves fertilization, the resulting embryo fails to survive. It is estimated that a large proportion of trisomy 21 embryos – which if they survived would have Down syndrome – also undergo miscarriage.

a *How could the tendency of trisomics to undergo miscarriage be discovered?*

Cells from the amniotic fluid surrounding an embryo may be drawn off, together with a small amount of the fluid, using a syringe (a technique called amniocentesis). These embryonic cells may be induced to divide in culture and their chromosome content (karyotype) determined. Trisomy 21 (and other abnormalities) can thus be detected during pregnancy and the mother can then choose whether to have the pregnancy artificially terminated.

STUDY ITEM

16.31 **Chromosome changes in a plant**

Polyploidy and aneuploidy sometimes occur together in flowering plants. The dandelion *Taraxacum officinale* is a triploid species (3n = 21).

a *What is the basic number of chromosomes in* **Taraxacum**?

Dandelion varieties are known in which the chromosome number is 20 rather than 21. These varieties can be distinguished by leaf shape, habitat preference, and other characters.

b *How many different varieties with a chromosome number of 20 could be produced by aneuploidy alone from an ancestral variety with a chromosome number of 21?*

Aneuploidy in a diploid species such as *Homo sapiens* is usually either lethal or results in an individual with severe handicaps. Aneuploidy in a polyploid species such as dandelion results in much less disturbance to normal development.

c *Would it be correct to argue from this that dandelion chromosomes carry fewer genes than human chromosomes?*

Examples of the effects of polyploidy on the growth of a plant species can be seen in *figure 32*.

Figure 32
Leaves and inflorescences of *Pelargonium zonale*. The plant on the left is haploid with 9 chromosomes; the central one is diploid with 18 chromosomes; and the plant on the right is tetraploid with 36 chromosomes.
Photograph, Dr C. G. Vosa, The Botany School, Oxford.

Dandelions are notorious for producing vast numbers of viable wind-dispersed fruits (often incorrectly called seeds). These do not result from sexual reproduction. The embryo results from an asexual process involving nuclear divisions resembling mitosis.

16.4 Meiosis and its significance

We have established in the introduction to this chapter that a special type of cell division is needed to reduce the chromosome number to the haploid state, so that fertilization may follow without doubling the chromosome number with each sexual generation. During meiosis, the nucleus divides twice while the chromosomes replicate only once. The result is four daughter nuclei, each with a single set of chromosomes, rather than the two sets of the original cell. The two nuclear divisions involved are called the first and second divisions of meiosis. Chromosome replication occurs in the interphase before division I, but the division of centromeres is delayed until division II. The events of meiosis are summarized in *figure 33*.

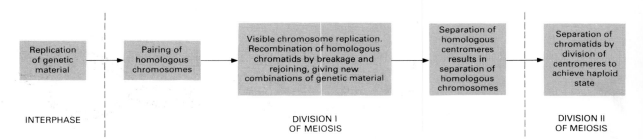

Figure 33
A summary of the events of meiosis.

Meiosis, like mitosis, is a continuous process which is arbitrarily divided for convenience of description into separate named stages. The first division of meiosis includes many significant events and the prophase of this division is therefore often divided into five named substages.

It may help you to understand meiosis if you watch a film of the process in which cultured cells from testes or anthers were used and photographed with phase contrast illumination. You will be able to see some stages at first hand if you carry out practical investigations 16A and 16B.

> **Practical investigations.** *Practical guide 5*, **investigation 16A, 'Gamete production by a plant', and 16B, 'Gamete production in an animal'.**

Meiosis, like mitosis, starts with an interphase during which the genetic material replicates. The chromosomes appear at the beginning of first prophase as long threads which are single and not visibly duplicated into chromatids. The long thin threads condense and, as they do so, each comes together with its homologue. (Remember that one chromosome of an organism is of maternal origin while its homologue, in the other set, is inherited from the organism's male parent.) Eventually each chromosome pair becomes closely associated, centromere with centromere, secondary constriction with secondary constriction, stainable band with stainable band. The pairs become very closely entwined and during their intimate association they become visibly duplicated into chromatids so that each pair consists of four strands.

Having become paired and duplicated, the chromosomes start to repel each other and move apart, but they remain attached at one or more points called *chiasmata* (singular: *chiasma*).

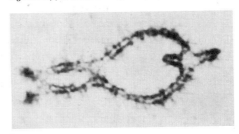

Figure 34
a A photomicrograph of chromosomes showing chiasmata.
b Diagram showing two hypotheses to explain the presence of chiasmata.
Photograph, Dr C. G. Vosa, The Botany School, Oxford.

Figure 34 shows two hypotheses which explain the presence of chiasmata. Hypothesis *a* suggests that maternal genetic material and paternal genetic material are conserved in their original form while *b* means that both maternal and paternal threads (chromatids) break and rejoin to form new composite structures. We shall examine the evidence for hypothesis *b* on page 81. Examine *figure 34* carefully and you will see that all four chromatids may be involved in chiasmata. If hypothesis *b* is correct and chiasmata are points of interchromosomal genetic exchange this is of great importance as it explains how variation may arise as a result of the formation of new combinations of genetic material (recombination).

As chromosome condensation proceeds and homologous centromeres repel one another more strongly, the position of the chiasmata is shifted towards the ends of the chromatids, and changes in the shape of the bivalents occur. These are explained in *figure 35*, which shows a situation with only one chiasma. Remember that several chiasmata are possible and lead to much more complex shapes.

Eventually, the nuclear membrane breaks down, and this marks the end of first prophase. First metaphase follows. The spindle forms and

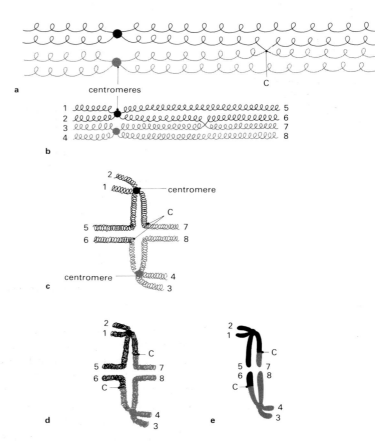

C = chiasma

Figure 35
A diagram showing the sequence of development of a single bivalent during 1st prophase of meiosis. The maternal chromosome material is black, the paternal grey. The distance between chromatids has been exaggerated to avoid confusion.
The whole bivalent would be curled up during the early stages of spiralization.
The change from configuration **b** to **c** is possibly caused by the mutual repulsion of homologous centromeres. The ends of chromatids have been numbered so that the two structures can be related.

spindle microtubules move the centromeres to the equatorial region (the metaphase plate). Each pair is orientated so that one homologous centromere will move to one spindle pole and the other to the opposite pole. It is important to notice that if one pair orientates with the maternal centromere towards pole A of the spindle, then a second pair may either orientate similarly (maternal centromere towards pole A) or in an opposite manner to the first pair (maternal centromere towards pole B). These possibilities are explained in *figure 36*.

a *What would be the genetic consequence if the first pair of chromosomes to orientate on the metaphase plate determined the orientation of the other pairs?*

In first prophase and first metaphase, homologous chromosomes are

Figure 36
Diagrams of two nuclei at 1st metaphase. There are two bivalents in each nucleus. The maternal chromosome material is black, the paternal material grey.

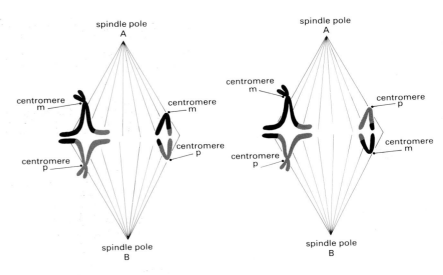

centromeres m and m = maternal centromeres
centromeres p and p = paternal centromeres

held together as pairs by the mutual attraction of sister chromatids and by the formation of chiasmata. The homologues pair closely at the start of 1st prophase but seem to repel one another quite strongly by the end of 1st prophase. The end of 1st metaphase and the start of 1st anaphase are reached when the terminal chiasmata part, and the homologous centromeres of each pair move apart towards the spindle poles with the chromatid arms trailing behind.

The resulting groups of chromosomes may despiralize and nuclear membranes may form around them, or each group may become immediately associated with a new, second division, spindle. Because of this variation, the end of the first division of meiosis, 1st telophase, is the most clearly artificial of the stages into which, for convenience, the process of meiosis is divided.

If daughter nuclei form at the end of the first division, there is a 2nd prophase during which the chromosomes spiralize. The nuclear membranes break down at the end of 2nd prophase and second division spindles form.

After second division spindles have formed (either during 1st telophase or after 2nd prophase) the centromeres become attached to the spindle microtubules and are drawn to the equator. This is called 2nd metaphase. The centromeres finally divide and sister chromatids move apart in 2nd anaphase. The chromatids despiralize, nucleoli appear, and nuclear membranes develop in 2nd telophase. First and second telophase are usually accompanied by the division of the cytoplasm. The complete process of meiosis thus results in 4 haploid cells being formed.

STUDY ITEM
16.41 Meiosis summarized

The events of meiosis are restated as a series in *figure 37* and illustrated by randomly arranged photomicrographs in *figure 38* (pages 45–46).

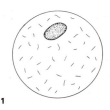

Interphase Genetic material replicates.

1

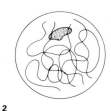

Start of 1st prophase Nucleolus becoming disorganized – chromosomes are long, thin, single threads,

2

Homologous chromosomes are pairing.

3

Chromosomes have shortened and thickened by spiralization and may be so closely paired that the bivalents (pairs) may appear as single threads.

4

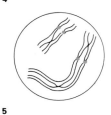

Bivalents fall apart slightly, to show chiasmata.

5

End of 1st prophase Condensation continues and this makes chromatids hard to see. Chiasmata become terminalized by repulsion of homologous centromeres.

6

1st metaphase Nuclear membrane has disintegrated. Centromeres, now repelling strongly, orientate on spindle. Bivalents are held together by terminalized chiasmata. Sister chromatids are closely associated and appear as a single thread

7

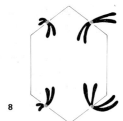

1st anaphase Association of sister chromatids ceases. Centromeres move towards poles of spindle, drawing chromatid arms behind them.

8

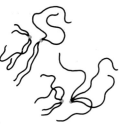

1st telophase Pairs of chromatids, joined at the centromere, may despiralize

9

2nd metaphase Two new spindles form. The chromatid pairs, now recondensed, orientate on the spindles by means of their common centromere.

10

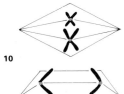

2nd anaphase Centromeres duplicate. Chromatids (now daughter chromosomes) move apart.

11

2nd telophase Daughter chromosomes of four haploid nuclei despiralize.

12

Interphase Chromosomes have completely despiralized and nuclear membranes have formed around the nuclei.

13

Figure 37
Diagrams showing the series of events in meiosis.

(Remember that there are more chromosomes in the photomicrographs.)

a *Study these two figures and make a table such as the one that follows, which best matches the numbered drawings of* figure 37 *with the photographs of* figure 38. *(Not every drawing has a corresponding photograph.)*

Drawing number (from figure 37)	Photograph number (from figure 38)
1	
2	
3	
etc.	

b *What is the haploid chromosome number in drawing 3 of* figure 37?

Figure 38
Photomicrographs showing the events of meiosis, arranged in random order.
Photographs **a**, **b**, **c**, **d**, **e**, **g**, **h**, *and* **i**, Dr C. G. Vosa, The Botany School, Oxford.
Photographs **f**, **j**, **k**, *and* **l**, Gareth Jones.

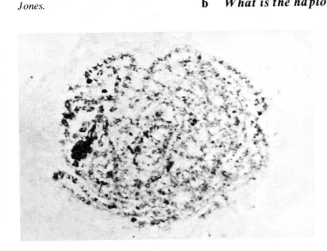

a

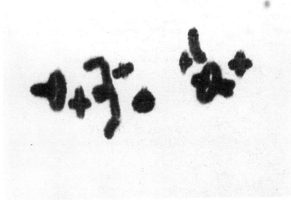

b

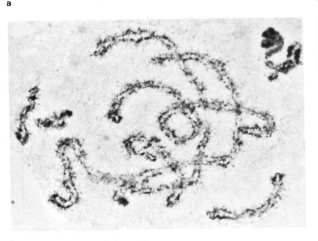

c

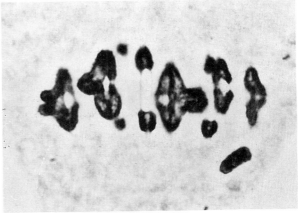

d

Chapter 16 The cell nucleus and inheritance

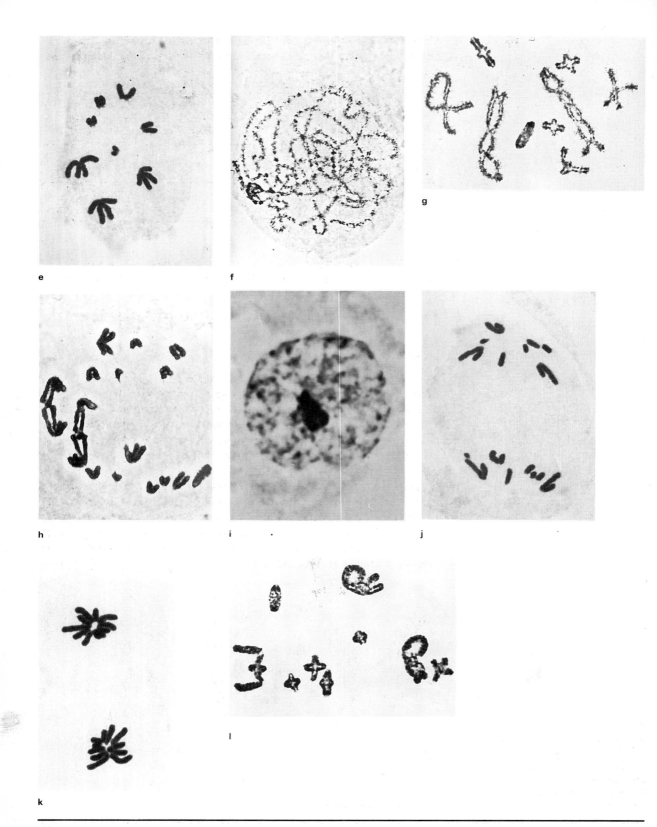

c Consider a chromosome with a centromere near the middle and a second, non-homologous chromosome with the centromere at one end. Copy and complete table 4 for these chromosomes.

Chromosome type	Number of chromatid arms trailing at anaphase of		
	mitosis	meiosis (division I)	meiosis (division II)
Centromere near middle			
Centromere at one end			

Table 4

d Give two reasons why it is important that chromosomes come together in their homologous pairs at meiosis and why this is not necessary at mitosis.

e For a diploid gamete to result, when would meiosis need to fail?

Meiosis in hybrids and polyploids

Meiosis has been called 'the dance of the chromosomes'. We have seen how each chromosome becomes aligned with, and pairs with, its homologous 'partner' and moves in an orderly manner on the spindle at anaphase so that one set of chromosomes segregates from another rather like two groups of dancers in an elaborate ballroom routine. The end product is four daughter nuclei, each with a complete and balanced complement of genetic material.

Imagine the chaos which would result if three or more people tried to carry out the movements of a waltz or some other formal dance normally performed by pairs of people. This is analogous to meiosis in a polyploid organism. In a triploid, each chromosome has two possible partners when pairing takes place during prophase I. *Figure 39* shows the possible pairing relationships in a triploid.

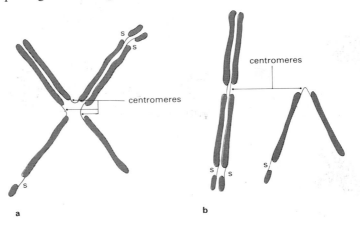

Figure 39
A diagram showing chromosome associations in a triploid.
a An association of 3 homologous chromosomes.
b A pair of chromosomes and a lone one.

s = secondary constriction

Single chromosomes and groups of more than two do not orientate properly on the spindle at 1st metaphase and there is no way of ensuring an orderly segregation into complete sets of chromosomes when meiosis has finished. Four nuclei with unbalanced chromosome complements result.

Tetraploidy can arise from a mitosis which has failed at some point after the duplication of the chromosomes. The tetraploid nucleus has four sets of chromosomes and, because this happened by chromosome duplication without chromosome separation, each chromosome has a genetically identical partner. If pairing could take place just with this sister chromosome, meiosis would be normal, but pairing is also possible with the homologous chromosome and its duplicate.

b *Figure 39 shows possible associations of chromosomes in a triploid at 1st prophase of meiosis. Make similar drawings to show the possible associations of four homologous chromosomes in a tetraploid. Show the two identical maternally derived ones in one colour and the two paternally derived ones in another colour. (Do not include a 'secondary constriction'.)*

The extent to which a tetraploid is fertile depends on a number of factors. In many cases meiosis is normal because the chromosomes pair in a regular manner, but chains of 3 and single chromosomes also form and this may result in frequent aneuploidy among the progeny of the tetraploid. Irregular pairing makes some tetraploids almost completely sterile.

Hybrids between species are often sterile and can only perpetuate themselves by some means of asexual reproduction where this exists. A hybrid zygote has a paternal set of chromosomes derived from one species and a maternal set derived from another species. The chromosomes may be so genetically different that they fail to form pairs at 1st prophase of meiosis and therefore fail to separate normally at 1st metaphase. Gametes with unbalanced chromosome complements result. When a hybrid is sterile because the paternal and maternal chromosome sets are too different for each chromosome to find a partner, the partner can be provided if the hybrid becomes tetraploid. These arguments are summarized in *figure 40*.

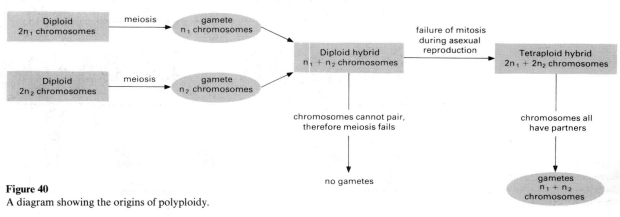

Figure 40
A diagram showing the origins of polyploidy.

Hybridization followed by polyploidy has been very important in the evolution of flowering plants and ferns and many species are known to be hybrid polyploids. Bread wheat (*Triticum aestivum*) is perhaps the most important example (see page 176).

c *Why has polyploidy been unimportant in the evolution of animals?*

16.5 The inheritance of sex

In the preceding sections, we have stressed that individuals of the same species usually have extremely similar chromosomes. Meiosis ensures that each gamete of a species normally has a set of chromosomes which appears similar to that of every other gamete of that species. The sex chromosomes which exist in the nuclei of many types of animal are an important exception. Humans have 46 chromosomes (n = 23) in their body cells and so each gamete has 23. One of these 23 is the sex chromosome, of which there are two forms, X and Y. Men have one of each type of sex chromosome in each of their body cells. These pair at meiosis and as a result sperms of two sorts are made, those that have an X and those that have a Y. Women have a pair of X chromosomes in their body cells and egg cells are thus all of one sort, that is, with an X chromosome. Human male and female karyotypes are shown in *figures 41* and *42*. In a man Y chromosomes are smaller than X chromosomes.

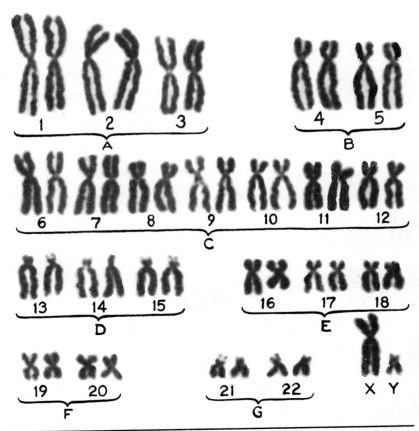

Figure 41
Human male chromosome karyotype.
Photograph, Biophoto Associates.

Figure 42
Human female chromosome karyotype.
Photograph, Biophoto Associates.

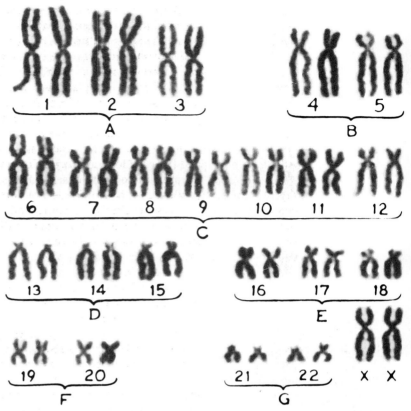

a   *Consider meiosis of a single human male cell. How many X-bearing and how many Y-bearing haploid nuclei would result?*

b   *If a man liberated 3×10^9 sperm cells at sexual intercourse, how many of these would you expect to be Y-bearing?*

c   *What ratio of male to female children will be expected in any population? Explain your answer and state what assumptions you have made.*

In humans and in other mammals, the male has two types of sex chromosome and can produce two types of gamete. The male is therefore called the heterogametic sex. In flies, grasshoppers, and spiders, the male is also heterogametic. In grasshoppers, there is no Y chromosome, the male having a single X chromosome which does not pair.

d   *What types of haploid cell would meiosis produce in a male grasshopper, and which of these types would produce male offspring after fertilization?*

In birds and butterflies, the female is the heterogametic sex. (Cocks have two X chromosomes and hens have an X and a Y.) It seems that a chromosome mechanism for determining the sex of offspring has arisen independently in several taxonomic groups.

Aneuploidy involving sex chromosomes has been observed in many species including humans. It results from faulty pairing of the X and Y chromosomes during meiosis or from failure of separation of chromosomes or chromatids (see section 16.2, page 33). Turner's syndrome occurs once in about 3500 female live births but is much more frequent in naturally aborted foetuses. In this condition all body cells have a single X chromosome. Zygotes producing Turner's syndrome probably arise from fertilization of a normal egg by a sperm lacking any sex chromosome. Turner's syndrome children are female but the ovaries, and consequently the secondary sexual characteristics, do not develop and puberty does not occur. Treatment with female hormones results in normal development of secondary sexual characters, but no eggs can be produced as the ovaries do not develop.

In Klinefelter's syndrome, body cells are XXY. Individuals with this syndrome are males with abnormally small testes which do not produce sperm, but the glands producing seminal fluid are active and sexual intercourse is possible. The extent to which secondary sexual characters are affected is variable.

XXX individuals occur once in 1000 female births. Such individuals have underdeveloped genital organs and limited fertility, and are often somewhat mentally retarded.

XYY males occur once in every 1000 male births. XYY men are significantly larger than normal XY men, though there is a great deal of overlap in size between the two groups.

Individuals that have only a Y chromosome are never found, presumably because the X chromosome has genetic material vital for development and zygotes lacking an X chromosome die.

While the genetic sex of the individual is determined at fertilization by the sex chromosome content of the gametes, the environment in which the embryo, and later the child, develops may influence the extent to which male or female characters develop. It would be a mistake to suppose that all genetic material relevant to sexual development is found on the sex chromosomes. The sex chromosomes probably act as a switch altering development in a male or female direction. From time to time individuals develop in whom the external manifestations of sex (penis, scrotum, or vagina) do not correspond to the genetic sex. A clue to genetic sex can be quickly obtained by taking a smear of cells from the lining of the inside of the cheek. The epithelium here undergoes continuous proliferation and large, flattened (squamous) cells are continually sloughed off and swallowed. These cells can be stained with dyes such as orcein (see *figures 43* and *44*). If two X chromosomes are present, one is invariably condensed in the interphase nucleus as a large blob of chromatin called a Barr body. The other X chromosome is completely unspiralized at interphase, and is presumably active in controlling cell development.

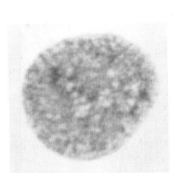

Figure 43
Human interphase nucleus from a male (× 2500).
Photograph, Dr C. G. Vosa, The Botany School, Oxford.

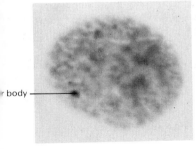

Figure 44
Human interphase nucleus from a female, showing the Barr body (× 2250).
Photograph, Dr C. G. Vosa, The Botany School, Oxford.

e *How many Barr bodies would you expect in nuclei of cheek cells from a girl with Turner's syndrome?*

Determining the genetic sex by the presence of Barr bodies is not entirely reliable and demands good staining. It is much more reliable to harvest white cells from a blood sample and induce them to undergo mitosis in culture when they can be treated with colchicine, and photographed when the chromosomes have dispersed in the cell. The genetic sex of very young embryos can be determined by a similar examination of cells from amniotic fluid.

The existence of sex chromosomes is in itself good evidence for the hypothesis that chromosomes carry genetic material, but we can go further than this. Male and female gametes have identical sets of non-sex chromosomes (autosomes), and this explains why reciprocal crosses usually produce the same result in breeding experiments (see section 16.1). As the X and Y chromosomes are visibly different, it is reasonable to assume that a unit of genetic material present on one might be completely absent from the other. In this case reciprocal crosses would not yield the same offspring. A good example of this is the inheritance of coat colour in cats. The patterns of inheritance found are summarized in table 5.

Cross	Character shown by male parent	Character shown by female parent	Characters shown by offspring
1	ginger	black	black males tortoiseshell females
2	black	ginger	ginger males tortoiseshell females
3	ginger	tortoiseshell	males are either black or ginger (in equal numbers) females are either tortoiseshell or ginger (in equal numbers)
4	black	tortoiseshell	males are either black or ginger (in equal numbers) females are either tortoiseshell or black (in equal numbers)

Table 5
(Note that 'ginger' cats have red markings on a pale red background; 'tortoiseshell' cats have an irregular pattern of black and red markings. All three colours may be associated with patches of white. This is inherited quite independently, and the presence or absence of white does not influence the pattern of inheritance shown in the table.) Tortoiseshell male cats are unknown.
Notice that in crosses 1 and 2 the male offspring always inherit the mother's coat colour.

f *From which parent must any male mammal inherit its X chromosome?*

g *Explain why tortoiseshell male cats are not found (with the exception of rare, infertile, XXY individuals).*

A fuller analysis of this interesting example must be delayed until we have mastered some concepts and procedures which are introduced in Chapter 17; we will return to it again in Chapter 19 when we consider the control of the expression of genes. The case clearly associates the inheritance of a character which has nothing to do with sex with the sex chromosomes. It provides very strong evidence that genetic material is carried on chromosomes and transmitted in an orderly and predictable manner as a result of meiosis. Non-sexual characters controlled by genes carried on the sex chromosomes are called sex-linked characters.

STUDY ITEM

16.51 The sex ratio in human beings

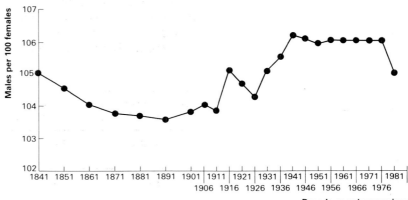

Figure 45
Sex ratio in live births in England and Wales 1841–1981.
Updated from Parkes, A. S., 'Sex ratio in human populations', in CIBA Foundation Symposium Man and his future, *Wolstenholme, G. (ed.), Churchill, 1963.*

a *What factors might change the expected ratio of the sexes, at birth, in a population?*

b *What are the advantages for humans (and other organisms) in having a 1:1 sex ratio? In what circumstances would it be a disadvantage?*

In most human populations more males are born than females, and the balance is tipped still more strongly in favour of males if the sex of spontaneously aborted foetuses and stillbirths is included in the data from which ratios are calculated.

c *There is a positive correlation between the excess of males over females at birth and the affluence of the group of mothers sampled. Put forward a hypothesis to explain this.*

In the past there has been a differential mortality between males and females, partly due to war and to accidental death, but also to different survival rates from a variety of causes. *Figure 46* shows how the ratio of males to females of various ages has changed between 1901 and 1983.

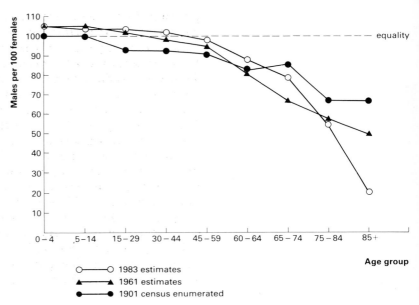

Figure 46
Sex ratio according to age (United Kingdom)
Data from the Central Statistical Office, Social Trends, *reproduced by permission of the Controller, HMSO, 1984.*

d *How might such changes in the proportions of males and females affect the size and composition of future generations?*

In a number of societies in different parts of the world, there has been a practice of female infanticide; that is, girl babies were sometimes killed at birth by the mother or by relatives. This practice allowed more than one man to be available to help support a woman and her children when conditions were very harsh. In Tibet, where some of the excess men become monks, they still contribute to the upkeep of the family.

In most societies, at least some families have a strong desire for a child of a particular sex, if only because they already have children of the opposite sex.

e *If the procedure of determining karyotype from cells of the amniotic fluid becomes routine, what might be the influence on the sex ratio and on the growth of populations?*

16.6 Inheritance in fungi

A large number of fungal species have been used in genetic investigations, and some of the most important steps in our understanding of genetic systems have been arrived at by using fungi as experimental material. They have a number of advantages over higher animals and plants and over insects such as fruit flies. These advantages include the following.

1 The existence of clear-cut biochemical variation; for example, one strain may require an organic substance such as a vitamin in its growth medium while a second strain is able to synthesize this substance from simple raw materials.

2 They are eukaryotes with meiosis similar to higher organisms.

3 The existence of a well-developed haploid phase in the life cycles; this means that the influence of a chromosome, or a gene situated on a chromosome, may be investigated without the need to take into account the effect of a homologous chromosome.

4 Extremely large numbers of spores are produced in a very small space, so that statistically improbable events can be detected.

5 Fungi grow quickly and cheaply in a small space and their environment can be tightly controlled.

6 Some species have a phase in the life cycle where two types of haploid nuclei, derived from different individuals, exist in a common cytoplasm but do not fuse. Such a heterokaryotic phase allows interaction of nucleus and cytoplasm to be studied (see page 12).

7 Some species develop specialized structures in which meiosis takes place and all four haploid products of a single meiosis remain together and can therefore be observed and isolated for study.

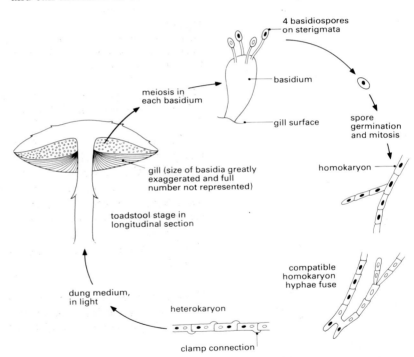

Figure 47
The life cycle of *Coprinus lagopus*. The fungus can produce asexual spores during the homokaryotic and heterokaryotic phases. These asexual phases have been omitted to simplify the diagram. (Note also that the diagram is not to scale.)

Figure 47 shows the life cycle of *Coprinus lagopus*. This is a small toadstool. Like all mushrooms and toadstools, it has a diffuse mycelium which digests organic matter and absorbs the products. Unlike fungi such as *Mucor*, with which you may be familiar, its mycelium is cellular. Some samples of mycelium have single nuclei in the cells. Such mycelia have grown by mitosis from a single uninucleate spore. The nuclei are haploid.

Other samples of mycelium have two nuclei in each cell. These heterokaryotic mycelia have characteristic clamp connections on the cells. Clamp connections are the result of a rather special type of cell division which ensures that each new cell has two genetically different nuclei. The details of this need not concern us here.

Both homokaryotic and heterokaryotic stages of *Coprinus* will grow on simple agar media. If two compatible homokaryons are put on the same culture plate they will fuse and form a heterokaryon. The heterokaryon grows more vigorously than either homokaryon. Only certain homokaryon strains will fuse in this way; other strains always remain separate. The ability of strains to fuse (compatibility) is under genetic control.

If a heterokaryon is transferred to a complex medium based on sterilized animal dung and kept in the light, it develops into toadstools. Underneath the caps of the toadstool are radiating gills similar to those on the mushrooms you can buy from a greengrocer. The gills are covered with large cells called basidia. Nuclear fusion takes place in the formation of the basidia and these then undergo meiosis. The four nuclei produced by the meiosis each migrate into one of the four spores, called basidiospores, which remain attached to the basidium that produced them (see *figure 48*).

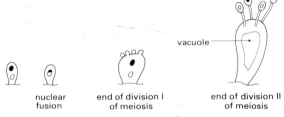

Figure 48
A diagram to show the formation of basidiospores.

If a heterokaryon is produced by fusion of a normal homokaryon and a second, compatible homokaryon which is unable to grow without the vitamin choline, the heterokaryon does not need an external supply of choline. If the heterokaryon is transferred to dung medium, it undergoes nuclear fusion after producing toadstools. A piece of gill can be cut from a toadstool and examined with a powerful dissecting microscope so that sets of basidiospores can be collected with a sterile needle. Each of the four basidiospores from a single basidium can be separately transferred to sterile, labelled Petri dishes and homokaryotic mycelia grown from each. These homokaryotic mycelia can then be tested for their ability to grow without choline.

It is invariably found that, on any basidium, two basidiospores grow into mycelia that can survive without choline in the growth medium, and two grow into mycelia that must have an external source of choline for significant growth, that is, two basidiospores take after each parent.

In order to interpret this result, two important genetic concepts must be introduced and defined.

First, the term *locus*: this is the position on a chromosome occupied by a specific unit of genetic material (gene).

Second, the term *allele*: alleles are different but alternative units of genetic material which occupy exactly the same locus on different but homologous chromosomes. By definition, alleles are inevitably separated when a diploid undergoes meiosis and forms gametes. The allele concept is illustrated by Practical investigation 16C.

> **Practical investigation.** *Practical guide 5*, investigation 16C, 'Inheritance of ability to synthesize starch in *Zea mays*' and investigation 16D, 'Spore colour in *Sordaria fimicola*'.

a   *Explain how the inheritance of an ability to produce choline in* Coprinus *illustrates the allele concept.*

b   *The pollen grains of a flowering plant are the haploid products of meiosis. Explain why pollen is much less useful than basidiospores as a means of establishing the allele concept.*

Staining techniques show that chromosomes have a very constant linear arrangement of bands and constrictions and it is therefore reasonable to assume that the genetic material has a constant linear arrangement on the chromosomes. This assumption may not be justified and needs testing. In most species there are many more genes than observable chromosome bands. Genes and their alleles cannot be seen, but their existence can be inferred from patterns of inheritance. Stronger evidence for a linear arrangement of loci is examined later (see page 80).

STUDY ITEM

16.61 Spore formation in *Neurospora*

One of the most famous fungi is the mould *Neurospora crassa*. Its life cycle is very similar to that of *Sordaria fimicola*. (See Practical investigation 16D.)

> **Practical investigation.** *Practical guide 5*, investigation 16D, 'Spore colour in *Sordaria fimicola*'.

The inheritance of biochemical characters in *Neurospora* was extensively investigated by two American scientists, G. W. Beadle and E. L. Tatum, who published their findings in the mid 1940s. We will consider some of their work in more detail in Chapter 19. The way in which spores are formed in the sexual cycles of *Neurospora* and *Sordaria* is summarized in *figure 81*. Notice carefully that the cell within which the spores are formed, the ascus sac, is so narrow that the nuclei must lie in single file and cannot slide past each other.

One set of results is shown in table 6. Two different haploid strains were crossed to produce the asci, one wild-type strain (able to synthesize the amino acid lysine from raw materials) and the other requiring lysine.

Figure 49
Wild-type *Neurospora crassa* on a petri dish.
Photograph, Dr Alan Radford, Department of Genetics, University of Leeds.

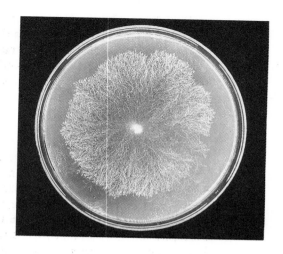

Table 6
N is normal growth on medium without lysine.
s is slight growth on medium without lysine (the spore contains a small store of lysine).

Spore number	Ascus number					
	1	2	3	4	5	6
1	s	s	N	N	s	N
2	s	s	N	N	s	N
3	s	N	s	N	s	N
4	s	N	s	N	s	N
5	N	s	N	s	N	s
6	N	s	N	s	N	s
7	N	N	s	s	N	s
8	N	N	s	s	N	s

a If the wild-type allele is represented by + and the lysine-requiring allele by L, what would be the genetic constitution of the diploid cell which produced each ascus?

b What would be the possible genetic constitutions of the nuclei in stage B of figure 50?

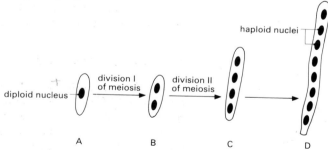

Figure 50
Stages in the development of an ascus.

c 'It is clear from these rather limited data that this inability to synthesize lysine is transmitted as it should be if it were differentiated from the normal by a single gene.' Explain this statement.

d Account for the difference in sequence between the spores in ascus number 3 and ascus number 4 in table 5.

e Between which two stages shown in figure 50 must this difference in sequence have arisen when the two asci developed?

In higher organisms there are no structures equivalent to the asci of *Neurospora*. The diploid phase of the life-cycle is the well developed one, the haploid phase being very restricted. The existence of alleles can be inferred from breeding experiments in higher organisms, but the number of assumptions that must be made are greater, as we shall see in the next chapter.

Summary

1. Male and female gametes contribute equally to inheritance because genetic material is found on their chromosomes (**16.1**).
2. Sexual reproduction involves fertilization which doubles the chromosome number. This must be halved again by meiosis at some stage of any life cycle (*figure 26*) (**16.1**).
3. Chromosomes have an individual structure, recognizable by staining. Each gamete contributes a similar set of chromosomes to a zygote which thus has two sets, in homologous pairs (**16.2**).
4. In meiosis the chromosomes replicate once and the nucleus divides twice. Each of the four haploid nuclei receives one of each type of chromosome (that is, one complete set) (**16.3**).
5. Chiasma formation allows recombination of maternal and paternal genetic material, as does the independent alignment of different pairs of chromosomes at metaphase I (**16.3**).
6. Changes in the chromosome content of an organism usually result in changes in development, and disturb the pairing of the chromosomes during meiosis (**16.2** and **16.3**).
7. Sex is determined by a chromosomal mechanism, and sex linkage is good evidence that genes occur on specific chromosomes (**16.4**).
8. Alleles are alternative units of genetic material found at the same point (locus) on homologous chromosomes (**16.5**).
9. In fungi it is possible to associate the inheritance of particular characters, involving biochemical products of cells, with the transmission of chromosomes to spores as a result of meiosis (**16.5**).

Suggestions for further reading

BOYCE, A. J. (ed.) Symposia of the Society for the Study of Human Biology Vol. 14 *Chromosome variations in human evolution.* Taylor & Francis, 1975. (An advanced text which describes the techniques used to study and compare chromosomes from different species.)

JOHN, B. and LEWIS, K. R. Carolina Biology Readers No. 65 *The meiotic mechanism.* Carolina Biological Supply Company, distributed by Packard Publishing Ltd, 1973.

KEMP, R. Studies in Biology No. 21 *Cell division and heredity.* Arnold, 1970.

MCDERMOTT, A. Outline Studies in Biology *Cytogenetics of man and other animals.* Chapman & Hall, 1975.

CHAPTER 17 VARIATION AND ITS CAUSES

17.1 Introduction

Organisms may differ from one another in an infinite number of ways. Sometimes the differences are obvious, but they can be subtle and may need careful measurements or special biochemical tests in order to reveal them. Similarities between individuals can also be striking as witnessed by the phrase 'as like as two peas in a pod'. Similarities and differences between individuals are considered under the general heading *variation*.

Variation has attracted the attention of scientists from the earliest times. People have asked such questions as:

1 What causes some kinds of organism to be very variable while others are relatively uniform?
2 Is variation produced by differences in the influence to which individuals are subjected as they grow and mature or do they inherit their differences?
3 What sorts of variable character are important for recognizing and naming different types or species of plant or animal?
4 How different must two individuals be before they are considered to be of different species; in other words, what is a species?
5 If variation is, at least in part, inherited, how does this come about?

STUDY ITEM

17.11 Continuous and discontinuous variation

> **Practical investigation.** *Practical guide 5*, **investigation 17A, 'Describing variation'.**

Some characters are said to vary discontinuously. This is the case if individuals can be sorted out into distinct groups (or sets) according to whether they show the character or its alternatives; for example, birds have feathers, mammals hair, some humans can roll their tongues into a U cross-section, others are unable to do this. There may be more than two alternative forms in a case of discontinuous variation, for example, humans may be blood group A or B or AB or O.

Continuously varying characters are those which do not allow individuals to be sorted out into distinct groups; for example, apples are not either sweet or sour (acid) but have every imaginable level of sugar and acid content within certain limits; humans are not either black or white, but show a range of skin shades from very pale to very dark.

a *Prepare two lists of characters which show variation in some familiar organisms. In one list, place the continuously varying characters and in the second, those showing discontinuous variation (or use the results of Practical investigation 17A, 'Describing variation').*

Variety	Height of plant from terminal bud to soil surface (cm)
Sutton's Purple Podded	44 46 43 49 57 55 38 55 52 44 55 51 50 53 65 54 53 61 62 63
Meteor	21 14 18 16 14 18 14 12 15 20 18 13 19 15 21 20 17 19 15 17

Table 7
Heights of young pea plants of two garden varieties grown in pots on a window sill (see figure 52).

Study table 7 and answer the questions that follow (you may need to refer to *Mathematics for biologists* to help you with some of these).

b **Prepare a table which groups the plants of each variety separately into 3 cm classes.**

c **Use this class frequency to show the size distribution of Meteor and Sutton's Purple Podded peas as a pair of histograms. Show these on the same set of axes, using different colours.**

d **Calculate and list the mean, the median, and the modal class values for each variety.**

e **In presenting data of this kind, explain why it is correct to use histograms and not continuous graphs.**

f **The data have been grouped into 3 cm classes. Give the advantages and disadvantages of using a smaller class interval. How can the class interval chosen for a histogram be related to sample size?**

g **What calculation would be needed to evaluate the hypothesis that the two groups of peas differed significantly in height?**

h **Is height a continuously or a discontinuously varying character in these varieties of peas?**

i **Making allowance for sampling error, is it probable that height is normally distributed in peas?**

j **Give two reasons why discontinuous rather than continuous variables are referred to in the construction of a key.**

17.2 The role of inheritance and environment

> Practical investigation. *Practical guide 5*, investigation 17B, 'The influence of environment on development'.

As an individual grows and develops, it is controlled by its genetic material and by the environment in which development takes place. All characters may therefore be under the influence of both inheritance and environment. Differences between individuals must arise from differences in the environment in which development took place or as a result of genetic differences, or from a combination of these influences. When it is stated that a character, such as coat colour in cats, is inherited, what is really meant is that variation results largely from genetic influences and that changes in the external environment influence the degree and type of variation very little.

The locust *Schistocerca gregaria*, on the other hand, shows marked variation depending on whether it is reared in isolation (low density) or in a group (high density). The differences are so great as to have led observers to assume that the two forms (swarm phase and solitary phase) were different species.

a *Design an experiment to test the hypothesis that the ratio of wing length to abdominal length of locusts is determined by the density at which newly hatched (first instar) hoppers are kept.*

A common waterside plant, amphibious bistort (*Polygonum amphibium*) (*figure 51*) can be found as two distinct forms, one growing in the water with elliptical floating leaves on distinct petioles, the other on land with narrower, erect, more hairy leaves on shorter petioles. The two plants appear strikingly different, yet if cuttings are taken from each and planted in the habitat of the other, new growth will occur which shows that the forms are interchangeable, the variation depending on the environment in which growth takes place. The examples mentioned above are of situations where the external environment produces striking variation.

Figure 51
Polygonum amphibium. Left, the terrestrial form; *right*, the aquatic form.
Photograph, Duncan Fraser.

b *Suggest some examples where variation usually depends on genetic differences and the environment seems to have little if any influence.*

Many characters, probably a majority, can be shown to vary as a result of environment and as a result of inheritance and the two types of influence may interact in complex ways. Height in humans is clearly inheritable since tall parents tend to produce taller offspring than short parents and these height differences are still apparent among children who have been fostered or adopted, that is, removed from the environment of their natural parents. Height is also influenced by environment since the mean height of a population of well nourished humans is significantly greater than that of a comparable population which has been poorly nourished.

A uniform environment may serve to prevent the expression of genetic variation; for example, a population of caged birds cannot show inherited variation in flying ability. Such hidden genetic variation may become suddenly important if a population encounters a new situation, such as a new type of disease.

The environment usually influences continuously varying characters more than discontinuously varying characters, and it is sometimes imagined that continuous variation results entirely from environmental influences. This is most misleading; very many vitally important continuously varying characters such as size, growth rate, and tolerance of environmental stress are strongly influenced by inheritance. Such things as the yield of fruit and seed by plants, the milk yield of cattle, and the egg production of poultry (all continuous variables) result from complex interaction of inheritance and environment.

There has been much argument among philosophers, politicians, and scientists about whether human qualities such as criminality, creativity, and intelligence are inherited or produced by environmental influences. There is a sense in which such questions are meaningless; clearly we inherit a brain and hence the physical basis of our personality and we feel, think, and act in an environment. In other words, neither inheritance nor environment can cause our personality by itself. What people are really arguing about is whether the difference between people (the variation) is inherited or environmentally produced.

Because it is usually impossible, or at least impracticable, to keep organisms in precisely controlled and identical environments throughout development, it is difficult to be certain about the cause of variation even of easily measured variables such as egg production in hens. In the case of variation in a character as difficult to define and measure as intelligence, the pitfalls are numerous.

In spite of the difficulties involved, investigation of the extent to which variation is produced by genetic differences is of great importance in developing more productive breeds of animals and plants.

c *Two groups of mice are found to differ significantly in mean mass. How would you seek to investigate whether this difference is totally or partially the result of inherent (genetic) differences between the groups?*

17.3 Mendel and his contemporaries

> **Practical investigation.** *Practical guide 5*, investigation 17C, 'Patterns of inheritance'.

On the 1st of July, 1858, Charles Darwin and Alfred Russel Wallace presented a paper to the Linnaean Society of London which was later published in the Society's Journal. In this paper and in the books which followed it, Darwin put forward his hypothesis of natural selection. In this, he suggested that new species of animal and plant arise by the spread of variants within existing populations. He showed that continuous and discontinuous variation occurs in all natural populations and claimed that some variants would have increased reproductive success at the expense of other variants. This would produce major changes in the population over long periods of time and would result in new species. (See page 163.)

Variation was thus of central importance to Darwin as it was to most biologists of the nineteenth century. Much scientific effort was devoted to systematics, and variation was important either because it made species difficult for the taxonomist to define or because it might provide evidence for the origin of new species.

A major difficulty for Darwin was the lack of a theory to explain how variation was inherited. It was widely believed that characters acquired by the influence of the environment during development could be transmitted to the offspring and also that characters inherited from the mother and father became blended so that the offspring and their descendants would show features intermediate between those of the original parents.

In 1868 Darwin put forward the idea of *pangenesis*. This hypothesis suggested that each organ of a body delivered genetic material into the blood and that this material became organized in the gonads into gametes after circulating in the blood. Because each organ contributed directly to the gametes, it was easy to imagine how characters acquired by organs could be transferred to the gametes and blended as they circulated in the blood.

a *How could the idea of pangenesis have been tested?*

Long before Darwin put forward his theory of natural selection, scientists and philosophers had been discussing the nature of plant and animal species and the possibility of new species arising. The Swedish botanist Linnaeus (see *Systematics and classification*) suggested in 1744 that entirely new species might arise by the hybridization of previously existing species and several other botanists took up and developed this idea (including Charles Darwin's grandfather Erasmus Darwin). Others strongly believed that species did not change and/or that hybrids could not give rise to new species.

The first to attempt systematically to cross plant species under controlled conditions was the botanist Koelreuter (1733–1806). Koelreuter showed that some hybrids were sterile and that those which were fertile did not breed true but gave rise to various types, some of which resembled one or other of the original parents.

A number of other workers followed Koelreuter and a mass of rather confusing data about the fertility and external form of plant hybrids resulted. One of the most prolific experimenters was Gärtner (1772–1850) whose work confirmed and extended that of Koelreuter. Gärtner found that some of his many hybrids gave rise to true-breeding varieties but he did not believe them to be species. The following brief extracts give a summary of some of Gärtner's main ideas.

'The formation of hybrid types has now become one of the most interesting and difficult of the subjects which are involved in the study of plant reproduction.

'The explanation of how the forms of hybrids originate and are constructed out of the elements and characters of the stem-parents is as important for the plant physiologist as it is for the systematic botanist; whilst for the latter a question of life is also involved. Are there stable species of perfect plants or have they been subjected to change or progressive development in the course of time, as some naturalists believe? This question has already come under discussion and we have given reasons for speaking in favour of the stability of plant species. Further clarification on this point will be furnished by the study of the origin and formation of hybrid types out of the characters of the stem-parents.

'The general similarity of hybrids with their stem-parents can be understood by thinking of the seeds as arising from the mixing which occurs in reproduction and not from pollen alone. However, since very few hybrids show an equal mixing of the characters of both types, but the one factor in the union often preponderates over the other, so the question arises: Which laws govern these modifications in the construction of hybrids? For these types are not vague or the result of chance; on the contrary, they always arise in the same manner and are of the same sorts....

'Other fertile hybrids, indeed the majority, give rise to various forms, deviating from the normal type, in the second and subsequent generations, *i.e.* varieties. These are either unlike the mother hybrid or they deviate from it to a greater or lesser extent, *i.e.* they degenerate in various ways. Koelreuter and Wiegmann also observed this...

'In many fertile hybrids these changes in the second and subsequent generations affect not only the flowers but also the entire habit, even to the exclusion of the flowers, whereby the majority of individuals from one reproduction retain the form of the mother hybrid, a smaller number have become like the stem-mother, and finally an individual here and there has moved closer to the stem-father.'

(*Sinnott, Dunn and Dobzhansky, 1958.*)

b *Did Gärtner believe in evolution?*

c *Judging from the above extracts, did Gärtner believe in blending inheritance?*

The work of Koelreuter, Gärtner, and others excited the interest of a young Austrian monk, Gregor Mendel, and he began to plan his experiments in 1854, before Darwin and Wallace put forward their idea of natural selection. One of Mendel's objectives was to discover whether new species could arise by crossing, but he was also interested in discovering laws regulating the inheritance of particular characters in his plants. Mendel realized that the researchers such as Gärtner who had preceded him had been successful only to a limited degree because of the poor design of some aspects of their experiments.

From his paper, it is clear that he had considered, in his planning, the following points:

1 Only true breeding stock should be used at the onset of the experiment and precautions should be taken throughout to exclude the possibility of unplanned fertilizations taking place.
2 Clear-cut discontinuous variation should be studied.
3 It is important to count the number of each type of offspring produced and calculate the relative proportions of the different types.
4 All offspring from a cross should be fertile so that the results of the cross could be followed for more than one generation.
5 The offspring of each individual must be separately examined since individuals which appear similar may transmit different characters to their offspring.

Because he knew that an experiment which completely fulfilled all these conditions would require a great deal of effort, Mendel gave a lot of thought to the choice of organism to use and eventually decided on the garden pea, *Pisum sativum*.

Besides his famous experiments on peas, Mendel made many crosses between plant species and also tried crosses between different varieties of honey bee, *Apis mellifera*. He found it very difficult to 'persuade' bees to mate under controlled conditions. This may have been fortunate as bees show an unusual pattern of inheritance. The males are haploid because they develop from unfertilized eggs. Their gametes are produced by mitosis. Fertilized (diploid) bee eggs develop into queens or workers. (See section 16.1, 'Do males and females contribute equally to inheritance?'.)

d *How many genetically different types of gamete can a male bee produce?*

Mendel devoted much effort to hybridizing species of the genus *Hieracium* (hawkweeds). The task was immensely difficult as these plants have minute florets. He persevered for many years, encouraged by the botanist Nägeli who seems to have been the only professional scientist with whom, as an amateur, Mendel had any contact.

We now know that hawkweeds have the ability to produce seeds by asexual means from diploid maternal cells. The sexual processes of meiosis and fertilization can thus be avoided. Both sexually and asexually produced seeds can occur in the same plant. Sexually sterile hybrids can still produce seeds asexually. Mendel believed that he had found true breeding hybrids. The results were inconsistent and disappointing. Nägeli showed no interest in the work on peas for which Mendel is now famous and he eventually gave up his experiments when his duties as Abbot of his monastery become more exacting and his eyesight failed.

It is often claimed that Mendel was extremely fortunate to have chosen characters in his peas which show simple patterns of inheritance. This is true, but his good fortune with peas was balanced by the extremely unfortunate choice of hawkweeds to which he devoted so much fruitless effort.

Mendel was unaware that genetic material was carried by chromosomes. He assumed that it existed somewhere in egg and pollen cells. The question of whether one or several male gametes participated in fertilization had not been completely settled and he carried out experiments in which single pollen grains were transferred to the stigma of flowers, thus satisfying himself that only one male gamete fertilized the egg.

His major effort was devoted to the pea because:

1 The pollen and stigma develop to maturity inside the flower bud and pollination occurs before the flower opens so that natural cross-pollination is extremely rare. Different varieties may therefore be grown in the same garden without shielding the flowers.
2 The flower bud is large and can be opened to remove the anthers without damaging the female part of the flower. Artificial cross-pollination can then be performed.
3 Many true breeding garden varieties were available with 'characters which are constant, and easily recognizable, and when their hybrids are crossed they yield perfectly fertile offspring'.
4 They are easily cultivated and need only a single season to complete their life cycle.

Mendel initially grew 34 garden varieties and from these selected 22 for further study. They were grown each year during the eight-year-long experiment and remained true breeding. The core of his experimental design is in the quotation below.

'If two plants which differ constantly in one or several characters be crossed, numerous experiments have demonstrated that the characters [shared in] common are transmitted unchanged to the hybrids and their progeny; but each pair of differentiating characters, on the other hand, unite in the hybrid to form a new character, which in the progeny of the hybrid is usually variable. The object of the experiment was to observe these variations in the case of each pair of differentiating characters and to deduce the law according to which they appear in the successive generations. The experiment resolves itself, therefore, into just as many separate experiments as there are constantly differentiating characters presented in the experimental plants...
'Furthermore, in all the experiments reciprocal crossings were effected in such a way that each of the two varieties which in one set of

fertilization served as seed bearer in the other set was used as the pollen plant...

'Some of the characters noted do not permit of a sharp and certain separation, since the difference is of a "more or less" nature, which is often difficult to define. Such characters could not be utilized for the separate experiments; these could only be applied to characters which stand out clearly and definitely in the plants.'

(*Sinnott, Dunn and Dobzhansky, 1958.*)

Breeding experiments concentrated on seven pairs of contrasting characters. In some crosses the parents differed only in respect of one pair of characters (we would call these monohybrid crosses); in other experiments dihybrid crosses, involving two pairs of characters, were carried out and there was one trihybrid cross.

Figure 52
Tall and dwarf peas (Sutton's Purple Podded and Meteor respectively) grown in pots on a window sill for 10 weeks. (See Study item 17.11.)
Photograph, Duncan Fraser.

As in many situations where great scientific discoveries have been made, luck played an important part. Each pair of alleles determining the characters Mendel chose to work with affected plant development independently and was inherited independently. Had this not been the case the results would not have demonstrated his laws so clearly.

STUDY ITEM

17.31 **Mendel's first experiment**

A seed contains a young plant and Mendel started with characters shown by the seed since he did not have to grow a plant in order to determine its type.

Before we consider results some terms must be defined.

1 Parental types are the true breeding varieties used to make an experimental cross.
2 The F_1 generation is the offspring directly resulting from a cross between true breeding types.
3 The F_2 is the offspring of F_1 individuals, either by self-fertilization or by crossing two F_1 individuals together.
4 The F_3 is the offspring of F_2 individuals.
5 A backcross is a cross between an F_1 individual and either of the parental types.

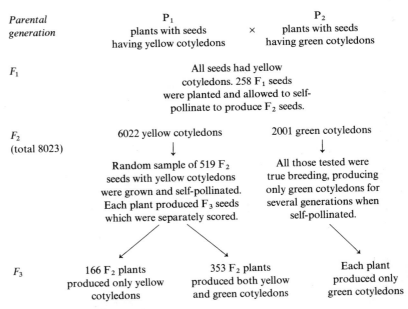

Table 8

Table 8 gives a summary of the results of one of Mendel's experiments.

a *What was the ratio of yellow cotyledons to green cotyledons in the F_2 generation?*

It is striking that all the F_1 plants resemble one parental type. A modern scientist would have concluded that a pair of alleles determined the difference between the plants in the experiment, and, that the allele (G_1), determining yellow cotyledons, was dominant to the allele producing green cotyledons (G_2), which is recessive. Mendel introduced the terms dominant and recessive and defined them clearly in the following extract.

'Experiments which in previous years were made with ornamental plants have already afforded evidence that the hybrids, as a rule, are not exactly intermediate between the parental species. With some of the more striking characters, those, for instance, which relate to the form and size of the leaves, the pubescence of the several parts, etc., the intermediate, indeed, is nearly always to be seen; in other cases, however, one of the two parental characters is so preponderant that it is difficult, or quite impossible, to detect the other in the hybrid.

'This is precisely the case with the pea hybrids. In the case of each of the seven crosses the hybrid-character resembles that of one of the parental forms so closely that the other either escapes observation completely or cannot be detected with certainty. This circumstance is of great importance in the determination and classification of the forms under which the offspring of the hybrids appear. Henceforth in this paper those characters which are transmitted entire, or almost unchanged in the hybridization, and therefore in themselves constitute the characters of the hybrid, are termed the *dominant*, and those which become latent in the process *recessive*. The expression "recessive" has been chosen because the characters thereby designated withdraw or entirely disappear in the hybrids, but nevertheless reappear unchanged in their progeny, as will be demonstrated later on.

'It was furthermore shown by the whole of the experiment that it is perfectly immaterial whether the dominant character belongs to the seed-bearer or to the pollen parent; the form of the hybrid remains identical in both cases. This interesting fact was also emphasised by Gärtner, with the remark that even the most practised expert is not in a position to determine in a hybrid which of the two parental species was the seed or the pollen plant.'

(*Sinnott, Dunn and Dobzhansky, 1958.*)

b *Does Mendel claim to have discovered dominance?*

You may have thought when reading these two passages from Mendel's work that he seemed to confuse the characters shown by his peas with the genetic material which determined these characters. A careful reading of his complete paper shows that he did not in fact make this mistake. He was the first biologist to realize that there is an important distinction between a character and the determinants of that character which are actually inherited via gametes.

We now use the term *phenotype* to refer to the character shown by a cell or organism and the term *genotype* to refer to the genes which its nuclei contain. Had Mendel introduced such terms, his paper might have been better understood by his contemporaries. The journal in which it was published in 1866, *The Transactions of the Brün Natural History Society*, was an obscure one but it was sent to as many as 200 scientific societies and institutions in Europe and America after 1866. The paper was referred to in other publications, in 1881 and again in 1892, but it was not until 1900, after Mendel had died, that three researchers, de Vries, Correns, and Tschermak traced and read the paper, finding it by means of the 1881 reference. They saw that Mendel had anticipated their own ideas and that his hypotheses explained their own results.

c *Are there any lessons about the international organization and communication of scientific research to be learned from the loss of Mendel's work between 1886 and 1900?*

STUDY ITEM

17.32 Analysis of monohybrid crosses

There is a variety of the domestic fowl called Andalusian which has rather attractive bluish coloured plumage. When two birds of this colour are mated, they never breed true. Some chicks develop the parents' blue colour but the rest are either black or else white streaked with grey. This very striking pattern of inheritance is the result of a pair of alleles which we can call a_1 and a_2. An analysis of the inheritance of the Andalusian character is a good introduction to Mendelian genetics. It is set out in tables 9a, b, and c. Experienced people often take short cuts in genetic

Parental generation			
phenotype	black	×	white
genotype	a_1a_1		a_2a_2
Genotypes of parental gametes	all eggs a_1		all sperm a_2
F_1 phenotype	All Andalusian		
genotype	all a_1a_2		
Genotypes of F_1 gametes	eggs either a_1 or a_2 (frequency of each 0.5)		sperm either a_1 or a_2 (frequency of each 0.5)

Assuming that the fertilization of each egg is a chance process and that both types of sperm have an equal chance of achieving fertilization,

		Frequency of F_1 sperm	
		0.5 a_1	0.5 a_2
Frequency of F_1 eggs	0.5 a_1	0.25 a_1a_1	0.25 a_2a_1
	0.5 a_2	0.25 a_1a_2	0.25 a_2a_2

∴ in F_2: Frequency of genotypes   0.25 a_1a_1 0.5 a_1a_2 0.25 a_2a_2 Total 1
Frequency of phenotypes   0.25 black   0.5 Andalusian   0.25 white   Total 1
Ratio of genotypes   1 a_1a_1 2 a_1a_2 1 a_2a_2
Ratio of phenotypes   1 black   2 Andalusian   1 white

Table 9a

Table 9b

Backcross 1
phenotype   black   ×   Andalusian
genotype a_1a_1 a_1a_2

Genotypes of gametes   all eggs a_1 sperm either a_1 or a_2
                                                             (frequency of each 0.5)

Assuming that fertilization is random,

		Frequency of F_1 sperm	
		0.5 a_1	0.5 a_2
Frequency of parental eggs	1.0 a_1	0.5 a_1a_1	0.5 a_2a_1

∴ in backcross 1:   frequency of genotypes   0.5 a_1a_1 0.5 a_1a_2 Total 1
                       frequency of phenotypes   0.5 black   0.5 Andalusian   Total 1
                       Ratio of genotypes   1 a_1a_1 1 a_1a_2
                       Ratio of phenotypes   1 black   1 Andalusian

Table 9c

Backcross 2
phenotype   white   ×   Andalusian
genotype a_2a_2 a_1a_2

Genotypes of gametes   all eggs a_2 sperm either a_1 or a_2
                                                              (frequency of each 0.5)

Assuming that fertilization is random,

		Frequency of F_1 sperm	
		0.5 a_1	0.5 a_2
Parental egg frequency	1.0 a_2	0.5 a_1a_2	0.5 a_2a_2

∴ in backcross 2:   Frequency of genotypes   0.5 a_2a_2 0.5 a_1a_2 Total 1
                       Frequency of phenotypes   0.5 white   0.5 Andalusian   Total 1
                       Ratio of genotypes   1 a_2a_2 1 a_1a_2
                       Ratio of phenotypes   1 white   1 Andalusian

analysis, especially when dealing with monohybrid examples, but it is best to be very thorough in each analysis you undertake until you are quite sure you have mastered the ideas completely.

Notice that there is a 1:2:1 ratio of genotypes and phenotypes in the F_2 and a 1:1 ratio of genotypes and phenotypes in the backcrosses. If such a result is obtained in a breeding experiment it means that a single gene locus controls the characters which are observed. There are two alleles segregating and neither is dominant.

A *homozygote* is an organism in which there are two identical copies of the same gene present, one on each of a homologous pair of chromosomes in each cell.

A *heterozygote* is an organism in which there are two different versions of a gene present; that is, there are two alleles, one on each of a homologous pair of chromosomes in each cell.

When one of a pair of alleles is dominant over the other, the heterozygotes resemble the dominant parental type. Crosses which show dominance, as did all those carried out by Mendel, can be analysed in exactly the same way as in table 9. The phenotype ratios are different but the genotype ratios are the same.

Table 10

		Grey body female	Grey body male	Ebony body female	Ebony body male			Grey body female	Grey body male	Ebony body female	Ebony body male
	Parental generation						Parental generation				
	F_1	4			4		F_1	5	5		
		62	104	0	0			121	128	0	0
	F_2	164	174	63	57		F_2	149	161	53	59
(Backcross 1)		338 : 120				(Backcross 2)		310		112	
	5 parental grey body			× 5F_1			3 parental ebony body			× 3F_1		
		Grey body female	Grey body male	Ebony body female	Ebony body male			Grey body female	Grey body male	Ebony body female	Ebony body male
	Backcross	83	78	0	0	Backcross		31	38	29	34
		161 : 0 / All grey						69		63	

The results shown in table 10 were obtained by crossing wild type and ebony bodied fruit flies (*Drosophila melanogaster*).

If the allele determining ebony body is e and the allele producing wild type (grey bodies) is e^+:

a Calculate the ratios and frequencies of the phenotypes in each F_2 generation and in the backcrosses.

b Carry out a full analysis of the inheritance of the ebony body character as in tables 9a, b, and c.

c Carry out a similar analysis of Mendel's result as shown in table 8.

In all the examples of monohybrid inheritance so far considered, the results of reciprocal crosses are exactly the same.

d Explain this on the basis of the chromosome theory of inheritance.

STUDY ITEM

17.33 Analysis of monohybrid crosses involving sex linkage

In the previous chapter (page 52) an example was introduced of the inheritance of the tortoiseshell colour in cats. It was shown that the pattern of inheritance of this character is consistent with the hypothesis that the gene involved is present in the X chromosome but not in the Y.

If we call the allele for black coat colour b_1 and the allele producing red (or ginger) colour b_2 then:

Males can be either b_1Y or b_2Y since they have one X and one Y chromosome. Females can be b_1b_1 (black), b_2b_2 (ginger), or b_1b_2 (tortoiseshell) since they have two X chromosomes and b_1 and b_2 do not show dominance. Tortoiseshell males are not found since the male cannot normally have two X chromosomes and so cannot be b_1b_2.

In writing the genotype of the heterogametic sex (see page 50) we have written a Y to remind us that the individual is diploid but that there is no allele on the Y. The presence of the X chromosome is understood. It would be possible to write the genotype of a male ginger cat as

$$X^{b_2}Y^{(-)}$$

to show this more fully but this is rather clumsy. A partial analysis is given in table 11.

Parental generation
phenotype   black female   ginger male
genotype b_1b_1 b_2Y

Genotypes of   all eggs b_1 sperm either b_2 or Y
parental gametes   (frequency of each 0.5)

Assuming that fertilization is random and that all sperm carrying b_2 are cells bearing X chromosomes,

	Frequency of sperm	
	0.5 b_2	0.5 Y
Frequency of eggs 1.0 b_1	0.5 b_2b_1	0.5 Yb_1

Frequency of genotypes   0.5 b_1b_2 0.5 b_1Y Total 1
Frequency of phenotypes   0.5 tortoiseshell female   0.5 black male Total 1
Ratio of genotypes   1 b_1b_2 1 b_1Y
Table 11 Ratio of phenotypes 1 tortoiseshell female   1 black male

a **Complete the analysis for other possible crosses, involving black, ginger, and tortoiseshell cats.**

b **Analyse and explain the results shown in table 12.**

	Wild type (red eye)		White eye			Wild type (red eye)		White eye	
	female	male	female	male		female	male	female	male
Parental generation	4		4		Parental generation	4		4	
F_1	16	12			F_1	22		17	
F_2	85	39	34	34	F_2	47	49	48	43

Table 12
Results of two breeding investigations of *Drosophila* involving wild type (red eyed) flies and white eyed flies. *Left:* wild type × white eye. *Right:* the reciprocal cross, white eye × wild type.

In pigeons there is a pair of alleles, r (wild type), which produces blue/black feathers and R (red), which produces reddish feathers. Red (R) is dominant to wild type (r). If a red female produces a wild type chick, this chick is always female, whatever the colour of its father.

c **Explain this (that red females produce blue female chicks, not blue males) and predict the results of crossing blue females with red males.**

17.4 Dihybrid crosses

> Practical investigation. *Practical guide 5*, investigation 17C, 'Patterns of inheritance'.

STUDY ITEM

17.41 Inheritance involving two different genes in *Drosophila*

In Study item 17.32, all the inheritance described could be explained by the existence of a single gene with two alternative forms (alleles). Table 13 shows a cross where two quite different pairs of characters, each controlled by a gene at a different locus, are involved. This is an example of dihybrid inheritance.

Table 13
A cross between normal winged, sepia eyed and vestigial winged, red eyed fruit flies.

Parental generation	normal winged, sepia eyed female flies	×	vestigial winged, red eyed male flies	
F_1	115 normal winged, red eyed flies			
F_2	151 normal winged, red eyed flies	47 normal winged, sepia eyed flies	58 vestigial winged, red eyed flies	9 vestigial winged, sepia eyed flies

The wild type of fruit fly *Drosophilia melanogaster* has red eyes and normal wings (*figure 53a*). Some strains of flies have dark bluish purple (sepia) eyes. Others have short and deformed (vestigial) wings (see *figure 53b*).

a

b

Figure 53
a wild type and b vestigial winged *Drosophila*. The males are on the left of each picture.
Photograph, Dr Nick Taylor.

Neither of the genes involved in the cross shown in table 13 is sex-linked so the sex of the progeny has been ignored.

We will adopt the following notation.

sepia eye = s; allele determining red eyes = s^+
vestigial wing = v; allele determining normal wing = v^+

a *Ignoring eye colour, what is the ratio of normal winged to vestigial winged phenotypes in the F_2 generation?*

b *Ignoring wing type, what is the ratio of red eyed to sepia eyed phenotypes in the F_2 generation?*

We are dealing with two straightforward cases of monohybrid inheritance showing dominance. These happen to have been observed at the same time in the same experimental family of flies.

It will be clear that the sepia parent is of genotype ss and that the vestigial parent is vv. The sepia parent is also true breeding for normal wild type wings and its fuller genotype is ss v^+v^+. By the same argument the full genotype of the vestigial parent is s^+s^+ vv. The cross is thus between the following genotypes:

ss v^+v^+ females × s^+s^+ vv males

The parents, being diploid, have a pair of alleles determining eye colour and a pair of alleles determining wing shape. Both pairs of alleles segregate at meiosis to produce gametes, so the gametes must have one of each pair:

ss v^+v^+ $\xrightarrow{\text{meiosis}}$ sv^+ s^+s^+ vv $\xrightarrow{\text{meiosis}}$ s^+v

These gametes must combine to give a doubly heterozygous F_1:

sv^+ + s^+v $\xrightarrow{\text{fertilization}}$ ss^+ v^+v

We could have written the double heterozygote as sv^+ s^+v but have put the alleles of each gene into pairs, ss^+ and v^+v.

When the doubly heterozygous F_1 generation produces gametes, meiosis could result in four possible allele combinations.

s^+v and sv^+ are the original combinations found in the parents and are called, therefore, parental types.

s^+v^+ and sv are new combinations and are called recombinant types.

c **Why are the combinations s^+s^+, ss, v^+v^+, vv, v^+v and s^+s not possible?**

A reasonable hypothesis would be that when the F_1 produced gametes, the original allele combinations of the parents would stay together. This means that s^+v and sv^+ gametes would be very frequent while the recombinants s^+v^+ and sv would be rare or absent. This prediction is made more specific in table 14 which assumes that no recombinant gametes are produced.

		Frequency of F_1 sperm	
		0.5 sv^+	0.5 s^+v
Frequency of F_1 eggs	0.5 sv^+	0.25 ss v^+v^+	0.25 s^+s vv$^+$
	0.5 s^+v	0.25 ss$^+$ v^+v	0.25 s^+s^+ vv

Predicted frequency of F_2 genotypes	0.25 ss v^+v^+	0.5 s^+s v^+v	0.25 s^+s^+ vv	Total 1
Predicted frequency of F_2 phenotypes	0.25 sepia eye, normal wing	0.5 red eye, normal wing	0.25 red eye, vestigial wing	Total 1
Predicted ratio of F_2 genotypes	1 ss v^+v^+	2 s^+s v^+v	1 s^+s^+ vv	
Predicted ratio of F_2 phenotypes	1 sepia eye, normal wing	2 red eye, normal wing	1 red eye, vestigial wing	

Table 14

There are in fact four phenotypic classes of fly in the F_2 so the hypothesis that no recombinant gametes are produced is false. Table 13 shows that the four classes are in the ratio:

| 9 normal winged red eyed flies | 3 normal winged sepia eyed flies | 3 vestigial winged red eyed flies | 1 vestigial winged sepia eyed fly |

The existence of the vestigial winged sepia eyed flies in the F_2 can only be explained by the fusion of two recombinant gametes:

$$sv + sv \xrightarrow{fertilization} ssvv$$

d *Assume that F_1 flies produce the four possible types of gamete with equal frequency ($\frac{1}{4}$ or 0.25 of each) and work out the frequencies and ratios expected in the F_2 generation on this basis.*

e *Is the prediction supported by the ratio actually found in the F_2 generation? (Note: we always expect departures from exact ratios in any group of offspring as a result of chance. Statistical tests must determine if the departure is significant.)*

F_1 males × vestigial winged, sepia eyed females

| Offspring | 188 normal winged, red eyed flies | 173 normal winged, sepia eyed flies | 181 vestigial winged, red eyed flies | 186 vestigial winged, sepia eyed flies |

Table 15
A test cross using the F_1 of the family shown in table 13.

We can further test our prediction that the F_1 generation produces four types of gamete in equal frequency by employing what is called a test cross. In a test cross, an individual is mated to a doubly homozygous recessive, that is, one homozygous for the recessive allele of both genes (in this case ssvv). The gametes from the doubly homozygous recessive must have the genotype sv and contain no dominant allele of either gene to mask the effect of any allele of either gene carried by the individual which is to be tested. Each different gamete type produced by the tested individual will produce a different phenotype of offspring. The ratio of phenotypes in the offspring will give an indication of the ratio of gametes produced.

f *Are the results of the test cross consistent with the hypothesis that parental and recombinant gametes are produced with equal frequency by F_1 individuals?*

g *Examine figure 36 in Chapter 16, page 43, and explain, in terms of the behaviour of chromosomes during meiosis, why four types of gamete in equal numbers are produced.*

A student, as part of an Advanced Level project investigation, mated a number of pairs of laboratory rats of various colours. Some of the results are shown in table 16.

The colours observed are shown in *figure 54*.

Figure 54
Fawn, black hooded, fawn hooded, and black rats.
Photograph, Matthew Dales.

	Black female			×	Fawn hooded male			
	Offspring phenotypes							
	FEMALE				MALE			
	Black	Black hooded	Fawn	Fawn hooded	Black	Black hooded	Fawn	Fawn hooded
Litter 1	2	0	3	1	1	1	1	0
Litter 2	3	2	1	0	0	4	0	0
Litter 3	1	1	0	2	3	1	0	2
Litter 4	1	2	0	2	1	0	1	2
Total	7	5	4	5	5	6	2	4

	Black female			×	Black male			
	Offspring phenotypes							
	FEMALE				MALE			
	Black	Black hooded	Fawn	Fawn hooded	Black	Black hooded	Fawn	Fawn hooded
Litter 1	3	1	1	0	1	2	0	0
Litter 2	1	1	0	1	2	0	2	0
Litter 3	4	0	1	0	4	0	0	0
Litter 4	2	2	1	0	2	1	1	0
Litter 5	3	2	1	0	2	1	1	0
Total	13	6	4	1	11	4	4	0

Table 16
Results of mating black and fawn hooded rats. Neither parent was from a true breeding strain

h ☐ *Assume dihybrid inheritance is involved and explain the results shown in table 16.*

Mendel was the first to discover and explain dihybrid ratios. He crossed peas true breeding for round seeds and yellow cotyledons with others true breeding for wrinkled seeds and green cotyledons, obtaining a 9:3:3:1 ratio of phenotypes in the F_2. He realized that this was a combination of two independent 3:1 phenotype ratios and so put forward the hypothesis known as his second law, that each pair of contrasting characters is inherited independently of all other pairs.

Let phenotypically round seeds = a
  phenotypically yellow cotyledons = c
  phenotypically wrinkled seeds = b
  phenotypically green cotyledons = d

The two monohybrid ratios in F_2 are:

(3a:1b) and (3c:1d)
$(3a + 1b) \times (3c + 1d) = 9ac + 3bc + 3ad + 1bd$ in a dihybrid.

17.5 Autosomal linkage

> Practical investigations. *Practical guide* 5, investigation 17C, 'Patterns of inheritance' and investigation 17D, 'Linkage and linkage mapping in tomato'.

The production of 9:3:3:1 ratios in dihybrid F_2 families and 1:1:1:1 ratios in test cross families is explained by the assumption that each pair of alleles is carried on a pair of chromosomes and that each pair of chromosomes orientates independently of other pairs on the spindle at 1st metaphase of meiosis (see *figure 36*, page 43).

Most organisms have a relatively small number of pairs of chromosomes so that each chromosome must carry many genes. If individuals heterozygous for two genes carried on the same chromosome were segregating we would not expect independent assortment. The combination present on the maternal chromosome would be expected to stay together, as would the combination on the paternal chromosome. This situation would reveal itself in breeding experiments where parental combinations were more frequent than recombinant types.

Examples where parental combinations of genes are conserved are very common. Genes which show this relationship and do not produce the characteristic 9:3:3:1 ratio in the F_2 are said to be linked. Experimenters usually detect linkage by means of test crosses, since this immediately reveals the proportions of the different types of gametes produced by an F_1 individual.

Table 17 (page 80) shows results obtained from experimental breeding programmes on *Drosophila*, involving the characters curled wing (c) and ebony body (e) and the wild type characters straight wing (c^+) and grey body (e^+).

The production of gametes with new combinations of alleles is called recombination but it can also be called crossing over. In a test cross, the

First investigation (A)

	P curled, grey female × straight, ebony male
Results of cross A	F_1 all straight, grey

	F_1 straight, grey female × curled, ebony male			
Test cross 1A	curled, grey 102	straight, grey 25	curled, ebony 29	straight, ebony 93

	curled, ebony female × F_1 straight, grey male			
Test cross 2A	curled, grey 147	straight, grey 0	curled, ebony 0	straight, ebony 129

Second investigation (B)

	P straight, grey female × curled, ebony male
Results of cross B	F_1 all straight, grey

	F_1 straight, grey female × curled, ebony male			
Test cross 1B	curled, grey 18	straight, grey 83	curled, ebony 90	straight, ebony 26

	curled, ebony female × F_1 straight, grey male			
Test cross 2B	curled, grey 0	straight, grey 67	curled, ebony 59	straight, ebony 0

Table 17

cross-over value between two genes is given by the formula

$$\text{cross-over value} = \frac{\text{number of recombinant offspring}}{\text{total number of offspring}} \times 100$$

a *What would be the cross-over value in a case where two genes were not linked, that is, where there was independent or random assortment?*

b *Calculate and list the cross-over values for each of the four test crosses reported in table 17.*

The first extensive studies of linkage were undertaken with *Drosophila melanogaster* and *Zea mays*. Many pairs of gene loci were tested for linkage in a large number of possible combinations. Some pairs of loci showed independent assortment while others showed linkage. The crossover value between any two loci proved to be constant and different pairs showed a wide variety of values between 0 and 50.

The following rules became apparent from studies of linkage:

1 In any species of organism the genes can be assigned to groups within which genes are linked with each other. These groups are called linkage groups.
2 The number of linkage groups is equal to the haploid number of chromosomes, for example, ten for maize, four for fruit flies.

3 Within each linkage group different pairs of genes have different cross-over values.

In table 17 you can see that the genes for wing shape and body colour show some recombination in the production of eggs by female F_1 flies but no recombination in the production of sperm by male F_1. This is an unusual feature of flies (Diptera). In most organisms there is an equal frequency of crossing over in males and females but recombination is totally suppressed in male flies.

In 1st prophase of meiosis it can be seen that chromatids break and rejoin (see Chapter 16, *figures 34* and *35*). When this occurs, maternal and paternal genetic material can be recombined and it is believed that crossing over as seen in breeding experiments is the result of chiasma formation in meiosis.

If crossing over is caused by the formation of chiasmata, it would be expected that genes close together on a chromosome would be tightly linked (low cross-over value) since a chiasma between them would be relatively infrequent. Conversely, genes far apart on a chromosome would have a cross-over value approaching 50 %. This idea means that genes may be mapped on each chromosome by means of their cross-over values within their linkage group.

Genetic maps have been prepared for a wide variety of organisms.

In *Practical guide 5*, investigation 17D, 'Linkage and linkage mapping in tomato', you will be able to follow through this technique used in mapping a chromosome. The sexual processes of bacteria are very different from those of eukaryotes as there is no meiosis, and the genetic material is much more simply organized. Bacterial genes can nevertheless be mapped by recombination, and the map turns out to be circular.

The circular map is clearly related to the circular organization of the bacterial genetic material (see page 108).

STUDY ITEM

17.51 An association between recombination and exchanges between chromosomes

In maize, two different chromosome abnormalities are known, both affecting chromosome number 9. In one case, a section of another chromosome has become attached (translocated) on to one end of chromosome 9. In the second case, there is a visible knob situated at the opposite end of chromosome 9 from that to which the translocated section is attached. These independent changes to chromosome 9 may occur in the same plant so that the chromosome is doubly affected. The types of chromosome 9 possible are shown in *figure 55*.

Figure 55
Different forms of chromosome 9 in maize.

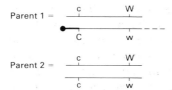

Figure 56
Genotypes of two maize plants crossed to produce the offspring shown in table 18.

Chromosome 9 of maize carries genes controlling two different seed characters:

seed colour: alleles C (coloured outer layer) and c (colourless)
storage material: alleles W (starchy endosperm) and w (waxy)

Two plants which have genotype and chromosome morphology as in *figure 56* were crossed.

Seed type	Number of seeds	Phenotype of seed	Knob present	Translocation present
A	3	coloured waxy	yes	yes
B	2	colourless waxy	no	yes
C	7	coloured starchy	yes	no
D	4	colourless starchy	no	yes
E	11	colourless starchy	no	no

Table 18
The offspring of the parents shown in *figure 56*.

The results are given in table 18. Assuming that W is dominant to w and C is dominant to c, answer the following questions.

a *Draw and label the pair of chromosomes with associated genes for each seed type (A to E) in the table, using drawings similar to those in* figure 56.

b *In which of the seed types (A to E) has no recombination taken place?*

c *In those seed types where chiasmata have resulted in the separation of the knob and translocation, is there also evidence of recombination of genes?*

d *Would the use of a cytologically normal parent of genotype ccww have been justified in this experiment as a substitute for parent 2?*

17.6 A model for the inheritance of continuously varying characters

> **Practical investigation.** *Practical guide* 5, **investigation 17C, 'Patterns of inheritance'.**

Mendelian inheritance, involving the segregation of alleles, explains the inheritance of discontinuously varying characters but what of continuous variation? Mendelian analysis puts forward hypotheses about the genotypes of individuals and their gametes and then makes predictions about the expected proportions of different types of offspring. If offspring cannot be assigned to one or more categories, that is, if the phenotype varies continuously, then Mendelian analysis is not possible.

This difficulty caused heated controversy among biologists in the first decade of this century. Many researchers who had become especially interested in the inheritance of continuous variation were following statistical methods pioneered by Francis Galton in 1889. These methods allowed the probability of a phenotype lying within a particular range of continuous variation to be predicted if the distribution of variation in the whole population and the phenotype of the parents were known. Researchers applying statistical techniques to continuously varying populations were called biometricians. Some of the biometricians considered Mendelian inheritance to be unimportant because so many highly significant variables such as size, productivity (milk yield, number of offspring produced), and vigour vary continuously.

Some Mendelians (as they were called) fell into the error of dismissing the work of biometricians because the biometricians often dealt with large populations of unknown or uncertain ancestry rather than with experimental families under controlled conditions. It was also believed by some Mendelians that continuous variation was entirely caused by environmental influence.

A most important series of experiments was conducted by Johannsen in Copenhagen and published in 1903. This research is outlined in *figure 57* in a simplified form.

a *What can be deduced from Johannsen's work?*

In 1908 Nilsson-Ehle published a paper reporting an investigation into seed colour of wheat. A variety with red seeds was crossed with one having white seeds. All the F_1 had pink seeds. The F_2 variation was almost continuous, with many different types of seed having shades from red through pink to white. Nilsson-Ehle suggested that his result was caused by the segregation of three pairs of alleles: A, a; B, b, and C, c. The red parent was a homozygote AA BB CC and the white parent a homozygote aa bb cc.

b *What was the genotype of the F_1 plants?*

c *What would have been the genotypes of gametes produced by the F_1 plants?*

The genes A, B, and C all allowed the formation of red pigment while their alleles a, b, and c all gave rise to white seeds. There was no dominance at all three loci and the effect of the three loci together on the phenotype was additive: that is, AaBbCc, AABbcc, AaBBcc, aaBbCC, and other combinations with three dominants produced similar phenotypes intermediate between red and white. Genotypes AAbbcc, AabbCc, AaBbcc etc. were lighter than those with three doses of red determinant and AABBcc, aaBBCC, AaBbCC, etc. were darker.

Nilsson-Ehle found one out of every sixty-four of his F_2 generation to be completely devoid of red colour: that is, it was like the white parent.

d *Explain how one in sixty-four white seeds in the F_2 generation was to be expected from the hypothesis outlined above.*

The way in which two or more genes with relatively small but parallel

All the beans were *Phaseolus vulgaris* (French beans) and were grown in a uniform environment. Only self-pollination was allowed.

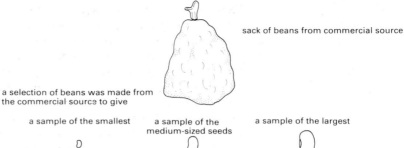

These samples were grown separately and the seeds of the resulting plants were weighed.
— — seeds from smallest parents; —— seeds from medium-sized parents; – – – seeds from largest parents.

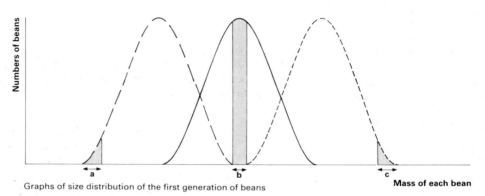

Graphs of size distribution of the first generation of beans

From each population of first generation bean seeds a further selection was made of those corresponding to the shaded areas of the graphs and having mass ranges **a**, **b**, and **c**. These samples were grown and the plants produced a second generation of seeds which were again weighed.

—— seeds from smallest-sized population   —— seeds from medium-sized population
– – – seeds from largest-sized population

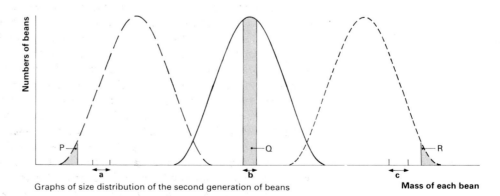

Graphs of size distribution of the second generation of beans

Selections were again made of bean seeds corresponding to the shaded areas P, Q, and R. Their progeny was even more distinct. After five generations, selection produced no further changes in size distribution and the largest and smallest seeds from each population produced offspring which were similar to each other and different from either of the other populations. Pure lines had been established.

Figure 57
The experiments of Johannsen.

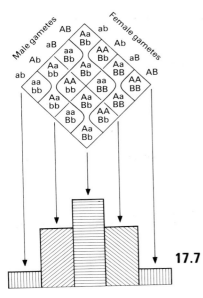

Figure 58
The distribution of phenotypes obtained in an F_2 generation with two genes of equal and additive effect but without dominance, neglecting non-heritable variation. The phenotypic expression is proportional to the number of capital letters in the genotype. There would be seven phenotypic classes with three such genes and nine classes with four genes.
From Mather, K. and Jinks, J. L., Biometrical genetics, the study of continuous variation, *Chapman & Hall*, 1982.

and additive effects can produce continuous variation is explained in *figure 58*.

Differences between genotypes are very small and are blurred by environmentally produced variation so the influences of the individual genes are difficult or impossible to determine. The work of Nilsson-Ehle helped to end the dispute between biometricians and Mendelians.

Statistical methods are very important in animal breeding programmes; for example, a prize bull may sire 5000 female calves by artificial insemination. His semen can then be stored in liquid nitrogen for several years. When his daughters grow up the mean and standard deviation of their milk yield can be compared with the mean and standard deviation of a random sample of heifers of the same breed to see if the bull has transmitted good milking qualities to his daughters. If he has, then his stored semen can be used for further breeding.

17.7 Mutation

> **Practical investigations.** *Practical guide 5*, investigation 17E, 'Mutation in yeast' and investigation 17F, 'The effects of irradiation in plant seeds'.

Throughout this chapter we have been concerned with genetic differences between individuals brought about by alternative forms of the same gene called alleles. How do new alleles arise?

It is one of the striking features of biological inheritance that once all the individuals in a population are homozygous for an allele they will breed true for very many generations. Only if a new allele appears is variation possible.

New alleles may often enter populations by crossing with individuals of different and genetically distant populations, but we cannot ultimately explain the origin of alleles in this way.

It has long been known to animal and plant breeders that entirely new inheritable variation may arise suddenly and quite unexpectedly in a true breeding population which has not been crossed with another variety. Breeders used to call such new variants 'sports' and have often preserved them as curiosities or because the new phenotype was valuable. The history of the domestic budgerigar provides a good example of a series of events of this kind. The wild type bird is green with a yellow face and characteristic black markings on the wings and under the beak. It is native to Australia and was domesticated as a result of a number of importations to Europe between 1840 and 1860.

By 1900 a breeding population was established in Britain and the bird was regularly exhibited at cage-bird shows. By then a yellow variety had appeared. In 1910 blue 'sports' arose spontaneously and were exhibited. These caused a sensation among bird fanciers and so great was the demand for blues (especially from Japan) that the price of a pair rose to over £100 during the 1920s.

Both blue and yellow are determined by recessive alleles of different genes responsible for pigment production in the developing feather. The green of the wild type is a mixture of blue and yellow pigments. A bird homozygous for yellow cannot produce the blue pigment and one homozygous for blue cannot make the yellow pigment.

The determinants of blue and yellow almost certainly arose spontaneously as a result of damage to genes or errors in the enzyme-catalysed replication of genes. Such a process is called *spontaneous mutation* and the new alleles and the individuals carrying them are called *mutants*.

Mutation at other gene loci controlling plumage colour has occurred since the 1920s, and by reassortment of these mutant genes a large range of colour phenotypes has been produced by budgerigar breeders. Other small domestic animals such as gerbils and hamsters provide similar histories.

Spontaneous mutation is a very rare process. Usually millions of individuals must be bred before a single mutant can be observed. Several different types of event can be called mutation. If a chromosome breaks at two points, broken ends may rejoin with the loss of the fragment between the two points of breakage (see *figure 59*). This produces a deletion mutation. A deletion is usually lethal when homozygous unless the region of chromosome lost in the process is very small.

DELETION

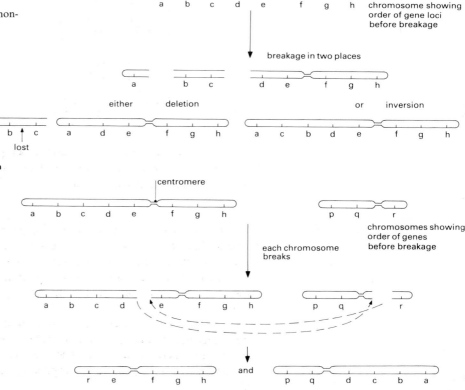

Figure 59
a Deletion mutation and inversion.
b Translocation between two non-homologous chromosomes.

Alternatively, either of the fragments without a centromere could be lost and the other attached to a new centromere to give a translocation and a deletion

Inheritance and development

If broken ends rejoin without loss of material but in a new arrangement, inversions or translocations result (see *figure 59*). When a cell is heterozygous for an inversion or translocation, the normal pairing of the chromosomes involved is disturbed (see *figure 60*). This leads to a reduction in fertility and one reason why hybrids between species are often of reduced fertility is that chromosome changes of this type have taken place since the evolutionary separation of the two species.

a   *In* **Drosophila** *the X chromosome sometimes breaks and then rejoins to give an inverted segment as shown in* **figure 60**.

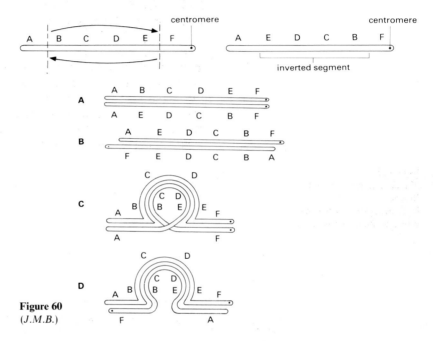

Figure 60
(J.M.B.)

During gamete formation in a female **Drosophila** *with one normal X and one inverted X chromosome, which way would these two chromosomes be expected to pair?*

(J.M.B)

Both inversion and translocation change the linkage between genes. If genes which were previously unlinked (on different chromosomes) or loosely linked (far apart on the same chromosome) become closely linked as a result of translocation or inversion, an advantage may result since favourable combinations of alleles may be held together by the linkage.

Chromosomal changes may result in loss or rearrangement of genetic material, but cannot by themselves produce real novelty unless a break point is actually within a gene. New alleles arise by rare errors in the enzyme-catalysed processes of replication of existing genes (see page 105).

Mutation involving the origin of a new allele is often called point-mutation to distinguish it from cytologically observable chromosome

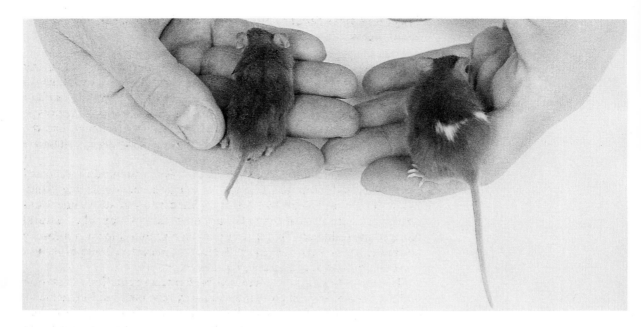

Figure 61
A short tailed and a long tailed mouse. The short tail allele (Sd) is dominant to Sd⁺ and lethal when homozygous.
Photograph, Duncan Fraser.

mutation. All types of mutation – chromosome and point-mutation – are random events. It cannot be predicted exactly which cell will suffer mutation or which gene or chromosome will be involved. Some genes seem to mutate more frequently than others and this greater probability of mutation may reflect a vulnerable arrangement or composition of the genetic material.

Mutation can occur in cells of the animal or plant body (the soma) as well as in those giving rise to gametes (the germ line). Somatic mutation can only be detected if the mutant cell divides to produce a patch of tissue of different phenotype from that of surrounding cells (see *Practical guide 5*, investigation 17F, 'The effects of irradiation in plant seeds'.) Mutation during the early development of an individual gives a body with cells of two distinct genotypes. This is called a mosaic.

A majority of point mutations result in alleles that are recessive to the 'normal' or wild type. There are numerous exceptions to this generalization (see *figure 61*).

If a mutant allele is recessive to that from which it was derived it cannot be detected until it occurs in the homozygous state. This can only happen if it is inherited from both parents. Since the parents must themselves inherit the mutant allele before they can transmit it, at least two generations must elapse between mutation and the appearance of a recessive mutant phenotype.

Research on mutation has usually concentrated on haploid organisms such as fungi or on prokaryotes, since the complication of dominant wild type alleles masking newly arisen recessive mutants is avoided. Mutation research in diploids such as mammals or flies has often been concerned with genes carried on the X chromosome.

b *Why should genes on the X chromosome be especially favourable for the detection of mutation?*

c *Yeasts and other fungi occur in the haploid state – what other advantage do they offer in mutation research?*

Alleles which have been frequent in a population for many generations are much more likely to produce an advantageous (successful) phenotype than are new alleles arising by the random process of mutation. The vast majority of mutants produce a phenotype which is less vigorous or less well adapted to the environment than the wild type. Many mutants do not allow survival if homozygous; these are called lethal mutants.

The first artificial or induced mutations were produced by exposing *Drosophila* and crop plant seeds to ionizing radiations (neutrons, β-particles, γ-rays, and X-rays). It is impossible to determine what proportion of spontaneous mutation is caused by the natural background radiation. This is quite unavoidable and comes from cosmic radiation, ultra-violet (UV) in sunlight, and radioactive isotopes such as those of uranium and thorium present in rocks.

Human activities have increased background radiation. Radiation pollution comes from such sources as X-ray machines used in medicine and industry, radioisotopes used in medicine and industry, waste from nuclear reactors (military reactors and those from power stations) and from nuclear weapons tests.

Natural background radiation varies greatly from place to place: for example it is greater in Aberdeen and other areas where granite rocks outcrop than in London. The radiation produced by synthetic sources is well within the limits of variation of the natural background but is still a matter of concern because it is not uniformly distributed and because it might increase greatly if nuclear weapons were used or if a reactor suffered a major accident.

It has been known for about forty years that certain chemicals are mutagenic. This was first discovered in research on poison gases known as nitrogen mustards during the Second World War. Since then, testing of large numbers of synthetic chemicals (products of the chemical industry) has shown many to be mutagenic. These include dyestuffs, food additives, food preservatives, and drugs. Mutagenic natural products are known also, the most active being aflatoxin, a product of the food spoilage fungus, *Aspergillus flavus*.

Different mutagens have very different mechanisms of action, producing in many cases different types of mutational change. For example, X-rays or the chemical bleomycin produce a high frequency of chromosome breakage and thus chromosomal rearrangements. Others, such as UV-radiation and the majority of chemical mutagens, give low frequencies of chromosome breakage but higher frequencies of point-mutation.

STUDY ITEM

17.71 The effects of radiation

A number of different types of high-energy radiation produce ionizations, that is, the generation of electrically charged chemical groups. Some types of radiation, such as neutrons and γ-rays are extremely penetrating and only large amounts of dense material such as lead or concrete provide effective shields. When a unit of radiation passes through a cell some of its energy is dissipated and it leaves a trail of charged particles. Some of the damage results directly from the ionization of parts of biologically important molecules, including genetic material, but most damage occurs indirectly as a result of chemical reactions between highly reactive ions and nearby molecules which were unaffected by the passage of the radiation.

Radiation is measured in units called rads (R). A rad is the amount of radiation producing 10^{-5} joule per gramme of material by which it has been absorbed.

Table 19 shows the result of an experiment in which barley seeds were irradiated at a dose rate of 895 R hr^{-1}. Different batches of seeds were removed from the source of radiation at different times so that they received different doses.

	Dosage (rads)				
	0	5	10	15	20
Germination (per 100 seeds)	92	86	84	81	80
Survival to 3-leaf stage (per 100 seeds)	85	80	83	58	8

Table 19
Germination of barley following irradiation.

a *Draw conclusions from these results.*

Figure 62 shows a summary of a number of investigations into the relationship between radiation and mutation in fruit flies.

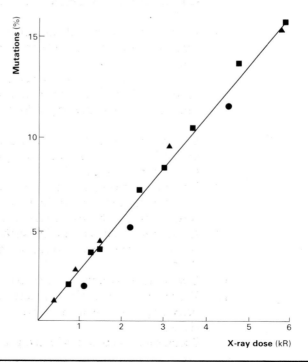

Figure 62
The direct proportionality between yield of recessive lethal sex-linked mutations in *Drosophila melanogaster* and dose of X-rays. Data from Timoféeff-Ressofsky; Oliver; Efraimson and Schechtmann, in Redei, G. P., Genetics, Collier Macmillan, 1982. Reprinted with permission of the publisher.

Such data have resulted in the generalization that there is a direct linear relationship between radiation dose and the yield of mutations, and that there is no safe lower limit of dose below which mutation can be avoided. (The same has generally been found to be true for chemical mutagens.)

b *Draw a graph to indicate what would be the relationship between dose and yield of mutations if the relationship were linear but a minimum (threshold) dose was needed to induce any mutations.*

☐

Mutation of human germ cells could result in the birth of genetically damaged children. A harmful recessive mutation could appear in a homozygote several generations after the mutation occurred. Many mutagens are also carcinogens (cancer-producing substances). This may be because development of cancerous tissue in some cases involves a somatic mutation. The carcinogenic effect of mutagens is a more serious hazard to humans than is mutation of germ cells since a mutagen must reach the gonads before it can affect the gametes while cancer can develop in any tissue.

Summary

1 Variation may be continuous or discontinuous (**17.1**).
2 Phenotype results from the interaction of genotype and environment (**17.2**).
3 Mendel was seeking to test ideas about the evolution of species when he discovered that characters of the whole organism are determined by pairs of factors which segregate in the production of the gametes (**17.3**).
4 Different generations of offspring in an experimental cross are described by the standard terms parental, F_1, F_2, F_3, and backcross (**17.3**).
5 The genetic constitution of a cell or individual is referred to as its genotype while the characters it displays are its phenotype (**17.3**).
6 The segregation of alleles produces a 1:2:1 ratio in a monohybrid F_2 (**17.3**).
7 Sex linkage means that a gene is present on one sex chromosome but absent from the other. The heterogametic sex cannot be heterozygous for a sex-linked gene (**17.3**).
8 If genes are located on different, that is, non-homologous chromosomes, they assort independently because each pair of chromosomes behaves independently during meiosis (**17.4**).
9 A test cross is a procedure in which an individual of unknown genotype is crossed with one known to be homozygous recessive (**17.4**).
10 Genes situated on the same chromosome are linked (**17.5**).
11 The extent of recombination between genes, measured by the percentage cross-over value, can be used to produce a genetic map (**17.5**).
12 Recombination between linked genes is accompanied by

recombination of variable, microscopically visible, features of a chromosome (**17.5**).
13 Inherited continuous variation can be explained by the additive effect of several gene loci. (**17.5**).
14 Mutation is the production of a new form of gene or a chromosomal change brought about by damage to, or faulty replication of, the genetic material (**17.7**).
15 Mutation rate can be artificially increased by chemical mutagens or irradiation (**17.7**).
16 There is no safe threshold dose of a mutagen (**17.7**).

Suggestions for further reading

(The references given for this chapter are mainly comprehensive textbooks on genetics and have sections relevant to Chapters 15–20 in this volume.)

CARTER, C. O. *Human heredity*. Pelican, 1970. (Ideas have developed and changed a good deal since this was written but it explains and describes many well known examples of Mendelian inheritance in humans very clearly. It is very readable.)

GARDNER, E. J. *Principles of genetics*. 6th revised edn. John Wiley, 1981.

MARSHALL, D. P. (ed.) *Advanced biology alternative learning project* (Abal) unit 8, *Genetics*. Cambridge University Press, 1984.

MILUNSKY, A. *Know your genes*. Pelican, 1980.

SINGER, S. *Human genetics: Introduction to the principles of heredity*. Freeman, 1978.

SINNOTT, E. W., DUNN, L. C., and DOBZHANSKY, T. *Principles of genetics* 5th edn. McGraw Hill, 1958.

CHAPTER 18 THE NATURE OF GENETIC MATERIAL

18.1 The search for genetic material

We have seen that inheritable variation is caused by inheritable units (genes) carried in the nuclei of cells and arranged in a definite order on chromosomes. Certain questions now present themselves:

1 Of what material are genes composed?
2 How does a gene function when it produces a character of the whole organism, the cells of which carry copies of the gene?

Although a long period of preliminary work was needed before biologists could formulate clearly such problems and develop the necessary materials and techniques for their solution, they have compressed most of the research into the second half of this century, when the progress has been extremely rapid. This has been a time of great activity and excitement among biologists, not only because it has brought them close to an understanding of the mechanism of inheritance, but also because it has thrown considerable light on the origin and nature of life itself.

In a school laboratory, it is difficult to do more than experience, to a limited extent, the methods used to study these problems. To understand fully the theoretical basis you will need to consider the experimental evidence from other sources. Histochemical tests that you can carry out are given in *Practical guide 5* investigations 18A, 'Testing for DNA, using the Feulgen technique' and 18B, 'Testing for DNA and RNA, using methyl green pyronin'.

Since genes appear to reside in or on chromosomes, a chemical analysis of chromosome material might help to discover the nature of genes. Chromosomes can be obtained by macerating tissue containing many actively dividing cells and then isolating the chromosome material by density gradient centrifugation. This process is very important in biochemistry. It involves pipetting a series of solutions of decreasing strength into a centrifuge tube to achieve a strong solution of high density at the bottom and a weaker solution of lower density at the top. The tissue extract is then placed on the surface and the tube spun at very high speed. Particles of different densities and shapes sink through the increasingly dense solution at different speeds and are thus separated.

Chromosome material (chromatin) turns out to be made up of proteins of several kinds and a high molecular mass substance called deoxyribonucleic acid (DNA). Any or all of these substances might contribute to the structure of a gene.

From observation of the inheritance of characters in organisms we can draw up a specification for a fully functioning gene.

1 It must replicate perfectly during cell division so that all the cells descended from a particular cell have at least one perfect copy.
2 Very rarely it must be able to undergo permanent alteration to produce a new allele; that is, it must be able to mutate.
3 It must carry information, that is, some sort of code which instructs

the cell or whole organism to develop a certain phenotype, (for instance, blood group A rather than group B).

4 It must be able to be controlled in its expression since a zygote contains many genes and the effects of most of these are manifested only in a limited number of organs or tissues of the organism that grows from the zygote (for example, only red blood cells produce haemoglobin).

For a long time scientists tended to favour the hypothesis that genes are made of protein. This was because protein structure is very complex. Each molecule can include up to twenty different types of amino acid. DNA was considered to be too simple a molecule as it contains only deoxyribose (a sugar with five carbon atoms) phosphoric acid, and four different nitrogen-containing organic groups called bases.

The apparent complexity of protein compared with DNA led to the hypothesis that genes consisted of protein and that DNA was involved in producing the structure of the chromosomes – a sort of scaffolding (or skeleton) to which the genes could be attached.

STUDY ITEM

18.11 DNA in the nuclei of different kinds of cells

Table 20 shows the mass of DNA contained in three different kinds of cells from seven different organisms. These data were obtained in 1949 by A. E. Mirsky and H. Ris at the Rockefeller Institute for Medical Research in the United States. They determined the number of nuclei in a given volume of a suspension in which the nuclei had been isolated. Determination of the mass of a known, very large, number of these enabled the mass of DNA per nucleus to be found.

Animal	DNA (mg $\times 10^{-9}$)			
	Nucleus of red blood cell	Nucleus of liver cells	Sperm	Amount in sperm column $\times 2$
Domestic fowl	2.34	2.39	1.26	2.52
Shad	1.97	2.01	0.91	1.82
Carp	3.49	3.33	1.64	3.28
Brown trout	5.79		2.67	5.34
Frog	15.0	15.7		
Toad	7.33		3.70	7.40
Green turtle	5.27	5.12		

Note:
Mammalian red blood cells do not possess nuclei but those of other vertebrates do.

Table 20
The amount of DNA in the nuclei of different cells.
After Mirsky, A. E. and Ris, H., 'Variable and constant components of chromosomes', *Nature*, **163**, 1949, pp. 666–667.

a From the information in table 20, what generalization can you make about the amount of DNA per nucleus in:
1 the red blood cells and liver cells examined
2 the somatic cells and gamete cells?

b *If DNA were the genetic material, would you expect to find data similar to that given in table 20?* Yes Somatic cells = 2x gamete

c *If a cell divides meiotically, what would you expect to happen to the amount of:*
1 genetic material ½ ed
2 structural (skeletal) material of the chromosomes?

d *From the data supplied by Mirsky and Ris, can we rule out the possibilities that DNA is both the genetic and the skeletal material, or just the skeletal material?* No 'cos if chromosomes ½ ed then both genetic + structural ½ ed

e *Would more DNA estimations from somatic cells and gamete cells of more kinds of animals (and plants) help to decide the role of DNA?* No - only make generalization more valid

f *Explain why the table contains blank spaces. Do the blanks significantly influence your conclusions?* Unknowns. It is personal scientific judgement that indicates if there is enough evidence here.

18.2 The evidence from viruses

Viruses are disease-causing agents which were first demonstrated by the associates of Pasteur in the nineteenth century. These particles presented great difficulty for the scientists who first put forward the hypothesis that infectious diseases are caused by 'germs'.

Viruses are too small to be studied with a light microscope, they can pass through filters designed to remove bacteria and other cells, and they can sometimes be crystallized. They have never been grown on non-living media and are only known to reproduce inside the cells of living hosts. They do not respire or metabolize compounds for themselves, and they rely entirely on enzymes in the host cell to convert energy and synthesize materials.

Much argument has arisen about whether viruses are living or non-living. More significantly, biologists debate whether they are similar to the earliest forms of life and so represent a link between chemical and biological evolution, or whether they have evolved from a component of a more complex cell or by the progressive loss of components from a cellular life form.

Each type of virus has a limited range of hosts in which it can reproduce. The 'life' cycle of a virus is often as follows. It is called the lysogenic cycle (*figure 63*).

Some types of virus have the ability to attach their genetic material to that of their host cell soon after they infect it. No virus protein coats are made and the virus genetic material is reproduced at each cell division and transmitted to all the descendants of the original host cell for many generations. It thus behaves just like any group of host genes. When a suitable stimulus, such as irradiation, is given to the host cell, the relationship between virus and cell breaks down, virus proteins are made and the host cell bursts to release many complete viruses.

The ability to enter into a stable relationship with the host cell is of very great interest for the following reasons:

1 It is believed that, in higher organisms, cells so infected may become

Figure 63
The lysogenic cycle of a virus. These diagrams are not to scale: the virus is very much smaller than the host cell.

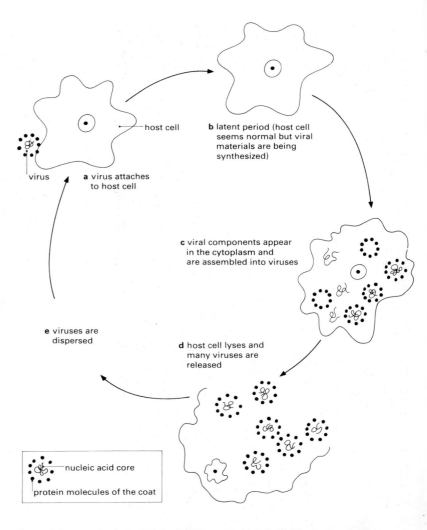

free of the controls on their development and divide to produce a tumour (cancer).

2 Viruses sometimes incorporate host genes into their structure and then transport these genes to the new cells they infect, thus altering the genotype of the new host cell. This can be made use of in so-called genetic engineering (see Chapter 21).

Viruses can be seen with an electron microscope and they are very diverse in structure. They always contain proteins and nucleic acids (either DNA or RNA). One group of viruses which have a fairly complex structure are the bacterial viruses or bacteriophages (see *figure 64*).

Bacterial viruses infect, and are reproduced in, bacteria (for example, *Salmonella*) which are found in mammalian guts. They were first studied in the hope that they might act as agents of control against bacterial diseases. This hope has never been realized but we have learned an enormous amount from them about the nature of genes and the role of DNA.

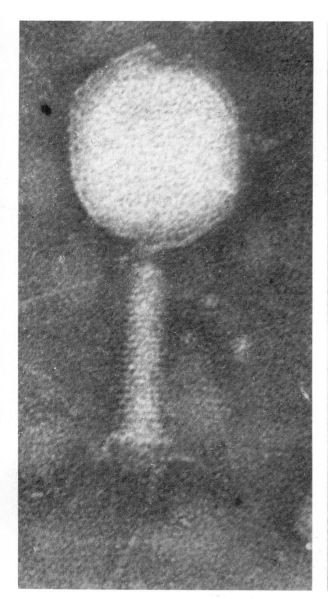
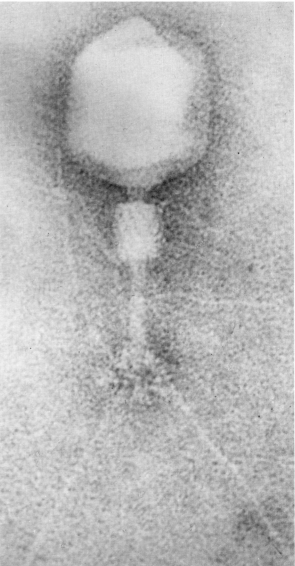

Figure 64
Electronmicrographs of bacteriophages. On the left is a normal phage with extended tail. On the right, the phage is 'triggered', showing contraction of the structure. ($\times 603\,500$)
Photograph, Dr R. W. Horne, Department of Ultrastructural Studies, The John Innes Institute, Norwich.

STUDY ITEM

18.21 Labelling viral DNA

The protein coat of a bacteriophage contains sulphur (two of the twenty amino acids found in protein have sulphur in their side chains) but no phosphorus. The DNA of the virus contains phosphorus but no sulphur.

The normal isotopes of phosphorus and sulphur are ^{31}P and ^{32}S. Both elements have readily available radioactive isotopes: ^{32}P and ^{35}S. If bacteria are grown with ^{35}S in the form of sulphate or ^{32}P in the form of phosphate they become labelled with the radioactive isotopes. Viruses produced in the labelled bacterial cells are themselves labelled. These labelled viruses can be used to test the hypothesis that DNA, and not

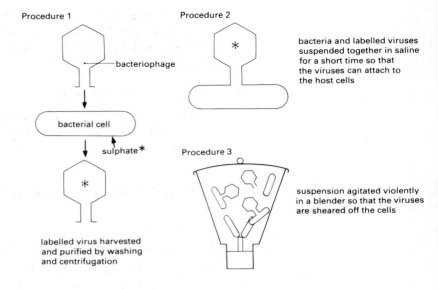
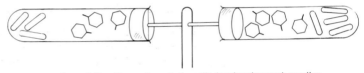
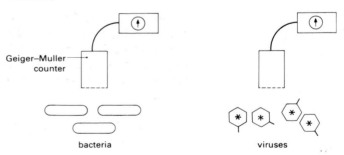

Figure 65
Procedures to demonstrate that the entry of DNA alone into a bacterium is enough to achieve complete and successful viral replication. (Not drawn to scale.)

protein, determines the structure of the next generation of viruses, that is, that DNA constitutes the viral genetic material.

Figure 65 outlines a set of experimental procedures using viruses labelled with radioactive sulphur but with normal (non-radioactive) phosphorus. The label is symbolized by a *.

The bacteria obtained after these procedures are free from radioactivity.

Some time after they have been freed from radioactivity, and therefore apparently from virus, the bacterial cells burst and release many new viruses similar to those which were removed by the blender from the surface of the bacteria.

a What does this result show? *Protein coat of bacteriophage need not enter the host cell. DNA must carry instructions for synthesis of coat*

b Explain how radioactive phosphorus (^{32}P) could be used to make more certain the roles of DNA and protein in the infection of cells by viruses. *R.A Phosphorous*

18.3 Direct evidence for DNA as genetic material

STUDY ITEM

18.31 Genetical transformation in *Pneumococci*

As early as 1928 Griffith showed that when he injected dead bacteria of one type into a mouse together with living bacteria of another strain, the living bacteria could acquire inherited characters from the dead ones.

The bacteria used were *Pneumococci*, which cause one type of pneumonia. They exist in two true breeding forms:

1 An infective (or virulent) form which produces disease symptoms in mice, and the cells of which develop a polysaccharide capsule. This type is called the smooth (S) form because its colonies look smooth on agar culture plates.

2 A non-infective (or non-virulent) form which cannot produce disease in mice, and the cells of which lack capsules. This type is called the rough (R) form because its colonies look rough when growing on agar plates.

Griffith injected heat-killed smooth cells and live rough cells into the same mice. Many of the mice developed disease symptoms, and he was able to isolate living smooth cells from them. Possibly some of the live rough cells had acquired the genetic information from the dead smooth cells to make a polysaccharide capsule, but other explanations are also possible.

1. heat-killed smooth cells might have survived + multiplied. This is ruled out by injecting heat-killed smooth cells without the living rough cells

2. Some living rough cells might have mutated to smooth. Ruled out by injecting living rough in no dead smooth

3. Mice could have been incubating pneumonia before start of exp. or acquired it by chance. This is ruled out by injecting a sterile saline

a What controls would be needed to rule out some of these alternative explanations?

To make sure that the true breeding smooth cells derived from the dead mice had originated from the rough cells he had used, Griffith employed the technique of genetic marking (see figure 66).

b Suggest two reasons why Griffith's result cannot be explained by mutation. *① Only small portion of cells transformed but the frequency + regularly too great to be mutation. ② Same mutation should be observable in pure rough culture*

Dead S cells might be able to mate with live R cells or the dead S cells might release a substance (genetic material) which is taken up by the R cells.

Figure 66
The experimental procedure of Griffith for genetic marking.

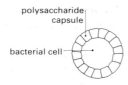

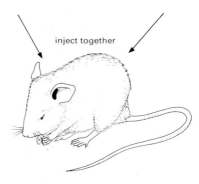

c **How could the latter hypothesis be tested?**

In 1944 O. T. Avery, C. M. Macleod, and M. McCarty of the Rockefeller Institute for Medical Research in the United States published an important paper which more or less settled the question of the nature of the hereditary material and identified it with DNA. They were the first to try to isolate from the S form a pure substance that was capable of transforming the R form into the S form in the same way as with dead bacteria. They showed that the active principle was DNA.

'The data obtained by chemical, enzymatic and serological analyses together with the results of preliminary electrophoresis, ultra-centrifugation and ultra-violet spectroscopy indicate that, within the limits of the methods, the active fraction contains no demonstrable protein, unbound lipid or serologically reactive polysaccharide and consists principally, if not solely, of a highly polymerized, viscous form of deoxyribonucleic acid.'

When you realize that it took $75\,dm^3$ of the bacterial culture to produce 10–25 mg of DNA, and consider the elaborate purification processes that were needed and the critical tests that were applied, you will be able to appreciate the effort behind the apparently simple statement that 'the hereditary material was identified as DNA'.

Once a highly purified extract had been obtained from the S type bacteria, there were still two further aspects to this important experiment. The first was to show that the extract could transform the living R type bacteria into the S type; the second that the extract was indeed DNA.

The first was fairly straightforward and you will be able to envisage the whole process and the necessary controls. Some details of the second may, however, be helpful.

Extracts of four samples of the DNA from the bacteria were analysed for carbon, hydrogen, nitrogen and phosphorus. The results are shown in table 21 together with the theoretical values for sodium deoxyribonucleic acid obtained from other sources.

Preparation number	Carbon (%)	Hydrogen (%)	Nitrogen (%)	Phosphorus (%)	N/P ratio
37	34.27	3.89	14.21	8.57	1.66
38B			15.93	9.09	1.75
42	35.50	3.76	15.36	9.04	1.69
44			13.40	8.45	1.58
Sodium deoxyribonucleate	34.20	3.21	15.32	9.05	1.69

Table 21
Analysis of DNA extracts and sodium deoxyribonucleate.
After Avery, O. T., MacLeod, C. M. and McCarty, M., 'Studies on the chemical nature of the substance inducing transformation of pneumococcal types'. J. exp. Med., **79 No. 2**, 1944, pp. 137–158.

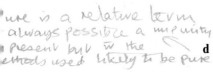

d *Do you think that the data in table 21 established that the substance extracted and isolated from the bacteria was chemically pure?*

e *What was the purpose of tabulating the ratio of nitrogen to phosphorus?*

In recent years much has been learned about the conditions which allow strains of bacteria to take up DNA and become transformed. DNA from higher animals and plants can be used and the bacteria may then acquire the ability to make polypeptides characteristic of the organism providing the DNA. In this way bacterial strains have been produced which make human insulin and human growth hormone. The cells of higher organisms can also take up DNA and incorporate it into their genetic material but not as easily as bacteria can. Research is in progress all over the world aimed at exploiting this to produce new types of plant by genetic engineering. Details of some of these developments may be found in Chapter 21.

18.4 The chemical structure of DNA

DNA may be extracted from the cells of animals, plants and micro-organisms. Soft tissues with small cells and a high density of nuclei, such as the thymus gland or root meristems, are particularly suitable. The tissue is macerated in ice-cold buffer solution and centrifuged to remove debris; the DNA is then concentrated and purified by chemical procedures.

The molecules of DNA extracted in this way are very long. If a solution is allowed to evaporate on a suitable surface and the surface is then coated with carbon, the DNA molecules can be individually resolved by the electron microscope. Many DNA molecules – especially bacterial DNA – viewed in this way appear circular.

DNA can be chemically analysed in a variety of ways. The simplest technique is to heat the solution with perchloric acid. This hydrolyses the molecule into simpler units which can be separated by chromatography.

Another important source of information about DNA has come from the technique of X-ray diffraction. Pure crystals of DNA must be produced (a difficult procedure) and X-rays are passed through a large, perfectly formed crystal. Atoms in any crystal are regularly arranged and they scatter the X-rays to produce a pattern which can be recorded on a photographic plate. From the diffraction pattern measurements are made which enable the position of atoms in the molecule of the substance to be deduced. This produces evidence for the molecular structure.

STUDY ITEM

18.41 **The relative quantities of nitrogenous bases found in various forms of DNA**

In 1951 E. Chargaff published values for the relative quantities of the nitrogenous bases found in DNA from a variety of organisms. He drew attention to some important generalizations which can be made when many different DNA samples are compared. Some data of the type used by Chargaff are shown in table 22. The molecular structure of the bases is shown in *figure 67*.

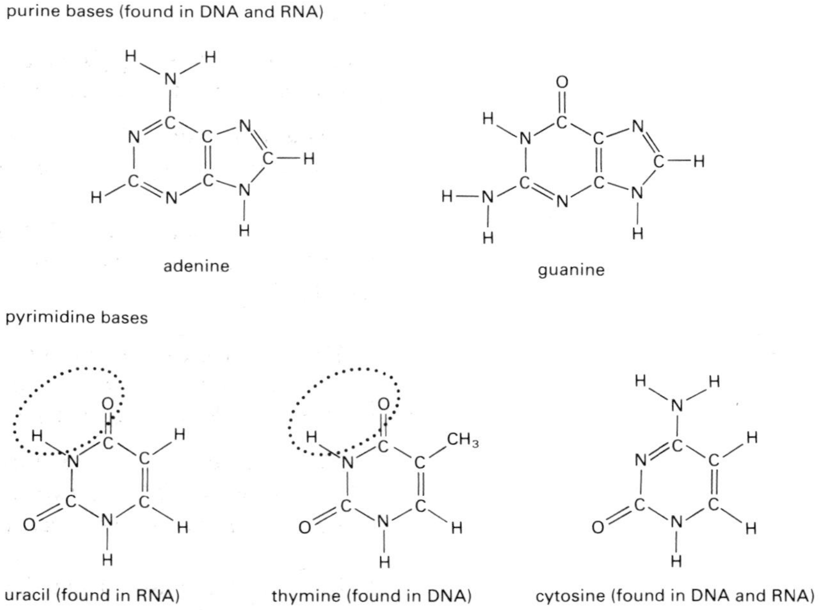

purine bases (found in DNA and RNA)

adenine   guanine

pyrimidine bases

uracil (found in RNA)   thymine (found in DNA)   cytosine (found in DNA and RNA)

Figure 67
The structure of the nucleic acid bases. The regions of thymine and uracil enclosed by dotted lines are involved in the pairing of bases – see *figure 68*.

Sources of DNA	Purine adenine	Purine guanine	Pyrimidine thymine	Pyrimidine cytosine	Purine/pyrimidine	Adenine/thymine	Guanine/cytosine	A + T/G + C
Bovine thymus	28.2	21.5	27.8	22.5	0.99	1.01	0.96	1.27
Bovine spleen	27.9	22.7	27.3	22.1	1.02	1.02	1.02	1.24
Bovine sperm	28.7	22.2	27.2	22.0	1.03	1.06	1.01	1.26
Pig liver	29.4	20.5	29.7	20.5	0.99	0.99	1.00	1.44
Pig spleen	29.6	20.4	29.2	20.8	1.00	1.01	0.98	1.43
Pig thymus	30.0	20.4	28.9	20.7	1.02	1.03	0.99	1.43
Pig thyroid	30.0	20.8	28.5	20.7	1.03	1.05	1.00	1.41
Rat	28.6	21.4	28.4	21.5	1.00	1.01	1.00	1.33
Mouse	29.7	21.9	25.6	22.7	1.06	1.16	0.96	1.24
Salmon	29.7	20.8	29.1	20.4	1.02	1.02	1.02	1.43
Shad	28.4	21.8	29.3	20.5	1.01	0.97	1.06	1.36
Wheat	27.3	22.7	27.1	22.8	1.00	1.01	1.00	1.20
Yeast	31.3	18.7	32.9	17.1	1.00	0.95	1.09	1.79
Escherichia coli	26.0	24.9	23.9	25.2	1.04	1.09	0.99	1.00
φX174 Bacteriophage	24.3	24.3	32.3	18.2	0.97	0.75	1.35	1.33
Vaccina virus	29.5	20.6	29.9	20.0	1.00	0.99	1.03	1.46
Arabacia lixula	31.2	19.1	30.5	19.2	1.01	1.02	0.99	1.61
Echinocardium cordatum	32.9	17.0	32.2	17.9	1.00	1.02	0.95	1.86

Table 22
Molar proportions of the nitrogenous bases in DNA from various sources expressed as moles for 100 moles of phosphorus in combination with them.

Examine the table and then answer the following questions.

a What can you say about the molar quantities of purine and pyrimidine present in any one sample of DNA?

b What do you notice about the ratio of the number of molecules of adenine to the number of molecules of thymine in any one sample of DNA?

c Is the same relationship found between the number of molecules of guanine and cytosine in any one sample of DNA?

d Are the data in table 22 compatible with the idea that the four bases combine together in pairs in the DNA molecule?

The last column in the table is calculated on the assumption that adenine and thymine form a pair and guanine and cytosine form another type of pair.

e Within the limits of experimental error, are the proportions of the two types of pair constant for different tissues in the same organism? If DNA were the genetic material, would you expect this?

f Within the limits of experimental error, are the proportions of the pairs different for DNA extracted from different species? Would you expect this if DNA were the genetic material?

Notice that the base proportions for rat bone marrow cells and bacteriophage ΦX174 are the same.

g Does this mean that the genetic information in rat bone marrow cells

and bacteriophage is the same? In answering this question it may help to consider the following pairs of English words:
mane and name
☐ cat and act

18.5 A model of the DNA molecule

In 1951 it was well known that DNA was made up of sugar, phosphate,

Figure 68
The structure of the DNA molecule.

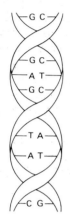

a The spiralized form of DNA

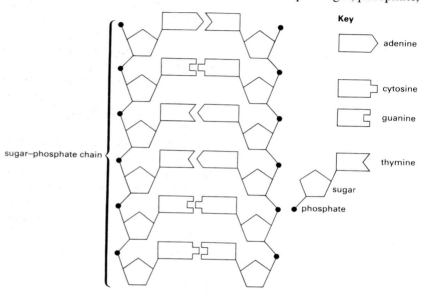

b A diagrammatic representation of a DNA molecule showing the two complementary strands

c The nucleotide units pair because a hydrogen atom is attracted to two other atoms. This is a hydrogen bond, shown thus: ···H—

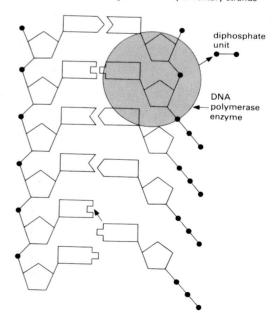

d A new strand forming on a pre-existing strand. Note that the nucleotides have three phosphates attached providing energy and four bases, but the way in which these subunits were linked together in the complete molecule was not understood, nor was it universally accepted that DNA was the genetic material.

A group of chemists was at that time researching in the Cavendish laboratory at Cambridge into the structure of biologically important molecules – especially proteins. The X-ray diffraction technique was an important tool in these researches. Two rather junior members of the team, Francis Crick and James Watson, became interested in DNA structure and obtained X-ray diffraction data from Maurice Wilkins of Kings College London. They hit upon the ingenious idea of making a very large number of models of the subunits of DNA, exactly to scale, using known atomic dimensions and chemical bond angles. The model subunits were then fitted together in a variety of ways to produce hypothetical DNA structures. Measurements were made of the relative distances between atoms in these models and the relative distances were then compared with Wilkins's X-ray diffraction data to see if any model fitted the data.

Several attempts ended in failure and Watson, in the account of the research in his book *The Double Helix*, hints that other members of the team regarded the whole exercise as a waste of effort! Eventually they hit upon the structure shown in *figure 68*.

After the model had been put together, its dimensions had to be checked against the X-ray diffraction data. Watson describes in his book how they went to lunch before the checks were completed – 'telling each other that a structure this pretty just had to exist'.

STUDY ITEM

18.51 DNA replication

When in 1953 the structure of DNA was published by Watson and Crick, they ended their paper with the telling words: 'It has not escaped our notice that the specific pairing we have postulated immediately suggests a possible copying mechanism for the genetic material'. The mechanism they suggested has been introduced in *figure 68* and is elaborated in *figure 69*.

Because each new double helix molecule would be half old and half new, such replication is called 'semi-conservative'.

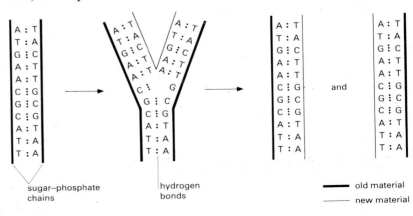

Figure 69
The replication of DNA.

Within five years, verification of this mechanism at the level of bacterial DNA and eukaryote chromosomes had been achieved.

The bacterium *Escherichia coli* was grown in a medium in which a heavy isotope of nitrogen, ^{15}N, had been incorporated. The bacterial cells synthesized the nucleotide bases (A, T, G, and C) using this heavy nitrogen and after several generations all the DNA in the cells was denser than normal as a result of the ^{15}N. The cells were then centrifuged, washed free of soluble ^{15}N compounds and transferred to a medium with normal ^{14}N nitrogen as the source for new growth.

Samples of cells were harvested after one and two cell generations. Their DNA was extracted and centrifuged in a density gradient (see section 18.1). The DNA sedimented until its density equalled that of the solution in the tube. Here it accumulated as a band of material. Samples of DNA from cells grown only in ^{15}N and only in ^{14}N were also centrifuged as controls.

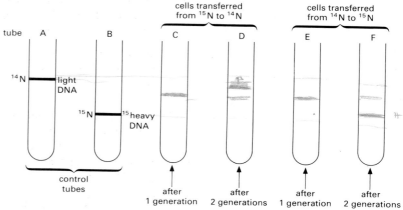

Figure 70
The labelling of replicating DNA with isotopes of nitrogen.

Figure 70 shows the DNA bands of the two types of control cells.

a   *Reproduce the figure by tracing or photocopying and use the semi-conservative hypothesis to predict the position of the DNA band or bands in the experimental tubes.*

b   *If cells are transferred from ^{15}N medium to ^{14}N medium, how many generations would elapse before a band of DNA containing ^{15}N disappeared?*

Isotope labelling, this time using the radioactive isotope of hydrogen called tritium, ^3H, can also be used to demonstrate semi-conservative replication in root tip cells. The technique used is autoradiography. Growing roots are supplied with thymine which has been labelled with ^3H. The thymine is specifically incorporated into the DNA of meristematic cells. After the cells have divided many times in labelled thymine they are transferred to non-radioactive thymine for one or two cell generations, after which they are spread on slides and covered with photographic emulsion. The slides are then stored in total darkness. Some of the ^{3}H atoms in the chromosomal DNA undergo radioactive decay and release very low energy electrons (β particles) into the emulsion above the radioactive chromosomes.

The emulsion is then fixed so that it can be exposed to light. The β particles from the radioactive areas of the cells produce silver grains in the emulsion which can be seen, and photographed, using a microscope. Because the emulsion is thin and transparent, it is possible to focus down and view the object, in this case a chromosome or chromatid, which has given rise to a group of silver grains.

Imagine a chromosome in late prophase of mitosis. The chromosome has replicated in an environment of non-radioactive thymine in a cell which is the end product of many cell divisions in labelled thymine.

c *Would one or both of the chromatids be labelled?*

d *If the chromosome had replicated once in non-radioactive thymine and had been killed and fixed during the second mitosis in unlabelled medium, would one or both of the chromatids be labelled?*

e *If this experiment is to be a fair test of the semi-conservative hypothesis, how many DNA double helices must there be in a chromatid?*

It is possible to distinguish between new and original DNA in chromosomes, not only by labelling with an isotope, but also by using a chemical, bromo-deoxy-uridine, that is incorporated into any newly synthesized DNA in place of thymine which is similar in chemical structure. DNA containing the chemical label stains differently from natural DNA. *Figure 71* shows a cell which has divided in a solution containing the label. One chromatid contains original DNA and is

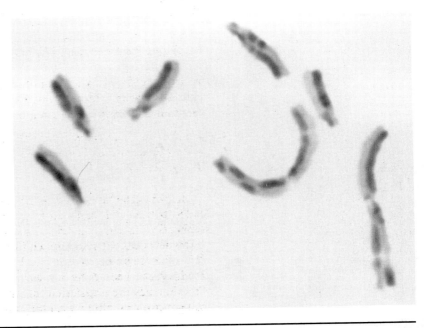

Figure 71
Chromosomes from a root tip cell of broad bean, *Vicia faba*, labelled to show semi-conservative replication.
Photograph, Dr C. G. Vosa, The Botany School, Oxford.

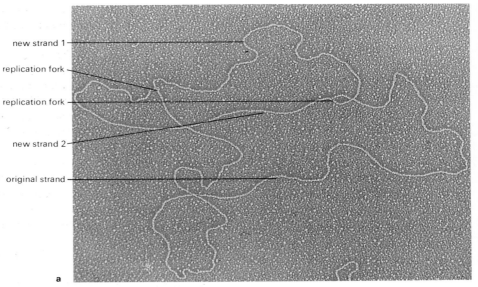

Figure 72
a Electronmicrograph of replicating mitochondrial DNA in *Plasmodium knowlesii*. Such a molecule is 6 μm in circumference. Bacterial 'chromosome' DNA is about 1200 μm in circumference and replicates in a similar way.
b This diagram shows how the bidirectional replication of a circular DNA strand begins at the origin and terminates at a diametrically opposite site. The replicating DNA appears similar to the Greek letter θ (theta) and so this type of replication is often called the θ type of replication.
Photograph **a**, *Dr Shirley McCready, The Botany School, Oxford. Diagram* **b** *is based on Redei, G. P.*, Genetics, *Collier Macmillan, London, 1982. Reprinted with permission of the publisher.*

darker since it has taken up more stain. The method shows that sister chromatids have broken and rejoined during mitosis. This may be caused by the bromo-deoxy-uridine – that is, it may represent abnormal behaviour and be an artefact – resulting from the experimental procedure.

Further experiments with viruses, bacteria, and eukaryotes showed that DNA replication originated at particular points (replication origins). Circular prokaryote DNA (see *figure 72*) only has one origin, but eukaryote chromosomes have multiple origins. DNA replication is a very complex biochemical process requiring a number of enzymes to carry out particular steps.

STUDY ITEM

18.52 A summary of the evidence that DNA is the genetic material

1 Gametes usually contain half the amount of DNA found in somatic cells of the same organism.
2 Most nuclei in a multicellular organism have the same amount of DNA, even if the nuclei are very different in size. Where this is not so, microscopic evidence often reveals the cells to be polyploid.

3 Pure DNA from one organism can be used to introduce a new gene into a cell so that the cell acquires an ability possessed by the cells from which the DNA was obtained.

4 When a virus infects a cell, only the nucleic acid portion needs to enter for the cell to make many perfect replicas of the whole virus.

5 All the tissues of a multicellular organism have the same types of DNA, as is revealed by base pair ratios. The DNAs of unrelated organisms often have very different base pair ratios.

6 The chemical structure of DNA appears to be well adapted to serve a genetic function.

7 In non-dividing cells, such as mature nervous tissue, there is evidence that DNA is not broken down and resynthesized while there is a constant turnover of the other cell constituents.

8 Studies in which DNA is labelled with a heavy isotope of nitrogen, ^{15}N, show that the replication of DNA is consistent with the model proposed by Watson and Crick.

9 Chemicals known to react strongly with DNA, such as nitrous acid or acriflavine, are mutagens.

10 The wavelengths of ultra-violet light (UV) which are the most efficient at producing mutation are those which are most strongly absorbed by DNA, that is, the DNA UV absorption spectrum matches the action spectrum for UV-induced mutation.

a *By means of a class discussion, a group discussion, or an essay, arrange the above list of evidence in order of importance. Which pieces of evidence are only circumstantial?*

b *The work of Watson and Crick is sometimes taken to herald the beginning of molecular biology. Is their research more crucial than that of others who are mentioned in this chapter?*

18.6 The breaking of the genetic code

STUDY ITEM

18.61 **The number of bases in the genetic code**

After they had worked out the structure of the DNA molecule, Watson and Crick went on to suggest how the information carried by it was used by the cell. Their hypothesis was that the sequence of bases determined the sequence of amino acids in a protein.

a *If only a single base coded for a single amino acid, how many of the latter could be coded?*

b *Calculate the minimum-sized group of bases which could code for 20 different amino acids.*

In making your calculations, remember that the direction in which a code is read is important. (In the code known as 'written English', 'cat' and 'tac' are not the same). Also remember that in the DNA code there are no gaps between 'words' as there are in written English, so it is simplest to assume that the group of bases corresponding to an amino acid is of constant size.

c 1 For how many different amino acids could, in fact, the minimum-sized groups of bases code?
2 If more than 20 amino acids could be coded, what might the solution to this problem be?

There is evidence that the DNA of bacterial cells is one very large molecule, yet the cell must require some hundreds of genes to specify all its structures and functions.

d *Does this introduce a fundamental difficulty for the hypothesis that the base sequence is a code? Explain your answer.*

STUDY ITEM

18.62 The site of protein synthesis

If an amino acid labelled with ^{14}C (or ^{35}S) is provided for protein synthesis by a population of cells, it will be incorporated into protein at the site or sites in the cells where protein synthesis takes place. If the cells are broken up and their organelles separated by centrifugation, the organelle at which synthesis is occurring will be labelled.

Protein synthesis can be extremely rapid and it has proved necessary to kill and analyse cells within a few seconds of supplying the labelled amino acids if the site of their incorporation into protein is to be detected. Longer periods of incubation with labelled amino acids have resulted in many cell constituents becoming radioactive (you should compare this with the use of $^{14}CO_2$ to detect the first product of photosynthesis; see *Study guide I*, chapter 7).

One of the first successful investigations was carried out by J. W. Littlefield and his co-workers and published in 1955. They injected ^{14}C leucine (an amino acid) into the veins of rats and showed that a component of the liver cell cytoplasm became more rapidly labelled with radioactive protein than any other part of the liver cell. The rapidly labelled component was the endoplasmic reticulum. When the phospholipid membranes of the reticulum were dissolved by detergent, most of the radioactivity remained bound to particles containing ribonucleic acid (RNA). These particles have become known as ribosomes (see section 18.9 later in this chapter).

The concentration of material at different levels in a centrifuge tube can be measured by the degree to which the different levels absorb light (optical density). Radioactivity can be measured with a scintillation counter. In 1959, K. McQuillen and his co-workers published their research concerning the incorporation of labelled amino acids into the cells of the bacterium *Escherichia coli*. Radioactive sulphate, $^{35}SO_4^{2-}$, was added to the culture. This was used by the cells to synthesize sulphur-containing amino acids which were incorporated into protein. If a large excess of non-radioactive sulphate, $^{32}SO_4^{2-}$, was added to the culture, the incorporation of labelled sulphate effectively ceased and the fate of the label already incorporated could be observed at various time intervals by sampling the culture, killing the cells and sedimenting the cell sap in a fast centrifuge (see *figure 73*).

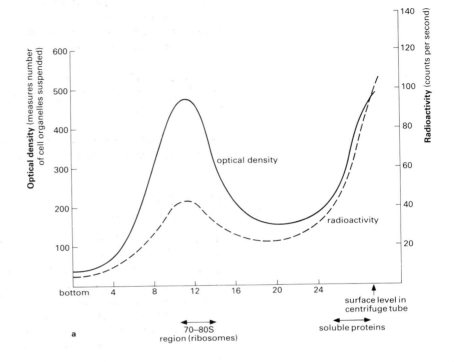

Figure 73
a The sedimentation analysis of bacterial cell juice. The bacteria were incubated with labelled sulphate for 15 seconds before analysis.
b Here the procedure was as for *figure 73a*, except that the bacteria were supplied with a large excess of non-radioactive sulphate for 120 seconds after the application of labelled sulphate. The radioactivity initially associated with ribosomes has passed to the soluble protein fraction.
Adapted from McQuillen, K., Roberts, R. B., Britten, R. J., 'Synthesis of nascent protein by ribosomes in Escherichia coli*' Proc. N.A.S., 1959.*

Chapter 18 The nature of genetic material   111

Particles sedimenting at different rates in a centrifuge can be identified by the rate of sedimentation under specified conditions. The rate of sedimentation is expressed in Svedberg units (S); for example, ribosomes have a sedimentation rate of 70–80 S.

The sequence of events suggested by McQuillen and his colleagues to explain their data was:

sulphate ⟶ amino acids ⟶ protein associated with 70–80 S particles ⟶ soluble protein

a *Explain how* figure 73 *provides evidence for this view.*

b ***Does the DNA of cells appear directly responsible for protein synthesis?***

c ***If the order of bases in a DNA molecule determines (codes for) the sequence of amino acids in a protein molecule, what problem is raised by the findings of Littlefield?***

When we considered evidence provided by experiments on *Acetabularia* (see chapter 15, section 15.3) we concluded that messages pass from the nucleus to the cytoplasm and determine the future pattern of development of the cytoplasm.

d ***With regard to the evidence provided by the work of Littlefield and McQuillen, are the messages DNA or protein, or are they likely to be another substance? Explain your answer.*** □

STUDY ITEM
18.63 The message linking nucleus and cytoplasm

Deoxyribonucleic acid (DNA) is only one type of nucleic acid found in organisms. There are also three types of ribonucleic acid (RNA). RNA differs from DNA in that the sugar ribose replaces deoxyribose (see *figure 74*).

The pyrimidine base uracil (U) replaces the thymine (T) found in DNA (see *figure 67*). Uracil and thymine have a very similar structure, and both form hydrogen bonds with adenine.

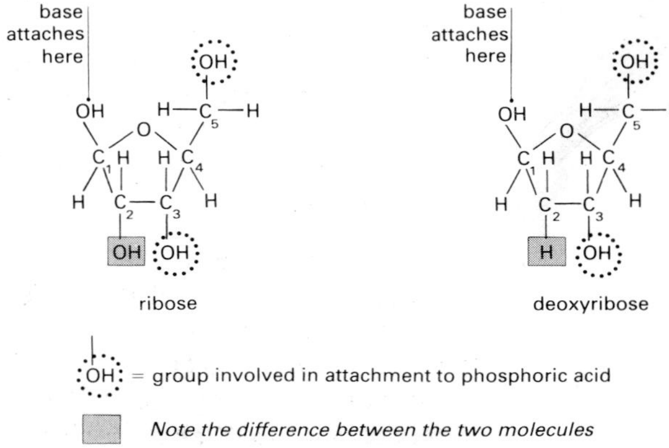
base attaches here
OH
H—C—H
O
C H
H C
H
C—C
H
OH
OH
OH
ribose
base attaches here
OH
H—C—H
O
C H
H C
H
C—C
H
H
OH
OH
deoxyribose
OH = group involved in attachment to phosphoric acid
Note the difference between the two molecules

Figure 74
The molecular structures of ribose and deoxyribose. (See also *Study guide I*, figure 112, page 153.)

As in DNA, the nucleotide subunits of RNA are linked by bonds between the sugar and phosphoric acid molecules. Notice that both types of sugar molecule have links to phosphoric acid at the carbon atoms numbered 3 and 5. Watson and Crick showed that a nucleic acid double helix will only form if the two strands are anti-parallel, with one sugar phosphate backbone running in a 3–5 direction and the other in a 5–3 direction. The 3–5 sugar phosphate linkage of nucleic acid molecules is very important because it provides a means whereby enzymes are able to 'read' the genetic code in the 'right' direction. All biochemical events in which the genetic code is used involve enzymes which move along a single nucleic acid strand. At one end of the strand is a sugar unit with carbon atom number 5 'free' (not linked to a neighbouring nucleotide). This is the 5 prime end. At the other end of the strand is a sugar with carbon atom 3 free – the 3 prime end. Enzymes 'read' the strand from any point towards the 3 prime end away from the 5 prime end.

The three different types of RNA are:

1 Messenger RNA (mRNA) – this is single stranded with no double helical regions. It usually has a high turnover; that is it is rapidly synthesized and broken down within the cell.
2 Transfer RNA (tRNA) – this is a single strand which folds back on itself so that the bases can pair between one part of the strand and another. The result is the clover-leaf shape shown in *figure 77*. tRNA is found free in the cytoplasm and there are at least as many types of tRNA as there are types of amino acid found in protein.
3 Ribosomal RNA (rRNA) – this is a long single strand which folds back on itself to form double-helical regions separated by single stranded regions. As the name implies, rRNA is associated with proteins to form cellular structures called ribosomes.

All three types of RNA can be shown by isotope labelling studies to be synthesized in the nucleus of eukaryote cells, after which they move into the cytoplasm.

a *Outline how such experiments might be carried out.*

Since protein is not made directly by DNA it was reasonable to suppose that the genetic message linking the code on the DNA with the site of synthesis (the ribosomes) was RNA. Messenger RNA was the prime suspect because of its rapid turnover.

Nirenberg and Mathaei (1961) extracted and purified mRNA from yeast and other sources. These mRNA types were added separately to tubes each containing the following:

bacterial ribosomes
bacterial enzymes and other soluble cell constituents of bacteria
bacterial tRNA
radioactive amino acids
adenosine triphosphate (ATP)
guanosine triphosphate (GTP)

Small amounts of labelled protein were produced in these cell free (*in vitro*) preparations. The protein was found to be characteristic of the mRNA source and not of bacterial protein.

b *Why would rapid synthesis and breakdown of a chemical message from nucleus to cytoplasm be an advantage?*

c *Why was it important to use mRNA from one organism and all the other constituents of the mixture from another, completely different organism?*

These experiments were followed by others in which single stranded RNA, similar to mRNA but of known and predetermined composition, was used to synthesize artificial polypeptides.

In the first of these experiments, the single stranded RNA that was used contained no other base but uracil. The structure can be represented by UUUUUUU...U repeated many times. This RNA produced a polypeptide which contained only the amino acid phenylalanine.

d *What amino acid does the sequence UUU (or TTT in the corresponding DNA sequence) represent in the genetic code?*

The use of synthetic mRNA, backed up by studies in which proteins produced by mutant cells were analysed, allowed the genetic code to be deciphered in the early 1960s. The code is as follows.

First base	Second base				Third base
	U	C	A	G	
U	phenylalanine	serine	tryptophan	cysteine	U
	phenylalanine	serine	tryptophan	cysteine	C
	leucine	serine	*chain terminates*	*chain terminates*	A
	leucine	serine	*chain terminates*	tryptophan	G
C	leucine	proline	histidine	arginine	U
	leucine	proline	histidine	arginine	C
	leucine	proline	glutamine	arginine	A
	leucine	proline	glutamine	arginine	G
A	isoleucine	threonine	asparagine	serine	U
	isoleucine	threonine	asparagine	serine	C
	isoleucine	threonine	lysine	arginine	A
	methionine (starts message)	threonine	lysine	arginine	G
G	valine	alanine	aspartic acid	glycine	U
	valine	alanine	aspartic acid	glycine	C
	valine	alanine	glutamic acid	glycine	A
	valine	alanine	glutamic acid	glycine	G

e *Give a set of three bases that codes for the amino acid proline.*

f *Give a base substitution (mutation) which would not produce a new polypeptide by substitution of a new amino acid.*

18.7 The synthesis of RNA

The following terms are used in relation to the code:

1 A codon – a group of three bases corresponding to a single amino acid.
2 Transcription – the production of a specific RNA, using a particular DNA sequence as a template.
3 Translation – the synthesis of peptides by ribosomes, using the information of a specific mRNA molecule.

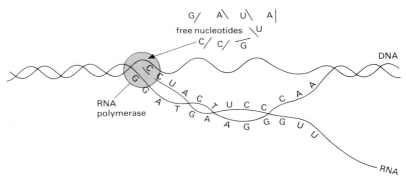

Figure 75
The production of an RNA molecule from a DNA template. In this diagram, RNA is shown as being synthesized only on one strand of the two in the DNA double helix.

Transcription is shown diagrammatically in *figure 75*. Notice that RNA is shown as being synthesized only on one strand of the two in the DNA double helix. This is logical as the RNA polymerase is only able to travel one way on the single stranded DNA, and the two DNA strands are anti-parallel. Also the mRNA copied from the two strands of the double helix would not be the same, and therefore would not give the same protein product when translated. However, can we verify that only one strand is transcribed?

If DNA solution is heated to a critical temperature under controlled conditions, the thermal motion of the molecules disrupts the hydrogen bonding between the base pairs, which may reunite if the mixture is cooled slowly. Each single strand will also form a double helical complex with other single stranded DNA or RNA molecules, provided these contain a complementary sequence of bases so that pairing is possible. DNA from some bacteriophages can be separated into two types of single strand by heat and then centrifuged so that the two types of strand separate into different bands. RNA produced by the same virus can be checked against the two bands of single stranded DNA and is found to attach by hydrogen bonding to only one type of DNA strand. This shows that the bacteriophage RNA is produced from only one strand of the DNA double helix.

Does transcription have specific start and stop points? The best evidence of this is cytological. *Figure 76* shows RNA being transcribed from genes for rRNA in the newt *Triturus viridiscens* and in the larva of the moth *Ephestia kühniella*.

Many RNA polymerase enzyme molecules are transcribing each of

Figure 76
The transcription of RNA from genes for rRNA:
a from a newt oocyte, *Notophthalamus* (*Triturus*) *viridescens* (× 18 900).
b from a moth larva, *Ephestia kühniella* (× 26 300).
Photograph **a** is by Dr O. L. Miller, Jr.; **b** is by Dr Andreas Weith, Med. Hochschule, Lübeck, W. Germany.

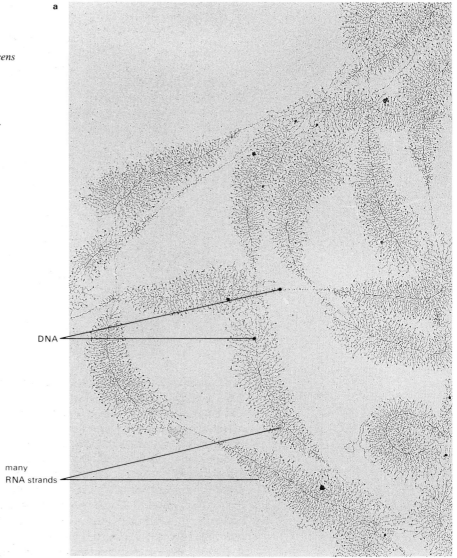

these genes. Each enzyme is producing an RNA transcript, and this increases in length as the enzyme passes along the gene. For each gene, therefore, transcription is starting at the 'apex of the Christmas tree' and stopping at the base of the tree. The enzyme and the RNA transcript then fall off the DNA.

Analysis of genes from a number of different organisms has revealed a common short DNA sequence – usually seven bases long – which is the initial recognition and attachment site for the RNA polymerase enzyme. Transcription actually is initiated a few nucleotides further along the DNA. Cells contain more than one type of RNA polymerase and this may be related to the control of the rate of production of different RNAs.

Transcription terminates at another specific short sequence of nucleotides distal to the coding region of the gene. As can be seen from

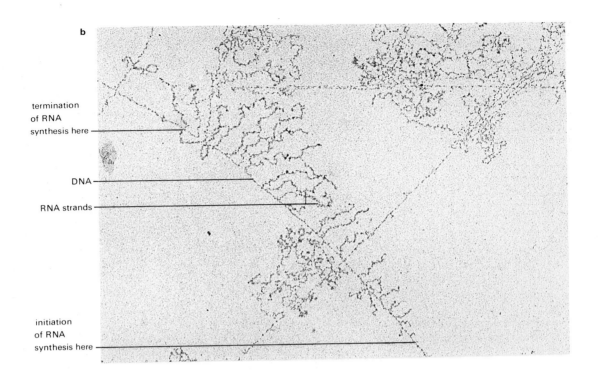

the genetic code (page 114) some amino acids are encoded by a single codon, others by more than one. The codon AUG codes for the amino acid methionine (a sulphur-containing amino acid). AUG also serves as a starting point in translation. The ribosome subunits are assembled at this codon and the first amino acid in all peptides is therefore methionine. This is often cleaved from the peptide after synthesis is complete. In bacteria, a modified form of methionine, *N*-formyl-methionine, is the initial amino acid for all peptides but elsewhere in the bacterial message AUG codes for methionine as in eukaryotes.

In micro-organisms some continuous mRNA strands may code for several polypeptides. In these cases one of the three codons UAA, UAG, or UGA serves as punctuation, causing translation to end after each gene. Translation is reinitiated by an AUG codon at the start of the next gene. These three termination codons do not code for any amino acid.

18.8 The mechanism of protein synthesis

Ribosomes are visible as small dots in electron microscope pictures of cell sections and are often associated with the endoplasmic reticulum in eukaryotes. In the bacterium *Escherichia coli*, ribosomes are about 20 nm in diameter. They consist of two unequal subunits which only come together in the presence of mRNA.

Amino acids cannot attach directly to ribosomes or to mRNA. The formation of a peptide bond requires energy and this is provided by ATP which activates the amino acid molecule. The amino acid is then transferred by a specific enzyme to the appropriate tRNA (see *figure 77*).

Amino acids are always attached to the 3-prime end of the tRNA which terminates with the codon CCA. The amino acid becomes linked to the terminal adenosine unit.

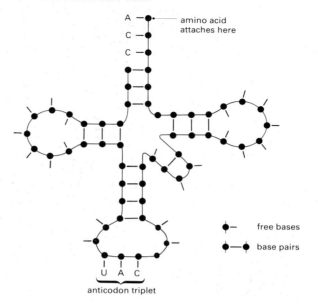

Figure 77
A representation of a molecule of tRNA. The molecule is not a flat structure but has a complex three-dimensional shape. It has about 80 nucleotides.

a   *Explain the term 'anticodon' as used in* **figure 77**.

Polypeptide synthesis begins by the attachment of the small subunit of a ribosome to a specific 6 or 7 base nucleotide sequence on a mRNA molecule. A methionine-tRNA (or *N*-formylmethionine in bacteria) then binds to the AUG start codon on the mRNA via its anticodon, in a special site in the ribosome's small subunit. The ribosome's large subunit then becomes associated (see *figure 79*).

The aminoacyl tRNA complementary to the second codon in the mRNA then binds to that codon at the ribosome, following which a peptide bond between the two amino acids is formed by the enzyme peptidyl transferase. This bond formation leads to the release of the tRNA, now cleaved from its amino acid.

The ribosome then proceeds along the mRNA, at each triplet adding on the aminoacyl-tRNA for each codon, forming the peptide bond, and releasing the free tRNA to which the growing polypeptide was formerly attached. These processes are provided with energy by guanosine triphosphate (GTP).

Just as there is a polarity to the mRNA (from 5-prime to 3-prime), there is a polarity to the polypeptide. The first amino acid is joined by its carboxyl end to the amino group of the next amino acid, which in turn is joined by its carboxyl group to the amino group of the next. As a consequence, the first amino acid retains its free amino group and the last ends up with a free carboxyl group.

Upon reaching one of the three codons UAA, UAG, or UGA, translation terminates, and the two halves of the ribosome dissociate from each other, from the polypeptide, and from the mRNA (see *figure 80*).

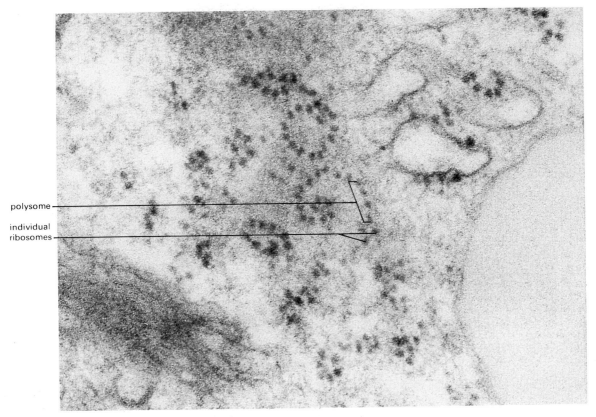

Figure 78
A strand of RNA with associated ribosomes (a polysome). (× 143 700).
Photograph, Biophoto Associates.

A series of ribosomes often becomes attached to the same mRNA molecule, each ribosome being at a different stage of translating the genetic code from RNA base sequence to amino acid sequence. The complex is called a polysome.

The events occurring in translation are summarized in *figures 79* and *80*.

During synthesis, the polypeptide folds up as a result of the interaction of amino acid side chains and consequently acquires its three-dimensional structure. It may also be modified by association with non-amino acid molecules (for example, the 'haem' group of haemoglobin). Enzyme proteins are often produced in an inactive form and are later activated by the removal of one or more amino acids from one end of the chain (as in the activation of trypsinogen to produce trypsin). (See *Study guide I*, chapter 6, page 193.)

Ribosomes are often associated with phospholipid membranes, especially those of the endoplasmic reticulum. The arrangement is such that the mRNA molecule lies on one side of the membrane but the growing polypeptide emerges on the other side. This gets over the problem of a large and complex protein molecule penetrating the membrane.

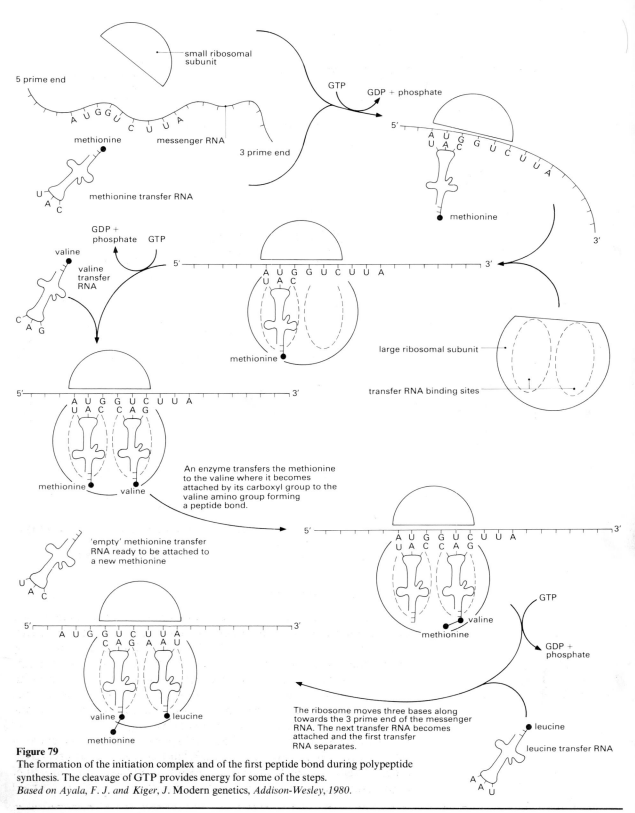

Figure 79
The formation of the initiation complex and of the first peptide bond during polypeptide synthesis. The cleavage of GTP provides energy for some of the steps.
Based on Ayala, F. J. and Kiger, J. *Modern genetics*, Addison-Wesley, 1980.

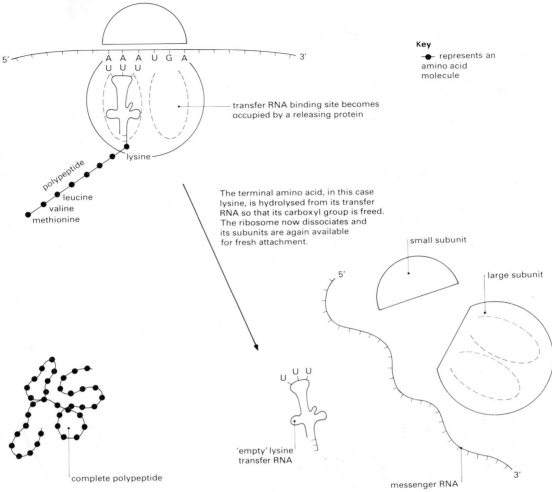

Figure 80
Steps in the termination of polypeptide synthesis when a termination codon (here UGA) enters the aminoacyl site on the ribosome.
Based on Ayala, F. J. and Kiger, J. Modern genetics, Addison-Wesley, 1980.

18.9 The structure of chromatin

Cytologists and biochemists have invested much effort in determining the relationship between DNA and the other constituents of chromatin. Typically this is composed of 40 % DNA, 46 % histone proteins, 13 % non-histone proteins, and 1 % RNA. It is clear that a DNA molecule must be coiled up, folded up, or packaged in some way since it is much longer than the chromosome of which it is a part. The changes in the length and appearance of chromosomes during the meiotic or mitotic cycle presumably reflect changes in this packaging.

Metaphase chromosomes show characteristic bands when suitably stained (see page 35) and it is thought that these bands also represent different states of DNA packaging. During interphase some regions of chromosomes or even complete chromosomes remain highly condensed and are visible as granules. This material is called heterochromatin and reflects yet another form of packaging. Only some of the genes in a cell are active in the synthesis of mRNA (see page 135) and there has been much interest in the organization of chromatin because it may provide information about how genes are regulated.

For many years the electron microscope proved singularly unsuccessful in revealing chromosome structure but much progress was made in the 1960s and 1970s. Scanning electronmicrographs, as in *figure 81*, show that a chromosome is a thread (or threads) which is so packaged that it forms loops which project outwards from the chromosome. In the so-called lamp-brush chromosomes of developing egg cells, these loops are large enough to be seen with a light microscope. Chromosomes may be made to swell or shrink *in vitro* by altering their environment and hence the manner in which their chromatin is packaged.

Very high magnification transmission electronmicrographs show that chromatin threads have a very regular structure consisting of beads of material of a definite size joined by finer threads. The beads are lengths of DNA about 160 base pairs long coiled around a core of histone proteins, with a particular type of histone, HI, acting as a spacer. Histones are proteins with many amino acids having basic side chains

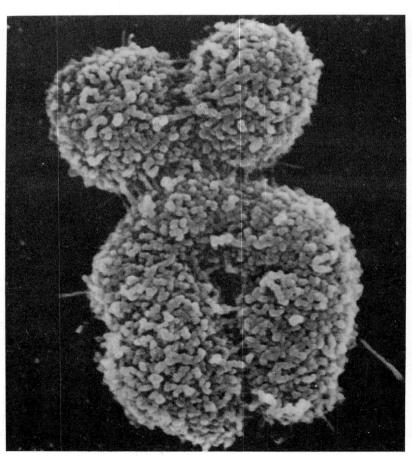

Figure 81
A scanning electronmicrograph of a human metaphase chromosome, showing the chromosome ultrastructure. ($\times 27\,500$).
Dr T. D. Allen, Paterson Laboratories, Christie Hospital and Holt Radium Institute, Manchester.

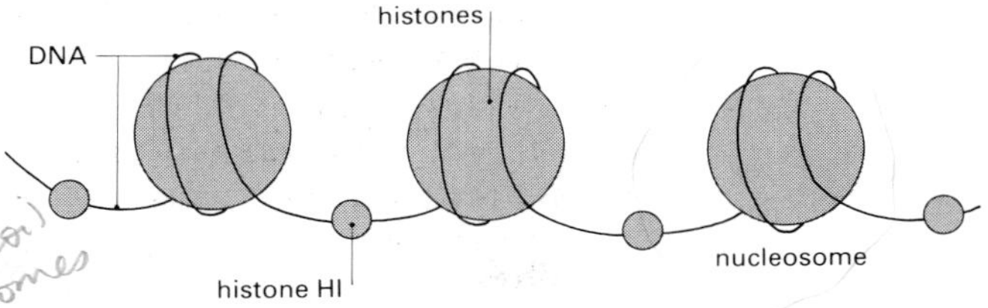

Figure 82
The relationship between DNA and histones in chromosome structure.

that combine with the phosphoric acid groups of DNA. The beads are called nucleosomes. (See *figure 82*.) Several amino acid sequences within histones have been conserved during evolution and show little variation when extracted from different organisms.

Organization into nucleosomes greatly reduces the length of the DNA. Non-histone proteins known as scaffolding proteins hold groups of nucleosomes together, further reducing the length of the strand. It is this level of organization which is responsible for chromosome banding.

Recognizably different protein structures are associated with centromeres and provide attachment for microtubules during division. Another group of proteins is associated with the specific pairing of homologous chromosomes during prophase I of meiosis. Almost all chromosomal protein seems to have structural functions and practically nothing is understood about the relationship between chromatin structure and the regulation of gene activity.

18.10 What is a genome?

'Genome' is a collective term for the genetic material of a cell (or a self replicating organelle within a cell such as a mitochondrion). In most organisms the genome is DNA but in some viruses, e.g. the AIDS virus and TMV, it is RNA. During fertilization the genomes of two haploid cells come together in a diploid zygote. It is therefore possible to refer to the haploid genome and the diploid genome of a species. In cells of hybrid origin (for example hexaploid wheat) one can refer to genetic material (information) originally contributed by one of the parental species as the genome of that species so the hybrid genome consists of two or more parental genomes.

It is known that large parts of the genome cells are neither transcribed nor translated. Some or all of this apparently redundant genetic material may have as yet undiscovered functions, for example in the regulation of the rate at which genes produce RNA. Many scientists believe that parts of the genome have no function. This raises the question of whether an organism exists to preserve and reproduce its DNA or whether the DNA exists to transmit the characters of an organism to its descendents.

Summary

1. Chromosomes contain DNA and protein (**18.1**).
2. Knowledge of patterns of inheritance and development allows us to draw up a specification for the genetic material (**18.1**).

3   Gametes contain half the DNA found in the somatic cells of the same organism (**18.10**).
4   The structure and replication of viruses can be used as evidence that DNA is the genetic material (**18.2**).
5   DNA extracted from one type of bacterium can be used to transform a second type – showing that it alone is able to specify the genotype of the cell (**18.31**).
6   DNA consists of a complex polymer containing deoxyribose sugar, phosphoric acid, and four nitrogenous bases. There are equal numbers of adenine and thymine units and also equal numbers of guanine and cytosine units (**18.41**).
7   The ratio of $A + T/G + C$ is constant for DNA from tissues of the same species but varies from species to species (**18.41**).
8   DNA molecules are able to replicate since they consist of two complementary strands and the bases on each strand are able to act as a template for the synthesis of a new complementary strand – this is semi-conservative replication (**18.51**).
9   A single strand of DNA can also act as a template for the synthesis of a complementary RNA molecule – this is transcription (**18.7**).
10 With the aid of a complex biochemical system involving tRNA and ribosomes, the sequence of bases on RNA can be translated into a sequence of amino acids in a newly synthesised polypeptide (**18.8**).
11 The sequence of bases in DNA is a code in which a triplet of bases (a codon) stands for a single amino acid (**18.8**).
12 Histones and other proteins associate with DNA to form chromatin which consists of repeating units called nucleosomes (**18.9**).
13 The genome of a cell, organelle, or virus is its complete, self-replicating nucleic acid content (**18.10**).

Suggestions for further reading

CLARK, B. F. C. Studies in Biology No. 83, *The genetic code*. 2nd edn. Edward Arnold, 1984.

FARNSWORTH, M. W. *Genetics*. Harper & Row, 1978. (A good and comprehensive account of genetics with a particularly clear section on the packaging of DNA into chromosomes.)

JACKSON, R. Carolina Biology Readers No. 86, *Protein biosynthesis*. Carolina Biological Supply Company, distributed by Packard Publishing Ltd, 1978.

ROBERTS, M. B. V. *Biology. A functional approach*. 4th revised edn. Nelson, 1986.

TRAVERS, A. A. Carolina Biology Readers No. 75, *Transcription of DNA*. Carolina Biological Supply Company, distributed by Packard Publishing Ltd, 1978.

WOODS, R. A. Outline studies in biology, *Biochemical genetics*, 2nd edn. Chapman & Hall, 1980. (The evidence for the genetic code is well explained.)

CHAPTER 19 GENE ACTION

19.1 Mutant complementation in *Neurospora*

This chapter is concerned with the way in which DNA – the genetic material – produces the phenotype. As is usual in science, any major discovery allows a whole host of new problems to be defined or redefined. The replication, transcription, and translation of DNA are processes which go a long way to explain how characters are transmitted unaltered for many generations and how specific polypeptides are produced. To what extent can the existence of complex phenotypes, like the shapes of wings or leaves or the colours of flowers and fur, be explained in terms of the translation of information coded in the base sequence of DNA?

We remain largely ignorant of how phenotypes are produced, even in the simplest prokaryote organisms, but some rather general hypotheses are available. They will help you to appreciate an area of biology where major research efforts are being made in laboratories all over the World and in which tremendous discoveries may be expected.

Much of our limited understanding of the way in which genes act begins with the researches of G. W. Beadle and E. L. Tatum in the United States. Their work was published in 1945 and concerned the fungus *Neurospora crassa*, already referred to in Chapter 16.

The life cycle of *Neurospora* is shown in a simplified form in *figure 83*. It has a vigorous, haploid mycelium which can reproduce asexually by means of spores called conidia.

If conidia or hyphae of different genotype meet, three outcomes are possible:

1 they may be vegetatively incompatible and the hyphae of the two types will grow as separate colonies;
2 they may fuse and grow as a heterokaryon;
3 if they are sexually compatible, hyphae or conidia from one mycelium will fuse with specialized reproductive hyphae, called trichogynes, from the other. This allows sexual reproduction to take place.

Neurospora is able to grow both on solid, agar containing media and in liquid cultures. The wild type fungus thrives on a carbon compound, such as sucrose, and a mixture of nutrient ions. It also requires the vitamin biotin. A growth medium which contains just these requirements and no extra nutrient substances is called a minimal medium. The fungus also grows well on complete media containing numerous amino acids, purine and pyrimidine bases, and vitamins. When growing on a complete medium it does not need to make for itself the various nutrients present in the medium.

Beadle and Tatum obtained suspensions of conidia from wild type *Neurospora* and irradiated them with a dose of X-rays sufficient to kill most of them. The spores that survived were likely to contain some nuclei which had suffered mutation.

Figure 83
a The life cycle of *Neurospora crassa*.
b Events within the perithecium – repeated many times.

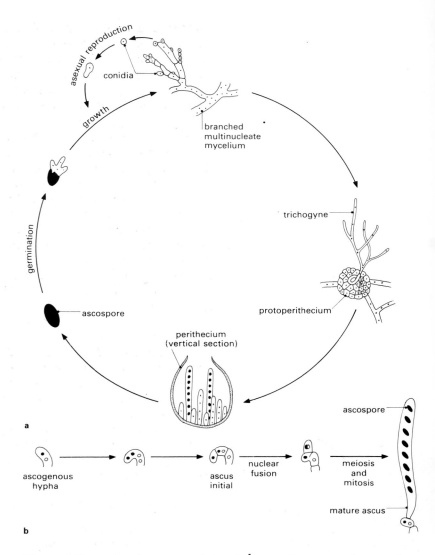

a *If a conidium contains two nuclei and a mutation occurs in one of them (or two different mutations occur, one in each nucleus) what type of mycelium would grow from the conidium?*

Irradiated conidia were used to provide nuclei to fertilize the trichogynes of sexually compatible wild type mycelia. In this way mutant nuclei gave rise to uninucleate ascospores. Large numbers of these were transferred individually to tubes of complete medium. Each ascospore produced, by mitosis, a mycelium which could be subcultured onto various types of media where its ability to grow could be observed.

By these laborious procedures, many haploid mycelia were isolated which had developed from single ascospores and which had inherited mutations produced by irradiation. Most of these looked perfectly normal but their mutant phenotype was revealed when they were tested on minimal medium. They would not grow on minimal medium because they required (from the environment) a nutrient substance which the wild type could make for itself. Each mutant was then tested on minimal medium which had been enriched with one or more substances and in this way the specific nutrient requirements of each mutant was determined. Some mutants required a vitamin, others a purine or pyrimidine base, and yet others particular amino acids.

Beadle and Tatum chose to work with mutants which needed the amino acid tryptophan. They isolated more than one hundred independently produced mutant mycelia which all required tryptophan. When spores from more than one of these mutants were transferred to the same agar plate with minimal medium, they were sometimes able to grow together as a heterokaryon while neither would grow alone. If two mutants are unable to grow alone but will grow together, either side by side or as a heterokaryon (or when both occur together in a diploid organism) the mutants are said to complement each other. Neither mutant will grow alone because some vital function is missing, but they will complement one another because each can supply a function missing in the other. If two mutants fail to complement it may be assumed that the same functional region of DNA (the same gene) has suffered damage in both mutants.

Figure 84 shows an example of complementation involving the nitrogen metabolism of a fungus. The wild organism can use nitrate as a source of nitrogen for protein synthesis. Nitrate is first converted to nitrite and then to ammonia by different enzymes. Mutants which cannot use nitrate as a source of nitrogen are unable to produce one or other of these enzymes. Very rarely a double mutant might occur which produces neither.

On a nitrate medium mutant B grows very slightly and produces nitrite which is excreted. Mutant A can utilize this nitrite so it grows better opposite B. The metabolic pathway involved is:

nitrate ⟶ nitrite ⟶ ammonia (used in amino acid synthesis)

Figure 84
Cross-feeding complementation in *Aspergillus*.
Photograph, Dr A. B. Tomsett, Department of Genetics, University of Liverpool.

b *Which steps do each of the mutants A and B lack?*

c *If two mutants, both requiring the same substance as in* **figure 84**, *are able to complement one another, what may be deduced about the biosynthesis of the substance in the cells of the organism concerned?*

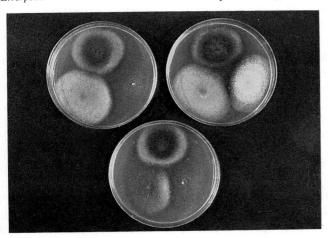

a

b

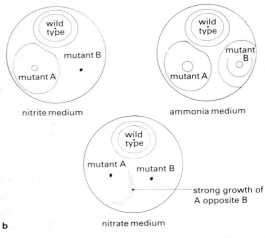

STUDY ITEM
19.11 Mutations and metabolic pathways

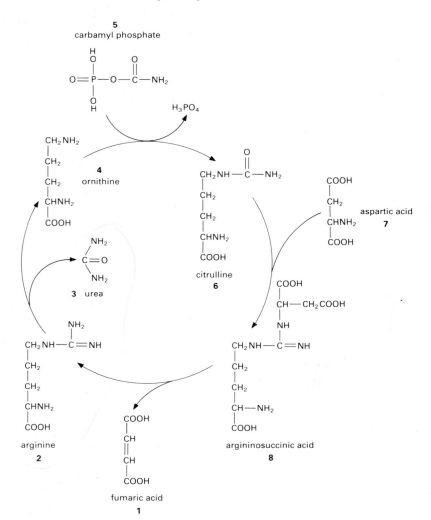

Figure 85
The synthesis of arginine and urea in the urea cycle.

Figure 85 shows a sequence of reactions which is widespread in organisms and which is used to synthesize the amino acid arginine. The same metabolic pathway is used in the synthesis of urea. The existence of metabolic pathways of this kind can be demonstrated using radioactive raw materials or sometimes by the use of inhibitors which block the action of a particular enzyme controlling one of the steps.

Mutants of fungi and bacteria can be isolated which require the amino acid arginine in their growth medium and in which one of the steps in the pathway of arginine synthesis is inactive. Such mutants are conveniently isolated and studied in yeast, using the replica plating technique. This is explained in *figure 86*.

The yeast cells are treated with a chemical mutagen or are irradiated. They are then spread onto a Petri dish of complete agar medium. Cells which survive the treatment divide to produce colonies. The wooden

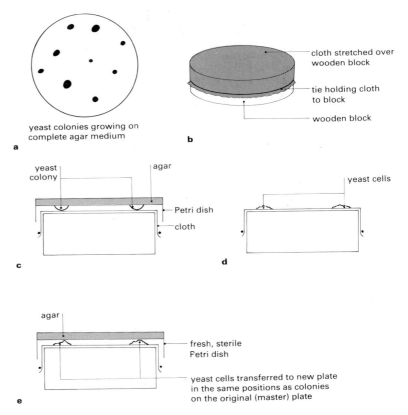

Figure 86
Replica plating technique.

block and felt cloth used in the technique must be autoclaved. The Petri dish with its yeast colonies is then pressed onto the cloth so that the cells stick to the cloth. These cells can then be transferred to several agar plates and will produce replicas of the pattern of colonies on the original or master plate.

If minimal medium is used in some of the plates, mutant colonies can be identified since they will develop on complete agar but not on minimal agar.

Figure 87 shows the results of an experiment in which yeast colonies derived from cells treated with a mutagen have been transferred to four other agar plates by replica plating.

Using a piece of polythene, tracing paper, or overhead projector acetate, make a tracing of the pattern of colonies on the master plate. Use this to identify the colonies which have developed on the four other plates.

a *Explain why four colonies have developed on minimal medium.*
b *How many mutants that require arginine have been produced?*
c *Explain why one colony has failed to grow on minimal medium but has grown on media containing both arginine and ornithine.*

Chapter 19 Gene action 129

Figure 87
Results of replica plating experiment. Not all the colonies on the master plate are able to grow on deficient media. All will grow on complete medium.

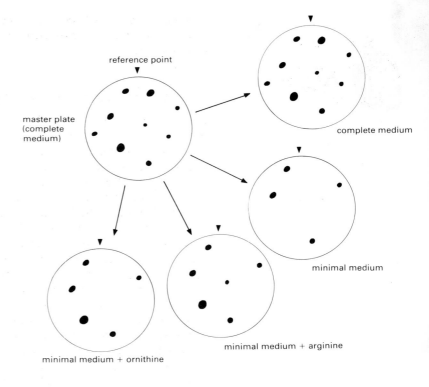

$$\xrightarrow{A} \text{ornithine} \xrightarrow{B} \text{citrulline} \xrightarrow{C} \text{arginine}$$

The sequence above is a simplified version of part of the metabolic pathway leading to arginine. Mutants could produce defective enzymes which failed to catalyse steps A, B, or C in the sequence.

d *Were any of the strains used wild type?*

e *Classify the mutants in table 23 as being type A, type B, or type C.*

f *Explain why a mutant which could grow on arginine or on ornithine but not on citrulline would be surprising.*

Table 23
The results of attempting to grow different strains of *Neurospora* on various media. The media were each inoculated with the same amount of *Neurospora* to begin with.

		Dry mass in mg after five days on medium			
	Mutant strain	Minimal medium, no supplement	Minimal medium + ornithine	Minimal medium + citrulline	Minimal medium + arginine
A	1	0.9	29.2	37.6	37.2
A	2	0.0	10.5	18.7	20.9
B	3	1.0	0.8	34.1	37.6
B	4	2.3	2.5	42.7	35.0
A	5	1.1	25.5	30.0	33.2
C	6	0.0	0.0	0.0	20.4

130 Inheritance and development

g *Pick out from table 23 two* **Neurospora** *strains that should show complementation, and two strains which would be unlikely to complement.*

19.2 Operators, repressors, and promoters

Lactose is a disaccharide of the subunits glucose and galactose and is found in milk. If it is to serve as a food for bacteria, the cells must synthesize a permease enzyme to transport it across their membranes and a β-galactosidase enzyme to cleave the lactose molecule into its glucose and galactose subunits. The bacterium *Escherichia coli* grows more vigorously on a medium that contains glucose than on one that contains lactose. In the absence of lactose it produces very little if any of the two enzymes needed to initiate the metabolism of lactose. Cells harvested from a glucose-containing broth culture by centrifugation and transferred to a lactose medium grow and respire very slowly at first and then more rapidly. More rapid metabolism is preceded by the synthesis of lactose permease and β-galactosidase.

Synthesis of enzymes only when their substrate is available is called *enzyme induction*. It has an obvious advantage to the cell since materials and energy are not wasted in making the enzymes when they are not needed. Enzyme induction is common in prokaryotes and also in yeasts.

The induction of lactose-metabolizing enzymes was intensively studied by researchers at the Pasteur Institute in Paris in the late 1950s and the hypotheses of two of these workers, Jacob and Monod (1960), have become famous.

Several mutants affecting the two enzymes were isolated and their position in the bacterial DNA mapped. Some mutants were analogous to those produced by Beadle and Tatum and abolished the function of either the permease gene or the β-galactosidase gene. Mapping of these mutants located them very close together on the DNA strand.

Some mutants were produced which could make neither type of enzyme, even in the presence of lactose, yet if those mutants were crossed with more 'ordinary' ones recombinants were recovered showing the apparently double mutant to contain intact copies of both genes. Clearly the mutation had caused both genes to be permanently switched off or *repressed* (see page 132).

A further class of mutants was found which produced large amounts of both types of enzyme even in the absence of lactose. These mutants were called constitutive mutants, and in them the genes were permanently switched on. Constitutive mutants and permanently repressed mutants could be mapped. They were found to involve either of two loci, one very near the genes coding for the enzymes and the other some distance away on the circular bacterial genetic map.

Jacob and Monod suggested that the genes coding for the enzymes were transcribed as a unit, starting at one end which they called the operator (o) locus. The operator region of DNA was able to interact with a repressor substance produced by another locus (r), see *figure 88*. If the repressor were attached to the operator, no transcription would take place and neither enzyme would be synthesized. Lactose, the substrate of

Figure 88
The lac operon of *E. coli*. A simplified map of the bacterial DNA showing the genes involved (not to scale).

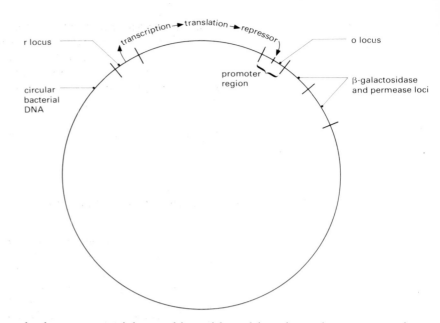

both enzymes, might combine with and inactivate the repressor, thus allowing transcription to proceed. In this way, enzyme synthesis is dependent on the presence of the substrate. Constitutive mutants and permanently repressed mutants could be mapped at either the o or the r locus.

a *Explain how the Jacob and Monod hypothesis accounts for both types of mutation at both loci.*

It has been found that if a micro-organism is given a rich supply of an amino acid which it must normally synthesize for itself, it stops making the enzymes which are specifically needed for the synthesis. The genes needed to make these enzymes continue to replicate and if the external supply of amino acid is interrupted enzyme synthesis restarts.

b *Suggest how the Jacob and Monod hypothesis could be adapted to explain the repression of enzyme synthesis by a product of their action.*

Genes coding for enzymes or other proteins needed by the cell are called structural genes. Genes such as o and r which control the rate of transcription are *regulatory* genes. A group of structural genes regulated by a single operator locus is an *operon*.
Transcription is carried out by an RNA polymerase enzyme of which several kinds are known. The polymerase enzyme binds to a sequence of DNA bases near the beginning of the transcribed sequence. This region is called a promoter and its function can be abolished by mutation. The promoter of an operon includes one or more operator loci and it is believed that gene regulation involves competition between an RNA polymerase and one or more repressor molecules. If the operator is free of repressor molecules the RNA polymerase binds to the promoter and

then transcribes the structural genes. The base sequences of some operators and promoters have been worked out. If 'foreign' DNA (e.g. human growth hormone) is to be functional in a bacterial cell, it must be attached to a bacterial promoter sequence that the bacterial RNA polymerase will 'recognize'. The bacterial cell will then make the 'foreign' RNA and the 'foreign' peptide.

c *Explain the possible value of constitutive mutants of bacteria in industrial processes.* ⇒ higher yield.

mutant may [not?] be not competitive

STUDY ITEM

19.21 An experiment demonstrating changes in the ability of yeast to ferment galactose

The experiment used a laboratory strain of brewers' yeast *Saccharomyces cerevisiae*. A stock culture of the yeast was established, using glucose as the carbon source. Cells were harvested from the stock culture by centrifugation and divided into two batches. The first batch (the control batch) was resuspended in a glucose medium. The second batch (the treated batch) was suspended in a medium in which glucose was replaced by galactose. The control and the treated cultures were aerated for 24 hours, then cells were harvested and used in the experiment.

The rate of fermentation was determined by measuring the rate of carbon dioxide production using Warburg respirometers (see *figure 89*).

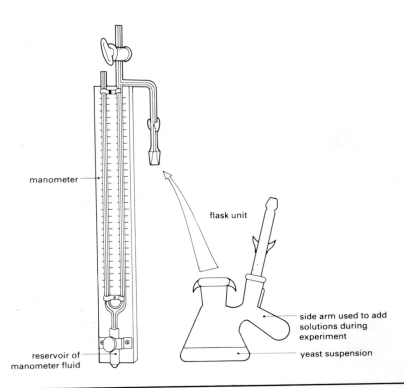

Figure 89
A diagram of the Warburg respirometer. The flask unit is continuously shaken in a thermostatically controlled water bath.

Either treated or control yeast was placed in the flasks. Both were suspended in potassium dihydrogenphosphate buffer solution to maintain a suitable pH. Either galactose or glucose solution was placed in the side arm of each flask. Two sets of apparatus containing only buffer solution were used to monitor changes in pressure and temperature which might produce random changes in the pressure of the gas in the respirometers. All the respirometers were flushed with nitrogen gas. After temperature and pressure had equilibrated in a water bath the sugar solutions were tipped from the flask side arms into the yeast suspensions. Changes in the volume of gas in each respirometer were recorded for 45 minutes. The results are given in table 24.

Minutes after tipping sugar from side arms	Increase in gas volume (mm^3) in each respirometer in successive five-minute periods			
	Respirometer 1 (control yeast with glucose)	Respirometer 2 (control yeast with galactose)	Respirometer 3 (treated yeast with glucose)	Respirometer 4 (treated yeast with galactose)
5	4.2	4.5	9.7	0
10	35.2	−2.2	21.9	4.7
15	−2.0	0	36.5	4.7
20	62.1	0	31.6	11.7
25	2.1	2.2	7.3	32.8
30	16.6	2.2	65.6	16.4
35	66.2	0	43.7	4.7
40	29.0	0	26.7	9.4
45	31.0	−2.2	19.4	7.0

Table 24
Carbon dioxide production by yeast.

a *Why were the respirometers flushed with nitrogen gas?*

b *List assumptions which must be made before the data can be interpreted.*

c *Display the data graphically.*

Changes in the thermobarometers (the sets of apparatus containing just buffer solution) were small and have been used to adjust the changes in volume shown by the experimental respirometers.

d *What indication is there that the thermobarometers have failed adequately to reflect temperature fluctuations in the experimental respirometers?*

The experiment was carried out by two students. One suggested that the results supported the hypothesis that S. cerevisiae can ferment galactose provided that there is sufficient time in the presence of galactose to allow induction of the appropriate enzymes.

The second student argued that 24 hours of aeration in a galactose medium had resulted in the multiplication of mutant cells able to ferment galactose and the death of genotypes unable to ferment galactose.

e *Explain how the results supported both these hypotheses.*

f *Suggest further experiments which might allow one or other of the hypotheses to be disproved.*

g *Does treatment with galactose affect the ability of yeast to use glucose?*

19.3 Control of transcription in higher organisms

In Chapter 15 we concluded from a number of lines of experimental evidence that genes are either activated (switched on) or inactivated (switched off) during differentiation and that this gene switching is at least partly the result of stimuli from the differentiating cytoplasm. Mutation studies allowed the detection of gene switching in bacteria and more recently the repressor substances involved have been isolated and shown to be proteins.

It is tempting to imagine that similar systems of gene control could account for differentiation.

Parts of the material in an interphase nucleus are in a highly condensed (heterochromatic) state while other parts are in a very much uncoiled (euchromatic) state. Is there evidence that heterochromatin consists of DNA which has been switched off?

Tortoiseshell cats are coloured with patches of ginger and black fur (see pages 52 and 73). They are heterozygous females in which one X chromosome has an allele determining ginger and the other an allele determining black. But why is one allele apparently 'dominant' over one patch of coat while the other appears 'dominant' over other patches?

The answer is that one of the two X chromosomes in a female mammal becomes heterochromatic while the other becomes euchromatic. The heterochromatic X reveals itself as a Barr body in interphase nuclei (see page 51). If it were a matter of chance which X chromosome in a particular cell became the Barr body, and if the transition took place early in embryonic development among a population of cells destined to become the ectoderm (skin) of the animal, this would explain patches of black and patches of ginger. Each patch would have in all its cells the same chromosome in a heterochromatic state. Cytological evidence does reveal that a heterochromatic chromosome remains heterochromatic over many mitotic cycles.

a *Give two reasons why this evidence is inconclusive.*

Gene activity in polytene chromosomes

Many species of flies and midges – including those in the genera *Drosophila* and *Chironomus* – have an unusual pattern of differentiation in some of their tissues which allows active and inactive genes to be observed microscopically. Certain cells in the larvae – those of the salivary glands are easiest to find – become very large. Homologous chromosomes in these cells pair and the DNA of each chromosome replicates repeatedly during cell growth but the centromeres of the

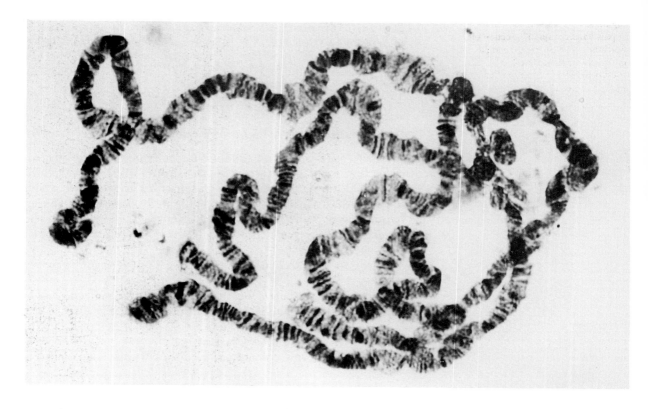

Figure 90
A photomicrograph showing a nucleus with polytene chromosomes (× 1600).
Photograph, Dr C. G. Vosa, The Botany School, Oxford.

chromosomes never divide. Very many copies of the chromosomes – they are not sufficiently separate to be called chromatids – remain in close alignment. The resulting structure can be called a giant chromosome but is more correctly referred to as a polytene chromosome (polytene means many threads). It is obviously very large when compared with its homologues in a normal diploid or haploid cell, but, more importantly from our point of view, it remains visible at all times and does not enter an interphase condition in which it becomes completely despiralized.

A conspicuous feature of polytene chromosomes is that they are banded (see *figure 90*). The bands are revealed by common staining techniques, such as orcein staining, and were found to be constant in number and position for each chromosome of a species. It has already been mentioned (on page 36) that deletions or duplications of certain of the bands in *Drosophila* are associated with mutation.

It is usually assumed that each band on a polytene chromosome represents a gene locus or group of loci and that it is visible because the many strands of the chromosome are in perfect alignment. If a section of a polytene chromosome which has been properly stained and well illuminated is examined under very high magnification with a light microscope, some of the bands can be seen to be in a 'puffed' state. (See *figure 91*.)

The puffed region is one in which the strands of the many stranded structure have become separated and more despiralized. We assume that chromosome puffs reflect genes or groups of genes which are being

Figure 91
A photomicrograph of a section of polytene chromosome showing a puffed chromomere. (× 4700).
Photograph, Dr C. G. Vosa, The Botany School, Oxford.

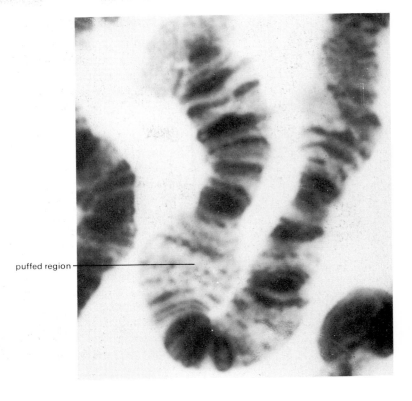

transcribed. Closer study of polytene chromosomes from a variety of *Drosophila* tissue has led to the following three conclusions:

1 the location of puffs is dissimilar in different tissues within *Drosophila*;
2 the pattern of puffing in a given tissue changes during its differentiation;
3 at any one stage all cells in a particular tissue contain chromosomes with identical patterns of puffing.

If puffs indeed represent active, switched on genes, it can be predicted that mRNA will by synthesized on puffed regions of a chromosome but not on more condensed chromosome bands.

b *How could this prediction be tested experimentally?*

Treatment of midge or fruitfly larvae with hormones which control ecdysis (the moulting of the exoskeleton during growth) has revealed that certain bands on the salivary gland chromosomes puff in a definite sequence in response to a particular hormone (see page 333). The hormone does not directly activate a gene but it stimulates changes in the cytoplasm of the cell which result in gene switching. In the case of the operon of *E. coli* which controls lactose metabolism, we understand the control of the structural genes and have identified a specific repressor substance which switches them off. We have no similar understanding of gene switching in higher organisms such as *Drosophila*.

19.4 Cell differentiation as a population phenomenon

Because cell differentiation reflects changes or events within the genetic material it might be assumed that it is only worth studying differentiation at the molecular level. That assumption is only partly justified, and it very soon becomes apparent that if one studies only isolated cells or subcellular fractions then one's understanding of cell differentiation will be somewhat limited. It was stressed in Chapter 15 that in all metazoans the cells are not autonomous units and that two fundamental aspects of multicellularity are cell diversity and the interdependence of the cells constituting the organism. Therefore, the ways in which we choose to study cell differentiation should not exclude that element of interdependence or interaction. We should remember that in addition to studying isolated cells and subcellular fractions it is important to consider the 'social' aspects of cells; that is to say the organization of cells into those discrete and integrated populations which we refer to as *tissues*.

What is the justification for such a claim? Remember that it is the particular organization of cells into tissue and the integrated organization of those tissues into an independent and functional organism that defines a species phenotype, and may even define individuals within that species. Furthermore, the homeostatic control at these levels of organization ensures the viability of the organism, and the breakdown of such homeostasis may result in the deterioration and even death of the organism. As an illustrative example consider the human species in relation to the lower primates. An orang-utan or a chimpanzee has the same range of basic cell types as the human animal. What separates and defines us from other primates is not necessarily any special cell type specific to humans, but the different organization of our cells. Chimpanzees and humans have the same range of cell types, but in building a human being those cell types have been deployed or organized in a slightly different fashion to give us the phenotype and abilities which characterize our particular species.

The message is that a phenomenon like differentiation (leading to cell diversity) needs to be studied in a number of ways each reflecting a different level of organization: molecular, subcellular, cellular, and 'population' level. In fact, isolated cells in culture, be they plant or animal, usually fail to display a differentiated state, whereas populations of cells generally do. Here is one brief example. It concerns the culturing of cartilage-forming cells. These cells are usually contained within individual cavities or 'lacunae' within the matrix they produce. In order to liberate them the tissue [*e.g.* sternum from an embryonic chick] is carefully incubated in the appropriate proteolytic enzymes to digest away the matrix but leave the cells intact. The result is a suspension of individual viable cartilage cells which can be cultured in such a way that we can monitor various parameters of cell behaviour such as division rate, movement, and synthetic activity. Let us suppose that we take two suspensions of such cells identical except for different population densities. On plating out these suspensions and leaving them to attach to the floor of the culture vessel we find that a number of parameters vary dramatically between the two populations. Firstly phenotype is different

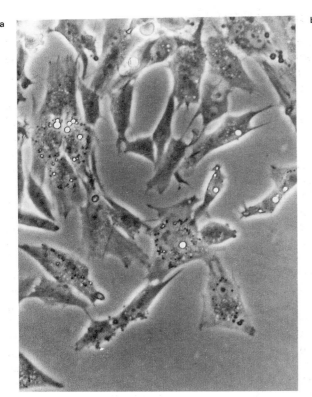 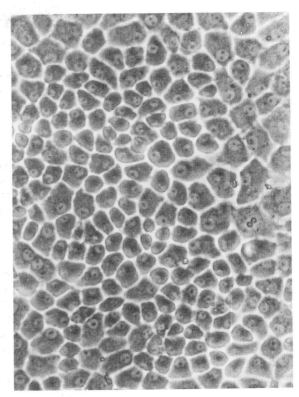

Figure 92
Photomicrographs of cartilage showing matrix and chondrogenic cells.
a Cultured cartilage cells at low population density where they appear like fibroblasts. ($\times$ 3100).
b Cells from the same source cultured at high population density where they form a matrix. ($\times$ 3100).
Photograph, Dr Peter Thorogood, Department of Biology, University of Southampton.

in that the cells in the low population density cultures are like fibroblasts in appearance and motile, whereas in the high population density cultures they flatten out to form a single layer of polygonal immobile cells.

However, one can learn only a limited amount from morphology – and perhaps this difference in phenotype might be attributed simply to the 'packing' of cells into available space. If we extend our observations by measuring some metabolic parameters to monitor the cells' synthetic behaviour we find further differences. The fibroblastic cells are proliferating rapidly, as judged by the incorporation of thymine labelled with tritium (^{3}H) into newly synthesized DNA, but making very little extracellular matrix, as indicated by the low level of incorporation of ^{3}H-proline and ^{35}S-sulphate into collagen and sulphated matrix polysaccharides respectively. In contrast, the polygonal cells proliferate very little but are making a relatively large amount of matrix. With time, the population density of the fibroblastic culture reaches that of the high population density cultures and cell behaviour and patterns of synthesis change accordingly; that is they begin to divide less but make more matrix. And if we make a subculture from cells from high population density cultures at a low density they, predictably, assume a fibroblastic phenotype and divide regularly, but make little matrix until their population density reaches the appropriate threshold.

So cells grown at a lower population density appear to lose many of their attributes which we regard as characteristic of their differentiated state and they might, in ignorance, be described as 'de-differentiated'. If we had considered only cells grown at the lower densities our interpretation about the synthetic ability of the cells and state of differentiation would have been very different from that gained by considering the high population density cultures. This demonstrates very clearly that if we wish to study cells fully we must consider them as interacting units within a population. Cells interact with their neighbours and may modify or even change their organization and their synthetic behaviour. This element of interaction within the cell population is important, not only in terms of developing tissue-specific organization, but also in terms of its maintenance, its regeneration, and also its breakdown during various disease states.

There is more to understand about the action of genes than simply finding out which are active and which are repressed.

19.5 Genes that influence metabolic reactions in humans

Figure 93 shows a simplified set of metabolic conversions, all requiring specific enzymes, relating to the amino acids phenylalanine and tyrosine. Homozygous recessive mutant genotypes and their associated phenotypes are known where one of the steps is blocked through failure to produce the required enzyme.

The C/c locus is involved in the synthesis of the pigment melanin. Melanin is laid down in the hair, in the skin, and in the choroid coat and iris of the eyes. It is a dark pigment and the more melanin present the darker the skin, hair, and eyes become. Very many loci influence the amount of melanin produced and its distribution. Individuals homozygous for the allele c (albino) produce none. Albinos have very fair hair, pale skin, and pink eyes. Because the normal protective function of the choroid coat and iris are missing, the eyes are very sensitive to glare and are easily damaged by bright sunshine. The skin is very easily sunburned. Albino mutants have been preserved by artificial selection in most species of domestic animal.

Mutants at the A/a locus were first recognized by the physician A. E. Garrod in 1902. Garrod came very near to the one gene–one enzyme hypothesis in his explanation of the mutant phenotype. Individuals homozygous for allele a produce urine which turns black on exposure to air. This is because of the oxidation of homogentisic acid in the urine. Blockage of the pathway for breakdown of excess tyrosine results in the excretion of homogentisic acid.

The third locus affecting tyrosine metabolism is the P/p locus. Individuals homozygous for p lack an enzyme which removes a hydroxyl (OH) group from the amino acid phenylalanine, thus blocking its conversion to tyrosine. This is the normal way in which excess phenylalanine in the body is removed. In pp individuals, the excess is deaminated and excreted in the urine as phenylpyruvic acid. The level of phenylalanine in the blood is much higher than normal and the physical and mental development of the individual is severely retarded. The syndrome is called phenylketonuria. The ill-effects of excess phenylalanine are not reversible.

Figure 93
A simplified set of metabolic conversions.

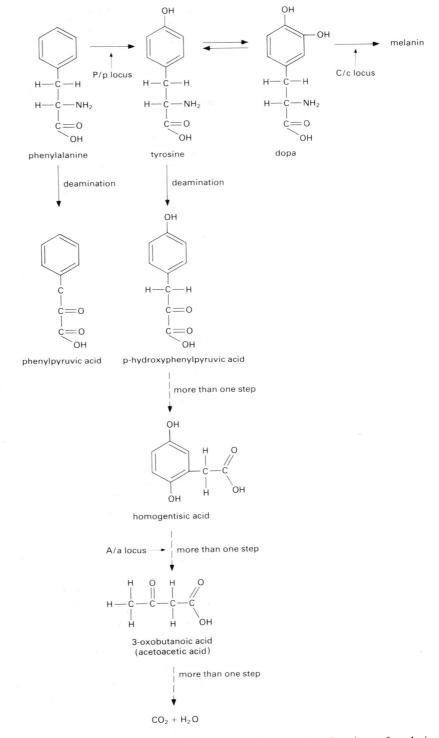

If pp individuals can be diagnosed within the first few days after their birth, they can be reared in normal health provided their diet is strictly controlled. Phenylalanine is an essential amino acid and must be present for protein synthesis and growth, yet the diet must not contain any excess. A routine blood test is used to detect the condition. By the time that phenylpyruvic acid appears in the urine, damage may already have started.

a What is the genotype of each of the parents of a baby who excretes phenylpyruvic acid?

b If these same parents choose to have a second baby, what is the probability that it will be affected?

It must be emphasized that only a minority of known mutations in humans and higher animals and plants can be understood in terms of the blockage of metabolic pathways. Many mutations probably involve loci which control the expression of other genes.

c In figure 93 tyrosine is shown as a product of phenylalanine. Explain why pp individuals are not albino in phenotype.

STUDY ITEM

19.51 The influence of thiamine on mutant tomatoes: an analysis of an experiment

Figures 94 and 95 show the results of treating a certain mutant of tomato with the thiamine (vitamin B_1) it requires.

The homozygous recessives (t1 t1) of a mutant known as t1 develop normally for the first week or two but seldom produce more than two or three true leaves, which are usually distorted. Although they are initially green, they rapidly bleach, die, and fall off, leaving a bare epicotyl and two nearly normal cotyledons. The heterozygotes (t1 T1) can usually be distinguished from the normal plants (T1 T1) in the early stages of growth by a slight but distinguishable chlorosis (yellowing) which disappears as growth continues.

The investigation showed that the mutant plants responded to thiamine whether it was applied as a general foliar spray to the cotyledons, the first true leaves, or the soil.

a Do the results shown in figures 94 and 95 indicate that the experiment was adequately controlled?

b From your study of the graph in figure 94 what can you say about the relationships between the concentration of thiamine used, the formation of chlorophyll, and the growth of the plants as measured by fresh mass?

c Why were the concentrations given as the square roots of parts per million?

d Why were the amounts of chlorophyll and fresh mass given as percentages of the normal genotype?

e What was the point of measuring chlorophyll content as well as fresh mass?

Figure 94
The effect of increasing concentrations of thiamine on fresh mass and chlorophyll content of the homozygous mutant (t1 t1) plants.
Based on Boynton, J. E., 'Chlorophyll-deficient mutants in tomato requiring vitamin B_1', Hereditas, **56**, 171–199, 1966.

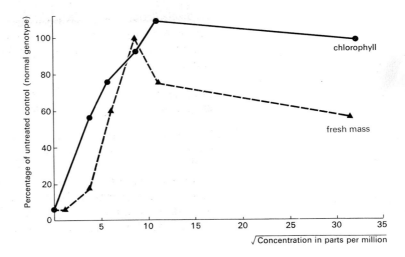

Figure 95
Growth curves of normal and mutant (t1 t1) plants untreated (continuous lines), and treated with 100 p.p.m. thiamine three times a week (broken lines).
Based on Boynton, J. E. 'Chlorophyll-deficient mutants in tomato requiring vitamin B_1', Hereditas, **56**, 171–199, 1966.

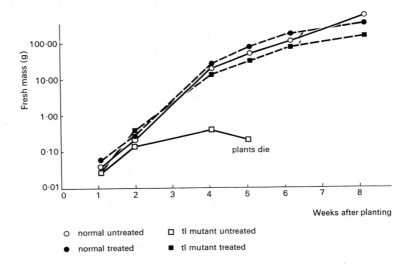

f What statistical techniques would you use to assess the significance of the results? SD

g How do you account for the fresh mass increase in the untreated mutants up to about the fourth week? Thiamine in cotyledons

h Thiamine-deficient mutant tomato seedlings grow normally if sprayed with a pyrimidine. Another substance, thiazole, is implicated in thiamine synthesis, but no growth occurs if mutant plants are sprayed only with this. In which enzyme-controlled step in the pathway below is there most likely to be a metabolic block?

precursor 1 —A→ pyrimidine —B→ precursor 3
                                              + —→ thiamine
precursor 2 —D→ thiazole —C→ precursor 4

Chapter 19 Gene action 143

You may be able to compare the results of this investigation with those of investigation 19A.

> **Practical investigation.** *Practical guide 5*, investigation 19A, 'Dwarfism in peas'.

19.6 Gene expression in heterozygotes

Much modern genetic research has concentrated on haploid organisms such as *Neurospora* and on prokaryotes. This is partly because gene expression in these organisms is not complicated by the possibility of a heterozygous condition, *i.e.* two different alleles of a gene in the same cell. If the cell contains two alleles – two similar but not identical genes having similar biochemical function – what consequences follow?

Mendel and the plant hybridizers who preceded him, observed that the heterozygote often resembled one of the parental, homozygous types so closely as to be indistinguishable. One allele is being expressed in the phenotype and the other is not and we refer to this situation as dominance.

One explanation for dominance is that the recessive allele is a deletion mutant. A section of DNA has been lost and the allele is therefore functionless. The single dominant allele in the heterozygote produces enough product by transcription and translation to give a phenotype like the homozygous dominant.

This hypothesis can be extended to explain some situations where there is no dominance and the heterozygote is distinctive. Table 25 shows an example using true breeding red-flowered *Antirrhinum* crossed with true breeding white-flowered *Antirrhinum*.

	red (true breeding)	×	white (true breeding)
F_1		100 % pink	
F_2	red	pink	white
	1	2	1

Table 25

The 1:2:1 ratio in the F_2 shows a clear case of monohybrid segregation, *i.e.* a single pair of alleles. The allele determining white might be a deletion mutant and a single red-determining allele in the heterozygote may not produce enough product to give the full red colour, so the heterozygote is a dilute red, *i.e.* a pink.

While some recessive alleles may represent functionless sequences of DNA, it would be very unwise to assume that this is always the case. The use of electrophoresis and immunological techniques in which antibodies are used to identify gene products often show that two substances are produced in a heterozygote and only one in each homozygote. This may be so even when one allele seems completely dominant. There is great interest in identification of heterozygotes by using such biochemical tests, since people carrying alleles producing

Figure 96
Two inbred plants of a species of tobacco, *Nicotiana rustica*. The plant in the middle is an F$_1$ hybrid produced by crossing the two inbred strains.
Photograph, Dr H. S. Pooni, Department of Genetics, University of Birmingham.

inherited diseases such as cystic fibrosis or haemophilia could be identified before they married and ran the risk of transmitting their disadvantageous allele. (See Chapter 20.)

If two true breeding homozygous individuals from different and unrelated populations are crossed, it often happens that the F$_1$ generation is larger and more vigorous than either of the parental populations. This situation has been observed in many domesticated animals and plants and is called *hybrid vigour* or *heterosis*. An example of heterosis in tobacco is shown in *figure 96*.

Heterosis is especially important in the culture of maize and chickens, and has been exploited to increase yields. If the true breeding parental strains are carefully chosen, desirable combinations of characters can be brought together in the F$_1$ as an additional bonus to the heterosis.

F$_1$ hybrids between two domesticated strains or breeds are likely to be heterozygous at several loci. There are two hypotheses which might explain heterosis:

1 Suppose one pure breed is of genotype aaBBCCdd and the other is of genotype AAbbccDD. The F$_1$ will be of genotype AaBbCcDd. If the recessive alleles at each locus are disadvantageous, the heterozygote will be more vigorous than either of its parents, since each recessive is masked by a dominant.

2 Suppose one pure breed is of genotype a_1a_1 and the other is of genotype a_2a_2. These homozygotes can only make one kind of enzyme protein each. The heterozygote a_1a_2 can make two variant types of the same protein since it has two types of DNA. The ability to make two types of protein gives it greater metabolic versatility and it is therefore more vigorous than either homozygote.

a   *Are hypotheses 1 and 2 as stated above mutually exclusive?*

Since heterozygosity is often an advantage we would expect to find that organisms have evolved mechanisms to maximize the chance that their offspring will be heterozygotes. Such mechanisms are discussed in Chapter 20.

> Practical investigation. *Practical guide 5*, investigation 19C, 'Gene expression in round and wrinkled peas (*Pisum sativum*)'.

19.7 Pleiotropy

If you have carried out investigation 19C you will know that a single gene may have several different effects on the phenotype. When this occurs, the gene is said to be *pleiotropic*. Before we decide that a gene is pleiotropic, we must consider the possibility that we are dealing not with a single gene but rather with a group of very closely linked genes which behave as a unit. *Figure 97* may make this distinction clearer.

It would be possible at this point for us to digress into a discussion about the definition of a gene; suppose, for example, that a region of DNA was transcribed as a single mRNA strand but was translated as two or more polypeptides – would we be dealing with one gene or two? The term gene was coined at a time when comparatively little was

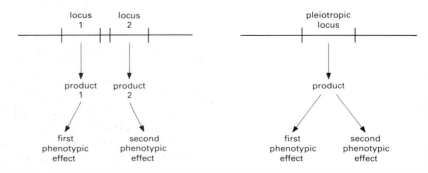

Figure 97
Pleiotropy and phenotypic effects.

understood about inheritance, and it can be used to mean rather different things by different geneticists. Many supposed cases of pleiotropy may turn out to be similar to case A rather than case B in *figure 97*. In only a few instances has the direct product of a region of DNA been identified and correlated with the expression of particular characters in the whole organism. The case of sickle-cell anaemia in humans is an example. This has been referred to in Chapter 15 and will be further considered in Chapter 20. A single base change in the DNA – a point mutation – results in the substitution of a single amino acid in one of the two types of polypeptide in the haemoglobin molecule. The result is a phenotype in which the red blood cells adopt a distorted (sickle) shape in conditions of low oxygen partial pressure. If the sickle-cell allele is homozygous, the cells sickle easily and blockage of capillaries can occur, the spleen is enlarged, and several other symptoms of disease result. The mean height of a sample of sickle cell homozygotes is slightly greater than that of a control group. Sickle-cell individuals are less likely to suffer attacks of malignant tertian malaria where this disease is endemic.

Another example of pleiotropy is shown by the vestigial wing mutation in *Drosophila*. The most obvious phenotypic expression of this allele is that when homozygous it results in small, functionless wings. Another effect on the phenotype is that the adult flies are less likely to move towards light, *i.e.* are less positively phototactic, than the wild type.

a *Design an experimental procedure which would demonstrate this.*

It has often been suggested that many alleles which occur with high frequency in human populations, *e.g.* alleles determining various blood groups, may have subtle pleiotropic effects which influence the life expectancy or fertility of those individuals that possess them. Individuals with blood group O have a slightly, but significantly, greater chance of developing a duodenal ulcer than do those with Group A or group B. Although it may be straightforward to demonstrate that particular genotypes are correlated with an increased risk of contracting a disease, showing a causal connection is rather more difficult.

An organism, even a simple prokaryote cell, is a highly complex whole. The genes influence different parts of the whole organism, but these parts all interact and one would therefore expect the majority of genes to be pleiotropic. This is the case for many genes which have been investigated. You will be able to study a clear example if you carry out investigation 19C.

19.8 Inheritance of comb shape in poultry

Soon after Mendel's paper was rediscovered in 1900, W. Bateson and R. C. Punnett of Cambridge published reports on the inheritance of comb shape in the domestic fowl, *Gallus domesticus*. *Figure 98* shows four forms the comb may take.

When true breeding rose-combed fowl or true breeding pea-combed fowl were crossed with true breeding single-combed fowl, the offspring in

a single b rose c pea d walnut

Figure 98
Comb shapes in poultry.
a Single. This is the familiar thin, high serrated comb normally found in the wild species. It is characteristic of the Leghorn, Minorca, Rhode Island, and many other breeds.
b Rose. This comb consists of a triangular mass of papillae, the apex of the triangle pointing backwards and extending into a free point. It is characteristic of the Wyandotte and some Dorkings.
c Pea. This is somewhat like the single but is lumpier, lower and closer to the head, with a lateral row of small tubercles on each side. It is found in the Indian Game and Brahma breeds.
d Walnut. This type of comb varies considerably but is characterized by a rounded, corrugated appearance, from which it takes its name. The only breed of which it is characteristic is the Malay.

both cases demonstrated a 3:1 ratio with respect to the shape of their comb.

When single-combed fowls were crossed with each other they always produced single-combed fowls. True breeding rose-combed fowl crossed with true breeding pea-combed fowl produced offspring all with walnut combs. When these were crossed the results in table 26 were obtained.

Parents	rose × pea
	(true breeding) ↓ (true breeding)
F_1	100 % walnut
F_1 cross	walnut × walnut
	↓
F_2	walnut rose pea single
	9 3 3 1

Table 26
The ratios of fowl with different shaped combs obtained from crossing walnut-combed birds.

When the single-combed fowls were crossed with F_1 birds with walnut combs the results in table 27 were obtained.

walnut × single
walnut rose pea single
1   1   1   1

Table 27
The ratios of fowl with different shaped combs obtained by crossing walnut-combed and single-combed birds.

The F_1 walnut × F_1 walnut result (table 26) of 9:3:3:1 is typical of a dihybrid cross. The F_1 walnut × single result (table 27) of 1:1:1:1: is typical of a dihybrid backcross to a homozygous recessive.

We may therefore conclude that:

1 The walnut phenotype is due to the interaction of the rose and the pea gene, and has a genotype which contains at least one R and one P.
2 The single genotype is a double homozygous recessive rr pp.
3 One possible rose genotype is RR pp.
4 One possible pea genotype is rr PP.

The F_2 generation of the F_1 walnut × walnut cross (see table 26) were then each crossed with single-combed fowl. The results are set out in table 28.

F_2 generation	Offspring per cent
all the single	100 single
$\frac{1}{3}$ of the pea	100 pea
$\frac{2}{3}$ of the pea	50 pea, 50 single
$\frac{1}{3}$ of the rose	100 rose
$\frac{2}{3}$ of the rose	50 rose, 50 single
$\frac{1}{9}$ of the walnut	100 walnut
$\frac{2}{9}$ of the walnut	50 walnut, 50 rose
$\frac{2}{9}$ of the walnut	50 walnut, 50 pea
$\frac{4}{9}$ of the walnut	25 walnut, 25 rose, 25 pea single

Table 28
The offspring resulting from matings between F_2 fowls and true breeding single combed stock.

Rose R p
Pea r P
Walnut R, P
Single r p

a ***From this information write down all the genotypes which result in***
1 the rose phenotype
2 the pea phenotype
3 the walnut phenotype.

You may find it easier to tackle this after referring to page 76.

b ***What do these data tell us about the way genes act?***

The comb types rose, pea, walnut, and single are so distinctive that they can be reliably distinguished in all poultry breeds and in the progeny of crosses between breeds. Within and between breeds, there is also continuous variation in comb shape. The single combs of an English Game hen and a Rhode Island hen are distinct and the 'wrong' type of single comb would be condemned by a judge at a poultry show. Breeders select birds which lie within a narrow range of variation for breeding show specimens.

c ***What does this tell us about gene action?*** More than R + P loci influence comb shape but only here do genes cause distinctive shape

In humans and in cats, a single mutant allele can produce the condition polydactylism. The hand (or forepaw) has an extra finger (or digit with claw).

d ***Does this indicate that a single gene determines the development of the hands of humans or the paws of cats?***
No many genes must be involved but Polydactyl gene has an allele → distinctive discontinuous variation

19.9 Epistasis

The work of Beadle and Tatum which we reviewed earlier in this chapter showed that each step in a sequence of metabolic reactions is controlled by a gene. We have also looked at some examples of mutations, such as phenylketonuria in humans, in which the phenotype of the mutant can be related to a blocked metabolic pathway. Most metabolic pathways in higher animals and plants are essential to life and mutants which seriously disturb the function of these pathways are lethal. There are some pathways that are less vital and these have been much studied in,

Figure 99
Genes which will show epistatic relationships.

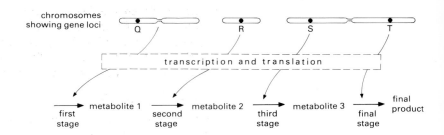

for example, coat colour of mammals, eye colour of fruit flies, and flower colour of ornamental plants. The wild type phenotype develops by the action of many genes, and mutants exist which influence the production of the pigments. They are often viable – at least under the sheltered conditions of an animal room or experimental garden.

Figure 99 shows diagrammatically the relationship between four genes and the metabolic pathway that they control. (The pathway is a hypothetical one and must not be related too closely to any actual example you may study.) Suppose metabolites 1 and 2 are colourless, metabolite 3 is a yellow pigment, and the final product a red pigment.

Gene S mutates to produce a recessive allele s. This mutant allele codes for an ineffective, functionless enzyme.

a *If an individual is of genotype QQRRssTT, can it produce either the red or the yellow pigment?*

b *What would be the colour of an individual of genotype QQRRSStt if allele t produced an ineffective enzyme?*

c *What would be the colour of a doubly heterozygous individual QQRRSsTt?*

We assume that all the substances involved in the production of the pigments are synthesized in the pathway shown in *figure 99* and cannot be absorbed from the environment of the organism.

d *If a mutation at locus Q blocks the pathway would mutation at loci R, S, or T make a difference to the phenotype?*

If one gene controls the ability of a different gene to influence the phenotype then the controlling gene is epistatic to the one which is controlled. In *figure 99* gene Q is epistatic to genes R, S, and T, while gene S is epistatic to T.

e *To which genes is R epistatic?*

Epistasis is a relationship between different genes which are often located on different chromosomes. It is very important to distinguish between epistasis and dominance. Dominance is a relationship between different alleles in a heterozygote. The dominant allele influences the phenotype and the recessive allele is not expressed.

In *figure 99* the wild type gene R may be dominant to a recessive mutant r. When r is homozygous the pathway may be blocked if r produces no effective enzyme. This means that QQRrSSTT and QQRRSSTT are both red in phenotype while QQrrSSTT is colourless (or white). The genotype qqRRSSTT may also be white if a q produces no effective enzyme. A cross between two true breeding individuals:

QQrrSSTT × qqRRSSTT

would give an F_1 all of which have the genotype QqRrSSTT. The F_1 would have a red phenotype since Q and R are dominant over q and r respectively. Because S and T are both homozygous wild type in both the parental white strains their existence can be ignored and we can rewrite the cross:

    QQrr × qqRR
(white) ↓ (white)
       Qq Rr
       (red)

The genes S and T can only be detected if they mutate to produce mutants which affect the pathway.

The Study items that follow explore some actual patterns of inheritance of pigmentation which can be explained by epistasis.

Practical investigation. *Practical guide 5*, **investigation 19B, 'The biochemistry of cyanogenesis in *Trifolium repens* (white clover)'.**

STUDY ITEM

19.91 An unexpected result from breeding sweet peas

The wild sweet pea, *Lathyrus odoratus*, bears flowers with a dark purple posterior petal and lighter coloured, almost red, lateral petals (see *figure 100*).

The cultivated sweet pea exists in many true breeding varieties which differ in height, scent, flower size, and many other characters. Sweet pea flowers are usually self-pollinated.

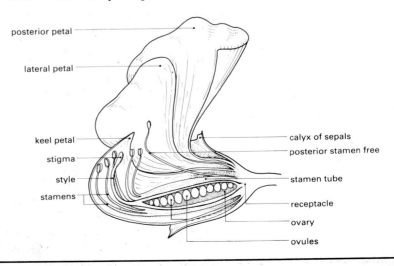

Figure 100
Longitudinal section of sweet pea (*Lathyrus odoratus*).
Based on Jepson, M. Biological drawings: part II, *John Murray*, 1959.

A cross was made between two true breeding white flowered varieties. All the F_1 were purple flowered like the wild plant. This result was shown to be repeatable and not the result of accidental pollination from some other variety.

Purple flowered F_1 plants were allowed to self-pollinate to produce an F_2. These latter consisted of 96 plants bearing coloured flowers and 68 bearing white flowers. Subsequent experiments gave the same ratio, within the limits of chance variation.

a Put forward a hypothesis to explain this result.

b Suggest a cross which would test your hypothesis and predict the outcome.

c Why is a result of this type important in commercial plant breeding?

STUDY ITEM

19.92 Breeding brown mice from black and white parents

In mice there is a mutant c which produces an albino phenotype when homozygous. The dominant wild type allele c^+ produces a coloured coat.

At a second locus, b^+ produces black fur and b produces brown fur.

A pair of mice, one black and one albino, which had been purchased from a pet shop, produced 10 litters. A total of 94 offspring were born, of which 49 were albino, 36 black, and 9 brown.

a What are the possible genotypes of the two parents?

b Work out the expected number of offspring of each phenotype on the basis of each hypothesis you suggested in a.

c Calculate the value of chi-squared (χ^2) for each hypothesis.

d Using probability tables, determine whether any of the hypotheses can be rejected as unlikely.

e Suggest crosses which would test the hypotheses experimentally.

Summary

1 Beadle and Tatum demonstrated by complementation tests and other experiments that the enzymes that catalyze stages in metabolic pathways are determined by genes (**19.1**).

2 In bacteria and some other organisms genes or groups of genes can be switched on and off by a feedback control process (**19.2**).

3 A group of structural genes controlled by a promoter, an operator, and a repressor is an operon (**19.2**).

4 Control of DNA transcription in higher eukaryote cells is not well understood but there is evidence that heterochromatin is inactive DNA complexed with proteins (**19.3**).

5 Polytene chromosomes in flies provide an unusual opportunity to observe groups of identical DNA strands as they are switched on and off during cell differentiation (**19.3**).

6 The behaviour of cells in tissue culture varies with their population density. The phenotype of a higher organism must be understood in relation to the organization of populations of cells as well as by an analysis of function within cells (**19.4**).

7 Metabolic pathways in humans and flowering plants may be blocked by mutation as can those of micro-organisms (**19.5**).

8 In heterozygotes one allele may influence the phenotype and the other not: this is dominance. A heterozygote may be a more vigorous phenotype than either homozygote: this is heterosis (**19.6**).

9 A single gene may have several different influences on the phenotype: this is pleiotropy (**19.7**).

10 Genes interact during development and the phenotype cannot be regarded as caused by any single gene (**19.6**, **19.7**, **19.8**).

11 If a gene has an influence early in a metabolic pathway it may control the expression of non-allelic genes acting at later stages in the same pathway. This situation is epistasis and is revealed by F_2 ratios such as 9:7 or 9:3:4 (**19.9**).

Suggestions for further reading

FINCHAM, J. R. S. Microbial and molecular genetics. 2nd edn. Hodder & Stoughton, 1976.

GURDON, J. B. Carolina Biology Readers No. 27, *Gene expression during cell differentiation*, 2nd edn. Carolina Biological Supply Company, distributed by Packard Publishing Ltd., 1978.

MANIATIS, T. and PTASHNE, M. 'A DNA operator–repressor system'. *Scientific American*, 238(1), 1976. Offprint No. 1333.

CHAPTER 20 POPULATION GENETICS AND SELECTION

20.1 Changes in population of *Drosophila*

There are many mutants of the fruit fly *Drosophila melanogaster*, some of which you may have handled. Most of those commonly available in school or college laboratories are hardy and breed almost as well as the wild type flies. Since thriving colonies of mutants are easy to maintain, the question arises: why are mutant individuals so rare among wild populations in such situations as dustbins or orchards? This question can be answered in part by setting up containers known as population cages in which mixed populations of wild type and mutant flies can be introduced and left to breed at random.

A good population cage is designed so that tubes of food medium, in which the flies lay eggs and in which the larvae are reared, can be removed and replaced without disturbing the population too much. Food tubes are replaced in rotation so that there is a constant and predetermined supply. At regular intervals a census of the population is made either by anaesthetizing and counting the adults or by removing a tube containing larvae and allowing them to develop into adults so that their phenotypes can be scored as a sample of the population as a whole.

Physical conditions within the cage, such as temperature and humidity, can be controlled, as can the quantity and quality of the food. The sample of flies from pure cultures used to start off the population cage can also be varied, for example they might be unmated (virgin) or mated within the pure culture before introduction to the cage. *Figure 101* shows the results of two population cage experiments lasting for several months.

Population I was started with 2 pairs of wild type flies from a pure stock culture and 22 pairs of vestigial-winged flies. Population II started

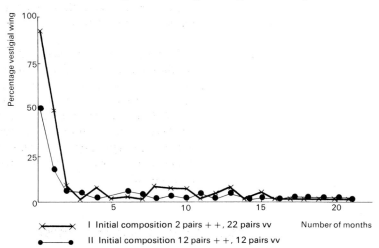

Figure 101
A graph showing changes in a population of *Drosophila*.

with 12 pairs of wild type flies and 12 pairs of vestigial-winged flies, *i.e.* equal numbers of the two types. Vestigial wings are the result of a single gene (point) mutation which is recessive to the wild type allele.

a *Explain why the rapid reduction in the proportion of vestigial-winged flies cannot be the result of the dominance of the wild type allele.*

The life cycle of fruit flies is approximately 21 days.

b *Roughly how many generations have elapsed during the experiment?*

Much research of this sort has been carried out into the fate of fruit fly (and flour beetle) mutants in mixed cultures of the mutant and the wild type. When the experiments involve mutants which have an easily observable effect on morphology or colour, the frequency of the mutant phenotype in the population usually declines to a low level, as in *figure 101* and the mutant is often totally eliminated. If adverse conditions, for example unsuitable temperature or food shortage exist, the mutant declines in frequency more quickly. The following hypotheses have been advanced to explain the decline in the frequency of mutants:

1 They are less fertile than the wild type.
2 They are less likely to mate successfully, with the result that wild type males inseminate a higher proportion of females.
3 There is a proportionally greater mortality among mutants at some stage in the life cycle.
4 The conditions in a population cage are unsuitable for mutants and they would decline in number in the absence of the wild type.
5 Wild type individuals attack mutant ones.
6 The decline in the mutant is due to chance (just as there are chance departures from Mendelian ratios).

c *Discuss and evaluate these hypotheses, suggesting experiments which would test some of them.*

It must not be imagined that all mutants are disadvantageous and are eliminated from mixed populations. Many *Drosophila* mutants, which are detected by microscopic examination of the salivary gland chromosomes or by biochemical investigation, persist at high frequency in mixed populations and are not disadvantageous. These mutants do not disturb the development of the flies enough to produce obvious changes – though they may influence physiology or behaviour. How beneficial or adverse the mutants turn out to be depends on the environment.

20.2 The Hardy–Weinberg model

Figure 101 shows the proportions of vestigial and wild phenotypes in a mixed population of flies. Since the vestigial allele is recessive, all the vestigial flies must have been homozygotes. Some of the wild type ones were also homozygotes but many – an unknown proportion – would have been heterozygotes.

One way of estimating the proportion of heterozygotes in a mixed population would be to obtain a sample of the dominant phenotype and test-cross each individual to a recessive homozygote. This would be easy to do for fruit flies but might be impossible if one were studying a wild organism which was difficult or expensive to breed under controlled conditions.

In 1908, two mathematicians, Hardy in England and Weinberg in Germany, independently published a mathematical model which would allow the proportions of genotypes in a population to be calculated from the frequencies of the phenotypes. The model is only valid if a number of assumptions about the population are true.

Suppose that a population starts with 100 homozygous females (AA) and 100 homozygous males (aa). The allele A is dominant to a. It is easy to see that in this first generation the frequency of each allele is $\frac{1}{2}$ (or 0.5 or 50%). In the next generation all the individuals will be of the dominant phenotype and will have a genotype Aa. There has been a drastic shift in favour of the dominant phenotype, but the frequency of each allele is still 0.5 since each cell in the F_1 individuals will be of genotype Aa and half the F_1 gametes will be of genotype A and half of genotype a.

When the F_1 mates and produces an F_2 by random fertilization the frequency of the F_2 genotypes and phenotypes will depend on the frequency of the two types of gamete A and a. If you refer back to table 9a (see page 71) you will see that the F_2 genotype ratio must be 0.25 AA, 0.5 Aa, and 0.25 aa, and if allele A is dominant to allele a the phenotype frequency will be 0.75 dominant and 0.25 recessive.

A situation where a population starts with equal numbers of males and females of different homozygous genotypes is a special case. In more general terms, if the frequency of the allele A in the founding population is p, the chance of an egg of genotype A meeting a sperm of genotype A is $p \times p = p^2$.

If the frequency of allele a is q:

the chance of getting a homozygous recessive (aa) is $q \times q = q^2$
the chance of an A egg meeting an a sperm $= p \times q$
the chance of an a egg meeting an A sperm $= q \times p$
the chance of getting a heterozygote is therefore $2pq$.

Since the total frequency of all genotypes in the population is by definition 100 per cent or 1:

$$p^2 + 2pq + q^2 = 1 \qquad \text{(equation 1)}$$

Similarly, the sum of the allele frequencies is by definition 100 per cent or 1:

$$p + q = 1 \qquad \text{(equation 2)}$$

The genotype and phenotype frequencies of the recessive homozygote (aa) in any population are the same (q^2). The square root of the frequency of the recessive phenotype in the population should be q.

This value of q can be substituted into equation 2 to calculate p. Once p and q have been calculated so can the frequency of the homozygous dominant and the heterozygote.

Provided certain assumptions are true, these frequencies should remain constant for all subsequent generations of the population – a situation known as Hardy–Weinberg equilibrium. The assumptions are:

1 The population is large enough for chance fluctuations in frequency to be unlikely.
2a Any genotype is as fertile as all the others; that is, the genotype does not influence the number of offspring an individual will have.
2b The genotype does not influence the chance of survival of an individual at least until after it has completed reproduction (death after reproduction cannot affect the next generation).
3a The genotype does not influence the choice of a mate from the population; that is, there is no assortive mating (see page 159).
3b Self-fertilization is no more probable than cross-fertilization, and there is no tendency for related individuals to mate.
4 Mutation of allele A to a or *vice versa* is so rare as to be negligible.
5 If individuals enter or leave the population by immigration or emigration their genotype does not influence their tendency to migrate; in other words, migration does not alter allele frequency.

In a real population, where the heterozygote can be distinguished from either homozygote, the frequency of heterozygotes predicted using the Hardy–Weinberg model can be compared with the actual observed frequency. Significant departure from the expected proportion of each genotype shows that at least one of the assumptions made by the model is false. The allele frequencies may be changing, and the rate of change can be assessed by comparing expected and observed frequencies.

Unfortunately, it is not possible to conclude, if allele frequencies in a real population match those predicted by a Hardy–Weinberg equilibrium, that all the assumptions on which the equilibrium depends are true. For example, an allele might improve the chances of survival of an organism but also increase the chance of that organism emigrating from the population and these influences might balance each other so that allele and genotype frequencies remained stable.

> **Practical investigation.** *Practical guide* 5, investigation 20A, 'Models of a gene pool'.

STUDY ITEM

20.21 Allele frequencies in human populations

A dominant allele R determines the ability of a human to roll the tongue into a U shape. The recessive allele r determines the absence of this tongue rolling ability.

In a survey of the population in an American city it was found that 64 per cent of the people could roll their tongues while 36 per cent could not.

a *What proportions of the population would be expected to have the genotypes RR, Rr, and rr, assuming the population is in Hardy–Weinberg equilibrium?*

Suppose that a population of several thousand humans has been isolated on an island for several generations and that in this imaginary population there are albinos, whose lack of pigment is due to a recessive allele of a gene controlling melanin synthesis. The members of the population choose their mates without reference to skin colour, and there is no difference between the fertility of the various genetic groups nor in the average age at which members of the various groups die.

b *If 4 per cent of the population is albino, what is the percentage of albinos in the population likely to be in 100 years' time?*

c *Give an important extra assumption that you have had to make in order to answer b. Do not question the facts you were given about the imaginary population but consider the special circumstances which must be fulfilled if the Hardy–Weinberg model is to be applicable.*

d *Complete table 29 for the imaginary population. It shows several frequencies of the albino allele.*

Frequency of the recessive allele c	Frequency of albinos (cc)	Frequency of heterozygotes (Cc)
0.9 (90 %)		
0.4 (40 %)		
0.1 (10 %)		
0.01 (1 %)		
0.0001 (0.01 %)		

Table 29
Genotype frequencies in a human population.

e *Does the frequency of the genotypes in the population depend on which of the alleles is dominant?*

The blood groups A, B, AB, and O are determined by a series of three alleles commonly designated I^A, I^B and i. I^A and I^B show no dominance to each other but both are fully dominant over i.

f *Tabulate all the diploid genotypes which are possible for these alleles and the corresponding phenotypes (blood groups).*

The Hardy–Weinberg equations for a situation where three alleles are involved are:

$p + q + r = 1$

and

$p^2 + 2pq + 2pr + 2qr + q^2 + r^2 = 1$

The frequency of the ABO blood group alleles in the English population is: I^A: 26 %; I^B: 6 %; i: 68 %.

g Calculate the frequency of the blood groups A, B, AB, and O in the English population. Assume it to be an ideal population in Hardy–Weinberg equilibrium.

h Are all of the assumptions that must be true for a population to be in Hardy–Weinberg equilibrium likely to be true for any real human population? *No — Such factors as assortive mating, migration, selection, small pop size*

20.3 Inbreeding and outbreeding

There are a number of ways in which sexual reproduction may be organized in a population.

1 Mating may be completely random, with each individual having the same chance of fertilizing any of the other individuals.
2 Self-fertilization may be possible, and, if this is so, it may happen either commonly or infrequently.
3 Mating may be common among related individuals.
4 Some phenotypes in the population may be more likely to fertilize each other than are other combinations of phenotypes. This is called assortive mating.

The existence of separate sexes is of obvious importance, since it makes self-fertilization impossible, but it does not prevent mating between relatives. Most types of animal have separate sexes and even those which are hermaphrodite, for example earthworms and snails, rarely achieve self-fertilization. Most flowering plants do not show separation of the sexes. There may be separate male and female flowers on the same plant, as in hazel, but more commonly each flower develops both pollen and ovules. Some species of flowering plants do have separate sexes, well known examples being hops, holly, dog's mercury and hemp. (The drug cannabis can only be obtained from female hemp plants.)

The fact that in some animal groups the male is the heterogametic sex while in others the female is heterogametic suggests that the separation of sexes has evolved independently a number of times in the animal kingdom.

Truly random fertilization is rare since there are so many factors which can influence the choice of mates. Geographical distribution is an important factor. If individuals are distributed in small groups, mates are likely to be chosen from within these groups and mating between relatives will be more frequent. Many organisms have a dispersal phase in their life cycle, and this is important for future breeding since unrelated individuals are more likely to meet after a phase of population dispersal.

Most species have special features which either restrict the possible choice of mates or which prevent self-fertilization and sometimes fertilization between related individuals. In sea lions a single male acquires a large harem of females and prevents other males from passing their genes to the next generation. Many sea lions in the next generation will be half-sisters and half-brothers (half-sibs) since they share a father in common. Young robins on the other hand disperse from the territory of their parents and eventually set up their own territory after a period of solitary wandering. They are monogamous, and choice of mates is much more random than in sea lions so half-sibs will be very rare.

Some species of plants have flowers which are so constructed that self-pollination and self-fertilization are almost inevitable. For example, in the garden pea self-pollination takes place in the bud before the flower opens. (As we have seen, Mendel exploited this feature of peas.) In radishes, white clover, and most varieties of cherries, pollen produced by a plant cannot fertilize ovules in the same plant or in any other plant of the same genotype. This makes cross-fertilization inevitable and also decreases the probability of related plants fertilizing each other since they are more likely to share the same genotype than are unrelated plants. The phenomenon is called self-incompatibility and is controlled by a locus at which there are several possible alleles, called s_1, s_2, s_3, $s_4 \ldots s_n$. As many as thirty alleles are known in some species. Being haploid, a pollen grain carries one of these alleles and it cannot succeed in fertilizing a plant which carries the same allele. For example, in the cross $s_1s_3 \times s_4s_6$ all pollen genotypes would be successful in fertilization, whereas in the cross $s_1s_3 \times s_3s_4$ pollen of genotype s_3 would be unsuccessful, and the cross $s_1s_3 \times s_1s_3$ would fail completely.

Very similar systems of self-incompatibility have evolved in many groups of fungi where the alleles are known as mating-type alleles.

a *What disadvantage might a plant which is self-incompatible suffer in reproduction?*

If a group of organisms has features which favour self-fertilization or mating between related individuals the group is said to be inbreeding. Groups where mating is random, or where self-fertilization or mating between relatives is less likely than could be expected by chance, are outbreeding.

b *Are humans inbreeders or outbreeders?*

c *What economic, cultural, and technological changes during the past 100 years may have changed the degree of outbreeding or inbreeding shown by the British population?*

The consequences of inbreeding and outbreeding

The Hardy–Weinberg model is able to predict genotype frequencies in a large randomly mating population because the rate at which homozygotes are produced by matings between heterozygotes is balanced by a similar rate of production of new heterozygotes by mating between the two homozygous types.

Aa × Aa	AA × aa
1AA 1aa 2Aa	Aa
(new homozygotes)	(new heterozygotes)

Outbreeding encourages the formation of heterozygotes, since under a system of random mating the probability of the two homozygous types mating is greatest.

Self-fertilization obviously steadily reduces the number of heterozygotes with each generation since under a system of self-fertilization heterozygotes produce homozygotes but not *vice versa*.

Mating between relatives also encourages the formation of a higher proportion of homozygotes than would be expected by chance. Relatives are more likely to share the same genotype than two individuals chosen at random. The chance of two different homozygous types mating becomes less and less with each generation of inbreeding so new heterozygotes are not produced. Eventually a number of homozygous pure lines develop.

If an individual is homozygous at many gene loci, meiosis will result in few new allele combinations among the haploid cells produced.

d *Under what circumstances would this be an advantage?* Constant environment

e *When would it be a disadvantage?* Fluctuating environment

In an outbreeding population most individuals are heterozygous at many gene loci and many new combinations of alleles are produced by meiosis in each generation.

f *How many haploid combinations could be produced by meiosis in an individual heterozygous at ten unlinked loci?* $2^{10} = 1024$

[each additional locus where there is heterozygosity doubles the possible no. of combinations]

Linkage reduces the rate at which new combinations of alleles are produced since recombination between linked genes can only occur if chiasmata form between them. Favourable combinations can be preserved by close linkage.

In the insect group *Hymenoptera* (ants, bees, and wasps) the male develops from an unfertilized haploid egg and produces many genetically identical gametes by mitosis. He transmits his own genotype to all his daughters (he cannot have sons). The female is diploid and produces new combinations by meiosis. One sex is specialized to conserve favourable combinations, the other to 'shuffle the pack' and produce new combinations which may be favourable under changed circumstances.

Each species of organism has evolved a balance between extreme inbreeding and homozygosity and extreme outbreeding and heterozygosity. This balance can change especially if changes in the environment alter breeding behaviour, population density, and patterns of dispersal.

Linkage and the frequency of chiasmata place a limit on the production of new genotypes. Chiasma frequency varies between different groups, for example male fruit flies have a meiosis in which chiasmata occur only rarely.

The degree of inbreeding shown by a species and the extent to which recombination is possible, provide an integrated system of control over variability. This system of control of variability is the breeding system.

g *Does asexual reproduction change the variability of a species?*

It must be remembered that in an outbreeding population the variation in phenotype shown, at least by adults, is often low because of selection acting in each generation. The population remains able to produce new variation in subsequent generations because not all the breeding individuals are of the same genotype and many are heterozygous.

The concept of a 'gene pool'.

If genetic material can be exchanged between individuals, either directly or as a result of mating between future generations, they share a common gene pool. If individuals share a gene pool, all the alleles and combinations of alleles which each contains will become mixed up and distributed among all the individuals in future generations. The speed at which mixing occurs will depend on the breeding system of the population.

> **Practical investigation.** *Practical guide 5*, **investigation 20A, 'Models of a gene pool'.**

New alleles appear in a gene pool either as a result of immigration from another gene pool or as a result of spontaneous mutation. If the assumptions upon which the Hardy–Weinberg equilibrium depends remain approximately true the frequency of new alleles will remain low but they will be constantly brought into new combinations by meiosis and fertilization.

20.4 Selection and genetic drift

Not all genotypes in a population are equally successful in reproduction. All populations are potentially capable of increase but can only do so within limits imposed by the availability of resources (see page 452). The difference between the reproductive potential and the actual production of new individuals in any period of time may be the result of mortality, reduced fertility, or a combination of these reasons. If a genotype contributes significantly more individuals to the next generation than would be expected, while other genotypes contribute proportionately fewer offspring, the genotype frequency of the next generation will be altered. If the population of breeding individuals is small, large deviations from expected genotype frequencies may occur by chance alone. One would not expect an exact 3:1 Mendelian ratio if one had an F_2 family of only nine individuals. In the same way, if a randomly breeding population is small, genotype frequencies will fluctuate unpredictably and one allele of a gene may be eliminated from the population by chance. This situation is known as genetic drift.

Some genotypes are at an advantage over others in the production of offspring. The advantageous genotypes will produce more offspring than the disadvantageous and are therefore more likely to be perpetuated in succeeding generations. This is known as *selection*.

Phrases such as 'struggle for existence' and 'survival of the fittest' have been used to describe selection. These are unfortunate since they suggest that organisms are aware of the process of selection and are consciously seeking to perpetuate their genes. This is quite untrue.

If an individual lives longer than its contemporaries, it is likely to contribute more offspring to the next generation. This means that genotypes which increase the chance of personal survival will be favoured. There are, however, many examples of situations in which individual reproductive success and individual survival are not correlated. Sterile female mammals may live longer than their fertile contemporaries because they do not suffer the strain of pregnancy and lactation under conditions of food shortage, yet sterility is unlikely to be selected. A male spider is likely to live longer if he does not seek out and court females since he may be eaten even in the act of mating, yet only by sacrificing his life will he perpetuate his genes.

Much attention has been focused in recent years on the possibility of kin selection. If all individuals in a group are closely related and share many genes, an individual may perpetuate his or her own genes by a behaviour which favours the reproduction of close relatives, even if this behaviour results in death or if it prevents the individual itself reproducing. Colonial ants, bees, and wasps probably evolved as a result of kin selection. A queen mates only once and to a haploid male so that all her offspring are genetically very similar (see page 161). Workers (sterile females) are almost as likely to perpetuate their genes by rearing the brood of a queen (one of their sisters or their mother) as they would be if they themselves laid eggs and reared their own offspring.

Darwin introduced the hypothesis of selection in order to explain increased variation and the origin of new species, yet the same mechanism also explains reduced variation and the maintenance of an unchanged phenotype over thousands or millions of years.

Figure 102
The frequency distribution of continuous variation in a hypothetical population.

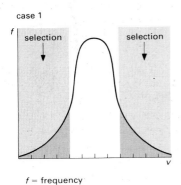
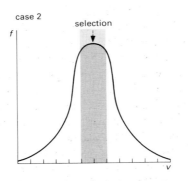
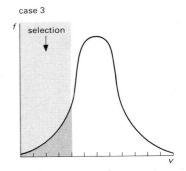

f = frequency
v = value of the variable measured
▨ = part of the population upon which selection is acting adversely

Figure 102 shows a hypothetical case of inherited continuous variation upon which selection is operating in three different environments (or at different times in the same place). You will find the situations easier to understand if you refer back to *figure 57* on page 84.

In situation 1, selection is occurring against the more extreme variants at both ends of the range of variation, *i.e.* individuals showing extreme variation are less likely to produce offspring which survive than are individuals showing a phenotype nearer the mode. In situation 2, selection is occurring against those of modal phenotype. In situation 3, selection is occurring against one extreme of variation.

    a *In which hypothetical case is selection encouraging variation?*

    b *In which case is selection tending to maintain phenotypic stability?*

    c *Will this process necessarily lead to stable frequencies of genotypes?*

    d *In which case is the gene pool being changed unidirectionally?*

STUDY ITEM

20.41 Interaction between strains of clover

Modern farmers frequently plant ley pastures. These are areas sown with grass seed to provide fodder for hay, silage, and grazing. The pasture is maintained for a few years, until it becomes colonized by less productive (wild) varieties of grass, after which it is ploughed and used for another crop.

Clover seed is often included in seed mixtures for ley pastures because clover is a legume and is able to increase levels of combined nitrogen in the soil, and because it is rich in protein.

Much effort has been expended in breeding varieties of clover which will persist in a pasture and which will provide high yields. In Britain and in New Zealand white clover *Trifolium repens* has been the most important species. (You may have studied variation in this plant in *Practical guide 5*, investigation 17A, 'Describing variation', and in investigations 19B and 20B in which different aspects of cyanide production in clover are explored.)

Subterranean clover, *Trifolium subterraneum*, has been an important crop in parts of Australia and the data presented here were gathered by J. N. Black of the University of Adelaide, using this species. He grew seedlings of two strains, Yarloop and Tallarook, at light intensities normally found amongst grass in pastures. The results are shown in *figure 103*.

    a *In what ways do the plants of the two strains differ in form under these conditions?*

    b *What might be the consequences if plants of the two strains were grown close together?*

Black sowed seeds of Yarloop and Tallarook separately in a number of seed boxes and in others he sowed a mixture of seed of the two strains. The seedlings were thinned to 3000 m^{-2} on emergence from the soil, and in the mixed populations a 50:50 mixture was achieved by selective thinning.

At intervals (19, 34, 48, and 62 days after emergence), Black took samples of the seedlings and discovered the yields of the strains in pure

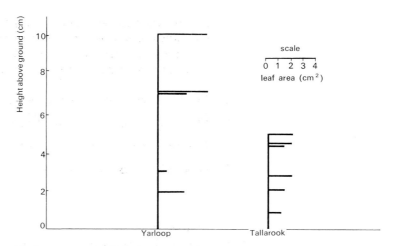

Figure 103
The leaf areas and height of typical isolated young plants of the Yarloop and Tallarook strains grown under reduced light intensity. (The height of a clover plant depends on the length of the petiole of its leaves.)
Based on Black, J. N. 'The significance of petiole length, leaf area and light interception in competition between strains of subterranean clover (Trifolium subterraneum L.) grown in swards', Aust. J. agric. Res., 11, No. 3, 277-291, 1960.

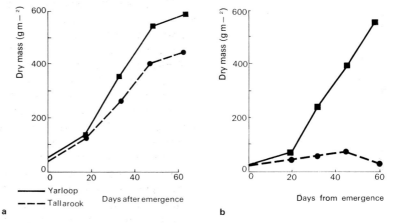

Figure 104
a The changes in dry mass of pure cultures of two strains of clover.
b The changes in dry mass of each type of plant when grown in mixed culture.
Based on Black, J. N. 'The significance of petiole length, leaf area and light interception in competition between strains of subterranean clover (Trifolium subterraneum L.) grown in swards', Aust. J. agric. Res., 11, No. 3, 277–291, 1960.

and mixed cultures. *Figure 104* shows the results. He measured yield by determining the dry mass of a sample of plants. Sufficient seedlings were grown so that sampling did not unduly influence the nature of the cultures.

    c *What were the advantages and disadvantages of thinning to a desired density rather than sowing seed at that density?*

    d *What was the combined dry mass of both varieties when grown in* mixed culture after 34 days? How does this combined mass compare with that of each variety when grown alone for the same time?

e *Assuming no plants had died after 34 days, was the mean dry mass of each Yarloop plant greater in the mixed or in the pure culture? Explain your answer.*

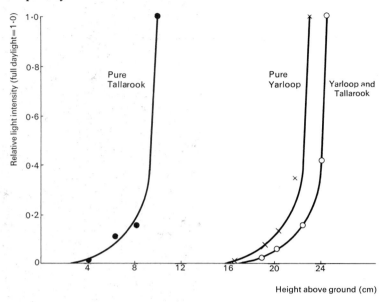

Figure 105
The absorption of light by the leaf canopy of the clover culture 62 days after emergence.

f *From figure 105, what was the relative light intensity 12 cm above the ground in the mixed culture after 62 days? Does this indicate that Tallarook is shaded in the mixed culture?*

No plants died in the pure cultures. Only Tallarook plants died in the mixed cultures. The death of 640 out of 1500 Tallarook plants is a typical result. The experiment shows strong selection against the Tallarook genotype when growing in competition with Yarloop.

g *Can it be assumed that similar selection would operate under field conditions?*

20.5 Polymorphism

If there are two or more alleles producing discontinuous variation in a population and the rarest of the alleles has a higher frequency than can be explained by spontaneous mutation, the population is polymorphic. Each different phenotype which constitutes the polymorphism is called a morph. The different morphs in a polymorphic population share a common gene pool and are found together in the same area.

A well known example of a polymorphic species is the snail *Cepaea nemoralis* (see *figure 106*) which is common in the chalk and limestone

Figure 106
The snail (*Cepaea nemoralis*), an example of a polymorphic species. Different morphs are shown in beech litter.
Photograph, Heather Angel.

areas of Britain. Your school or college may well have a collection of the shells of this species or you may be able to collect living specimens and record the frequency of morphs.

This example of polymorphism is a complex one which must have evolved long ago because it is shown by shells buried under Neolithic barrows. The background colours – brown, pinkish-brown, or yellow – are determined by a series of three alleles. (Brown is dominant to pinkish-brown and yellow, pinkish-brown is dominant to yellow.) A second locus controls the presence or absence of bands on the shell (unbanded is dominant to banded). If bands are present, other loci determine the number and width of the bands. There is yet another locus controlling the colour of the lip of the shell.

Snails are obviously rather immobile animals and there is likely to be a restricted flow of genes between neighbouring populations. Each population has a characteristic frequency of the different morphs and much research has been undertaken to explain their occurrence. One factor of importance is predation. Some morphs are better camouflaged than others in particular types of vegetation. Banded phenotypes blend well into hedgerows and unbanded pinkish-brown and brown types are well camouflaged among dead leaves in beech woodland.

The importance of camouflage may be tested by comparing the frequency of morphs marked and released into a population with the frequency of the morphs eaten by song thrushes. Song thrushes break the shells open on stones in their territory and marked shells can be collected round these 'anvils' some days after releasing the live marked individuals into the territory.

[margin note: not change appearance]

a *What precautions would need to be taken when marking and releasing the snails?*

In some parts of Britain it can be shown that thrushes take more unbanded snails in hedgerow habitats than would be expected by chance, and more banded and yellow individuals in beechwood habitats. These opposite selection pressures ensure that several morphs maintain a high frequency in each locality. In downland habitats in southern Britain, and more especially in France, there is no correlation between the frequency of morphs and the degree of camouflage likely to be afforded by the local environment. In these areas the polymorphism is not being maintained by thrush predation.

In hot, dry summer weather snails cement themselves to a leaf or stem with mucus and withdraw into the shell where they remain dormant until cooler, rainy weather returns. During these periods, called aestivation, they may become heated by solar radiation. If thermocouples are inserted into the shell openings of a live snail of dark colour and into a pale morph of the same size, the dark snail can be shown to warm up significantly quicker in direct sunlight.

[margin note: diurnal animal – warm up earlier + so be active earlier, pale forms may reflect heat during hot weather]

b *Under what circumstances might the ability to absorb radiant energy be an advantage and when might it be a disadvantage?*

c *The locus controlling the presence or absence of bands in* **Cepaea** *is closely linked to the locus controlling the background colour of the shell. How is this linkage an advantage to a population of snails?*

[margin note: Favourable combination selected for]

In previous chapters (see pages 29 and 147) we have mentioned a human polymorphism involving the production of sickle-cell haemoglobin. The common type of haemoglobin all over the World is haemoglobin A, but in certain areas, notably in Africa and parts of the Arabian peninsula, S haemoglobin occurs in a substantial minority of individuals. Those of genotype Hb^AHb^A are more likely to die of malaria than are Hb^SHb^S individuals or the heterozygotes Hb^AHb^S. Many sickle-cell homozygotes die as a result of the anaemia from which they suffer and from complications which attend the unusually high rate of turnover of their red blood cells. The polymorphism continues because the heterozygotes $Hb^A Hb^S$ have a selective advantage over either homozygous type; in other words, they enjoy the greatest chance of survival. In the Yemen, sickle-cell homozygotes $Hb^S Hb^S$ are less disadvantaged than are similar African homozygotes, and this has been shown to be associated with high levels of foetal haemoglobin which is often present in the people of the Yemen. The foetal haemoglobin stabilizes the shape of the red blood cells so that they sickle less easily under conditions of low oxygen partial pressure. The tendency to produce foetal haemoglobin in adult life is under genetic control. In other countries besides the Yemen the expression of the condition is variable, some homozygotes becoming much more ill than others.

d   *Malaria has been eliminated in many parts of the World where it was once common. What will happen to the sickle-cell anaemia polymorphism in an area where malaria has ceased to be common?*

Hb^s a disadvantage + will be reduced to a frequency where elimination ≡ new mutation

Evolution in the peppered moth (*Biston betularia*)

Many species of European moths are active at night and rest immobile on tree trunks and branches during the day. Such moths are very well camouflaged on the bark of the species of tree on which they usually rest. During the first half of the nineteenth century, industrialization based on coal increased greatly in many parts of Britain. Smoke from coal burning causes soot and tar to be deposited on buildings and on the bark of trees. Most coal contains some sulphur and this appears as sulphur dioxide in smoke. Lichens are extremely sensitive to sulphur dioxide pollution and are rapidly eliminated near areas where coal is burned and in areas to which prevailing winds carry the smoke.

The camouflage of bark-resting moths depends on their resemblance to a patch of lichen. They become much more conspicuous when lichen is killed and still more so when the bark becomes dark with deposits of soot. During the nineteenth century, collectors began to capture increasing numbers of dark-coloured or melanic mutant forms of many species of moth. The first melanic form of the peppered moth was taken near Manchester in about 1850. Melanic mutants have become more frequent in some seventy species of bark-resting moth during the past century. There is good evidence for the spread of these polymorphisms since the Victorians were keen collectors of butterflies and moths. Collectors seek out new and/or unusual variants, and many extensive Victorian collections are still preserved.

The phenomenon of industrial melanism has been most closely studied in the peppered moth by E. B. Ford and later by H. B. D. Kettlewell at the University of Oxford. There are three recognizable morphs: the pepper-coloured form *typica*, a much darker but still mottled form *insularia*, and a black form *carbonaria*. These forms are produced by three alleles – *carbonaria* is dominant to *typica* and *insularia*, and *insularia* is dominant to *typica*.

STUDY ITEM

20.51 Polymorphism in the peppered moth *Biston betularia*

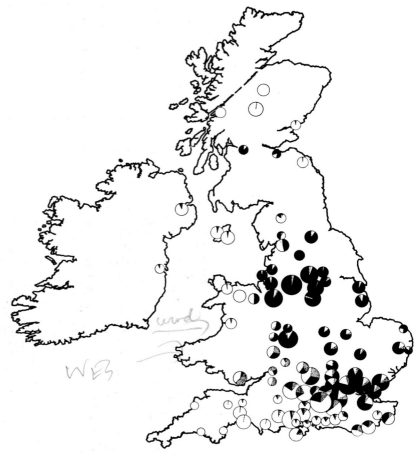

Figure 107
Sketch map showing the frequencies of the three forms of the peppered moth, *Biston betularia*, in Great Britain. The circles show the localities in which collections were made and their size indicates the number of moths in the sample. The largest circles represent collections of more than 4000 moths and the smallest (Dundee) down to 17 moths. The size of the black area of each circle gives the percentage of *carbonaria*, the stippled area the percentage of *insularia* and the white area the percentage of the typical form. Note that the proportion of *carbonaria* is high in industrial regions and to the east of them but decreases rapidly to the west, south and north. The proportion of *insularia* compared with typical behaves in much the same way, showing that it also is an industrial melanic.
Based on Sheppard, P. M. *Natural selection and heredity*, 4th edition, Hutchinson, 1975.

Figure 107 shows a map of Britain which indicates the distribution of the three morphs.

a In which areas is **typica** most frequent?

b Explain this distribution of **typica**?

c What ratio of offspring would be expected in the F_2 of a cross between a homozygous **carbonaria** and a homozygous **typica**?

When such an F_2 is reared, there are often significantly fewer *typica* offspring than would be expected. This is probably because the heterozygotes are hardier than either homozygote, and *carbonaria* is hardier than *typica*.

d If such selection took place in the wild, what would be expected to happen to the **typica** allele frequency?

In 1956 Kettlewell reared many *carbonaria* and *typica* moths in captivity and released equal numbers of the two forms in a woodland near Birmingham with dark, lichen-free tree trunks, and in a second wood in Dorset with trunks covered by a rich lichen flora. The data of two experiments are in table 30.

	Typica	Carbonaria
Dorset	26	164
Birmingham	43	15

Table 30
Moths seen to be taken by birds (equal numbers of both morphs released).

e *What inference can be drawn from table 30?*

Figure 108 shows a bird in the act of capturing a moth.

Figure 108
A bird capturing a moth.
Photograph, Dr H. B. D. Kettlewell.

f *Does the photograph confirm that predation is an important influence in maintaining the polymorphism of the peppered moth?*

g *Suppose 99 per cent of moths in an area are* **typica**, *1 per cent are* **carbonaria**, *and the* **insularia** *allele is absent from the local population. Calculate the expected frequency of homozygous and heterozygous* **carbonaria**.

h *Is it likely that* **carbonaria** *individuals captured at a period before this form became common would be homozygotes?*

In populations of moths captured from the Midlands, heterozygotes are indistinguishable from homozygotes. When a *carbonaria* individual

Chapter 20 Population genetics and selection 171

is mated with a *typica* captured in an area such as west Dorset where wild *carbonaria* are never found, the heterozygotes are rather lighter than homozygous *carbonaria*; that is, dominance is not complete. *Carbonaria* specimens from early collections are often lighter than modern *carbonaria*. This is because in areas where *carbonaria* has been a rare mutant, as in Dorset today and in the Midlands before the industrial revolution, the allele is not fully dominant over *typica*. Complete dominance evolves by selection for genes which modify the effect of *carbonaria* on the phenotype. This happens after the polymorphism becomes established in an industrial area.

Since the 1950s smokeless zones have spread throughout Britain and the use of coal as an energy source in the home and in small factories has declined. Large modern power stations have very high chimneys and the sulphur dioxide they produce is carried into the atmosphere to fall as acid rain over Norway and Sweden.

i *What changes in the distribution of industrial melanic forms would be expected as a result of these developments?*

STUDY ITEM

20.52 Evolution of Warfarin resistance

The use of poisons to kill pests has resulted in the evolution of important polymorphisms and this has greatly reduced the efficiency of the substances concerned. Resistance of agricultural pests to DDT is referred to in Chapter 28. Many pathogenic bacteria have evolved resistance to antibiotics such as penicillin.

Warfarin is a poison which prevents blood from clotting. It is used as a drug to prevent thrombosis (the blockage of arteries by clots of blood) in humans. Rats are easily killed by Warfarin since it results in internal haemorrhages similar in effect to those suffered by haemophiliacs. Because rats are much more sensitive to Warfarin than are domestic animals such as dogs and cats, it is very widely used.

Resistance to Warfarin first developed in a rural area in mid-Wales. It is controlled by a single dominant allele. Animals with this allele for resistance need large quantities of vitamin K in their diet.

Figure 109 shows a map of the area where resistant rats originated and the concentric lines show the extent of the resistant population in the year indicated.

a *Explain how a small Warfarin-resistant population might have developed initially.*

b *How would the fact that resistance is determined by an allele dominant over the pre-existing allele affect the increase in numbers and spread of the resistant population?*

c *Account for the rate of spread of resistance to Warfarin between 1967 and 1970.*

d *Studies have shown that resistant rats rarely constitute more than 50*

Figure 109
A map showing the area where rats resistant to Warfarin originated. The lines show the spread of the resistant population.
Data from Greaves, J. H. and Rennison, B. D. 'Population aspects of Warfarin resistance in the brown rat, Rattus norvegicus', Mammal review, 3, 1973.

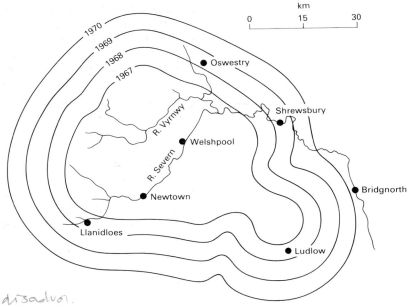

per cent of the population. Put forward some hypotheses to explain this.

e Explain the likely implications of this genetic situation for the selection of the population in the central area of the map.

Alcohol dehydrogenase in *Drosophila* – polymorphism at the molecular level

Fruit flies and their larvae depend on the sugars present in decaying fruit and on the yeasts which ferment these sugars. Ethanol is produced by yeast fermentation and flies must be able to metabolize the ethanol present in their diet. Their tissues contain the enzyme alcohol dehydrogenase which is part of the system involved in ethanol metabolism.

Using the technique of gel electrophoresis (see Study item 15.42 on page 27) the proteins in fly body fluid can be separated in an electric field. Each type of protein migrates through the gel at a specific rate and is present as a band in the gel. Those proteins that are able to act as alcohol dehydrogenase are revealed by a staining technique. *Figure 110* shows a photograph of the results of gel electrophoresis of the bodies of flies which are a sample from a wild population.

The example shown in *figure 110* is a simple one since the phenotypes can be explained by the action of one gene having two alleles. Many examples of biochemical variation are known where several alleles of a single gene exist and can be found in several possible paired

Figure 110
The results of gel electrophoresis of the bodies of flies. Heterozygotes produce three bands, homozygotes only one. The three bands can be explained by the association of peptide molecules in pairs. Since two types of peptide are produced we have: (A + A), (A + B), and (B + B).
Photograph, Dr J. A. Stewart, Department of Genetics, University of Leeds.

combinations in a diploid. (The inheritance of the A and B blood group antigen system is like this.)

In other cases there may be more than one gene able to produce proteins with similar enzyme function. Each of these genes may have alleles producing slightly different versions of each enzyme.

e   *Examine figure 110 and explain why a heterozygote results in three bands and a homozygote in only one band.*

f   *Count the number of flies of each phenotype shown in figure 110 and calculate the allele frequency. Is the sample large enough for the frequencies to provide reliable estimates of the true allele frequencies?*

g   *Why might the ability to produce two slightly different types of alcohol dehydrogenase be of advantage:*
   1 for an individual fly
   2 for a fly population?

Biochemical investigations have revealed that polymorphisms are even more common and important than one would have suspected by studying those which can be seen with the naked eye (such as colour variations). Subtle variation in the protein products of genes results in subtle variation in the physiology of the whole organism. Only in a few cases, the best known of which is sickle-cell anaemia in humans, do we have any understanding of the chain of cause and effect, starting with variation in the sequence of bases in DNA and ending with variation in the functioning of the whole organism.

Very many highly disadvantageous human mutations are known. The majority of these, cystic fibrosis and phenylketonuria, for example, are recessive. As death from infectious disease and malnutrition has become relatively rare in the more affluent nations, inherited malfunctions have become relatively more important. In the past it was impossible to diagnose healthy heterozygotes in any population. Close relatives of any person with a severe disorder caused by a recessive mutation have been faced with the moral and practical dilemma of whether they might be heterozygotes (carriers) and many have asked for genetic counselling. The counsellor was until recently only able to advise whether or not it was possible that the mutation had been inherited and estimate the probability of it being transmitted to children or grandchildren. Biochemical investigation will in the future be able to reveal which of the healthy relatives of a sufferer are heterozygotes.

Many couples who have produced a child suffering from a disorder caused by a recessive mutation have decided not to have any more children because of the high probability of producing a second affected child.

h   *What is the probability of such parents producing a healthy child?*

Direct identification of genes or gene products in tissues taken from a foetus (by amniocentesis) allows parents at risk to start a pregnancy knowing that it can be terminated by a legal abortion if the foetus is identified as a homozygous mutant which would develop into a child suffering from the inherited disorder.

20.6 Artificial selection

Since Neolithic times, that is, from about 8000 B.C., increasing numbers of animal and plant species have been domesticated. The environment of a domesticated organism is very different from that of its wild counterpart. For example, it may be transported as a result of trade or colonization to a different continent, as happened when European explorers and settlers took wheat and horses to North America. Even under the most primitive of farming methods, a domestic population will be protected from predators and obvious competitors (weeds) will be eliminated.

Wild populations respond to changes in their environment with changes in allele frequency so that they become more closely adapted to the new conditions. Domesticated populations evolve in a similar manner and it is not necessary to suppose that humans deliberately selected certain individuals or certain genotypes. Wild species of wheat have spikelets which are brittle when the ear is ripe. They become easily detached from the stalk, for example by the wind, and fall to the ground. When early humans harvested wheat, genotypes which had spikelets remaining more strongly attached to the stalk would be selected since they would have an increased chance of being gathered and sown again.

Many species of wild plant have seeds which remain dormant, even under conditions favourable for germination, until a time when the resulting seedlings are most likely to survive to maturity. Some species show variable dormancy, a few seeds germinating as soon as they fall from the plant while the germination of others is delayed. This acts as an insurance that at least some seedlings will survive.

When humans started to sow seeds and harvest crops, a new set of selection pressures were generated. Seeds which germinated late matured late, after the crop had been harvested. Only those that germinated early were selected as parents of the next crop. This was not as a result of conscious effort on the part of the farmer, but simply because he could only harvest and thresh the crop once. Genotypes conferring quick, uniform germination were at a selective advantage.

Apart from unconscious selection of this sort, farmers and gardeners have deliberately chosen certain individuals or populations for reproduction while using others for food. The operation of castration has been known from very early times and was of great importance in animal breeding since it prevented undesired males from transmitting their genes.

A combination of unconscious and deliberate selection acting over several centuries has resulted in the development of domestic varieties of animals and plants with phenotypes quite different from their wild ancestors. (See *figures 111, 112,* and *113.*)

In wheat, polyploidy and interspecific hybridization have been of

Figure 111
Ears of wheat. Cultivated wheat (*Triticum aestivum*) is shown on the left, and wild (*Triticum monococcum*) on the right.
Photograph, Biophoto Associates.

Figure 112
a The domestic Middle White pig.
b The wild boar.
Photographs: **a** *by permission of the National Pig Breeders' Association;* **b** *by permission of the Royal Zoological Society.*

Figure 113
a Cultivated lettuce.
b Wild lettuce. (The number is the plant breeders batch number.)
Photographs, The Glasshouse Crops Research Institute, Littlehampton, Sussex.

vital importance in the development of the modern cultivated species (see *figure 114*.). Most modern varieties belong to the hexaploid *T. aestivum*. This is unknown in the wild and developed under cultivation in Mesopotamia around 6000 B.C. *T. monococcum* has been cultivated as

Figure 114
A chart showing the origins of modern bread wheat.

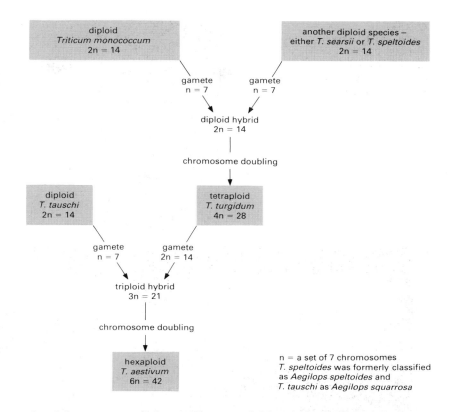

n = a set of 7 chromosomes
T. speltoides was formerly classified as *Aegilops speltoides* and *T. tauschi* as *Aegilops squarrosa*

the einkorn group of wheats. The tetraploid varieties include the emmer wheats and durum wheats (used to make pasta).

High yielding modern varieties have replaced traditional varieties which were grown by peasant farmers. Many traditional varieties have become extinct but others are preserved by plant breeders in gene banks and are an important international resource.

The dog has been domesticated for longer than any other organism and variation between breeds is especially striking. *Figure 115a* shows a

Figure 115
Modern dog breeding.
a Smooth-haired dachshund.
b Rottweiler.
Photographs: **a** *Frank Lane Picture Agency;* **b** *Marc Henrie.*

breed selected for creeping into badger sets and *Figure 115b*, one used for guarding cattle.

Modern genetic theory has provided a tremendous stimulus to the selective breeding of animals, plants, and some micro-organisms, especially yeasts. Unconscious selection, or selection of phenotypes without an understanding of inheritance, has been replaced by carefully planned breeding programmes. Selective breeding will give quick results in species such as rabbits or chickens which produce many young and take only a few months to mature. Cattle only produce one calf at a time, and this takes years to mature. Many breeds have been developed in different parts of the World and most of these have valuable inherited qualities, for example the Jersey cow has milk with high cream content, while cattle originating in India and Africa are able to withstand drought and high temperatures. In these latter breeds the fat store is deposited as a hump, rather than uniformly under the skin of the back as an insulating layer.

Improving breeds of cattle by selection (always a slow process) has been aided by two technological advances – artificial insemination and the longterm storage of semen at the temperature of liquid nitrogen.

A young bull is selected because his pedigree suggests that he has a high probability of carrying very desirable combinations of genes. Semen from the bull is used to inseminate a good sample of cows while the rest is stored. When the sample cows have calves, the quality of the daughters as milkers can be assessed. If it turns out that the bull produces many good milkers among his daughters, his stored semen can be used for further artificial insemination, even after he has been sold or slaughtered.

It has been claimed that five bulls could provide enough semen to service the entire British dairy herd (several million cows) by artificial insemination.

Why would a geneticist regard this as undesirable?

Figure 116
Breeds of cattle produced by different selection pressures.
a Jersey bull.
b Zebu bull.
Photographs: **a** *Gordon Cradock/Farmers Weekly;*
b *Walter Jarchow/Frank Lane Picture Agency.*

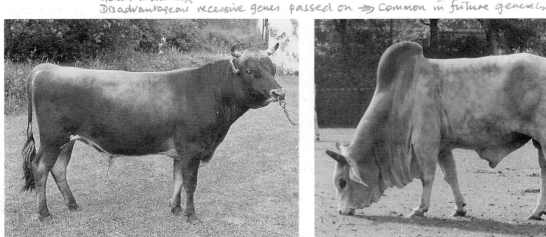

a   b

The semen of human males could be stored and used for artificial insemination of thousands of women. It has been suggested that especially well favoured men, for example sportsmen and Nobel prizewinners, could donate semen for storage so that the human race could be improved by widespread artificial insemination.

b *Ignoring aesthetic or moral objections, what reservations might a geneticist have about this idea?*

Most important variables which are manipulated by breeders, for example milk yield, are influenced strongly by both inheritance and environment, and vary continuously. Inheritance of such qualities is by the small additive effect of many loci (see page 85). Before embarking on a long and expensive programme of selective breeding it is usual to try and estimate the extent to which the variation shown in the population is a result of genotype rather than environment. This can be done by comparing the variance

$$\frac{\sum (x - \bar{x})^2}{n}$$

(see *Mathematics for Biologists*) shown by groups of related individuals with that shown by similar groups chosen at random. If variation turns out to be produced mainly by environmental influence, selection will not be worthwhile and the breeder must seek a new strain as a starting point for the programme.

A study of inherited resistance to disease

Diseases caused by fungi, protozoa, bacteria, and viruses affect not only humans but also cultivated plants and domestic animals. They result in serious losses of yield and quality in crops and livestock. The economic and effective control of such diseases is an important goal for applied biologists if the World's present and future population is to be adequately fed. Populations of wild plants and animals are also subject to attack by infective organisms. Such attacks may regulate the size and density of natural populations and also influence their gene pools.

STUDY ITEM

20.61 Inheritance of resistance to tobacco mosaic virus (TMV)

TMV is a rod-shaped, RNA-containing virus. It can infect a number of plant species including tobacco and tomato. The symptoms vary with the plant used and with the strain of virus. They include stunting of growth and yellow blotches on the leaves (see *figure 117*). An infected tomato plant is commercially useless as it produces little fruit. The virus is readily transmitted on hands, clothes, and implements or by plant to plant contact.

There is a gene located on chromosome 9 of tomato, alleles of which confer resistance to TMV. The wild type allele, $tm-2$ is for susceptibility. This is recessive to $TM-2$ which confers resistance and to $TM-2^2$ which also confers resistance. Some virus strains overcome the resistance of $TM-2$ but none have so far been able to infect $TM-2^2$ plants.

Figure 117
A resistant tomato leaf (left) and one infected with TMV.
Photograph, The Glasshouse Crops Research Institute, Littlehampton, Sussex.

Table 31 shows the relationships between two strains of TMV and three true breeding types of tomato.

Virus strain	Genotype of host plant		
	tm–2 tm–2	TM–2 TM–2	TM–2^2 TM–2^2
0	susceptible	resistant	resistant
2	susceptible	susceptible	resistant

Table 31

In a breeding programme, two tomato plants of unknown resistance genotype were crossed. Cuttings from plant A were susceptible to virus strains 0 and 2. Cuttings from plant B were resistant to both strains of virus.

a *Why were cuttings of the plants tested for resistance, rather than the plants themselves?*

A sample of seedlings produced as a result of the cross were tested for resistance, using the two strains of virus.
Of these, 59 seedlings proved resistant to both strains of virus. The remaining 63 seedlings were resistant to strain 0 but susceptible to strain 2.

b *Explain this result and give the genotypes of the two parents.*

c *If resistance of the type discussed above was first discovered in a wild tomato species of no commercial value, suggest how a plant breeder might incorporate the newly discovered character into an existing commercial variety.*

Plant breeders recommend that precautions against TMV infection, such as washing hands and changing overalls, should be continued even when a resistant variety of tomato is being grown.

d *Explain the reason for this precaution.*

Breeders usually hope to discover alleles at more than one locus which confer resistance to a pathogen and then try and incorporate resistance at both loci into commercial varieties.

e *Explain why it is desirable for a plant to be resistant at more than one locus.*

Conservation of gene pools

Modern farmers have been immensely successful in increasing yields of crops and livestock products – maize, wheat, rice, milk, egg, and meat production have all been greatly improved during the past few decades. Yield is one aspect of the phenotype and as such is affected by the genotype and the environment. More favourable environments for crops and stock have been provided by control of pests and diseases and by improved nutrition. Animal and plant breeders have succeeded in producing strains which are able to respond to the special environment created by mechanized intensive agriculture.

Some poultry farms keep more than a million birds, either for meat or for eggs. The birds are housed at very high density and can only remain alive as a result of mechanized feeding, watering, and ventilation systems. Intense selection has been applied to a rather small number of poultry breeds, especially to the Rhode Island Red and the White Leghorn. New strains have been developed which grow fast and show behaviour suitable for the very intensive systems in which they are kept. Varieties of wheat have been produced which grow and mature more quickly than older varieties, allowing two crops per year to be obtained in some areas where only one crop was previously possible. The newer wheat varieties have short stalks, a phenotype produced by alleles of genes controlling height (see *Practical guide 5*, investigation 19A). Short stalks allow more of the total production of the plant to be used for developing grain, make mechanized harvesting easier, and prevent the plants being damaged by wind in the very large fields. Modern wheat varieties respond well to large fertilizer inputs.

The requirements of mass harvesting and mass marketing have encouraged farmers to grow varieties which develop a highly uniform product. Tomatoes for example are only in demand by large chain stores if they are round and of uniform size and colour. The British public is believed to require a fairly small round red tomato. There are many varieties available which are irregular in shape or plum shaped and others which show a wide range of colour – yellow, tangerine and even striped. Such varieties are almost unknown to the general public.

Because the development of an improved variety involves intense selection from a fairly small population over a number of generations, it is inevitable that the variability of the new strain will be less than that shown by the founding population. The gene pool of modern varieties is smaller than that of the breeds or varieties used as the foundation stock in the breeding programme. With the spread of modern varieties there has been a corresponding decline in the numbers of traditional varieties of crops and stock. Many breeds and varieties have become extinct. Animal and plant breeders are becoming increasingly alarmed at the threat to the future progress which this implies. Successful selection demands that a large gene pool is available.

One motive for conserving traditional varieties is that they may be able to provide alleles of disease resistance genes which can be incorporated into modern commercial strains. A programme of crossing followed by repeated backcrossing to the commercial strain is accompanied by selection for the desired allele.

Farmers have increasingly used F_1 hybrids between inbred strains of sheep, cattle, poultry, and maize because of the heterosis they show and because the good qualities of two strains can be combined in the F_1 hybrid without loss of uniformity. If rare traditional varieties are conserved it is hoped that the unique combination of characters which their gene pools determine will be expoited in the future by using them as parents to produce F_1 hybrids of great commercial value. One problem faced by those wishing to conserve gene pools is to decide which traditional varieties to conserve. The problem is eased where the breed is attractive, such as Highland cattle (see *figure 118*).

Pig breeds have become extinct more frequently than those of other livestock species because wealthy landowners have not maintained ornamental herds of pigs!

Figure 118
Highland cattle.
Photograph, Scottish Tourist Board.

Another problem is that conservationists often have to 'rescue' the few surviving members of a breed. (The Gloucester cattle were represented by only eleven pure bred individuals when they were saved by the Rare Breeds Survival Trust.)

> *loss of variability + genetic drift results in undesirable characters become fixed in pop*

c *What is likely to be the result of maintaining a small population of a rare breed of animal or plant (or a population of a rare wild species such as the mountain gorilla) for many generations?*

20.7 Reproductive isolation

Mutation, recombination, and selection can produce very large changes in the genotype and phenotype of individuals in a population. Such changes take place in the wild but can be most clearly observed in domesticated or laboratory populations. In domesticated species such as dogs the different breeds are so different from one another and from their wild ancestors that we must question whether they should be called separate species. The species is the fundamental group in nearly all systems of classification (see *Systematics and classification*). Most of us think we have a pretty good idea what we mean by the term 'species' yet it is notoriously difficult to define.

Many species – humans are a good example – which are distributed over a very wide geographical range have developed distinctive phenotypes characteristic of populations from different regions. These are called geographical races. Both naturally occurring geographical races and varieties produced by artificial selection are able to interbreed and produce vigorous and fertile offspring.

In contrast to the situation where distinctive forms are interfertile, there are other cases where very similar forms cannot breed together. A famous example of this is a group of *Anopheles* mosquitoes found in Europe and the Near East. These mosquitoes are all very similar in appearance and were classified as members of the species *Anopheles maculipennis*. When researchers studied the group in captivity and investigated their ecology (motivated by the need to control mosquitoes because they transmit malaria), it was realized that what had been considered to be a single species was in fact a group of at least five species. They do not interbreed and require different habitats. Some of the species are able to transmit malaria while others are not. The adults of all the species have a very close physical resemblance and can be confused even by experienced observers.

If two organisms are able to breed together and produce fertile offspring under natural conditions they are by definition members of the same species. This definition cannot be applied to all groups which are referred to as species. In many cases breeding behaviour has never been observed and it is therefore impossible to decide whether distinctive forms are species or merely races. Experienced taxonomists must take all the available information into account and make a judgment which can later be upset. Palaeontologists must clearly define their species entirely in terms of morphological differences since one can hardly test the hypothesis that two fossils were interfertile.

Figure 119
Polish bantam cock.
Photograph, Duncan Fraser.

Figure 120
Old English Game cock.
Photograph, Poultry World.

Figure 121
Light Sussex cock.
Photograph, Poultry World.

Figure 122
White Leghorn cock.
Photograph, Barnaby's Picture Library.

When domesticated species are selectively bred to produce distinctive varieties the breeder often has a particular purpose in mind. Fowls, for example, have been selected for ornamental purposes, for fighting ability, for meat, and for egg production (see *figures 119, 120, 121,* and *122*). Some of these objectives may be combined, but they cannot all be met efficiently by one genotype.

The gene pools of domestic breeds are kept artificially separate by restricting the choice of mates available to highly prized specimens (artificial reproductive isolation) and by preventing cross-bred animals from breeding.

The most obvious mechanism which can reproductively isolate two wild populations of the same species is geographical separation. The desert of central Australia provides a barrier to the migration of most species of mammal and bird. It has developed as a result of climatic changes in the (geologically) recent past and isolated populations around its margins have evolved into new species.

The most famous example of geographical isolation resulting in the formation of new species is that of the Galápagos Islands lying in the Pacific Ocean west of the South American continent. The fauna of these islands were observed by Charles Darwin and provided him with very good evidence for his evolutionary theory. The islands are isolated from the mainland and from each other and each of the larger ones has distinctive races or species of birds, tortoises, and marine iguanas.

STUDY ITEM

20.71 An isolated population of house mice

Many of the small islands and island groups lying off the coast of Britain, for example the St Kilda group in the Outer Hebrides, the Orkney Islands and the Isles of Scilly, have distinctive races of small mammals. These have evolved because the gene pools of the mammals are entirely separate from those on the mainland, and the selection pressures on the islands are different from those operating on the mainland.

One island population which has been studied very intensively is the population of house mice (*Mus musculus*) found on the small and windswept bird sanctuary island of Skokholm lying off the coast of Dyfed in West Wales. Historical records left by lighthouse keepers and others suggest that the mice were introduced in a pile of sacks brought to the island by rabbit catchers. During winter, only the coast of the island is habitable. The mice survive the winter in crevices among rocks and feed on marine animals around the high tide mark. In the spring the whole island can be colonized and young mice disperse over it. In the autumn there is 80 to 90 per cent mortality.

The frequency of the mutant allele of the gene 'diffuse haemoglobin' was estimated in a house mouse population in a Dyfed corn rick and in the Skokholm mice during the spring and autumn of 1968. At the same time, the number of mice heterozygous at the locus of 'diffuse haemoglobin' was determined in samples from each population. The estimates are given in table 32, where N is the number of mice.

	Pembrokeshire corn rick N = 75	Skokholm Spring 1968 N = 87	Skokholm Autumn 1968 N = 82
Frequency of the mutant allele (%)	15.8	65.7	57.3
Number of heterozygotes observed	19	35	19
Number of heterozygotes estimated	19	39	40

Table 32
(J.M.B.)

a *Assuming dominance, how would the number of heterozygotes be determined?*

b *Were the mainland and island populations both polymorphic?*

c *Advance two possible hypotheses to explain why the number of heterozygotes observed in the autumn population on Skokholm was less than expected.*

d *Suggest two possible reasons for the different frequency of the mutant allele in the Skokholm mice compared with the mainland mice.*

A distinctive variety of the house mouse formerly existed on the St Kilda Islands, west of the Outer Hebrides. These mice lived in the cottages of the crofters and became extinct within a few years of the crofters abandoning the islands. The St Kilda Islands have a race of field mouse (*Apodemus sylvaticus*, sub species *hirtensis*) which still exists there.

e *Suggest why house mice were able to survive in the absence of human habitation on Skokholm but were unable to do so on St Kilda.*

Ecological and behavioural isolation

If two populations become geographically isolated it is likely that each will become adapted to a different type of environment and the genetic constitution of each will become more and more different as a result of chance (genetic drift) and because of contrasting selection pressures. After a long period of isolation a climatic change and the expansion or migration of the population may bring them into contact. The two gene pools may then merge. It may also happen that the differences have become so great that the mating between the two types is infrequent and that when it occurs, the hybrid offspring are poorly adapted and do not survive. Under such circumstances there would be great selective advantage for genotypes which chose only their own kind for mating and avoided those of the other race. The races would then evolve into distinct species.

In cases where two populations have come to differ in karyotype as a result of polyploidy, aneuploidy, inversion, or translocation, the chromosomal differences may result in reproductive isolation because hybrids would be totally or partially sterile. The formation of new species of plant by polyploidy has been mentioned in section 16.3.

STUDY ITEM

20.72 Isolation in North American crickets

Professor R. D. Alexander of the University of Michigan has made a close study of several related species of field cricket found in the eastern and central United States. We will concentrate on two species *Gryllus fultoni* and *G. vernalis*.

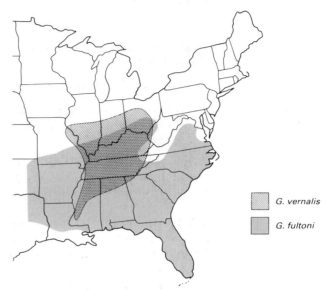

Figure 123
The geographic distribution of *Gryllus vernalis* and *Gryllus fultoni* in North America.
Based on Alexander, R. D., *Singing insects*, Patterns of life series, Rand McNally, 1967.

Figure 123 shows the geographical distribution of the two species. *G. fultoni* has two geographical races. The southern race inhabits oak–pine woodland while the northern race inhabits abandoned fields and prairies. *G. vernalis* occurs in oak – hickory and other woodlands – a habitat similar to the southern race of *G. fultoni*. Figure 124 summarizes the differences between the three types of cricket.

a *Is the northern race of G. fultoni likely to compete with G. vernalis? Explain your answer.*

b *Is the southern race of G. fultoni likely to compete with G. vernalis? Explain your answer.*

Male crickets vibrate their wings together and produce rhythmic noises by means of a chitinous file and scraper known as the stridulating apparatus. The male has three types of song:

1 a sound pattern which threatens other males;
2 one which attracts receptive females;
3 one used to court the female after she has approached.

The songs can be recorded on tape and analysed visually by means of

Figure 124
G. vernalis and *G. fultoni*.
Based on Alexander, R. D., Singing insects, *Patterns of life series*, Rand McNally, 1967.

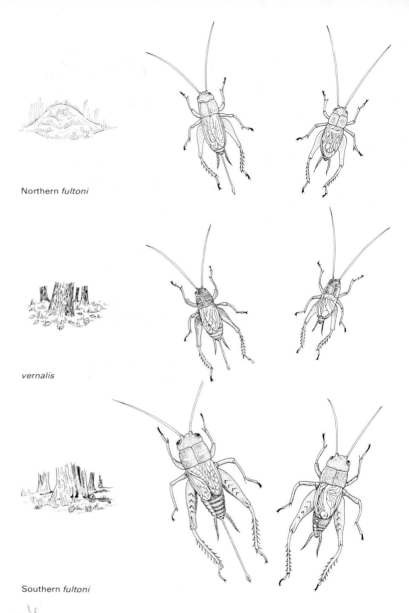

Northern *fultoni*

vernalis

Southern *fultoni*

an audiospectrograph. *Figure 125* shows the audiospectra of *G. fultoni* and *G. vernalis*.

The two species *G. fultoni* and *G. vernalis* can hybridize in the laboratory but have only very rarely been observed to hybridize in the field.

c *Use the information provided above to suggest two reasons why hybrids are rare in the field.*

The European cricket *G. campestris* has songs which are extremely similar to the north American species *G. veletis*. Crickets which have an overlapping distribution and which are adult at the same time of the year never have similar songs.

Figure 125
The audiospectrographs produced by G. vernalis and G. fultoni. Both species produce bursts of sound separated by silent phases. G. vernalis has approximately five bursts per second, each burst consisting of three peaks of amplitude. G. fultoni produces seven bursts per second, each burst consisting of three much less distinct peaks.
Based on Alexander, R. D., Singing insects, *Patterns of life series*, Rand McNally, 1967.

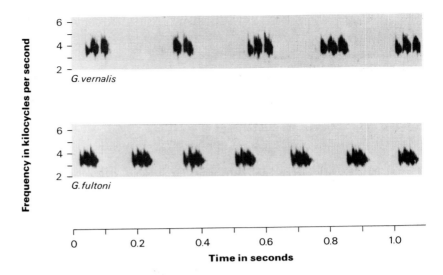

d *Does this information support your hypothesis produced in answer to c?* Yes since it suggests that species that are likely to meet have distinctive songs

Figure 126 shows the wing length and body length of many individuals of G. vernalis and G. fultoni plotted on the same scattergraph.

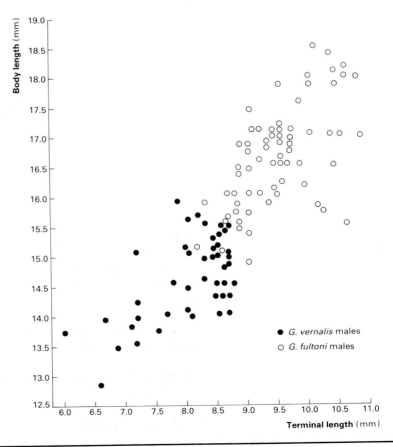

Figure 126
The wing-length and body-length ratios in G. vernalis and G. fultoni.
Based on Alexander, R. D., Singing insects, *Patterns of life series*, Rand McNally, 1967.

Chapter 20 Population genetics and selection

Before the work of Alexander showed that these forms differ in ecological requirements and in song, they were regarded as being members of a single species.

☐ e *What lesson should this scattergram provide for taxonomists?*

[handwritten: behav. + ecology as well as morphology important]

Summary

1. If two alleles exist in a mixed laboratory population, their frequency usually changes with time and one may be eliminated (**20.1**).
2. The Hardy–Weinberg model is explained by reference to hypothetical and real human populations. It is used to show that allele frequencies are stable from generation to generation unless disturbed by chance or by an outside influence (**20.2**).
3. Populations vary in their breeding system, that is the extent to which self-fertilization, mating between relatives, and recombination at meiosis takes place (**20.3**).
4. Outbreeding populations show much variation and variability and are well adapted to a changing or fluctuating environment. Inbreeding populations are well adapted to stable conditions (**20.3**).
5. 'Gene pool' is a term used for all the genes and combinations of different alleles of these genes which are able to recombine in a population as a result of sexual reproduction with some measure of outbreeding (**20.3**).
6. Selection is a situation in which the frequencies of genotypes change because not all are equally fitted to survive and reproduce in a particular environment (**20.4**).
7. Selection may come about through competition between genotypes for scarce resources (**20.4**).
8. Polymorphisms (situations in which two or more alleles exist at frequencies too high to be explained by mutation) are common in many species. They exist because either no allele has a selective advantage, and existing allele frequencies reflect the past history of the population, or because there is a dynamic equilibrium between several selective agencies (**20.5**).
9. Electrophoresis and other biochemical tests can detect variation in natural populations and may be able to distinguish between heterozygotes and homozygous dominants (**20.5**).
10. Several examples of polymorphism illustrate different types of selective agent (**20.5**).
11. Humans have both deliberately and unconsciously selected certain genotypes in populations of animals and plants. This has resulted in the development of domesticated varieties (**20.6**).
12. Animal and plant breeders are able to use knowledge of genetic systems to breed superior strains much more efficiently than our ancestors were able to by trial and error. Breeding of new commercial strains has been so successful that we need to conserve traditional varieties (**20.6**).
13. A population must have a separate gene pool, isolated from other gene

pools, if a distinct variety or species is to develop. This can occur by geographical isolation. Other isolating mechanisms may then develop (**20.7**).

Suggestions for further reading

ALEXANDER, R. D. *Singing insects.* Rand McNally, 1967.
BERRY, R. J. *Inheritance and natural history.* Collins, 1977.
BERRY, R. J. Studies in Biology No. 144, *Neo-Darwinism.* Edward Arnold, 1982.
BOWMAN, J. C. Studies in Biology No. 46, *An introduction to animal breeding.* Edward Arnold, 1974.
LAWRENCE, W. J. C. Studies in Biology No. 12, *Plant breeding.* Edward Arnold, 1968.

CHAPTER 21 THE PRINCIPLES AND APPLICATIONS OF BIOTECHNOLOGY

21.1 What is biotechnology?

Biotechnology is big business. Governments and industrial corporations around the world are investing heavily in research, and hundreds of small, specialist companies have been created. The boom in biotechnology began in university laboratories during the 1970s with advances in genetics and immunology which attracted much attention from the media, and finance from investors. At the same time, the oil crisis, and the subsequent realization of the finite nature of the Earth's energy resources, led to a search for new, non-fossil fuels, and for energy-efficient manufacturing methods. Furthermore as most biological processes occur at low temperatures and consume renewable raw materials or waste products, they are attractive in economic and ecological terms. This combination of factors has resulted in a dramatic upsurge in commercial activity in biotechnology.

Biotechnology can be defined as the application of biological organisms, systems, or processes to manufacturing and service industries. The agents involved may be whole organisms (mostly micro-organisms), single cells or biological substances such as enzymes. Biotechnology is therefore multidisciplinary, harnessing the techniques of microbiology, biochemistry and engineering to a rapidly increasing number of applications. Its impact has been felt in a wide range of industries, which includes those concerned with foodstuffs, medical products, fuel, waste treatment, and chemicals. There have been many recent developments which show enormous potential and could herald far-reaching changes in science and society comparable with the microchip revolution.

21.2 Fermentation

Economically, the fermentation industry is the most important area of biotechnology, with a worldwide turnover of around $15 billion (excluding beverages). Louis Pasteur described fermentation as 'the consequence of life without air'. In biochemical terms, fermentation can be defined as a catabolic (breaking down) process in which the formation of the end products from the substrate takes place anaerobically (without oxygen). In industrial jargon, however, the term 'fermentation' is applied to all production of commodities by microbial culture, whether it is by aerobic or anaerobic means. The products are many and varied, and include organic chemicals, fuels, antibiotics, enzymes, foods, and food additives.

Control of the environment and development of the organism are two methods by which maximum yield of product at minimum cost is achieved. Fermenter vessels are used to maintain the organism in an environment in which temperature and pH are controlled, a suitable concentration of nutrients (including dissolved gases) is provided, inhibitory substances are excluded or removed, and contamination by non-productive or destructive organisms is prevented. A system which contains only a selected desired organism is said to be 'aseptic' and this state is achieved by heat-sterilizing the entire system followed by inoculation with a small culture of the appropriate organism.

Development of the organism is achieved by mutation and selection. Random genetic mutations are induced by exposing the organism to a mutagenic agent, such as ultra-violet radiation or a chemical mutagen, and the survivors are cultured and tested for ability to synthesize the desired product. Occasionally a mutation will result in an increased product yield; organisms altered in this manner are selected and cultured and the process is repeated. Product yields have been increased many times over by the use of this technique, and most fermentation processes use 'overproducing' strains.

STUDY ITEM

21.21 Lysine production by 'overproducing' strains of bacteria

The pathway illustrated in *figure 127* shows the production of the amino acids lysine, threonine, and methionine by the wild type strain. It is controlled by feedback inhibition: lysine and threonine act together to inhibit enzyme 1.

Two kinds of mutant strains have been developed which produce larger amounts of lysine than normal.

a   *With reference to the pathway shown here, suggest possible mechanisms for this increased production of lysine.*

A brief history of fermentation

Humans have used micro-organisms in the preparation of food and drink for thousands of years. The Egyptian and Mesopotamian cultures are known to have fortuitously discovered wine, ale, vinegar, and leavened bread as long ago as 5000 B.C. The sugary juices of crushed grapes can be fermented to alcohol and carbon dioxide by wild yeasts on the skins of the fruit. Similarly, a barrel of wine or ale left open is susceptible to airborne infection by ethanoic acid bacteria, forming vinegar. In analogous ways, fermented products evolved around the World; examples are soy sauce in the Far East and yoghurt in the Balkans. All these processes became improved and controlled through purely 'trial and error' methods, in hindsight a remarkable achievement.

Microbes were first observed by Antonie van Leeuwenhoek in 1680, but their significance in fermentation was not realized until much later, through the work of Pasteur in the 1860s. Until then, the predominant view was that formation of new products in fermentation was the result of chemical instability. Pasteur demonstrated the relationship between the growth of micro-organisms and chemical changes in fermentation processes, and showed that each type of fermentation is accompanied by the growth of a specific organism. Microbiology thus became an exact science and laid the foundations for much of today's biotechnology.

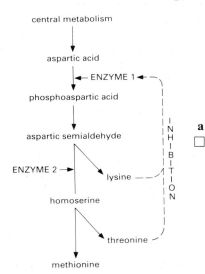

Figure 127
The pathway for the production of amino acids by the wild type strain of bacterium.

Figure 128
Operation of a laboratory scale fermenter (capacity 5 dm^3) in continuous culture.
Photograph, Biotechnology Centre, School of Industrial Science, Cranfield Institute of Technology, Bedford.

The birth of industrial fermentation

The work of Pasteur and others led to the development of new fermentation products and techniques. The production of baker's yeast in large aerated vats was initiated at the end of the last century, and microbial production of industrial solvents and chemicals began in the early 1900s. During World War I (1914–1918), a serious shortage of propanone (then known as acetone) for the manufacture of explosives was resolved by the fermentation of corn starch to propanone by *Clostridium acetobutylicum*. Other valuable products of this process were butanol (a solvent used in paints) and riboflavin (a vitamin). The first commercial plant for the production of citric acid was opened in 1923, utilizing the mould *Aspergillus niger*. Many other industrial chemicals have been produced commercially by similar means, including ethanedioic acid (oxalic acid), itaconic acid, gluconic acid and trans-butenedioic (fumaric acid).

Asepsis was not a problem in these early fermentations, which proceeded in open tanks. In these processes, prevention of contamination is a natural consequence of the fermentation conditions. Examples include ethanol and propanone production, where the high concentration of product is toxic to potential contaminants, and organic acid production, where the low pH suppresses the growth of other organisms.

Antibiotics

Since World War II (1939–45), the fermentation industry has been revolutionized by the development of the antibiotic industry. Many micro-organisms synthesize and secrete complex compounds that are selectively toxic to other micro-organisms, a phenomenon first noted by Alexander Fleming in 1928, when a plate culture of *Staphylococcus aureus* became contaminated with a mould. Fleming observed the inhibition of bacterial growth in the vicinity of the mould colonies and reasoned that a secreted chemical was responsible. He identified the contaminant as *Penicillium notatum* and named the secreted chemical penicillin. The practical significance of this discovery was not widely recognized at the time, and the isolation of pure penicillin eluded Fleming. After several years, he abandoned his work on the problem.

The limitations of antibacterial drugs of the time were cruelly obvious during the Second World War, and the search for alternative therapeutic agents led to renewed interest in penicillin. Two Oxford scientists, Howard Florey and Ernst Chain, succeeded where Fleming had failed, and demonstrated the effect of purified penicillin against a wide range of infections caused by Gram-positive bacteria (see note on table 33). A programme of research and development in the United States led to the large-scale production of penicillin within three years, using a new high-yielding organism, *P. chrysogenum*, which was suitable for cultivation in deep, aerated tanks.

The success of penicillin led to an intensive search for new antibiotics. Thousands of such compounds have now been isolated from a wide range of bacteria and fungi. Many are insufficiently selective in toxicity for medical use, but a number of important therapeutic agents have been found, including streptomycin, chloramphenicol, erythromycin, and the tetracyclines. In addition, numerous chemically modified (semi-synthetic) antibiotics have been developed, notably the penicillins and the chemically related cephalosporins.

The production of antibiotics led to a completely new set of problems for the fermentation engineer. Contamination by resistant bacteria caused considerable losses in early penicillin production, emphasizing the need for pure culture methods. The development of the aseptic fermenter during the post-war years was a most important step in the antibiotics industry, and led directly to new products in biotechnology, such as enzymes (see section 21.4), amino acids, polysaccharides, and single cell protein. In addition, the evolution of new fermented products through genetic engineering would have proved impossible without the exclusion of competing organisms.

STUDY ITEM

21.22 Selective toxicity of antibiotics

The success of antibiotics relies on the exploitation of biochemical differences between the host cells and the target cells. An antibiotic which is highly toxic to an infecting organism but relatively harmless to its host is said to have a high specificity or selective toxicity.

a *There are several major types of pathogenic agents – viruses, bacteria, fungi, and protozoa. From your knowledge of the cellular structure of these organisms and of the mammalian host cells, say which types will be most and least susceptible to selectively toxic antibiotics.*

Antibiotic	Target organisms	Mode of action
Penicillin	*Gram +ve bacteria	Inhibits Gram +ve bacterial cell wall synthesis
Rifampicin	Gram +ve bacteria Mycoplasmas	Inhibits bacterial RNA polymerase
Tetracycline	Gram +ve bacteria Gram −ve bacteria Rickettsia Mycoplasmas	Inhibits bacterial ribosomes – prevents tRNA binding in protein synthesis
Chloramphenicol	Gram +ve bacteria Gram −ve bacteria Rickettsia Mycoplasmas	Inhibits bacterial ribosomes – probably prevents peptide bond formation in protein synthesis
Ketoconazole	Fungi	Inhibits fungal cell membrane synthesis
Polyoxin D	Filamentous fungi	Inhibits fungal cell wall synthesis

* The Gram stain separates almost all kinds of bacteria into two groups – Gram-positive and Gram-negative. Gram-positive bacteria stain blue to purple; Gram-negative bacteria do not retain the stain.

Table 33
The action of some antibiotics.

Refer to table 33 in order to answer the following questions.

b *Why is penicillin toxic to Gram-positive bacteria and not mammalian cells?*

c *Penicillin is ineffective against Gram-negative bacteria. What can you deduce from this?*

d *Penicillin and Polyoxin D are both known to inhibit cell wall growth, but are effective on different organisms. What can you deduce about the cell walls of Gram-positive bacteria and filamentous fungi?*

e *Rifampicin is selectively toxic towards some bacteria but does not harm mammalian cells. Can you suggest a reason for this?*

f *Tetracycline and chloramphenicol both inhibit bacterial ribosomes. Why are they ineffective against eukaryotic ribosomes?*

g *Ketoconazole inhibits the synthesis of sterols (large insoluble molecules) in fungal cell walls. What can you say about the presence of sterols in mammalian or bacterial cell membranes?*

Amino acids

Amino acids have been commercially produced by fermentation for approximately forty years, with two-thirds of the World market supplied by Japan. A number of amino acids have important medical applications, particularly in the treatment of dietary deficiencies. Their dietary importance is also reflected in their use as supplements in animal and human foodstuffs. Most of the fermentation capacity in the food industry, however, is devoted to the production of L-glutamic acid, which is converted into monosodium glutamate for use in taste enhancement.

Polysaccharides

Polysaccharides are used as thickening and gelling agents, particularly in the food, adhesive, and paint industries. Although the majority of industrial polysaccharides are plant products, such as alginate and agar from seaweed, the use of microbial polysaccharides is increasing. Some have properties that make them suitable for specialist applications, such as xanthan gum, a polymer of unusual chemical composition produced by *Xanthamonas campestris* and used in oil recovery, where its viscosity enhances the effect of water pumped into the well, thereby increasing the volume of oil displaced. Other microbial polysaccharides, for example alginate produced by *Azotobacter*, may eventually replace some plant products, since the latter tend to vary in their availability.

Single cell protein

Single cell protein (SCP) is a term used to describe microbial cells grown for human and animal food. As in other areas of biotechnology, the concept is not new. Single-celled organisms have formed part of the human diet since ancient times. In Africa and Central America, the cyano-bacterium *Spirulina* is harvested for use as a major protein source. Yeasts are consumed in leavened bread, as dietary supplements and, in times of shortage, as foodstuffs. The yeast *Candida utilis* was produced as a food in Germany during World War II.

The world requirement for protein-rich food has led to renewed interest in microbial protein. In theory, the process is attractive for several reasons. Micro-organisms produce protein far more efficiently than any food crop or farm animal, with many species doubling biomass in an hour or less under optimum conditions. The protein has high nutritional value (generally superior to plant proteins, some of which are deficient in essential amino acids), and can be produced from a variety of substrates, including waste material. The method of cultivation is independent of climate, giving production all year round and great flexibility in the choice of production site. A wide range of bacterial, fungal, and algal species are suitable for SCP production, with a correspondingly large range of possible substrates, including methanol, ethanol, oil-derived hydrocarbons, agricultural wastes, and even sewage.

Whilst business economics, political considerations, and concern about potential health hazards connected with the carcinogenic compounds in oil-based substrates have caused the abandonment of SCP projects in some countries, ICI's SCP product, Pruteen, has been accepted for use in several animal feed applications. Pruteen is made by the fermentation of *Methylophilus methylotrophus*, using a methanol substrate. (See *figure 129*.)

Figure 129
ICI's Pruteen plant at Billingham.
Photograph, Imperial Chemical Industries plc, Agricultural Division.

Chapter 21 Principles and applications of biotechnology

Since the economic feasibility of SCP for animal feed has yet to be demonstrated, several companies including Phillips Petroleum and ICI are exploring the use of SCP in human food. ICI has a joint venture with Rank Hovis MacDougall to make myco-protein, a microfungus from the same family as mushrooms and truffles. The substrate used is glucose syrup extracted from wheat starch. Myco-protein has a threadlike shape, which makes it possible to simulate animal protein textures. It was sanctioned as safe for human consumption by the British government in 1980, and can be used in a variety of food products.

Fuel production

The dramatic increase in the price of crude oil during the 1970s led to renewed interest in microbially derived fuels. By far the leading exponents of biotechnological energy production are the Brazilians in the 'green petrol' or 'gasohol' programme. Ethanol, fermented from sugar cane or cassava, is used as a petrol extender by being blended with petrol for use as motor fuel. Although the cost per unit volume of ethanol is currently greater than that of petrol, its production makes political sense, decreasing oil imports and providing a significant safeguard against fluctuations in the price and supply of oil. However, the practice of devoting food crops to fuel production at a time of worldwide population growth and food shortage is likely to provoke continuing controversy.

On a smaller scale, the treatment of organic pollutants (sewage and agricultural and industrial wastes) by anaerobic fermentation yields methane as a by-product. The relative abundance of natural methane gas precludes the development of this process into a major source of energy. However, small-scale methane production by this method is common throughout the world, and it helps to supply the energy requirements of many farms and villages in Asia.

21.3 Genetic engineering

Humans have influenced the genetic composition of many different organisms. Animals have been selectively bred for desired characteristics, such as docility and meat yield in cattle, or attractive physical features in dogs. Improvements in crop yield and disease resistance have been achieved by selective breeding in plants. A variety of mutation and selection techniques have been used on micro-organisms to increase product formation. However, the advent of genetic engineering – more accurately, *gene manipulation* – has taken this process into a new era.

The complexity of higher organisms currently prevents genetic manipulation of plants and animals, although research is continuing here, and may produce the most spectacular results of all in biotechnology. At present, genetic engineering is only effective in single-celled organisms.

It has been known for some time that genetic information can be transferred from one micro-organism to another by *plasmids* (small, closed circular loops of DNA (*Figure 130*) and *phages* (bacterial viruses).

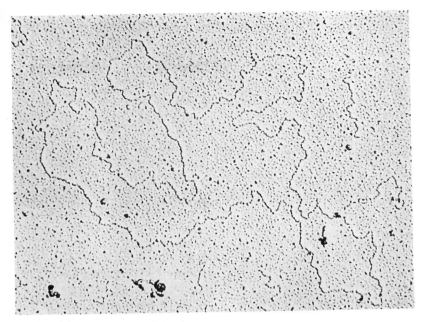

Figure 130
Electronmicrograph of plasmid DNA from *Pseudomonas* ($\times 18\,600$).
Photograph from Warner, P. J., Higgins, I. J., and Drozd, J. W., 'Examination of obligate and facultative methylotrophs for plasmid DNA, FEMS Microbiology Letters, *1, 6, 1977.*

These are both termed *vectors*. Many bacteria contain plasmids, which replicate independently and code for many functions in the bacterial cell, notably antibiotic resistance and the ability to grow on unusual nutrients. The ability of plasmids to pass from one cell to another is one reason why resistance to antibiotics can spread rapidly throughout a bacterial population. However, transfer of plasmids only occurs naturally between closely related species of bacteria, usually by a sexual process called conjugation. Similarly, infection by phages only affects certain groups of bacteria. In other words, the transfer of genetic information is limited in the natural world. The techniques of genetic engineering can overcome these limitations, permitting transfer of foreign genes for specific desired goals.

The four stages in gene manipulation are:

1 the formation of fragments of DNA, including the gene desired for cloning (replication),
2 the insertion (splicing) of the DNA fragments into vectors,
3 the introduction of vectors into the organism to be transformed, and
4 the selection of the newly transformed organisms for cultivation.

Formation of DNA fragments

DNA fragments of appropriate size are generated using one of a group of enzymes known as *restriction enzymes*. Each restriction enzyme recognizes a specific base sequence on the DNA molecule and cleaves the molecule within that sequence. Some restriction enzymes cleave both

Figure 131
The action of some restriction enzymes.

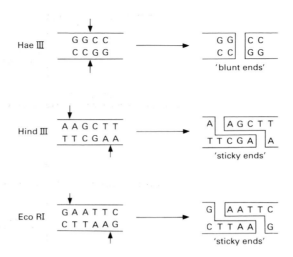

strands of DNA opposite one another, leaving a pair of 'blunt ends' (see *figure 131*) which are not suitable for genetic manipulation without further modification. However, most restriction enzymes cleave in staggered fashion, forming cohesive termini or 'sticky ends'. Each DNA fragment resulting from digestion by a restriction enzyme will have identical 'sticky ends'.

This method of generating DNA fragments is suitable for most bacterial genes. However, higher organisms have extra sequences of DNA (introns) within the genes which are excised from the mRNA transcript before translation into a protein. This excision is not possible in the bacterial host, so the presence of the eukaryote introns must be prevented. For a short DNA sequence, such as that for the peptide hormone somatostatin, which is involved in growth, it is possible to synthesize the gene chemically from its constituent nucleotides, complete with 'sticky ends'. An alternative method, used for longer eukaryote genes, is an enzymatic synthesis using the corresponding mRNA. The mRNA, which is present in multiple copies in the cell, is purified and treated with two enzymes: reverse transcriptase, which forms a single stranded DNA molecule corresponding to the mRNA, and DNA polymerase I, which completes the second strand of the DNA molecule. The 'sticky ends' can be added enzymically or chemically, giving a DNA fragment ready for insertion into a vector (see *figure 132*).

Splicing

The DNA fragments containing the gene for the desired product can now be inserted into a vector, either a phage or a plasmid. The choice between the two depends on several factors, such as the size of the DNA fragment and ease of monitoring the insertion.

In early experiments in gene manipulation, naturally occurring plasmids were used, but these have some disadvantages and were superceded by new 'tailor made' vectors constructed in the laboratory by splicing together sequences from other plasmids. The most versatile and popular vector is the plasmid pBR322 (see *figure 133*) which is used to

Figure 132
The enzymatic synthesis of a gene.

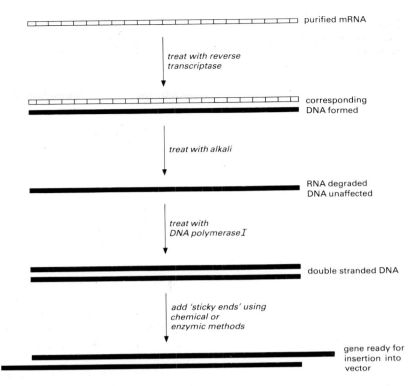

transform *Escherichia coli*. pBR322 carries genes for resistance to the antibiotics tetracycline and ampicillin, and thus can confer a selectable phenotype on the transformed bacteria. It is small (an important property in plasmids) and is 4362 base pairs long; the exact sequence of these base pairs is known. The existence of unique cleavage sites for a number of restriction enzymes allows great flexibility in its use.

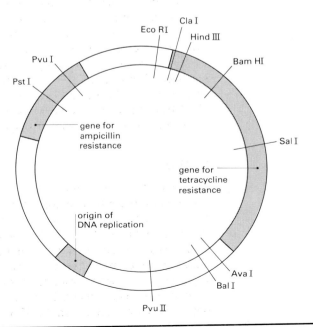

Figure 133
The plasmid vector pBR322 with antibiotic resistance genes and unique cleavage sites. Restriction enzymes (Pru I and Hind III etc.) cleave the plasmid at specific sites as shown by the lines.

Figure 134
Insertion of DNA into a plasmid vector.

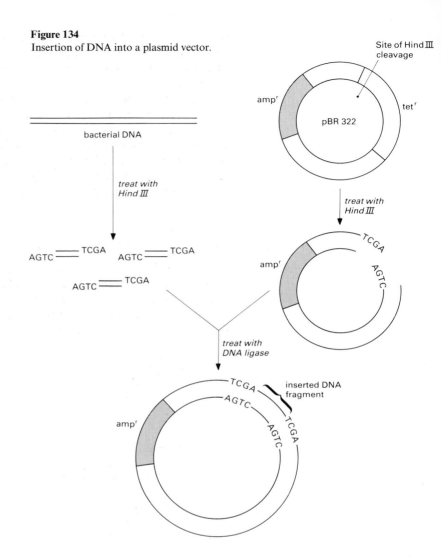

Figure 134 shows the steps necessary to insert a gene into a plasmid vector. The plasmid pBR322 and the DNA containing the desired gene are treated with the same restriction enzyme (Hind III in this case). The resulting 'open' plasmid and DNA fragments have complementary 'sticky ends' which are joined by mixing and adding the enzyme DNA ligase. This process is relatively inefficient – some plasmids will be rejoined without a DNA fragment, some DNA fragments will be joined together, and of those plasmids with a fragment correctly inserted, only a small proportion will contain the desired gene. However, it is possible to transform the host cells and later select those organisms containing the gene.

Transformation and selection

The new recombinant DNA can now be introduced into the host bacterial cell, usually by uptake of the DNA from the surrounding medium. The host cells are pretreated with calcium chloride which renders them 'competent', that is, able to take up DNA. The mechanism of entry of DNA into the cell, and the role of calcium chloride are unknown.

Selection of transformed bacteria is achieved by plating out on agar containing an antibiotic (ampicillin in this case). Those bacteria containing the plasmid are ampicillin resistant (amp^r) and will form colonies, whereas the untransformed cells are susceptible (amp^s) and will not. Further selection for the desired gene can take place by various techniques. The simplest, where applicable, is to plate out the bacteria on a medium, where only those organisms containing the gene can grow.

STUDY ITEM

21.31 The selection of a transformed organism

It has already been mentioned that the plasmid pBR322 is used to transfer genetic information from one micro-organism to another. For example, bacteria which do not normally grow on starch can be made to do so by using the plasmid to insert DNA containing the gene coding for the enzyme amylase, as illustrated in *figure 135*.

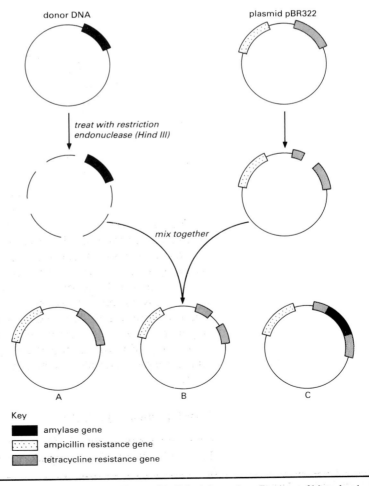

Figure 135
The insertion of DNA containing the gene coding for the enzyme amylase from plasmid pBR322 into the DNA of a micro-organism which does not normally have the gene.

Following transfer of the plasmid into the recipient organism, selection of the bacteria is achieved by growing them on specific media, containing different components.

a **Which of the newly formed plasmids (A, B, and C) would allow the bacteria containing it to grow on media containing**
  1 ampicillin?
  2 tetracycline?
  3 amylase?

Production of genetically engineered biochemicals

The techniques just described have led to the production of hormones and other polypeptides which are difficult and expensive to extract from mammalian sources. The most important products include somatostatin, insulin, and interferon (an anti-viral agent).

The expression of the gene for somatostatin in the bacterium *Escherichia coli* was one of the early successes in cloning mammalian genes. It is impossible to provide adequate quantities of the hormone by the traditional means of purification, since an extract from one and a half million sheep brains will only yield 5 mg of pure somatostatin. It is now possible to obtain this amount from a 9 dm^3 culture of genetically engineered *E. coli*. The first step was the chemical synthesis of the somatostatin gene from its constituent nucleotides, and its insertion into pBR322. The bacteria were transformed as previously described and cultured by using fermentation techniques similar to those used for antibiotics.

Insulin, a two-chain (A + B) peptide hormone, is used daily by millions of diabetics. For many years insulin has been extracted from pig or cattle pancreas, but the increasing demand from a growing number of diabetic patients has led to fears of a shortage. In addition, these insulins are chemically slightly different from the human hormone, which may account for the side effects suffered by some diabetics. Genetically engineered 'human insulin' is now being produced in quantity, and it is hoped to resolve the problems of supply and side effects. The genes for human insulin A and B chains were chemically synthesized and cloned separately in *E. coli*, resulting in two different strains, one producing A chain and the other B chain. The two peptides were purified separately and subsequently linked to form the active hormone.

The production of a human interferon from genetically manipulated *E. coli* is one of the most exciting and significant achievements of biotechnology. The group of glycoproteins known as interferons are natural anti-viral agents, present in most vertebrates, which function by making cells resistant to viral attack. This property, and the possible use of interferons as anti-cancer drugs, have stimulated great interest in their clinical application, but it has proved most difficult to extract the minute quantities present in animal tissues, and interferon purified in this manner is prohibitively expensive. A number of companies are now using genetically engineered *E. coli* to produce interferons more economically, and these are now undergoing clinical trials. While interferons have yet to live up to their 'wonder drug' reputation, it is hoped that they will prove important in fighting cancer and viral diseases.

21.4 Enzymes

Enzymes are biological catalysts; that is, they are protein molecules that mediate the specific chemical reactions of living cells. Enzyme technology is concerned with the isolation and purification of enzymes and their application in industry and biochemical analysis.

Enzymes perform reactions rapidly and at moderate temperatures, usually acting on a specific substrate (such as glucose isomerase, which converts glucose to fructose) or a specific group of substrates (such as subtilisin which is a protein-degrading enzyme or protease). It is this degree of specificity, which is only found in biological systems, that makes enzymic methods attractive. Over two thousand enzymes are now known, and it is thought that many more may yet be discovered.

Enzymes, like fermentation, were first used unwittingly, in the preparation of foods. Some ancient oriental processes use enzymes secreted by fungi to hydrolyse starch to sugar before fermentation to produce rice wine and saki. Rennet, prepared from calf stomach, has long been used to coagulate the milk protein casein, the first step in cheesemaking.

The existence of enzymes was a controversial issue in nineteenth-century science, and the debate was not settled until 1897, when Buchner demonstrated the fermentation of sugar to alcohol and carbon dioxide by an extract of yeast cells. Subsequent research established the nature and role of enzymes, and laid to rest the doctrine of vitalism (which postulated that a 'vital force' resided in the cell, mysteriously guiding its chemical activities). Buchner's discovery marked the beginning of modern biochemistry.

Enzyme purification

Most enzymes used in industry are derived from micro-organisms (see section 21.2), though important exceptions are the plant proteases and animal proteases (such as rennin). Some enzymes are by nature extracellular, that is, they are secreted by the cell into the surrounding medium. Others, which are resident in the cell, must be released by physical, chemical or enzymic disruption of the cell wall and/or membrane.

A typical purification might proceed as follows. Once freed from the cell, the required enzyme must be separated from the medium, which will include other proteins as well as broken cells and other cell constituents such as nucleic acids. Solid material is removed by centrifugation or filtration, and nucleic acids are precipitated using specific reagents or hydrolysed by the addition of specific enzymes (nucleases). The proteins are precipitated, usually by the technique of 'salting out', in which ammonium sulphate is added to increase the ionic strength of the medium. Proteins are less soluble in high ionic strength solutions, and form a precipitate that is easily harvested.

The required enzyme can be concentrated and purified from the protein mixture by chromatography. Various techniques are used, but in each the principle remains the same: the enzyme is retarded during its passage in solution through a column packed with solid material. The basis of separation may be difference in molecular size or charge, or affinity for a specific constituent of the column material.

Not all enzymes can be purified in this way; different enzymes require different techniques. New methods will continue to emerge as novel enzymes are discovered.

Immobilized enzymes

Enzymes may also be used in soluble or immobilized forms. Soluble enzymes are convenient in applications where the separation of the enzyme from the reaction mixture is not required (for example in biological detergents), but in other instances the recovery and re-use of the enzyme can be economically important. This can be achieved by immobilizing the enzyme to an insoluble, chemically inert support, such as cellulose, thus conferring easy recovery and in most cases increasing the stability of the enzyme.

A number of immobilization techniques are available to the enzyme technologist (*figure 136*). The enzyme may be chemically bound to a polymer, entrapped in gel capsules or membranes, or simply cross-linked to form insoluble proteinaceous aggregates. Whole microbial cells can also be immobilized, and in some processes are preferred to immobilized enzymes as biocatalysts.

Industrial applications

Although enzyme preparations were commercially available by the turn of the century, sales remained relatively small until the 1960s, when demand from the detergent and food industries increased, and enzyme production multiplied rapidly.

The use of enzymes in detergents accounts for about one-third of all enzyme production. Biological washing powders contain protein-degrading enzymes (mainly subtilisin from *Bacillus subtilis*) which dissolve protein-containing materials such as clotted blood. This type of dirt is very difficult to remove by conventional washing. Efforts to develop enzymic methods for other biological stains, including lipids and vegetable colours, have so far proved unsuccessful.

A number of enzymes are used in starch and sugar processing, both in soluble and immobilized forms. The most important enzymes are the amylases, which convert starch into glucose, and glucose isomerase, which converts glucose to the much sweeter sugar, fructose. These processes have replaced the chemical methods previously employed, and are used worldwide to yield huge quantities of their respective products.

Other applications of enzymes are found in the chemical industry, notably in amino acid production, cheesemaking (formation of whey from milk protein using rennet), textiles (removal of starch employed in processing by amylases), leather treatment (removal of hair from hides

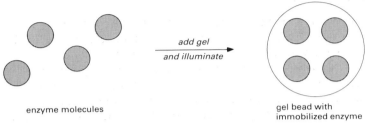

Figure 136
Techniques of enzyme immobilization.

by protein-degrading enzymes), and food processing (prevention of discoloration using glucose oxidase).

Analytical applications

The analysis of the composition of biological materials is of great importance in clinical chemistry and food quality control. Many compounds can be detected and quantified by adding the appropriate enzyme and measuring the formation of product. For instance, the enzyme urease catalyses the following reaction:

$$NH_2CONH_2 + 2H_2O + H^+ \longrightarrow 2NH_4^+ + HCO_3^-$$
urea

The concentration of the products can be determined using ion selective electrodes, and the original concentration of urea determined

Chapter 21 Principles and applications of biotechnology 207

by comparison with known standard solutions. Many enzymes are now available in diagnostic test kits; glucose oxidase, for example, is used to detect high levels of glucose in the blood of diabetics. Others are available in automated analytical systems. (*Figure 137.*)

Biosensors

The analytical application of enzymes, combined with immobilization techniques, have given rise to a new type of analytical device, the *biosensor*. The most common type is the enzyme electrode, where a thin layer of immobilized enzyme is held in close proximity to an electrochemical detector. The chosen enzyme catalyses the breakdown of the specific compound under analysis, and the electrochemical detector measures the concentration of a product or reactant. This method combines the specificity of the enzyme with the speed and simplicity of electrode measurement.

The first biosensor was the glucose electrode, proposed in 1962, and based on the enzyme glucose oxidase, which catalyses the following reaction:

$$\text{glucose} + O_2 \longrightarrow \text{gluconolactone} + H_2O_2$$

Early devices measured oxygen depletion, using an oxygen electrode. Since then, a number of different glucose electrode designs have been suggested. The most effective are those based on the electrochemical detection of hydrogen peroxide liberated during the reaction, and those which rely on direct electron transfer between the enzyme molecule and the electrode via a mediator. Examples of these three glucose electrodes are shown in *figure 138*.

Biosensors are available for a wide range of compounds. Not all use

Figure 137
An enzyme-based probe connected to a BBC microcomputer via an Artek programmable interface.
Photograph from Poole, R. J. and Dow, C. S., Microbial gas metabolism, *Academic Press, 1984.*

Figure 138
Some enzyme-based biosensors. *After Aston, W. J. and Turner, A. P. F., 'Biosensors and biofuel cells',* Biotechnology and genetic engineering reviews, I, *Intercept, 1984.*

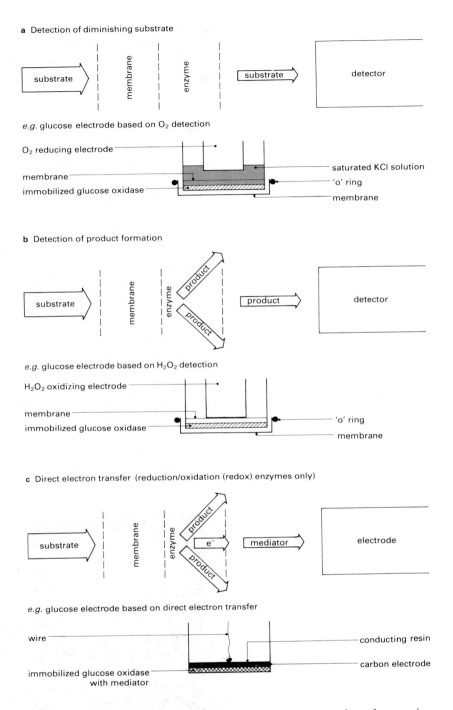

electrochemical methods of detection; some can respond to changes in temperature, mass, or optical properties. The applications of biosensors are numerous: in hospital and food science laboratories (sample analysis), in medicine (personal glucose monitors for diabetics), in environmental control (monitoring toxic compounds), and in industrial processes (fermentation control).

21.5 Immunology

In recent years there have been great advances in our understanding of the complex processes of the immune system in humans and animals and this is described in detail in chapter 24. The body responds to invasion by foreign material (such as bacteria, viruses, foreign proteins or tissues, and synthetic chemicals) by initiating a set of reactions in order to combat the invasion. The foreign particles are collectively known as antigens, and the molecules which combine with antigen, produced by the body in response to invasion, are called antibodies.

Antibodies are glycoproteins made by specialized cells. A vast number of different antibodies can be made, each binding very selectively to a specific antigen structure. When an antigen is recognized, antibodies bind specifically and facilitate its removal from the body by ingestion by phagocytes (specialized white blood cells). Disease occurs when the immune response is inadequate to cope with the antigen, for instance a rapidly multiplying micro-organism. Once induced by an antigen, the specific antibody will remain present in the body for many years, conferring immunity from that antigen.

Vaccination has long been used to stimulate the natural production of antibodies. A vaccine is a preparation of an attenuated (weakened) pathogenic micro-organism that cannot cause disease but still elicits the immune response from the host system. The antibodies raised against the attenuated organism can in most cases recognize and neutralize a subsequent infection by the active pathogen (see section 24.8). In some diseases, however, the process of attenuation changes the antigenic determinants on the cell surface, and vaccination is ineffective. In others, the disease is caused by a number of different strains of the pathogen. There are still no effective vaccines against influenza, cholera, syphilis, and malaria.

Monoclonal antibodies

Antibodies are of great use in the diagnosis and treatment of disease, in detection and quantification of chemicals, particularly drugs, in the body, and in purification of small amounts of rare compounds. Until recently, obtaining antibodies was a time-consuming and inefficient process, where samples of blood from a vaccinated animal were partially purified to give a mixture containing different antibodies. Because of the small amounts present and the structural similarity of the different antibodies, it is impossible to purify a single antibody by this method. Attempts to culture antibody secreting cells *in vitro* have failed, since the cells are not viable for long under these conditions, and a mixture of antibodies are produced.

The problem was solved in 1975 by Cesar Milstein working at Cambridge, who first produced monoclonal antibodies. He fused the antibody-secreting cells with tumour cells called *myelomas*, using the technique of cell fusion (see *figure 139*). The resultant hybrid myeloma cells, called *hybridomas*, secrete antibodies, and are immortal (a property of the parent tumour cells) and, following isolation, can be grown in bulk

Figure 139
Preparation of monoclonal antibodies.

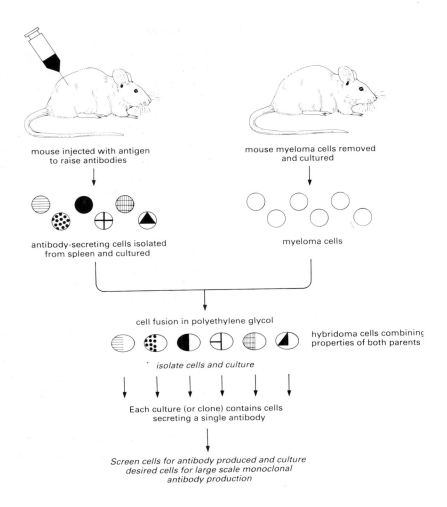

culture as a pure clone (that is all the cells are derived from a single hybridoma cell). For this reason, the products are known as monoclonal antibodies. Milstein's work was rewarded with a Nobel Prize in 1984.

Immunological techniques are standard methods in scientific analysis; in the past, mixtures of antibodies derived from vaccinated animals were used. Monoclonal antibodies are replacing these mixtures for many procedures, and are available in kit form for specialist diagnosis. A monoclonal antibody assay for alpha fetoprotein levels in amniotic fluid is used to reveal possible brain defects in the developing foetus. A rise in the levels of alpha interferon is commonly found in AIDS sufferers (see page 370), and a monoclonal assay is available for this purpose. A refined pregnancy test, based on the monoclonal antibody to a placental hormone, gives a fast and reliable diagnosis of pregnancy. A number of other diagnostic kits are now coming on to the market, and many more are expected to follow.

The use of antibodies in therapy is under investigation, with several possible applications. It is hoped to use the properties of antibodies in cancer therapy, exploiting the specific binding of the antibody to cancer cell antigens. One method is the targeting of drugs by attaching the drug molecule to an antibody, thus concentrating its effect at the site of the tumour. Alternatively, it may be possible to develop antibody cells that directly attack and destroy the tumour.

21.6 Future trends in biotechnology

The achievements discussed in this chapter could be dwarfed by developments in the next few decades if the potential of biotechnology is fulfilled. In this section we review briefly research that may come to fruition before the end of the twentieth century, and speculate on more longterm possibilities.

Health care

We have seen that biotechnology already makes considerable contributions to medical science in the prevention, diagnosis, and treatment of disease. All aspects of medical biotechnology promise further progress, none more so than genetic manipulation. New therapeutic products, like hormones and interferons, will continue to emerge as genetic engineering techniques improve. New vaccines, made by cloning the genes for antigenic determinants of pathogenic organisms, should prevent diseases where current immunological techniques are ineffective. Genetically engineered vaccines are already under development, including a malaria vaccine, a priority in tropical medicine.

These innovations are concerned with genetic manipulation of micro-organisms, but more sophisticated therapeutic techniques are envisaged in which the human genome (see page 123) could be altered. For instance, genetic defects could be detected and corrected at an early stage of foetal development, and deficiency diseases such as diabetes and thalassemia could be treated by the insertion of cells containing the genes for production of insulin and haemoglobin respectively. The applications of human genetic manipulation are not confined to medicine, and concern over the ethical implications has prompted a ban on recombinant DNA experiments on humans at present.

Biotechnology is involved in the development of artificial organs. It may be possible to treat liver failure by implanting a device containing immobilized liver cells or enzymes to perform the functions of the ailing organ. This method could also be used with insulin-producing cells to create an artificial pancreas for diabetics. However, as in organ transplants, the introduction of foreign biological material could cause problems with rejection. Alternatives could use biosensors to monitor body functions and artificial devices to effect a response, for example an artificial pancreas based on an implanted glucose sensor linked to an insulin pump.

Agriculture

Livestock farmers will undoubtedly benefit from the many advances in medical biotechnology. Genetically engineered growth hormones are now available, and may be used in the future to increase meat yield. Foot and mouth disease, trypanosomiasis (sleeping sickness) and other livestock diseases could be eradicated if new genetically engineered vaccines succeed.

However, the greatest potential for biotechnology in agriculture lies in the improvement of food crops by genetic manipulation. As in mammalian systems, plant genetic techniques are still in their infancy, and there are enormous problems yet to be surmounted, but knowledge is advancing rapidly. One of the aims of plant genetics is the improvement of cereal proteins, which often lack certain essential amino acids. It should eventually be possible to introduce the genes coding for the essential amino acids into the plant genome, thus improving its nutritional value.

All plants require nitrogen in order to make the proteins, nucleic acids, and other metabolites necessary for growth, but they are unable to utilize atmospheric nitrogen. A number of micro-organisms possess the ability to fix nitrogen gas, a process catalysed by the enzyme nitrogenase. Leguminous plants form symbiotic relationships with these organisms to derive nitrogen fixed in this manner, but other crops cannot be cultivated without expensive nitrogenous fertilizers. A major longterm goal in biotechnology is the introduction of nitrogen fixation into crop plants by the insertion of the nitrogenase gene.

Photosynthesis in most plants is an inefficient process owing to a lack of specificity in a key enzyme, ribulose bisphosphate carboxylase (or rubisco), which fixes atmospheric carbon dioxide into cellular carbohydrate. Under illumination, the enzyme acts as an oxygenase and incorporates oxygen instead – the phenomenon of photorespiration (see *Study guide I*, Chapter 7). This reaction has no apparent value to the organism and competes directly with the carbohydrate synthesis. Genetic engineering techniques might be used to modify rubisco to eliminate the oxygenase activity, thus greatly increasing photosynthetic yield of carbohydrate.

Energy

Biotechnology is likely to play an increasingly influential role in future energy production. The generation of fuels from renewable resources will become more important as supplies of fossil fuels diminish. Ethanol production from biomass is planned to expand in many countries, with vehicles running on pure ethanol becoming more common. As the economics of energy production change, other biological sources may become viable, for instance the bacterial generation of hydrogen.

Alternative biological systems could be examined in the search for new energy sources. The power of photosynthesis has yet to be harnessed directly to provide energy, though the use of plant biomass in fuel production can be regarded as an indirect exploitation of photosynthesis, in which the reduced carbon compounds formed are converted into fuels suitable for use by humans. In the future, direct conversion of solar energy by biological means may be possible, using the photosynthetic proteins of the chloroplast in artificial systems to incorporate carbon dioxide into carbohydrates, or to generate a current of electricity directly via the excitation of electrons during illumination.

Industry

Many chemical conversions using biocatalysts are possible, but not economically viable at present. However, the use of enzymes and microbes in industry is likely to increase as economies of production change and biotechnological techniques develop.

New enzymes are continually being discovered, some of which have properties with potential for industry. For example, the recently discovered extreme thermophilic bacteria may yield enzymes capable of withstanding the high temperatures and pressures used in some manufacturing processes. Biotechnologists intend to speed the process of finding new enzymes by improving on nature, using genetic manipulation to alter the amino acid sequence of the enzyme and thus change its catalytic properties. The specificity of enzymes could be improved, or changed to utilize the different substrates. The stability to extremes of temperature or pH, or to organic solvents, could also be improved. Such 'tailor made' enzymes could be designed for specific industrial processes.

Enhanced commercial availability of enzymes should also result from the continually improving techniques of protein purification. The search for new methods took biotechnology into space in 1983, when the American company McDonnell Douglas installed enzyme purification equipment aboard the space shuttle Challenger. The technique of electrophoresis, in which the protein is concentrated by movement in an electric field, suffers from gravity-related complications; great improvement was obtained from the gravity-free environment. The high cost was justified by the enhanced purity of the high-value product. Not all advances in enzyme technology are as dramatic, but the experiment emphasizes the progressive and ambitious nature of biotechnology.

The direct transfer of electrons between redox proteins and electrodes has already been exploited in biosensors (see section 21.4), and has led to the suggestion that proteins could be used in 'molecular electronics'. It is envisaged that protein molecules could be specifically immobilized onto an electron-conducting support, to form a network which could act in analogous ways to the silicon chip. The so-called biochip would probably be far smaller than today's microchips and could have far more complex functions. Other speculative biochip designs incorporate organic molecules, which can exist in two different states, interchangeable by the application of an electric field. Such molecules are envisaged as components in an artificial biological memory, with the two states representing 0 and 1 in binary arithmetic. However, such innovations are unlikely to be with us for many years to come. The biochip may be the ultimate in biotechnological speculation, but if research proves successful it could have more impact on the way we live than any other aspect of biotechnology.

Summary

1. Biotechnology is not new; biological processes and organisms have long been exploited by humans with great success (**21.1**).
2. The evolution of biotechnology follows a growth in the understanding of the underlying science and technology involved.
3. This can be divided up into three phases; firstly, one of descriptive observation (up to 1940) which corresponds to the period before and some time after the contribution made by Pasteur. Products such as alcoholic beverages, cheeses, yoghurt, vinegar and organic acids came from this period (**21.2**).
4. Secondly, there was a phase of greater insight and understanding (1940–1975) which brought about a period in which many new products, largely medicines such as antibiotics, emerged (**21.2**).
5. Thirdly, we are now witnessing a phase in which greater control of the science is possible. The use of genetic engineering techniques and monoclonal antibodies have led to products such as human insulin and diagnostic test kits (**21.3**).
6. In spite of these modern developments, it is interesting to note that in terms of money, classical products such as wine, beer, and foods represent more than 75 per cent of the turnover of all biotechnological products.
7. For the future, there is every indication that humans will continue their efforts to exploit biological organisms, systems, or processes to produce useful goods and services. In the near future, biotechnology is likely to contribute most to the pharmaceutical and fine chemical industries, while in the longer term this will extend to agriculture, energy, and more chemicals (**21.6**).
8. It should be remembered, however, that the future development of biotechnology may be influenced as much by social, political, and economic considerations as by scientific and technical progress (**21.6**).

Suggestions for further reading

DAY, M. J. Studies in Biology No. 142, *Plasmids*. Edward Arnold, 1982.

A Scientific American Book, *Industrial microbiology and the advent of genetic engineering*. W. H. Freeman & Co., 1981. (A useful and comprehensive collection of reprints on the subject of biotechnology.)

SMITH, JOHN E. Studies in Biology No. 136, *Biotechnology*. Edward Arnold, 1981.

WARR, J. ROGER. Studies in Biology No. 162, *Genetic engineering in higher organisms*. Edward Arnold, 1984.

WISEMAN, A. (ED) *Principles of biotechnology*. Surrey University Press, 1983.

CHAPTER 22 **METHODS OF REPRODUCTION**

22.1 **Why organisms reproduce**

Reproduction is the production of new individuals. In unicellular organisms all cells are capable of reproduction under favourable conditions. They reproduce by division of a fully grown cell to form two smaller daughter cells which can repeat the process, resulting in an increase in numbers (*figure 140*). Only lack of resources or adverse conditions, causing cells to die before they have reproduced, prevents this process from continuing indefinitely with disastrous consequences for other living things.

In multicellular organisms, only some of the cells are concerned with reproduction and the individual has a finite natural lifespan determined genetically by the controlled senescence of other cells. The lifespan may be only a few hours, as in the Mayfly imago, or over a hundred years, as

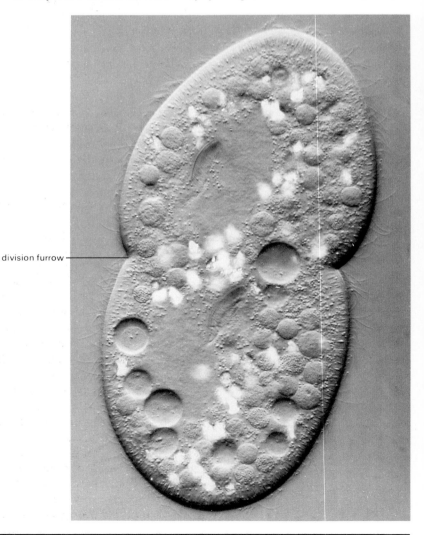

Figure 140
Simple fission in *Paramecium aurelia* (× 6600). The cytoplasmic structures replicate themselves before division to form two daughter cells.
Photograph, Dr David J. Patterson, Department of Zoology, University of Bristol.

Figure 141
The bristlecone pine, *Pinus aristata*, growing in the White Mountains of California. *Photograph, Charlie Ott/Bruce Coleman.*

in the giant tortoise, *Testudo gigantea*. The oldest recorded tortoise died as the result of an accident when it was known to be aged over 152 years. Individuals of some tree species, including *Pinus aristata*, the bristlecone pine of California (*figure 141*), are believed to have lived for up to 5000 years.

a *How do scientists estimate the ages of such trees?*

Eventually, even under the most favourable conditions, these trees die, like individuals of all other species, and only if reproduction replaces the adults which die will the species survive. Some individuals die as the result of environmental influences before completing their natural lifespan and frequently before reproducing themselves. Thus, additional progeny are required in order to maintain a constant population size.

Environmental factors that are favourable for some individuals but not others will vary in intensity with the time and place. When such pressures are low the population will increase, and when they are high the population will decrease. Which progeny survive to reproduce depends to a large extent upon how well the genotype, and consequently the phenotype, is adapted to the prevailing conditions. In stable uniform conditions those offspring most resembling the parental type are most likely to be successful, so reproductive mechanisms which conserve the parental genotypes are likely to be favoured. In contrast, in a changing or heterogeneous environment, parental genotypes may be at a disadvantage and new gene combinations may be more successful. In these circumstances successful reproduction must incorporate mechanisms to generate genotype variations.

Longterm evolutionary survival, then, requires a combination of, or a compromise between, two factors: the conservation of the parental genotype and the generation of new genotypes able to adapt to changing environments. Present day species must have met this requirement in the past, but not all will necessarily continue to do so in the future.

Forms of reproduction

There are essentially two forms of reproduction. *Asexual* reproduction depends entirely on mitotic cell division (see Chapter 15). *Sexual* reproduction also involves mitosis, but in addition includes two special events, meiosis (a modified form of nuclear division – see Chapter 16) and fertilization (the fusion of two specialized gamete cells), each occurring once at particular points in each sexual cycle. This apparently straightforward distinction between sexual and asexual reproduction is complicated by cases in which either fertilization or meiosis has been eliminated from the sexual cycle in one or both sexes. Examples of this are found throughout the plant kingdom and among the lower groups of animals. These life cycles, which form a very heterogeneous group, are sometimes classified as sexual, because they are modifications of complete sexual cycles; alternatively, they may be classified as asexual, because they do not have all the essential features of a sexual life cycle, especially when fertilization is omitted, or the consequences are genetically the same as in asexual reproduction. They may be placed in a third category, termed partially sexual or *subsexual*. Because of differences between plant and animal life cycles, botanists' classification of this intermediate group usually differs from that of zoologists. As these life cycles include some familiar, readily accessible, and instructive examples, they will be discussed later.

22.2 Asexual reproduction

Asexual reproduction depends upon the retention during development of embryonic cells with the full development potential, or *totipotency*, capable of giving rise to entire new individuals following mitotic cell divisions in the mature organism. In unicellular organisms, where the cell is the individual and usually all cells are capable of division, cell division *is* reproduction and thus mitotic cell division is asexual reproduction. In multicellular organisms, most mitotic divisions are part of the process of growth and development which produces cells of diverse types and specialized functions, not all of which retain the ability to regenerate complete organisms. Thus asexual reproduction is generally restricted in nature to fungi and plants, including the higher seed plants with their permanently embryonic root and shoot meristems (tissue formed of cells all capable of dividing and differentiating to form the plant body – see section 23.9), and to certain animals in the lower groups.

> **Practical investigation.** *Practical guide 6*, investigation 22A, 'Reproduction in fungi'.

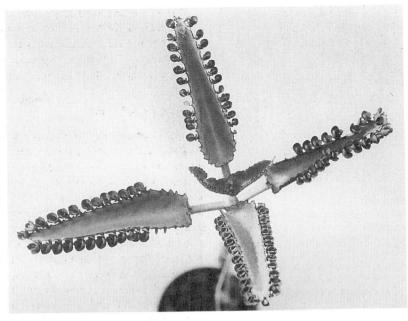

Figure 142
Three examples of vegetative reproduction in plants.
a Canadian pondweed, *Elodea canadensis*.
b *Bryophyllum* with plantlets on its leaf margins.
c *Fragaria*, the strawberry, with runners.
Photographs, Malcolm Fraser.

Two forms of asexual reproduction can be distinguished. The first, often referred to as 'vegetative' even in animals, occurs when new individuals are regenerated from somatic tissues of structures which are never directly involved in sexual reproduction. In some fungi, vegetative reproduction by asexual spores is common. Asexually produced spores are also found in some plant groups, such as the pteridophytes. Sometimes parts of the plant body may break off and establish themselves separately. Canadian pondweed (*Elodea canadensis*), of which only female plants are known in the U.K., has colonized much of the country by means of fragmentation (*figure 142a*). Some bryophytes, pteridophytes, and higher plants reproduce vegetatively by separating off a multicellular propagule. In angiosperms of the genus *Bryophyllum* (*figure 142b*) vegetative reproduction occurs through the presence of meristematic cells in the leaf margins. These generate small plants capable of independent existence when they fall off the leaf. In other angiosperms, such as the strawberry (*Fragaria*), lateral buds form runners which grow out from the parent, producing buds and roots and eventually independent plants (*figure 142c*).

Asexual reproduction is common in other plant groups, such as the grasses. A clump of grass may contain one original plant and a number of units, rooted tillers, more or less connected to it.

a *In such a case, how would you define an individual plant?*

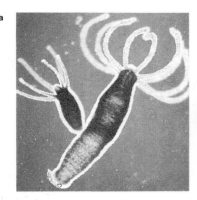
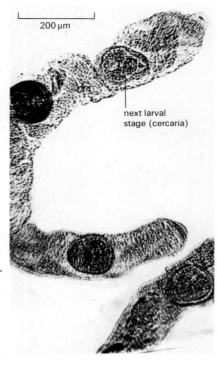
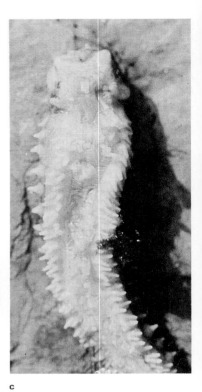

Figure 143
Asexual 'vegetative' reproduction in animals.
a *Hydra* with a bud.
b Larval stages (redia with cercaria) of *Clonorchis sinensis*, a liver fluke parasitic in humans.
c The starfish *Linckia* regenerating itself from one arm.
Photographs: **a** *and* **c** *Biophoto Associates;* **b** *Shaw, A. C., Lazell, S. K., and Foster, G. N.,* Photomicrographs of invertebrates, *Longman, 1974.*

Asexual reproduction is found in only a few animal groups. It occurs in the coelenterates, such as the freshwater *Hydra* (*figure 143a*) which, when food is abundant, reproduces by producing a bud which grows into a small replica of itself and breaks off. The Oriental liver fluke (*Clonorchis sinensis*), a parasitic platyhelminth belonging to the Trematoda, is an important parasite in humans in parts of the East. The fluke has a multiplicative phase when one type of larva, called a redia, produces, by asexual means, another type of larval form, the cercaria (*figure 143b*). A number of polychaetes, a group of marine annelid worms, also form new individuals by a process of budding. Some echinoderms either split into two, or, as in the starfish *Linckia* (*figure 143c*), the arms are cast off and each grows into a complete new individual.

In higher animals asexual reproduction of any kind is much rarer. An example is the production of more than one embryo from the products of a single zygote. Identical offspring produced by this means (see section 23.2) are comparatively rare in humans, but are the norm in some mammals. The nine-banded armadillo (*Dasypus novemcinctus*) normally produces four identical young (*figure 144*), and the twelve-banded armadillo may produce as many as eight.

The second type of asexual reproduction involves the formation of embryos and some of the structures normally involved in sexual reproduction, but without meiosis or fertilization. Thus in some *Citrus* species, including grapefruit, tangerine, and orange, each seed contains several embryos derived mitotically from maternal tissue. Such a system

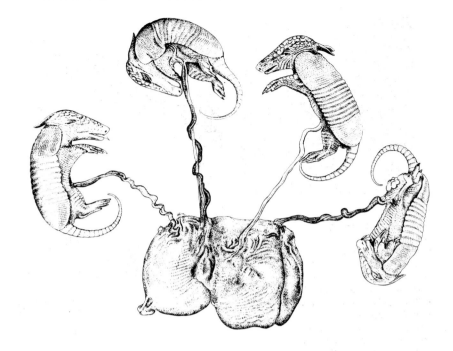

Figure 144
Four identical foetuses, each attached to its own placenta, derived from a single fertilized egg of the nine-banded armadillo, *Dasypus novemcinctus*.
From Benirschke, K. 'Fetal homeostasis', (ed. R. M. Wynn) N.Y. Acad. Sci. **1**, 237, 1965.

retains the dispersal mechanism usually associated with sexual progeny. In animals this type of asexual reproduction involves *parthenogenesis*, that is, the development of an egg without fertilization. In these asexual mechanisms, the egg is produced mitotically from maternal tissue, as in *Apis* (see section 22.4). In many species where all other somatic cells have lost their totipotency, this is the only mechanism for asexual reproduction.

STUDY ITEM

22.21 Asexual reproduction

a *Consider the examples of asexual reproduction illustrated in figures 142 and 143. What do all examples of asexual reproduction have in common?*

b *What are the advantages and disadvantages of asexual reproduction in the wild?*

c *Some plants reproduce asexually by vegetative means, as shown in the examples in figure 142. Others, such as Citrus, reproduce asexually by means of seeds with asexual embryos. What are the advantages and disadvantages of these two different types of asexual reproduction?*

d *Some domesticated crop plants are propagated for commercial purposes by means of vegetative propagation which occurs naturally in the species. Name three examples, and describe how they are propagated.*

e *Many other crop and ornamental plants do not normally reproduce vegetatively, but are propagated asexually by horticultural techniques.*

Name as many of these techniques as you can, in each case giving an example of a species which is commercially propagated by that means. What do all these techniques have in common?

[margin note: *all retain genetic info*]

f *More recently, some plants have been successfully propagated in vitro from meristem tissue, or cell cultures. In some cases, several identical plants have been regenerated from a single cell isolated from a differentiated tissue. What does this tell you about these plant cells?*

[margin note: *lge scale / slow seeds / hybrids*]

g *Why, and under what circumstances, are plants commercially propagated by asexual rather than sexual means?*

[margin note: *very early (identical twins)*]

h *Asexual reproduction occurs in vertebrates, including humans, but only rarely and at only one stage of development. When does it occur and why does it not occur at other stages?*

STUDY ITEM

22.22 The life cycle of aphids

Figure 145 outlines the complex life cycle of the black bean aphid (Aphis fabae). In the spring, dormant fertilized eggs on the spindle tree (Euonymus europaeus) hatch into wingless females. These produce, parthenogenetically from unfertilized eggs, winged females which migrate to secondary hosts, such as beans and sugar beet. On these hosts, parthenogenetic production of females continues, producing first wingless and later winged forms, the latter migrating to new hosts. During this period aphids cause serious damage to crop plants.

In the autumn, one generation consisting of males as well as females is produced, again parthenogenetically. The males are produced following the loss of one of the two sex chromosomes during the development of some of the eggs. As in the preceding generations, eggs retaining both sex chromosomes of the parent develop into females. Both males and females of this generation are winged and they migrate to spindle trees where wingless sexual females are produced which mate with the winged males. Both sexes produce gametes after meiotic cell division, but in the male only the sperm cells inheriting the sex chromosome develop; those without a sex chromosome degenerate. In the female, meiosis is regular in these sexual forms, and all eggs contain one sex chromosome. All zygotes thus contain two sex chromosomes and are therefore destined to become females. The female lays the fertilized eggs which remain dormant over winter.

[margin note: *economic way to produce lge no. of offspring to exploit an environment*]

a *What is the major advantage of reproduction by parthenogenesis in the black bean aphid's life cycle?*

[margin note: *reduces variation*]

b *Some species of aphid reproduce only by parthenogenesis. What might be the disadvantage of such a life cycle?*

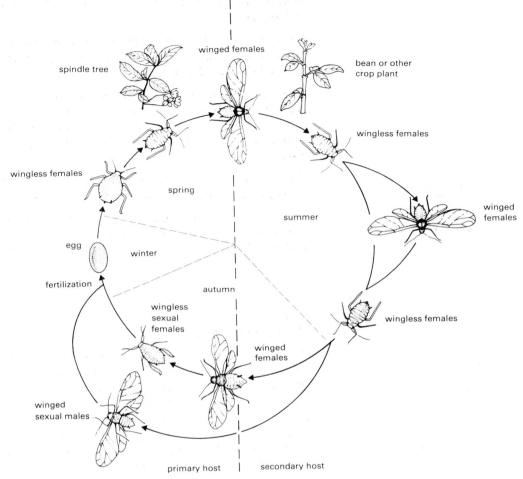

Figure 145
Life cycle of the black bean aphid, *Aphis fabae*. Apart from the production of fertilized eggs, all other generations are produced by parthenogenesis.
Based on Dixon, A. F. G., Aphid ecology, *Blackie*, Glasgow and London, 1985.

Isolate eggs

    c *How would you try to prove that an insect, such as **Aphis fabae**, was reproducing parthenogenetically?*

drop in temp / day length /

    d *What factors might cause sexually reproducing forms to appear in September?*

Quality of host / overcrowding
temp / day length / mature plants

    e *After a wingless aphid population has been living on a host plant for some time, winged forms begin to appear. What factors might promote the production of winged forms?*

    ☐ **f** *How might knowledge of the aphid's life cycle help in planning control measures?*

    The essential feature of mitosis, upon which all asexual reproduction is based, is that, except in the rare instances when mutation occurs, both daughter cells inherit the same nuclear genotype as the parent cell. This is ensured by the exact duplication of the genetic information when the DNA replicates, followed by the precise distribution on the spindle of equivalent complements of daughter chromosomes containing identical copies of the DNA. New individuals derived from mitotically produced cells are thus genotypically and, given the same environmental conditions, phenotypically identical (see section 15.2).

A population of individuals produced by processes which result in their possessing the same genotype is known as a *clone*. Clonal propagation may be successful in the wild. However, it is possibly a disadvantage for evolutionary survival in the long term (see section 22.7).

b Why do you think that there may be such a disadvantage?

c Why might it be, at least in the short term, a definite advantage?

Figure 146
Pioneering plant species.
a Bindweed, *Convolvulus* spp.
b Couchgrass, *Elymus repens*.
c Bracken, *Pteridium aquilinum*.
Photographs, Malcolm Fraser.

Asexual reproduction allows a successful individual to propagate rapidly to form a clonal population with a relatively restricted distribution. This enables selected genotypes to exploit an available site to the maximum. As a consequence, this is one of the reproductive strategies found among 'pioneering' plant species of open, disturbed habitats, including cultivated ground. Some of these colonizing species persist into later stages of vegetational succession and even invade established plant communities nearby. Examples of these particularly troublesome weeds include bindweed (*Convolvulus* spp.), couch grass (*Elymus repens*), and bracken (*Pteridium aquilinum*) (*figure 146*).

Asexual reproduction also provides a temporary reprieve for sexually sterile hybrids and polyploids which would otherwise not survive naturally beyond the life span of one individual. There are several examples of this, particularly in the plant kingdom. Among these are the *Citrus* hybrids such as tangelos, including the 'Ugli' and other hybrids of tangerine and grapefruit, and citranges, hybrids of citron and orange, which produce viable seeds with asexual embryos, like the parent species.

Clonal propagation has been exploited by humans, especially in the economically important angiosperms. Natural methods are used when available and easily manipulated, as with strawberry runners, potato tubers, and onion 'sets', but in many other plants, vegetative propagation can be induced by a variety of artificial methods. These usually involve removing part of the plant and either stimulating it to produce the missing parts, as when a stem cutting produces adventitious roots from the cut end, and when a root cutting produces adventitious shoots, or attaching it to an established root stock, as in grafting or budding. There are now laboratory techniques by which cultures of parent plant tissues can be induced to form large numbers of new independent plantlets (*figure 147;* see also Study item 23.21).

Figure 147
Artificial production of a clone; oil palm, *Elaeis guineensis*.
Photograph, Unilever Research.

22.3 Sexual reproduction

All sexual cycles have in common the alternation of fertilization with a meiotic division, together with many mitotic divisions. The mitotic divisions give rise to the multicellular phases or, in the case of unicellular organisms, increase the number in the population. Fertilization and meiosis have an essential role in generating and determining the level of genetic variation (see Chapter 16).

Fertilization is the fusion of two gametic cells. In most organisms, the fusing gametes are distinguishable, one being larger than the other. This type of fertilization is called *anisogamy* when both gametes are motile. However, the larger gamete in most plant and animal groups is non-motile and is called an 'egg cell' or 'ovum'; this type of fertilization is called *oogamy*. After fertilization by the smaller, and, except in most seed plants, motile, male gamete, the egg cell becomes the unicellular zygote. In some algae, such as *Fucus*, and aquatic animals, including many fishes and amphibia, the egg cell is released and fertilization occurs in the water. In some other algae, including *Spirogyra*, and in many animals, including most insects, reptiles, and birds, it is the zygote which leaves the protection of the parent organism.

In all other plant groups, and in viviparous sexually reproducing animals, such as scorpions, the adder (*Vipera berus*), the common lizard (*Lacerta vivipara*), sharks, and the common aquarium fish, the guppy (*Lebistes reticulatus*), and all placental mammals, the progeny are released at an advanced stage of development. In these, the zygote is retained and develops into an embryo protected by the tissues of the parent organism (see section 22.8).

a *What different meanings do biologists give to the term 'egg'? Which is the correct definition?*

Fusion of the gametes is a process called *syngamy*. The consequent fusion of their nuclei brings two sets of chromosomes together within the same nucleus, thus combining the genetic information. Frequently, but not always, the two fusing gametes are derived from different original zygotes, a process of cross-fertilization. As the gametes, like the individuals which produced them, are likely to differ genetically by having different alleles at many of the gene loci, a zygote produced by cross-fertilization will be heterozygous for those loci.

Later in the same cycle, meiosis will provide the mechanism for recombining these allelic differences in a range of new combinations, none of them perhaps identical with those which combined at fertilization. However, meiosis can achieve nothing in the way of genetic recombination if the parental zygote is homozygous at all loci. This can happen as a result of the fusion of genetically identical gametes, as after self-fertilization (see section 22.6).

avoids halving the chromosome complement in each generation. At the mitosis immediately preceding meiosis, the two anaphase chromosome groups reform into a single 'restitution' nucleus, now containing twice the chromosome complement, each chromosome present previously being now represented twice. At the subsequent meiosis, each chromosome pairs with its exact replica. The events of meiosis then follow normally, but as each bivalent consists of a pair of identical chromosomes, no recombination is achieved and the reduction of the chromosome number merely restores the original complement. In the animals, the eggs then develop parthenogenetically to form a new individual with the same chromosome complement as the parent. In plants, a gametophyte is formed which has the same chromosome complement as the sporophyte (see section 22.5). In the flowering plants, parthenogenetic development of an unfertilized egg usually occurs to give a maternal embryo in an otherwise normal seed. In the ferns, it is other cells of the gametophyte which, in the total absence of any female gametes, develop by mitotic divisions into a new sporophyte, a process known as apogamy. A relatively easily obtained example which shows such a cycle is *Dryopteris affinis*, a close relative of the male fern, *Dryopteris filix-mas* (*figure 149*).

In all these examples, the results of this modified cycle are the same. No recombination occurs at meiosis and no new heterozygosity is generated by fertilization. The genetic consequences are the same as in asexual reproduction but the organisms appear superficially to go through the processes of sexual reproduction, and in the case of plants retain the natural seed or spore dispersal mechanism. Many of these examples are interspecific hybrids, often triploid, which would otherwise be sterile.

22.5 Variations in life cycles

Leaving aside the obvious differences in structure between the various plant and animal groups, there are three basic types of sexual cycle, differing in the distribution of mitoses in relation to fertilization and meiosis (*figure 150*).

In one type, the *haplontic* life cycle (*figure 150a*), the cells of the haplophase, possessing the gametic chromosome complement, occupy every stage except the zygote. In the *diplontic* cycle (*figure 150b*), the

Figure 150
Basic types of sexual life cycles.
a Haplontic life cycle.
b Diplontic life cycle.
c Haplo-diplontic life cycle.

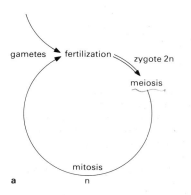
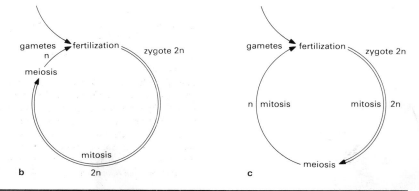

cells of the diplophase, possessing the zygotic complement, form the major part of the life cycle, excluding only the gametes. The final type is the *haplo-diplontic* cycle (*figure 150c*). Here, during the life cycle, the organism is found in both a haplophase and a diplophase. In plants the haplophase is known as the *gametophyte* and the diplophase as the *sporophyte*.

STUDY ITEM

22.51 Life cycles

Consider the haplontic cycle (*figure 150a*).

a When do mitoses occur? HAPLOPHASE / zygote divides meiotically

b By what type of cell division are gametes formed? MITOTIC

c *Chlamydomonas*, an alga, exhibits this type of life cycle. Which other organisms fall into the group of haplontic organisms? You will need to consult texts to find out. MUCOR

Now consider the diplontic cycle (*figure 150b*).

d When do mitoses occur in the life cycle? DIPLOPHASE zygote divides mitotical

e What type of cell division gives rise to the gametes? MEIOTIS

f What organisms exhibit this type of life cycle? Consult texts if necessary.

Finally, examine the haplo-diplontic cycle (*figure 150c*).

g When do mitoses occur? in both. zygote div. mitotically

h By what type of cell division are gametes formed? MITOTIC

i This type of life cycle is said to show an alternation of generations. Why? 2 W

j In some algae, such as **Cladophora** (*figure 151*), the two phases of the cycle appear exactly alike. In such a case, how would you set about deciding to which phase a given organism belonged? *presence of reproductive structures*

k Which groups possess this type of life cycle?

l What type of life cycle do the following organisms possess? Consult suitable textbooks.

Blatta orientalis	Marsilea vestita	Pinus sylvestris
Dryopteris filix-mas	Mucor hiemalis	Polytrichum commune
Homo sapiens	Pelargonium crispum	Rana temporaria
☐ Lumbricus terrestris	Pinnularia viridis (a diatom)	Spirogyra adnata
		Ulva lactuca

In some cases, such as *Cladophora* where the two phases of the haplo-diplontic cycle are morphologically indistinguishable, the alternation is

Figure 151
The green alga, *Cladophora* (× 350).
Photograph, Biophoto Associates.

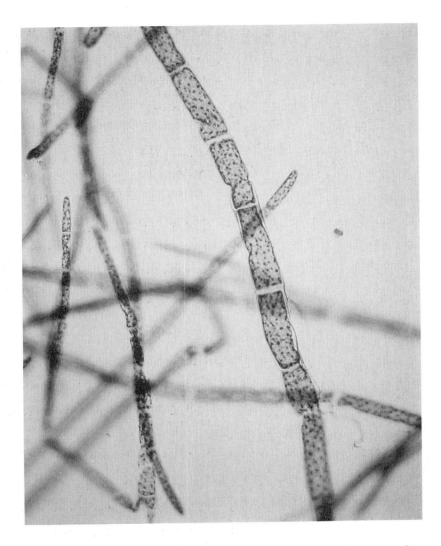

described as *isomorphic*. Most haplo-diplontic cycles are *heteromorphic*, meaning that there are conspicuous differences in size and morphology between the generations.

> **Practical investigation.** *Practical guide 6*, **investigation 22C, 'The life cycle of *Dryopteris filix-mas*, the male fern'.**

This raises fascinating questions about the control of development. The genotype must include information for controlling two distinct developmental sequences and one sequence must be suppressed, and its effects erased, when the alternative sequence is activated at the start of development of the next generation. It is perhaps significant that the two generation switches occur at the two unicellular stages of the cycle, the egg/zygote and the spore, and that both of these become isolated from surrounding cells and show major cytoplasmic changes at the time of the switch.

STUDY ITEM
22.52 Alternation of generations

In plant groups which have a haplo-diplontic cycle one or other of the phases is usually dominant.

a *Which is the dominant phase in each of the following groups? Consult suitable texts to check your answers.*

1 Angiosperms — flowering plant
2 Bryophytes — moss
3 Gymnosperms — conifer
4 Phaeophytes
5 Pteridophytes — fern

b *On the basis of what you already know about reproduction, what evidence is there to indicate whether the change in chromosome number, which usually accompanies the alternation of generations, is a causal factor, or even an essential feature, of the phase change?*

Electron microscope studies of several sexual angiosperms and ferns at meiosis and gametogenesis/fertilization have shown that three striking features are characteristic of the events accompanying the phase change:

1 The formation of an impermeable wall surrounding the unicellular spore or zygote produced.
2 A reduction and then a restoration of ribosome numbers.
3 A de-differentiation followed by a re-differentiation of plastids and mitochondria.

c *How might these features be involved in the phase change?*

Apogamy can occur naturally (see section 22.4), but it can also be induced in ferns which are normally sexual, like bracken, *Pteridium aquilinum*, by culturing gametophytes on an appropriate synthetic medium. The converse phase change from sporophyte to gametophyte by mitotic regeneration without meiosis, known as *apospory*, can also be induced experimentally in some ferns. Thus the first-formed leaves of bracken sporelings, if removed and placed in contact with the culture medium, produce outgrowths which develop into gametophytes. It is not known whether apospory occurs in ferns under natural conditions but it can occur spontaneously in some angiosperms.

d *Are the three features we have listed essential to the alternation of generations, or are they merely some of the several unique events involved in meiosis and gametogenesis? Use the information you have been given to plan investigations which would help to answer* ☐ *this question.*

Some pteridophytes and all seed plants are *heterosporous*, that is, they produce two types of spore from different types of meiotic cell. Usually they are in separate spore cases, or *sporangia*, often in separate cones or flowers, or even on separate individuals. The smaller spore, or *microspore*, develops into a male gametophyte, producing male gametes, while the larger spore, or *megaspore*, produces a female gametophyte, which in turn develops one or more egg cells. The change from homospory to heterospory has important implications for the breeding system (see section 22.7).

> **Practical investigation.** *Practical guide 6*, investigation 22B, 'Reproduction in *Marsilea vestita*, the shamrock fern'.

The most advanced plant cycles resemble most animal cycles, because animals are usually unisexual, producing gametes of only one type.

Another evolutionary trend has provided the reduced gametophyte with increased protection by retaining it within the megaspore, and then, in seed plants, within the *megasporangium* (the structure in which the megaspores develop). This has important consequences for plants because, unlike most animals, they are sedentary and require a dispersal phase which can exploit external agents in order to colonize new sites. The dispersal unit needs to be relatively small for easy transport, resistant to irradiation and desiccation in transit, and dormant; otherwise the plant may develop and rapidly lose viability before favourable growth conditions are available after landing in a suitable site. In the lower land plants, such as bryophytes and most pteridophytes, the main or only dispersal unit is the meiotically formed unicellular spore of the haplophase. Where, as in all flowering plants, the megaspore is retained within the sporangium, the spores cannot contribute to dispersal. This is because even the microspores, in this instance the pollen grains, though released and still dispersed by external agents, can only function if they land on, or near, the megasporangium of another plant of the same species. In these plants the role of dispersal has been taken on by an entirely new structure, the *seed*. This is relatively massive compared with a spore and consists of a young multicellular embryo developed from a zygote, surrounded by any remains of the gametophyte and the protective structures of the previous sporophyte generation. At the same time, the pollen grain has taken over from the male gamete as the agent for cross-fertilization. In most gymnosperms and all angiosperms, the male gamete is not motile; an external pollen vector transports the male gametophyte from one plant to another and the pollen tube then carries the gamete to the egg cell. This mechanism releases the plant from the dependence on free water for fertilization which occurs in bryophytes and pteridophytes.

> **Practical investigation.** *Practical guide 6*, investigation 22E, 'Events leading to fertilization in angiosperms'.

STUDY ITEM

22.53 **The evolution of life cycles**

The following plant groups are thought to represent steps in the evolution of green plants:

1 Angiosperms
2 Pteridophytes
3 Gymnosperms
4 Chlorophyta

a *Place the groups in sequence from primitive to advanced.*

b *What were your criteria for deciding the order?*

c *What evolutionary trends are revealed by this sequence in relation to:*
 1 *the development of the gametophyte relative to that of the sporophyte?*
 2 *the retention of the female gamete, the zygote/embryo, the gametophyte, and the spore within the structures in which they were formed?*
 3 *the requirement for water for the male gamete to reach the egg?*

d *What are the genetical and other implications of these trends?*

Examine the data from all the life cycles that you have studied.

e *What evidence is there that diplontic cycles have evolved from haplo-diplontic ones and haplo-diplontic cycles from haplontic ones?*

22.6 Gametogenesis and fertilization

In the sexual life cycle, gametogenesis always immediately precedes fertilization but meiosis, as we have seen, does not always immediately precede gametogenesis.

In diplontic organisms, including ourselves and other mammals, the products of meiosis develop directly into gametes. There is no intervening somatic development; thus, in heterozygous individuals, all the gametes may well be genetically different. Gametes fusing at fertilization will have come from different meiotic cells and in many cases from different individuals. This is because *meiocytes* (reproductive cells before meiosis) produce either male or female gametes, but not both, and most diplontic organisms have two distinct sexes.

In haplontic and haplo-diplontic organisms, including angiosperms and almost all other plants, the direct products of meiosis are not gametes but spores which divide mitotically and develop into multicellular gametophytes before some cells differentiate into gametes.

In the ferns and other plants which produce more than one gamete per gametophyte, this means that all the gametes from one gametophyte, and thus from one meiotically produced spore, are genetically identical, barring a rare mutation. Moreover, in a plant with hermaphrodite gametophytes, there is the possibility that those gametes which fuse at fertilization will also be identical (see section 22.4).

In angiosperms, the gametophyte has been reduced to the point where the female produces only one gamete per gametophyte and, although the male produces two, only one contributes to a zygote. Effectively, therefore, each functional spore gives rise ultimately to a single gamete. In pollen (male) meiosis, all four products of each meiosis usually develop into functional spores, but in the female meiosis, three spores usually abort and only one embryo sac (see page 239) is produced. There are therefore parallels with the majority of animals where only one of the four products of each female meiosis gives rise to an egg cell, but all four products of male meiosis form sperm. In both higher animals and plants there are many more male meioses than female. Wastage of the protected female gamete is much less than for the male gametes which have to transfer from one individual to another.

Gametogenesis in humans

Before you study the development of the fertilized egg to give a mammalian embryo and ultimately a new adult individual, it is most valuable to have studied the reproductive system of a mammal, in order to place the event in a proper context.

> **Practical investigations.** *Practical guide 3*, investigation 10A, 'The relation of the urinary system of a mammal to other systems of the body' (includes dissection of male and female reproductive systems); and *Practical guide 6*, investigation 22F, 'The structure and function of the mammalian gonads'.

Male
In the testis of the human male, each *primary spermatocyte*, or sperm mother cell, produces two secondary spermatocytes at the end of the first division of meiosis. Each secondary spermatocyte then divides at the second division of meiosis to form two *spermatids* (*figure 152*). There are many spermatocytes in each of the one-metre long seminiferous tubules and there are about 800 seminiferous tubules in each testis, accounting for much of their volume. This enormous output of spermatids, over 200 000 per minute, is maintained over many years because new spermatocytes are continually being differentiated from mitotically dividing spermatogonia. The spermatids, once formed, differentiate into mature spermatozoa (sperm). This takes place with the head of the developing sperm embedded in large Sertoli nurse cells which nourish them. When mature, about two days after meiosis commenced, the sperm are released into the cavity of the seminiferous tubule. The journey to the urethra then takes at least 12 days. A system of tubules leads from the seminiferous tubules to the coiled ducts of the epididymis lying at the back of the testis, where sperm can be stored for up to a month; then, if not ejaculated, they are absorbed. In the epididymis, cell debris and fluids accompanying the sperm are removed. During ejaculation the sperm travel from the epididymis along the vas deferens to the urethra. As they leave the vas deferens, seminal fluid is added. This fluid has a complex composition and is produced by the seminal vesicles, the prostate gland, and Cowper's glands. The transport of the sperm during ejaculation is achieved by peristaltic contractions of the smooth muscle in the testes and vasa deferentia as well as in the erect penis. The 200 to 500 million sperm ejaculated can be replaced in 24 hours.

Female

In the human female, only one of the four meiotic products of each *primary oocyte* develops into a female gamete (*figure 153*). At birth there

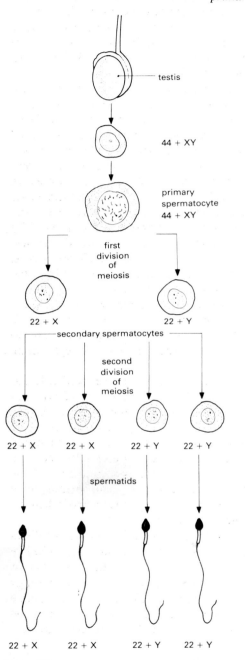

Figure 152
Diagram showing the stages of human spermatogenesis. The number of chromosomes at each stage is given, without the sex chromosome(s), which are shown after the number.
Based on Moore, K. L., The developing human, 3rd edn., W. B. Saunders, 1982.

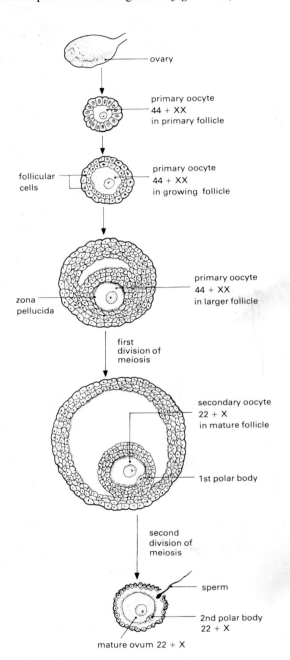

Figure 153
Diagram showing the stages of human oogenesis. The number of chromosomes at each stage is given, without the sex chromosome(s), which are shown after the number.
Based on Moore, K. L., The developing human, 3rd edn., W. B. Saunders, 1982.

are about two million oocytes present in the two ovaries, reducing to some 300 000 at maturity. Each oocyte, in its protective follicle of surrounding epithelial cells, enters the first division of meiosis during foetal development but stops during prophase before birth. Some of these oocytes and their follicles will resume development at some time between puberty and menopause. Throughout this period, all stages of development are present as new follicles begin to grow. Each month several follicles enlarge and vacuolate, and one, or occasionally two, of these eventually rupture while the rest degenerate. Just before the follicle ruptures, the primary oocyte completes the first division of meiosis in the nucleus, with an accompanying very unequal division of cytoplasm. This division produces a large *secondary oocyte* and a small 'polar body' with a complete set of chromosomes but little cytoplasm. The polar body does not develop much further and may not even complete the second division of meiosis. The sister cell, the secondary oocyte, in a protective jelly with a surrounding layer of cells, is released when the follicle bursts at 'ovulation', half-way through the menstrual cycle. Usually called the 'egg' (although strictly speaking, the female gamete does not mature fully until after fertilization), the secondary oocyte is carried the few millimetres to the Fallopian tube, down which it then slowly travels. Only if a sperm penetrates the surrounding protective layers and into the cytoplasm is meiosis completed. The second division of meiosis takes place as the sperm enters; another unequal distribution of cytoplasm then produces a large egg and a small second polar body, which is extruded and lost like the first. Shortly after the female gamete matures, the nucleus fuses with the male nucleus already present in the cytoplasm, and the zygote is formed.

Thus, oocytes which are eventually fertilized will have been held up in mid-prophase of meiosis I for at least 9–17 years prior to puberty and perhaps as long as 45–50 or more years if released just before menopause. It has been suggested that this interruption of meiosis might sometimes be implicated in the increased incidence with maternal age of certain congenital disorders due to chromosomal anomalies, such as Down syndrome, where an extra chromosome (number 21) is transmitted (see section 16.3). Perhaps the centromere/spindle interaction during the meiotic divisions is more prone to error when the cell has been held for a long time in first prophase.

Another feature of female meiosis is that the recombination of alleles of genes on homologous chromosomes due to chiasmata occurs before the mother's birth; in contrast, recombination of genes on different chromosomes occurs a few hours before fertilization (at 1st anaphase), or a few minutes afterwards (at 2nd anaphase). These events are unlike those in the male where all the genetic segregation and recombination events occur in rapid succession about two weeks before fertilization.

Gametogenesis in flowering plants

Male
In each developing anther of a flower bud, each of hundreds or thousands of *pollen mother cells* undergoes both divisions of meiosis in

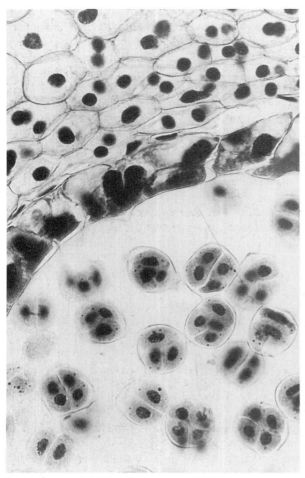

Figure 154
The tetrad stage of pollen formation in a lily anther (× 350).
Photograph, Biophoto Associates.

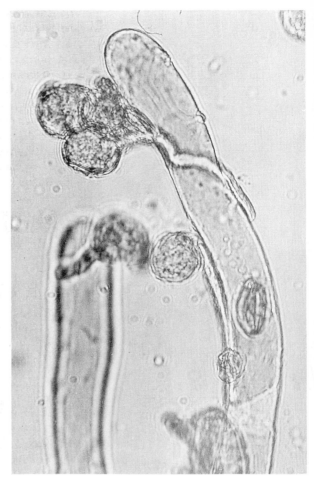

Figure 155
A photomicrograph of pollen grains germinating on a stigma (× 770).
Photograph, R. H. Noailles.

rapid succession to produce four similar microspore cells in a usually spherical tetrad (*figure 154*). After release from the mother cell and separation, each microspore or pollen grain develops into the male gametophyte generation, consisting at maturity of only three cells. The pollen grain then undergoes division by mitosis, pollen grain mitosis, with an unequal distribution of cytoplasm, while still in the closed anther within the bud. The larger 'vegetative cell' does not divide again but the smaller 'generative cell' does. The generative cell consists of a condensed nucleus surrounded by a thin layer of cytoplasm, a cell membrane, and sometimes a thin wall, and it eventually lies suspended in the cytoplasm of its sister vegetative cell. The generative cell divides mitotically once more to produce two similar male gametes with genetically identical nuclei. Like the generative cell, the gametes are not naked nuclei but complete cells with cytoplasm (though it is often limited in quantity and lacking conspicuous organelles), a membrane, and even a cell wall.

The generative cell division, like pollen grain mitosis, often occurs within the anther of a bud, as is the case in the buttercup (*Ranunculus* sp.), but in many other species, including members of the lily family, Liliaceae, it is delayed until the pollen has produced a pollen tube growing down a style after pollination (*figure 155*). Mature pollen grains are thus released when the flower opens, either as two-celled gametophytes (often misleadingly referred to as 'binucleate pollen grains'), or as three-celled gametophytes ('trinucleate pollen grains') including two male gametes. The pollen grains are not motile and depend on external agents to carry them to the receptive stigmas of other flowers. The gametes also show no signs of motility and appear to be carried passively by the tube cytoplasm from stigma to egg cell.

Practical investigation. *Practical guide 6*, investigation 22E, 'Events leading to fertilization in angiosperms'.

Female
In each developing ovule (immature seed) within the protective ovary (immature fruit) of a young flower bud is an *embryo sac mother cell*. This undergoes the divisions of meiosis, and cell wall formation, in rapid succession to form a usually linear tetrad of four cells, known as megaspores. In most species, three of these degenerate and the remaining

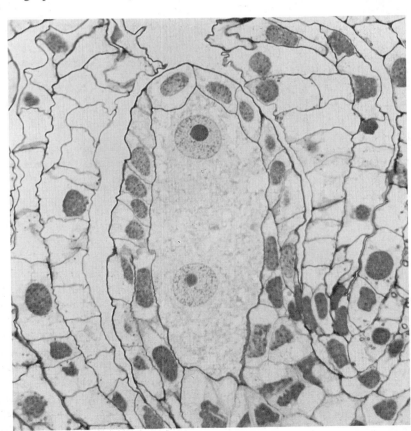

Figure 156
A photomicrograph of an embryo sac at the two-nucleate stage in a young lily ovule (× 1000).
Photograph, Biophoto Associates.

end cell undergoes three synchronous mitoses to form, in succession, a two-, a four-, and finally an eight-nucleate immature *embryo sac* (*figure 156*). Three of these nuclei remain at the anterior (micropylar) end, near the surface of the surrounding maternal sporangium tissue called the *nucellus*. They become separated by cell walls, to form a single egg cell and two, similar, synergid, or 'helper' cells. At the other end of the embryo sac, similar events produce three 'antipodal' cells. The remaining two nuclei migrate to a point between the egg cell and the centre of the embryo sac and remain, without individual cell walls, in the vacuolate cytoplasm of the 'central cell' which occupies much of the embryo sac volume. The mature embryo sac, the female gametophyte, is now ready for fertilization. The protective integuments have grown up round the nucellus leaving only a small pore, the *micropyle*, at the tip. The style and stigma have developed and become receptive, and the flower has opened to allow pollination.

STUDY ITEM

22.61 **The life cycle of angiosperms**

Figure 157 summarizes the whole sexual cycle of angiosperms, including the events shown in more detail in *figures 154–156*.

a *Comment on each of the numbered stages in figure 157. What other stages must occur between 10 and 0?*

b *In which phase of the life cycle does vegetative reproduction also sometimes occur? Is this the same in pteridophytes and bryophytes?*

c *Why is it incorrect to refer to a pollen grain as a male gamete?*

d *What happens to each of the components of the embryo sac after fertilization? You will need to consult suitable texts.*

e *In what ways is the endosperm unlike other tissues? Can you suggest any possible reasons for its unusual features?*

f *In evolutionary terms, what features of the angiosperm's life cycle reflect its position as the most advanced plant group?*

Sexual cycles in mammals

We have seen in the flowering plant that it is essential that pollen is liberated and then transferred to a mature receptive stigma of the same species. Similarly with animals, gametes must mature at the correct time with the development of associated structures. Thus, in mammals, changes in the reproductive tract must be synchronized for the implantation of the embryo into the uterine wall.

Most mammals have definite and controlled breeding seasons. They do not usually breed all the year round. Humans, as we know, are one of the exceptions to this.

a *What are the advantages of having a breeding season?*

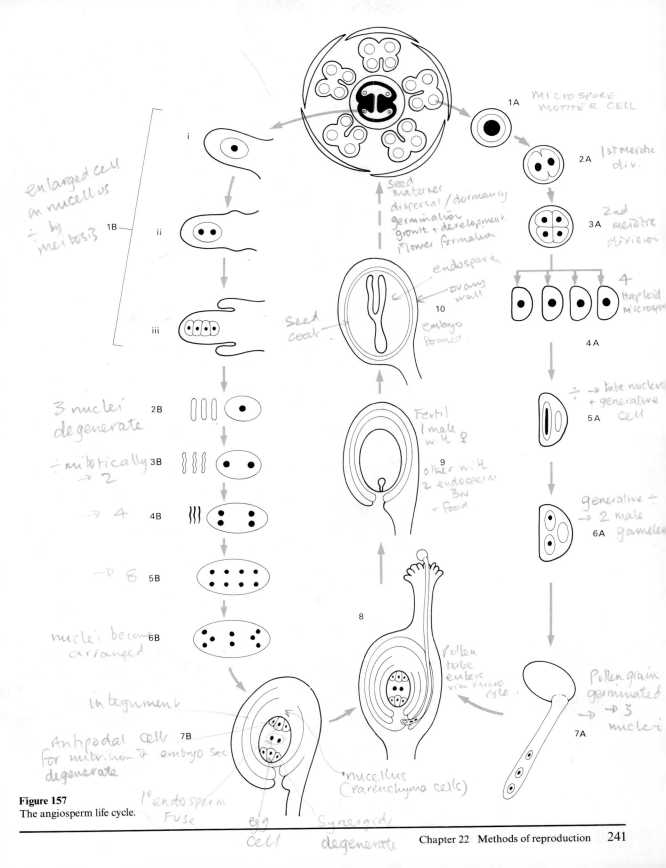

Figure 157
The angiosperm life cycle.

Chapter 22 Methods of reproduction 241

The major sign of the onset of the breeding season is that the female shows willingness to allow copulation to take place. This is the time when she is said to be 'on heat' or 'in oestrus'. Oestrus is the time when ovulation normally occurs, when one or more eggs are released from the ovary and begin their journey down the Fallopian tube. Thus, if copulation occurs at that time the likelihood of fertilization is high. Other internal changes must also take place in the female. For example, the lining of the uterus must be prepared to receive the fertilized egg, so that it can implant into the uterine wall and continue its development.

Some mammals, such as deer, wild sheep, and lions, normally only come into oestrus once a year, but this is often altered in captive mammals. Others, such as the domestic cat and dog, have two or sometimes three periods of oestrus in a year. Many rodents come into oestrus several times during the annual breeding season. Thus in the Mongolian gerbil (*Meriones unguiculatus*) the birth of a litter is normally followed within 24 hours by oestrus and mating. During the breeding season this results in a succession of litters at regular intervals.

Normally, ovulation is closely associated with copulation as sperm are usually only viable in the female reproductive tract for a day or so. Indeed, in rabbits, copulation stimulates ovulation, which follows about ten hours later. No wonder there is the saying 'to breed like rabbits'. An interesting exception to this association is found in many bats. Here copulation takes place in the autumn; the sperm are stored in the uterus during hibernation over winter, and ovulation and fertilization take place in the spring. In the badger (*figure 158*) and some other mammals there is another variation. Badgers mate in midsummer and the embryo

Figure 158
Two new-born badger cubs, eight days old, born at the normal time in February 1980, and a six-month-old badger born in September 1979, after its mother had been subjected to an artificial winter in a climate-controlled chamber. The mothers of these differently aged young were all mated at the same time in February 1979.
From Austin, C. R. and Short, R. V., Reproduction in Mammals. Book 2, 2nd edn, Cambridge University Press, 1982 (*redrawn from photograph by R. Canivenc and M. Bonnin*, J. Reprod. Fert., Suppl. **29**, 1981, p. 27).

grows to form a ball of cells or blastocyst (see page 270), but it remains at that stage until the end of the year. Only then does the blastocyst implant into the uterine wall and recommence development. The young badgers are born about two months later. This phenomenon is called delayed implantation.

b ***What are the advantages of the unusual reproductive cycles of the bat and badger?***

It is not yet known exactly what controls implantation, but a knowledge of the mechanism might have important implications. One of the major problems in the successful development of a 'test-tube' baby (see page 251) is the correct implantation of the embryo in the mother's uterine wall. In contrast, the so-called 'morning-after' contraceptive pill works by preventing implantation of the embryo.

STUDY ITEM

22.62 **The effect of males upon oestrus in female mice**

Mice are social animals and, as with other social mammals, the presence of adults of the opposite sex seems to have an effect on their reproductive physiology.

An experiment was carried out with twenty mature female mice of a strain called 'Q', which were caged in a stock box that had previously been scrupulously cleaned and left in the open air for 24 hours. In the first part of the experiment smears were taken daily from the vagina of each mouse. The smears were examined for the type of cells that they contained. The day on which oestrus occurred was identified because the smears on that day contained cornified epithelial cells (*figure 159*).

In the second part of the experiment, on the twelfth day, two males were housed separately in an inner compartment of the stock box and fed

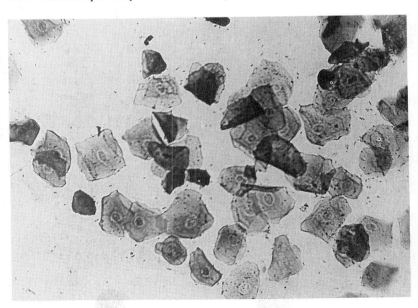

Figure 159
Vaginal smear of a mouse consisting of only cornified epithelial cells indicating occurrence of oestrus.
Photograph, M. R. Young.

Figure 160
A mouse stock box. At the top is the compartment in which the males were housed, with a separate water bottle and food supply. The area where the females were kept is in the foreground.
Photograph, Malcolm Fraser.

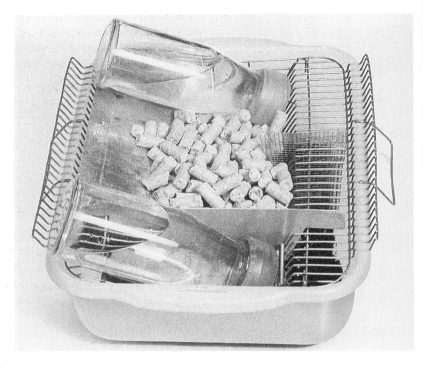

separately from the females (*figure 160*). The mice were given a day to adapt to the change in the experimental conditions, and then smears were taken again for a further 12 days.

In the third and last part of the experiment, the first set of conditions was restored by removing the males and recording the incidence of oestrus for a further 12 days.

The results of the whole experiment are summarized in *figure 161*.

Figure 161
Effects of social circumstances on oestrus. The horizontal axis of each chequered table represents a period of 12 days. The 20 females studied are represented on the vertical axis. A day when their vaginal smears were found to be cornified, indicating oestrus, is shown by a grey square.

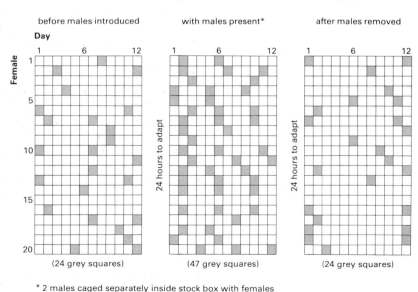

244 Inheritance and development

a What do you think was the value of the last period of the experiment when the males were removed and conditions were the same as at the beginning? Calculate the percentage of time for which the mice were in oestrus.

b What do these figures suggest is the effect of having:
1 a number of females together without a male?
2 males in the immediate vicinity of females?

c When males are present what do you notice in the majority of females about:
1 the timing of oestrus?
2 the frequency of oestrus?

d What hypothesis can you suggest to explain the effect of the males on the female oestrus?

e What modifications in the experimental procedure would you make to test your hypothesis?

f What practical benefit could the social effect on the incidence of oestrus have in managing a stock of mice for studies of development?

It is interesting that evidence has been found that synchronization of oestrus also occurs in humans. In the USA, college girls living together in a residence block were found to have much closer synchronization of menstruation (which follows oestrus) than would be expected by chance. This suggests that there may also be a female/female interaction.

In some mammals, apart from behavioural changes in the female, the time of ovulation is marked by other effects. Thus in some primates, such as the baboon, there are striking changes in the coloration of the female's face and buttocks. Another effect is seen, for example, in the female rhesus monkey (*Macaca mulatta*) which produces a chemical that acts as a strong sexual attractant to male rhesus monkeys. A chemical such as this, which is produced by an animal and affects the behaviour of another, is known as a *pheromone*. The use of scent by women appears to provide an obvious analogy, but almost certainly the phenomenon does not have such a simple explanation. In fact it has been suggested by Janzen (1983) that such fragrances, which are sometimes related to the smell of ripe fruits, may be a past memory of the time when certain plants evolved fruit fragrances which attracted humans to them as a source of food, and thus aided their seed dispersal. Further, it is possible that the recipient of one signal system, from the plant, might also be attracted to the signal system of the other, the female human, or as Janzen suggests 'a perfumed and painted member of our society is quite unconsciously a fruit mimic' – well, it's worth thinking about!

Menstrual cycle

In a major group of primates known as Old World monkeys or catarrhines (because they possess narrow noses), which includes apes and humans, oestrus occurs about every 28 days, assuming that the animal is not pregnant. Between each oestrus the uterus is prepared for the possible reception of an embryo. This process, repeated monthly, is known as the oestrous cycle. However, with humans, it is more common to speak of the menstrual cycle, since it is not usually easy to tell the time of oestrus, only of menstruation (see below).

During the menstrual cycle a number of physiological changes take place. First, one egg (rarely more) matures and is released from one of the ovaries. Second, changes occur in the uterine wall so that it is in a suitable state to receive a zygote. Third, should fertilization not occur, the lining of the uterine wall breaks down. The loss of the uterine wall lining, together with a certain amount of blood and mucus, constitutes menstruation.

These changes are mediated and controlled by hormones.

The major hormones involved in maintaining the menstrual cycle and the effects that they have, are summarized as follows.

1 A region of the brain, the hypothalamus, produces a hormone, the releasing hormone (RH). This hormone causes the anterior pituitary gland to secrete two hormones, follicle stimulating hormone (FSH) and the luteinizing hormone (LH).

2 The FSH affects the ovary and causes the development of special cells, called follicle cells (see page 237), which surround a primary oocyte. As a result, a structure called a Graafian follicle develops (*figure 162*).

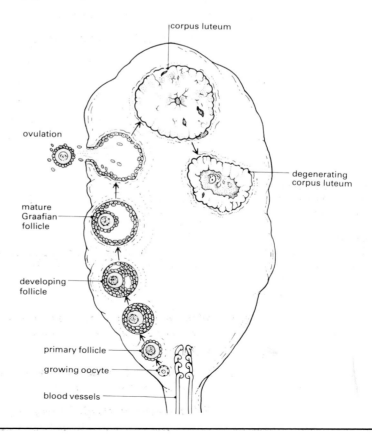

Figure 162
Diagram of a mammalian ovary showing the development of a primary follicle to ovulation, the production of the corpus luteum, and its degeneration in the absence of pregnancy.
Based on Austin, C. R. and Short, R. V., *Reproduction in Mammals*, Book 1, 2nd edn, Cambridge University Press, 1982.

246 Inheritance and development

3 From puberty to the menopause, the ovary secretes a background level of a fourth hormone, oestrogen. At this stage, some of the follicle cells secrete an additional amount of this hormone. The hormone's target organ is the uterus where it stimulates the development of the uterine wall, so that it is ready to receive a *blastocyst*. This is a hollow ball of cells formed by division of the original zygote.

4 As the concentration of oestrogen circulating in the blood reaches a critical level, it causes a surge in the production of both FSH and LH from the pituitary. This surge stimulates the Graafian follicle to release the oocyte, the process of *ovulation*.

5 After ovulation, LH stimulates the inner cells of the Graafian follicle to expand and fill the cavity left by the oocyte. They form a yellow body called the *corpus luteum*. LH causes the corpus luteum to produce a fifth hormone, progesterone, as well as more oestrogen.

6 Progesterone inhibits the production of both FSH and LH. The reduction of the latter hormone in turn causes the corpus luteum to cease functioning and to degenerate.

7 If a blastocyst implants in the uterine wall, the cells forming the outer wall of the blastocyst, called the *trophoblast*, secrete a hormone, chorionic gonadotrophin. This maintains the corpus luteum and its hormone production.

STUDY ITEM

22.63 **Control of the menstrual cycle**

Consider the information already given, together with that in *figure 163*.

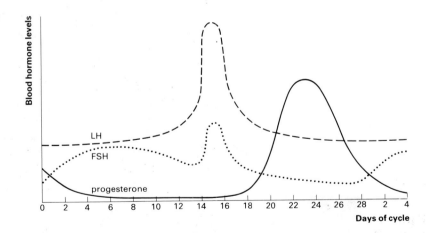

Figure 163a
Graph showing the changes in concentration in the blood of three hormones during a menstrual cycle.
Diagram a based on Tribe, M. A. and Eraut, M. R., 'Hormones', Basic Biology Course, Book 11, Cambridge University Press, 1979.

a *What may prevent the maturation of further eggs during:*
   1 an individual menstrual cycle?
   2 pregnancy?
b *If pregnancy does not occur, what may explain the drop in progesterone level at day 25?*
c *What are the effects of the drop in progesterone level?*

Figure 163b
Graph showing the body temperature changes in a woman over a menstrual cycle.

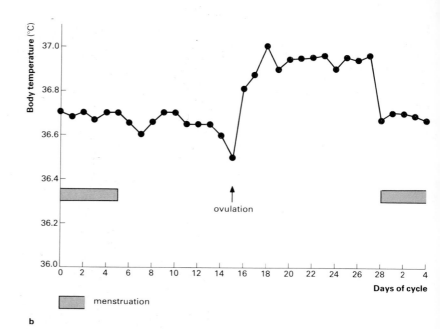

b

d **What information about the menstrual cycle could a daily record of body temperatures provide?** drops av start ovulation rises until menstruation then starts again

e **The hormone FSH has been used as a 'fertility drug'. How might it act?** Stimulates follicles to release eggs

f **Hormone preparations, taken orally, are used as contraceptive pills. What hormones might you use in developing a suitable pill? Why would they possess a contraceptive effect?**

progesterone + oestrogen stop production of FSH + LH

g **Since such hormone preparations can be taken orally, what does this fact tell you about the chemical structure of the hormones?**

h **At menopause, the ovaries cease the production of oestrogen (although they continue to produce a hormone, testosterone, which in the male affects the development of secondary sexual characteristics). What effects does this change have on the female? How could such effects be reduced?** male characteristics take oestrogen

i **Explain the menstrual cycle in terms of a feedback system.**

Events leading to fertilization in humans

The reproductive system of all higher mammals, besides producing gametes, also provides a suitable internal environment for fertilization and the consequent development of the embryo(s).

> **Practical investigation.** *Practical guide 3*, **investigation 10A**, 'The relationship of the urinary system of a mammal to other organs of the body' (includes dissection of male and female reproductive systems).

248 Inheritance and development

Figure 164
A human egg cell, a primary oocyte showing (top right) a polar body and the zona pellucida surrounding the oocyte.
Photograph, from the film 'The first days of life' by Guigoz, distributed by Boulton-Hawker Films Ltd. Copyright Guigoz-Claude Edelmann–Jean-Marie Baufle.

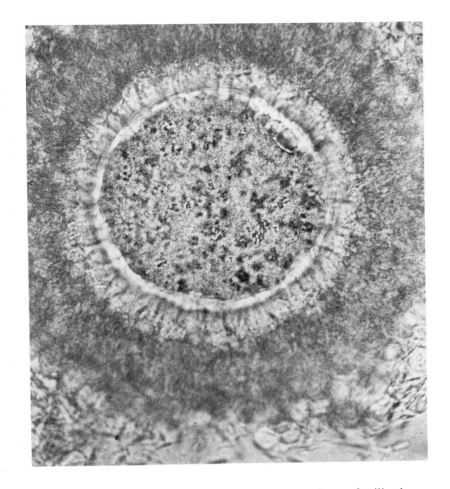

As in all mammals, every human being develops from a fertilized egg. This egg is the result of the fusion of a female gamete, the egg, and a male gamete, the sperm. The human egg is a large cell, about 150 μm in diameter (*figure 164*), and is enclosed in a protective capsule, the *zona pellucida*. The egg contains food reserves which nourish the embryo during the earliest stages of development. As we have seen, the egg is released at ovulation. Under normal circumstances one egg is released from the ovaries of a woman every month. Occasionally two eggs are released in the same month and, if both are fertilized, they may develop to give non-identical twins.

For fertilization to occur, sperm must be introduced into the genital tract at, or very close to, the time of ovulation. It is estimated that, on average, sperm can remain viable in the female genital tract for 24 hours after ejaculation, although some may live for as long as three days. Likewise, the lifetime of an egg after ovulation is limited. The egg can survive for 12–24 hours. If it is not fertilized within this time, it will die.

The scene for a successful fertilization is set as follows. Millions of sperm have been deposited in the vagina after ejaculation during copulation. The secondary oocyte ('egg') is travelling slowly down the Fallopian tube. The sperm travels a distance of some 15 cm to fertilize the

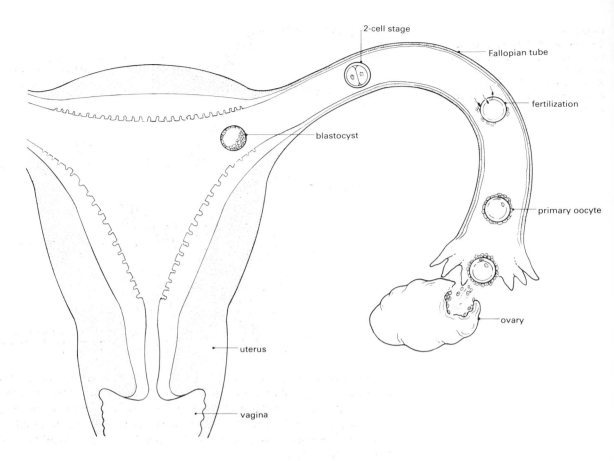

Figure 165
Diagram of the female genital tract showing the passage of the egg down the Fallopian tube and its position at fertilization.

egg. You can trace the route of this journey in *figure 165*. Travelling at speeds of up to 8 cm an hour, the journey can be completed in less than two hours. Of the 200–500 million sperm deposited in the vagina it has been estimated that only 200–300 reach their goal. However, if there is a reduced concentration of sperm in the ejaculate, infertility is likely.

c *Why do you think this is so?*

It is estimated that if, instead of the normal 100 million sperms cm^{-3}, less than 20 million sperms cm^{-3} are produced, then the man is likely to be sterile.

Normal fertilization involves the fusion of one sperm with the egg. During their stay in the female genital tract, the sperm are prepared for fertilization. This involves a change in the plasma membrane of the sperm and is called *capacitation*; it takes about seven hours. Capacitated sperm are ready for fertilization on accomplishing the journey to the egg. On contact with the zona pellucida and any cells from the ovary that are stuck to the zona, changes occur in the sperm head (see *figure 166*). The acrosome membrane fuses at several points with the overlying sperm plasma membrane and releases its contents extracellularly. The acrosome contents are enzymes that digest away the cement between the surrounding cells and also the zona pellucida and allow the sperm to

Figure 166
Diagram illustrating the acrosome reaction and sperm penetration of an oocyte.
Based on Moore, K. L., The developing human, *3rd edn, W. B. Saunders, 1982.*

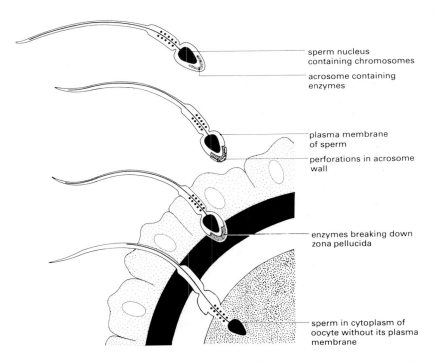

come into contact with the plasma membrane of the egg. Once contact is made between the plasma membrane of the head of the sperm and the plasma membrane of the egg, the two membranes fuse. Following fusion, the head and tail of the sperm enter the egg and the plasma membrane of the sperm is left as a patch incorporated into the plasma membrane of the egg (*figure 166*).

Immediately following the fusion of one sperm with the egg, changes take place in the egg plasma membrane to prevent further sperm fusing with the egg (polyspermy). Most of the information about the changes in eggs that block polyspermy has been obtained from studies on sea urchin eggs, which can be easily fertilized *in vitro*. These studies have shown that about one second after the fusion of the sperm with the egg, a weak block to fertilization, involving a change in the electrical potential of the egg membrane, has been established. This is followed after about 25–30 seconds by a permanent and strong block brought about by the fusion of granules in the outer, cortical, region of the egg and the egg plasma membrane.

Fertilization triggers a number of other events. In the human egg, the second division of meiosis is completed to give a fully mature egg and a tiny polar body which later degenerates. The nucleus of the mature egg is the female pronucleus and this fuses with the nucleus of the sperm, the male pronucleus. Thus the diploid set of chromosomes of the new individual is established.

Recent advances have made it possible to fertilize human eggs *in vitro*. The fertilized eggs are maintained in culture during the earliest stages of development. Then the very young embryos can be introduced into the uterus of the mother. If all goes well, such an embryo will implant and develop into a baby. By these techniques, women who were previously sterile due to problems such as blocked Fallopian tubes have a chance of having a baby.

d   *Write an essay describing, from a sperm's point of view, the process from its formation in the testis until it achieves union with an egg in the oviduct.*

Events leading to fertilization in flowering plants

Fertilization and the events immediately preceding it have been studied in much less detail in flowering plants than in animals, and little is known about their physiology. Fertilization has probably been studied most closely in the seaweed, *Fucus*, which releases large eggs into the water for external fertilization, and in the heterosporous 'fern', *Marsilea*, where fertilization in the single clearly visible archegonium on each small mega-gametophyte occurs soon after placing dormant sporocarps in water.

> **Practical investigation.** *Practical guide 6*, **investigation 22B, 'Reproduction in *Marsilea vestita*, the shamrock fern'.**

In angiosperms, after pollination on compatible stigmas, the pollen grains germinate to produce a pollen tube which elongates rapidly by growth at the tip. The vegetative cell cytoplasm is concentrated near the tip of the tube, where Golgi bodies synthesize the new wall material. The tube penetrates the surface of the stigma and then grows down through the tissues of the style, probably directed by a chemical gradient towards the ovules.

e   *How would you set up an investigation to find evidence for a chemical attractant in the stigma?*

> **Practical investigation.** *Practical guide 6*, **investigation 22E, 'Events leading to fertilization in angiosperms'.**

A short distance behind the tip of the pollen tube, and embedded in cytoplasm, lie the vegetative cell nucleus and the generative cell, or two male gametes if generative cell mitosis has already taken place. These move down with the cytoplasm until the tube tip reaches the micropyle of an ovule, enters the embryo sac within, and then releases the tube contents near the egg cell and polar nuclei (*figure 157*).

One of the two apparently identical male nuclei then enters the egg and moves across the cytoplasm to the egg nucleus and fuses with it. It is often assumed that the male parent contributes a nucleus and no cytoplasm to the zygote and hence cytoplasmically determined characters are said to be only maternally inherited. However, although it is still uncertain whether the male gamete cytoplasm regularly enters the zygote with the nucleus, there is convincing genetical evidence in a few angiosperm species, including *Pelargonium*, that chloroplasts are inherited from both parents. For most other species examined there is no unequivocal genetical or cytological evidence either way. The seedlings contain no paternal chloroplasts, but, unlike the case of *Pelargonium*, there are no detectable plastids in the male gamete's cytoplasm to be transmitted even if the cytoplasm does enter.

STUDY ITEM

22.64 **Inheritance of plastids in *Pelargonium***

Breeding experiments have been carried out on two *Pelargonium* cultivars, 'Flower of Spring' and 'Paul Crampel'. Both are variegated, with a striking white border to each leaf. This is due to mutant plastids which are found in the cells of the subepidermal layer of the shoot apex. These later fail to develop into green chloroplasts in the tissues, including the leaf margins, which develop from those cells of the shoot apex. The result is an orderly arrangement of mutant white tissues in an otherwise normal plant. Such a plant is known as a chimera (*figure 167*).

In table 34, W indicates the typical variegated form of each cultivar. Because the subepidermal layer of the apex gives rise eventually to the gametes as well as the leaf border, the variegated forms are equivalent to entirely white plants for breeding purposes. True albinos do not survive beyond the seedling stage. G indicates an entirely green plant produced as a cutting from one of the pure green shoots which arise spontaneously from green sectors of the cultivar. W and G plants are therefore genetically identical (isogenic) in all respects other than the chloroplasts.

Figure 167
Pelargonium cultivars.
Photographs, Harry Smith Horticultural Photographic Collection.

Cultivar	Cross (female parent first)	Seedling frequencies G	V	W
'Flower of Spring'	G × G	172	—	—
	W × W	—	—	5
	G × W	60	11	50
	W × G	88	24	2
'Paul Crampel'	G × G	97	—	—
	W × W	—	—	8
	G × W	108	139	83
	W × G	127	50	3

Table 34
Inheritance in *Pelargonium*. The frequencies of green (G), white (W), and variegated (V) seedlings after controlled crossings within cultivars of *Pelargonium* × *hortorum*. All crosses were made under similar conditions.

Germinated seedlings are scored as green (G), white (W), or variegated (V). Variegated seedlings have a mosaic of green and white tissues. Some variegated seedlings in time 'sort out' into pure green, pure white, or chimeras. Microscope studies reveal that some cells in variegated plants have both normal and mutant plastids. Further studies have shown that all three embryo types had an equal likelihood of survival in the seedling stage.

a   *Why do albino plants not survive beyond the seedling stage?*

b   *What can you conclude from the data in table 34 about the plastids and how they are inherited?*

c   *What can you deduce from these results regarding the inheritance of cytoplasm in* **Pelargonium**?

The transfer of male gametes to the female

In algae and the lower plants (bryophytes and pteridophytes), the motility of the liberated sperm cells in the surrounding water is sufficient to convey them to female gametes on different free-living gametophyte plants in the immediate vicinity. In some aquatic animals with external fertilization, gametic motility is again sufficient for fertilization of egg and sperm from different parents, once both kinds of gamete come into close proximity. In the higher plants and animals, the male gametes are either non-motile, or motile for only short periods and able to cover only short distances; thus they can function only within the maternal parent. This removes the need for free water for fertilization, an important step in the evolution of terrestrial organisms, but it creates the need for an alternative mechanism to achieve the cross-fertilization of gametes from different individuals. Cross-fertilization is necessary to generate the heterozygosity required to produce variability in the progeny.

For animals, this presents few problems. The male introduces active sperm directly into the female without exposing the gametes to damaging external conditions. In higher plants, immobile rooted individuals are usually at a distance from each other, and, even when they are close, there is no way that male cells can be transferred from one plant to another without the aid of an external agent. This means that the cells transferred are liable to be exposed to dry conditions, ultra-violet irradiation, and temperature extremes. The structure which has evolved to withstand these conditions during transfer is not the gamete itself but the pollen grain containing the reduced male gametophyte, quiescent within a thick and often pigmented wall. The external vector which carries the pollen grain passively from one plant to another may be wind, animals, or, in a few aquatic species, water. Although the spores of angiosperm ancestors were wind-dispersed and so too, as a general rule, are the modern representatives of these groups, the early angiosperms were probably pollinated by early insects as the two groups began to co-evolve. Subsequently, some angiosperms have reverted to wind-pollination, while others have evolved a closer relationship with more advanced insects of several different groups. In yet other angiosperms, more recently evolved animal types have been exploited, particularly the birds and bats, which are capable of flight and thus of rapid movement over considerable distances and between flowers on tall plants.

Wind is a reliable vector in certain open habitats but there is no way of influencing the direction of pollen movement. As pollen is only effective if it lands on the specialized receptor, the stigma, of a closely related plant, much is wasted, and therefore much has to be produced.

For animals to be exploited as vectors, they must be attracted to the flower. In most cases the attractant is food, in particular carbohydrate-rich nectar and/or excess, protein-rich, pollen. These are advertised by conspicuous colouring and sometimes also by scent. In a few plants, with highly specialized pollination mechanisms, for example the cuckoo pint, *Arum maculatum* (figure 168a), female insects of particular species are

Figure 168
a A section through the inflorescence of the cuckoo pint, *Arum maculatum*, which attracts insects with its scent.
b The orchid, *Ophrys cretica*, which mimicks a bee.
Photographs, Biophoto Associates.

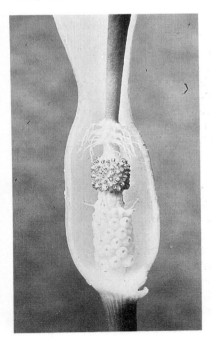

a b

attracted, not by food, but by the flowers mimicking the smell of the rotting carrion on which the insects would lay their eggs. Insects, especially certain species of midges, are attracted to the *Arum*'s flower spike by the dung-like odour it gives out and by its warmth, produced by a high rate of respiration of stored starch. On reaching the basal chamber of a newly opened spike where the female flowers are open, they are trapped, unable to climb out because of the hairs and the very slippery surface of the epidermal cells at the neck. On the following day, the stigmas become non-receptive and the upper, male flowers on the flower spike open and release pollen with which the flies become covered. The hairs at the entrance wither and the flies escape and may then visit another flowering spike where the female flowers are ready for pollination. Self-pollination is prevented by this mechanism. Other types of flowers adopt the remarkable strategy of attracting male insects by mimicking the female. This has been investigated in the orchid genus, *Ophrys*, where it was found that the flowers of each species attracted males of one or two particular species of bee or wasp which apparently attempted to copulate with the flower, acquiring pollen in the process (*figure 168b*).

However it is attracted, in moving from flower to flower, the animal vector unwittingly transfers pollen from the anther of one flower to the stigma of another. Some of the nectar and pollen-feeding animals feed indiscriminately on a wide range of relatively unspecialized flowers. In such cases, many pollen grains will finish up on stigmas of the wrong species, as well as on structures other than stigmas. In many other cases, close relationships have evolved between interdependent animal vectors and plant food sources. In these, more discriminating feeding behaviour, incorporating particular food preferences, is associated with a characteristic foraging behaviour and response to specific colours and scents. This has been reflected in a more specialized flower with corresponding size, shape, structure, colour, scent, and pollen or nectar output. Such specializations increase the chance of pollen being carried to the right part of the right flower. In the most highly evolved relationships, including those involving egg-laying and pseudo-copulation rather than feeding, a species of plant may be pollinated by no more than one or two species of animal visitor. In the American *Yucca* plant, for example, all species found east of the Rockies are pollinated by a single species of moth, *Tegeticula yuccasella*. The female moth transfers pollen to the stigmas while laying her eggs in some ovaries. These relationships are very efficient in that transported pollen has a very high chance of being conveyed to the stigma of another flower of the same species.

f *Why are these species with such specific mechanisms vulnerable? What modifications would reduce this vulnerability?*

Most of the visible features of a flower are associated with the pollination mechanism. The co-evolution of a plant with a particular vector has given flowers a characteristic *pollination syndrome* which often indicates the main type of vector involved.

Practical investigation. *Practical guide 6*, investigation 22D, 'Pollination mechanisms in angiosperms'.

STUDY ITEM

22.65 Pollination syndromes in *Trichostema* species

Table 35 provides information on five Californian species of *Trichostema*, a genus in the mint family (Labiatae). Two of these species are automatically self-pollinated and are not pollinated by any animal visitors. A further two species are obligate out-breeders, cross-pollinated by bees and other hymenopterans. The fifth species is cross-pollinated, mainly by humming birds.

Table 35 Floral characteristics of Californian species of *Trichostema*.

Species of Trichostema	Floral structure	Nectar volume (cm^3 flower^{-1} day^{-1})	Ratio of numbers of pollen grains: ovules per flower	Percentage seed set	
				with pollinating visitors	without pollinating visitors
T. lanatum [hummingbird ×+ self]	A lanatum	5.37 ± 0.33	2471 ± 159.1	51.9	45.8
T. micranthum [self]	B micranthum	0.0	137 ± 7.5	78.0	88.6
T. laxum [× obligate]	C laxum	0.95 ± 0.06	1422 ± 88.5	63.6	—
T. austromontanum [self]	D austromontanum	0.0	233 ± 15.0	90.2	87.1
T. ovatum [× obligate]	E ovatum	0.78 ± 0.01	1035 ± 41.2	93.2	—

Using the data provided, answer the following questions:

a How is each of the species **T. lanatum, T. micranthum, T. laxum, T. austromontanum,** and **T. ovatum** *pollinated? Give reasons for your conclusions*

b *How are the flower size, nectar volume, and pollen:ovule ratio related to*
  1 *breeding system (self-fertilization or cross-fertilization), and*
  2 *pollinating agent?*

c *What do your conclusions imply about the commitment of resources to pollination mechanisms?* [Selection favours economy of resources]

Successful pollination brings viable pollen to the receptive stigma of a compatible plant. The stigma has to be exposed, or at least readily accessible, in order to be pollinated by a visiting animal. The female gamete, on the other hand, has to be protected and is deep within the tissues of the ovule, surrounded, in angiosperms, by the ovary. This may itself be protected, for example, by being situated at the base of a long tubular flower or embedded in the flower stem. It is the style, which is also unique to angiosperms, which links the accessible stigma with the hidden ovule and provides the route for the pollen tube as it carries the non-motile gametes along the last short but vital stage of their journey to the egg.

Species recognition

If integrity of species is to be maintained, successful cross-fertilization in sexual reproduction must involve two individuals of the same species. Hybridization between different species is usually either unsuccessful or the hybrid produced is, like the mule, only very rarely capable of further sexual reproduction (*figure 169*). Exceptions only occur when species are very closely related and, though distinguishable, are not yet separated biologically, or when species have diverged during evolution because of geographical or ecological isolation rather than genetically determined barriers.

Unsuccessful hybridization represents wasted reproductive effort, and, in particular, when the barrier operates after zygote formation, a waste of female gametes. Mechanisms which prevent hybrid fertilization avoid this wastage. Such mechanisms exist and in many cases involve discrimination against individuals, tissues, or cells of another species. This implies the existence of a recognition process. In many higher animals, such a process operates at the level of the individual. Behavioural responses leading to mate selection mean that, at least in natural populations, it is rare for a male to mate with a female of another species. As a consequence, hybrids are uncommon.

In plants, there is little except distance and specialized pollination

Figure 169
The horse and donkey will mate, producing a mule. The mule is, however, almost always sterile.

mechanisms to prevent pollen of one species from reaching the stigma of another. This may explain, at least in part, why hybrids are more common in plants. There are, however, severe limits to the range of successful plant hybrid combinations. The fact that hybrids are not more abundant is in some cases due to rejection of foreign pollen by the stigma and style. In such instances, the growth of the alien pollen is halted, or at least slowed down, so that it fails in competition with pollen of the same species. This response implies the ability of the stigma or style to recognize and discriminate against foreign pollen. The underlying mechanism, which may resemble that of self-incompatability in some ways, is not well understood (see section 22.7). Nevertheless, it has sometimes proved possible to overcome barriers of this sort by artificial manipulations in order to make crosses between distantly related crop species, for example between wheat and rye, for the purpose of gene transfer.

STUDY ITEM

22.66 **Speciation in campions**

Red campions are locally abundant throughout Britain, growing on rich soils amongst established plant communities in woodland margins and clearings. White campions are common colonizers of open habitats on waste ground, cultivated land, and disturbed roadsides. Red and white campions have a similar geographical range through Europe and beyond, but their ecological ranges only overlap in wayside hedgerows, where they sometimes grow in mixed populations. In some of these, plants with pink flowers also occur. In a few populations, almost every shade from red to white is represented by at least one individual.

a *What possible explanations are there to account for the populations with pink-flowered individuals?*

b *How could you distinguish between your suggested explanations?*

The red campion flowers in early summer. It has no obvious scent and is usually pollinated during the day by long-tongued bumble-bees, butterflies, and flies. In white campions, flowering continues into mid- and late summer after the red campions have finished. When first produced, these evening-scented flowers are closed during the day, but open to be pollinated by night-flying moths in the evening. Both red and white campions are dioecious (they have separate male and female individuals). The two forms are interfertile, but the 'alien' pollen grows more slowly than it would on the stigma of a flower of the same colour. Further, although the progeny of the cross produce seed, non-viable pollen grains are more common than in the parent types.

Further information obtained in a study of campion populations in south-west England is given in *figure 170*.

c *What conclusions about the relationship between red and white campions can you draw from these data?*

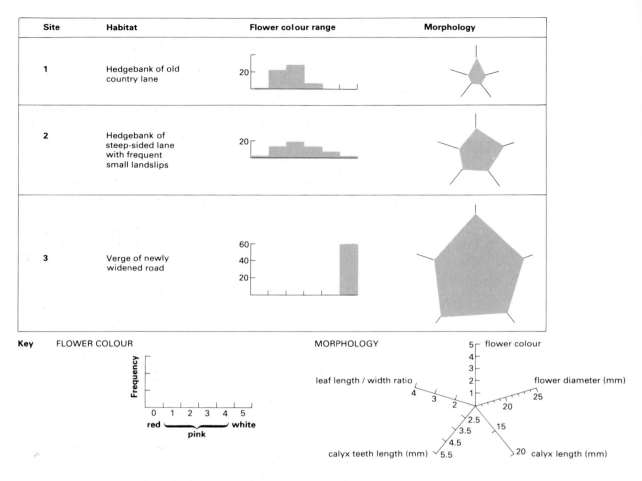

Figure 170
Data on campion population in south-west England.

d *Do you consider that red and white campions are two separate species, or two extreme forms of one species? Give your reasons.*

Similar phenomena, operating at the cell level but influencing events at the species level, also occur in other groups – for example, when fungal hyphae of different species meet in mixed culture, as in soil. Fusion of different strains of hyphae of the same species is often a necessary preliminary to sexual reproduction.

> **Practical investigation.** *Practical guide 6*, investigation 22A, 'Reproduction in fungi'.

The final example of a recognition phenomenon in sexual reproduction underlies all the previous examples. It is fertilization itself. Most cells formed during the development of an organism do not fuse with other cells, however close they may be to each other. Convincing examples of male gametes fusing with cells other than female gametes are rare. Clearly mutual recognition by the male and female gametes is a necessary precursor to fusion.

22.7 Breeding systems

The recombination of alleles made possible by sexual reproduction is the basis of variation and adaptation in later generations. However, wholly unrestricted recombination would not be advantageous. This is because, firstly, some of the novel genotypes would have unsuccessful combinations of alleles. Secondly, unrestricted recombination would produce a successful new genotype but would then dismantle it at the next generation. So conservation of much of the genotype must be necessary to maintain success. The chromosomes, each of which contains a large number of genes in a linear sequence, provide the means of preserving many advantageous combinations of alleles at functionally related gene loci. The maximum amount of recombination made possible by the events of sexual reproduction is, therefore, strictly limited, and even this level is frequently not achieved. The extent to which recombination is restricted varies with the species. The optimum amount of recombination, which is when the balance between conservation and change allows the maximum exploitation of available resources, will depend on the type of organism and its habitat.

Where conservation of the genotype is favoured, there are several ways in which variation can be severely restricted, or even eliminated. Thus asexual reproduction gives rise to clones of individuals which remain identical, apart from the consequences of accumulating mutations. In addition, asexual reproduction preserves any existing heterozygosity, with its attendant vigour. However, asexual reproduction is not found in most animals and in many plants with restricted life spans, such as annuals.

In sexual reproduction, reduction of variation can be achieved in two main ways: by reducing the potential for recombining heterozygous loci during meiosis and by reducing the level of heterozygosity established at fertilization. Recombination at meiosis is restricted when chiasmata are localized and few in number. It is also reduced when the total number of chromosomes in a haploid complement is reduced by structural rearrangements, thus cutting down the number of bivalents assorting randomly at meiosis.

Inbreeding, that is breeding between related individuals who therefore have at least some alleles in common, also reduces the level of heterozygosity. The low level of heterozygosity then, in turn, limits the amount of variation which can be generated subsequently at meiosis.

An extreme form of inbreeding occurs when two genetically identical gametes fuse.

a *What can you say about the genetic make-up of the zygote produced by such inbreeding?*

Such inbreeding happens in some species of fern. Subsequently, at meiosis there is no recombination.

b *What will be the characteristic of the genetic make-up of the gametes of the next generation?*

Continual inbreeding of this type will produce homogeneous populations of genetically identical progeny, as in asexual clones.

c *In what way will the products of such inbreeding be different in their genetic make-up from clonal individuals?*

This is what happens in homosporous fern plants when an egg is fertilized by a sperm from the same hermaphrodite gametophyte. Both gametes are formed mitotically from the same meiotic spore and are therefore genetically identical, thus producing 100 per cent homozygosity in one generation. Such occurrences need to be limited if any variation is to be produced in later generations.

In some species, strict conservation of the genotype is not advantageous, and the generation of some variation by recombination is favoured. The organism therefore has to be capable of outbreeding. This requires cross-fertilization between gametes of distantly related individuals which, while sharing the essential features of the species, will differ allelically at many loci and thus produce highly heterozygous zygotes. As most higher plants are hermaphrodite, a balance of both self- and cross-fertilization will allow for both the generation of new genotypes and the conservation of the most successful ones. The relative contribution of self- and cross-fertilization varies between species. It may favour outbreeding, even to the point of completely excluding self-fertilization (or selfing). This would occur in species where highly heterozygous individuals in heterogeneous populations are advantageous.

STUDY ITEM

22.71 Reproductive structures in flowering plants

Flowering plants can be grouped according to the type of reproductive structures they possess. The plant may be dioecious, monoecious, or hermaphrodite.

a *Find the meaning of these terms by consulting suitable texts.*

b *What is the effect on the frequency of selfing of:*
1 *being dioecious?*
2 *being monoecious?*
3 *being hermaphrodite?*

c *What characters or mechanisms found in hermaphrodite plants would reduce the likelihood of self-pollination?*

In practice the various factors involved may not prevent selfing, but they do influence the breeding system by affecting the relative proportions of self- and cross-pollination.

An entirely different mechanism exists in some plants in which physiological interactions between substances in the stigma and substances incorporated in the pollen grain determine whether germination and tube development will take place. The interacting substances are genetically determined, and when the substances in the pollen grain and the stigma are determined by the same allele, pollen growth is suppressed and fertilization prevented. When different alleles are involved, growth is normal. It is therefore another example of a recognition response. Such mechanisms provide a very effective means in a hermaphrodite flower of favouring cross-fertilization at the expense of self-fertilization, regardless of the pollination mechanism. In many outbreeding species where both cross- and self-pollination can occur, such as cabbages (*Brassica* spp.), it is the only means of suppressing self-fertilization. In such cases only carefully controlled pollinations will reveal the true breeding system.

STUDY ITEM

22.72 Self-incompatibility in plants

Table 36 shows the results of some breeding experiments carried out with marrowstem kale, a cultivated fodder-crop of *Brassica oleracea*, a species which also includes the cabbage.

Female as seed parent	Male as pollen parent					
	A 1	A 2	B 1	B 2	C 1	C 2
A 1	0.6	1.3	0.0	0.1	15.8	20.9
A 2	1.1	3.0	0.5	0.8	26.9	23.0
B 1	0.2	0.3	1.3	1.2	24.4	28.5
B 2	0.7	0.5	0.1	1.4	26.6	27.1
C 1	28.0	30.7	33.7	31.0	4.4	3.4
C 2	37.0	34.8	34.2	35.0	7.7	2.5

Table 36
The numbers of ripe seed per fruit produced after controlled pollination in *Brassica oleracea*. Each value is the mean of ten fruits. A, B and C are different genotypes, differing allelically at the S and other loci. 1 and 2 are replicate plants of each genotype.

The incompatibility response of the pollen and stigma is determined by alleles at one locus, S. In *B. oleracea*, this occurs in the parent sporophyte, so all pollen grains from one plant have the same incompatibility response regardless of the particular allele each grain inherits. Within the species there are many alleles, usually classified according to the dominance expressed in pollen and stigma development. Thus S_1 is dominant to S_2, S_2 to S_3, and so on. When the incompatibility phenotype is the same in the pollen and the stigma, most pollen tubes fail to penetrate the stigma, and fertilization occurs rarely, if at all. Thus self-fertilization (e.g. S_1S_2 female × S_1S_2 male) is prevented and most plants are heterozygous for S alleles. Crosses involving the same dominant allele (e.g. S_1S_2 × S_1S_3) are also incompatible. Occasionally, dominance is incomplete (codominance), or dominance relationships are different in pollen and stigma.

Each ovary contains 35–40 ovules. Compatible crosses yield 15–40 seeds; incompatible combinations produce 0–4 seeds. Pollination of stigmas before the flower opens with pollen of the same plant (= bud-selfing) yields 7–30 seeds.

Answer the following questions, using the data provided.

a Why does bud-selfing result in a good seed-set?

b How might bud-selfing be useful?

c Which of the genotypes included in table 36 are self-incompatible?

d In the crosses involving genotypes A, B, and C, which are the incompatible combinations?

e What can you say about seed-set in self-incompatible (SI), incompatible (I), and compatible (C) combinations?

f If only three S alleles are involved in total in genotypes A, B, and C, what are the three genotypes with respect to the S locus?

In some plants, such as the primrose (*Primula vulgaris*), more than one breeding system is used. In most populations of primrose about half the flowers are of one type and half of the other (see *figure 171*). In the first type, called pin-eyed, the style is long with the stigma at the mouth of the corolla tube. In the other type, thrum-eyed, the style is short and the anthers are at the top of the corolla tube. The reason for the name of the first type is obvious. The analogy behind the second name can be discovered by looking in a good dictionary. Other distinctions between the two types are morphological differences in the stigma surface and size differences in the pollen grains.

All these modifications would appear to ensure that only pollen from one type of flower is transferred to the other type. However, the real barrier appears to be a physiological one. Pollen placed on a mature stigma of the same type of flower only grows a few millimetres into the style before growth ceases.

Angiosperms are, and probably always have been, predominantly

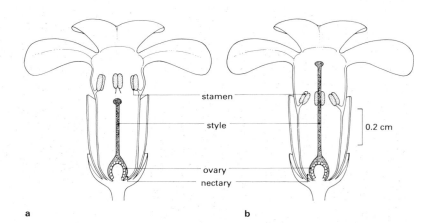

Figure 171
Sections of two types of primrose
(*Primula vulgaris*) flower:
a thrum-eyed.
b pin-eyed.
Based on Bell, P. R. and Woodcock, C. L., The diversity of green plants, 3rd edn, Edward Arnold, 1983.

hermaphrodite, and physiological self-incompatibility is thought to have arisen early in their evolution and provided the initial means of ensuring outbreeding. This in turn produced the variability essential for the development of the present diversity within the group. As they have evolved, some species have reinforced or replaced self-incompatibility with other outbreeding mechanisms affecting pollination, while others have lost it and become self-fertile, and even predominantly self-fertilized, as an adaptive response to selection pressures favouring conservation of the genotype.

In almost all animals, because they are diplontic, selfing between gametes originating from the same cell is impossible because each meiotic product produces only one gamete. Selfing between different gametes of the same individual occurs, but only rarely, because most animals are unisexual, male or female, and even hermaphrodites mostly cross-fertilize. In that respect, most animals are like either dioecious or self-incompatible plants, but the further regulation of the degree of inbreeding and outbreeding is achieved in a completely different way in animals. The closest degree of inbreeding possible in most species is sib-mating, and this is not uncommon in captive and domesticated animals. In natural conditions in many animals, both vertebrate and invertebrate, there is a social structure in the population which determines which are the breeding males, which are the breeding females, and what are the pairing relationships and family bonds, if any.

Most birds, for example are monogamous, although with many species, especially the passerines, such as sparrows, the relationship only lasts for one breeding season. Amongst other birds, including many seabirds, the pair bond may last much longer, perhaps for life. Turning to mammals, in the baboon (*Papio* spp.) the animals live in social groups (*figure 172a*) that are organized around a dominant hierarchy of males. The large dominant male of the group tends to mate exclusively with the high ranking females, but permits other males to mate with lower ranking females. In contrast, in the red deer, *Cervus elaphus*, the sexes live in separate herds for most of the year, except for the period in autumn,

Figure 172
Different types of social grouping.
a A group of baboons with male, females and young, grooming.
b A red deer stag with hinds.
Photographs: **a** *Clem Haagner/Ardea London.*
b *Hans Reinhard/Bruce Coleman.*

the 'rut', when the mature females are in oestrus (*figure 172b*). At that time the adult stags take up territories which they defend against rivals. The territorial stag attracts females in oestrus by roaring and by adorning his antlers with debris from vegetation. These different types of social structure affect the type of mating that occurs. The breeding system is thus determined, within the limits imposed by a diplontic cycle of unisexual individuals, by the influence of mobility and behavioural responses on mate selection, factors not present in plants.

Studies of plants have shown that there is a relationship between the level of recombination achieved by a species and the type of habitat or community in which it grows. Thus, there are circumstances in which a reduced recombination level confers an adaptive advantage by limiting variation in the progeny and thereby allowing maximum exploitation of an available habitat by a successful genotype.

22.8 Embryo protection and nutrition

Plants

An embryo, a young dependent multicellular sporophyte, protected by parental tissues, occurs in plants. It forms by the development of the zygote within the tissues of the gametophyte, which in the seed plants is in turn retained within the protective structures of the maternal sporophyte.

In bryophytes and pteridophytes, embryo development begins within the archegonium, the structure in which the female gametes are formed. There the embryo receives protection, water, and nutrients from adjacent tissues of the usually independent and photosynthetic gametophyte (*figure 173*).

In bryophytes, the small sporophyte remains attached to the gametophyte and continues to depend upon it for at least some of its water and nutrient supply. In the pteridophytes, the embryo is initially colourless and rootless and dependent on the gametophyte, but it develops into a complete, though small, green sporophyte by the time the

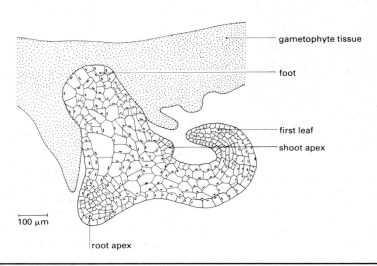

Figure 173
Pteridium aquilinum (bracken). A vertical section of a developing embryo attached to the gametophyte.
Based on Bell, P. R. and Woodcock, C. L., The diversity of green plants, 3rd edn. Edward Arnold, 1983.

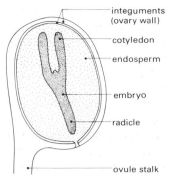

Figure 174
Vertical section of a dicotyledonous seed in the ovary.

gametophyte dies. In both groups, development of the embryo into a young sporophyte is uninterrupted under good growing conditions.

In the seed plants, the embryo develops within the gametophyte, the embryo sac, which in turn is protected within the ovule by tissues of the sporangium (*nucellus*) and ovule wall (*integuments*). In angiosperms, further protection is provided by the ovary wall surrounding the ovule (*figure 174*). Embryonic development continues until seed dispersal. When seeds ripen for dispersal, and all metabolism is reduced to a very low level, embryonic development temporarily ceases at a stage when rudimentary organs are present but independent existence is not yet possible. Development into an independent sporophyte is resumed when conditions are suitable for germination.

Compared with the unicellular spore, the dispersal unit in the other plant groups, the relatively massive seed requires the commitment of more resources. However, the number of seeds produced is, in general, less than the number of spores from spore-producing plants. The larger fern species may produce 10^7 to 10^{10} spores per year, depending on the size of the plant, whereas a herbaceous perennial angiosperm produces 10^2 to 10^5 seeds, depending on size and habitat. The highest numbers occur in large species growing on open sites, such as foxglove (*Digitalis purpurea*) which may have up to 100 000 seeds per plant.

a   *The fact that fewer seeds are produced per flowering plant than spores per fern of similar size suggests that the seed possesses significant advantages over the spore. What might be the advantages?*

The transition from an immature embryo to an independent seedling occurs without direct contact with the parent plant, but the embryo is nevertheless supported by a parental food source, as we are about to see. There are some parallels here with egg-laying animals. Moreover, just as some animals are viviparous (see section 22.3), giving birth to live young after eggs have developed within the parent, there are even a few viviparous plants, in which dispersal is delayed until after germination. One such is *Rhizophora*, a tree which is a principal component of many mangrove swamps. The embryo germinates while still attached to the tree and may grow to a length of a metre before it drops off the parent plant, into the swamp below (*figure 175*). Other examples are usually also associated with other extreme environments, such as high altitudes and latitudes.

Nutrition of the plant embryo
After fertilization, the zygote forms a more or less distinct filament of large cells, the *suspensor*, which carries the apical cell away from the narrow micropylar end and towards the centre of the embryo sac, in which the endosperm is developing. The developing embryo is directly dependent upon the parent plant for water and other nutrients. These are brought to the basal region of the ovule through the very small vascular strand, often a single xylem vessel and sieve tube, which runs through the ovule stalk. This vascular strand links with the vascular system of the

Figure 175
Stage in the germination of a seed of mangrove (*Rhizophora mangle*), whilst still attached to the parent plant.
Photograph, Philip Kelly/Oxford Scientific Films.

flower stem and thence the rest of the plant through the vascular bundles of the placenta of the ovary. These connections are clearly visible in a seed which has been specially treated (called clearing) for microscopical examination.

The embryo proper develops from the apical cell at the end of the suspensor. There is no direct vascular connection between the embryo and the small vascular strand ending at the base of the ovule. Studies of the physiology of embryogenesis have been few but it appears that, initially, nutrients reach the developing embryo via the suspensor and at least sometimes also directly through the nucellus and embryo sac or endosperm. At a later stage of development additional reserve materials may accumulate in the by then massive multicellular endosperm, as in the seeds of plants such as the onion (*Allium cepa*). Alternatively, the endosperm tissue may degenerate while the embryo is growing, when the reserve materials accumulate in the cotyledon(s) of the embryo itself, as in the seeds of the pea (*Pisum sativum*) and the broad bean (*Vicia faba*).

Animals

In egg-laying animals the eggs are often protected. Thus the eggs of the common frog (*Rana temporaria*) are covered with a thick layer of jelly which only disintegrates after the tadpole has hatched, probably as a result of enzymes secreted at that time. The desert locust (*Schistocerca gregaria*) is usually found in dry environments, and only lays its eggs when the soil conditions are right – damp, but not waterlogged. The eggs are further protected by being laid in batches and buried up to 13 cm below the soil surface, the female's abdomen being adapted to stretch and dig downwards. In the hen the egg is protected by a complex structure (*figure 176*), which in one sense acts as an 'artificial pond' in which the embryo can develop. The shell and egg membranes permit the exchange of gases, while at the same time filtering out bacteria and reducing the loss of water. Mechanically, a hen's egg is a very strong structure: try crushing an egg in your hand. It is virtually impossible to break a normal egg by this means unless you push the shell with one of your fingers.

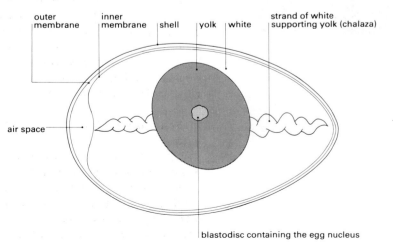

Figure 176
A section through a hen's egg to show its structure.

Figure 177
Some unusual parents.
a The Surinam toad, *Pipa pipa*.
b The African mouthbrooder (*Haplochromis burtoni*).
c A male sea horse (*Hippocampus erectus*) with brood pouch.
Photographs: **a** *Heather Angel;* **b** *Jane Burton/Bruce Coleman;* **c** *Des Bartlett/Bruce Coleman.*

In many animals, the eggs, after being laid, are cared for by one or both parents. This is almost universal in birds, except where, as in the cuckoo, the female utilizes foster parents, by laying her eggs in the nest of another species. In some animals the eggs, immediately on laying, are placed in a cavity of the body of a parent, there to continue development under the parent's protection. Sometimes this can be a most unusual situation. Thus in the Surinam toad (*Pipa pipa*), the eggs develop in honeycomb-like pits in the back of the female (*figure 177a*). In the fish, the African mouthbrooder (*Haplochromis burtoni*), the eggs develop in the female's buccal cavity (*figure 177b*), while in the sea horse (*Hippocampus* spp.), it is the *male* who retains the eggs in a brood pouch beneath his tail (*figure 177c*). Perhaps the most unusual of all is the gastric brooding frog (*Rheobatrachus silus*) from Australia where the whole of larval development takes place in the stomach of the female.

b *What are the advantages of such behaviour?*

The most important group of viviparous animals are the mammals. Apart from a small group, the Monotremata, which includes only the duck-billed platypus (*Ornithorhynchus anatinus*) and the spiny ant-eater (*Tachyglossus* spp.), all mammals are viviparous. However in the marsupials, such as kangaroos, wallabies, and opossums, the young are born at an early stage of development. They then migrate from the opening of the female reproductive tract to the pouch, or marsupium, firmly attach themselves to a nipple of a mammary gland, and complete their development. At this stage the embryo is tiny. In a wallaby, for instance, it may have a mass of one gram, about 0.00008 per cent of that of the adult female. Compare this with the human where the mass of a newborn baby would be 5 to 6 per cent of that of the mother.

In the advanced group of mammals, the Eutheria, which includes humans, the zygote contains very little food reserve, so it is essential that contact is rapidly made with maternal tissues to obtain nutrient. In humans, when the fertilized egg reaches the uterus the original single cell has already divided to form a hollow ball of cells, the blastocyst. The outer blastocyst wall, or trophoblast, is thin over most of the surface. Lying inside the wall at one point, is a group of cells, the inner cell mass, from which the embryo will develop (*figure 178a*).

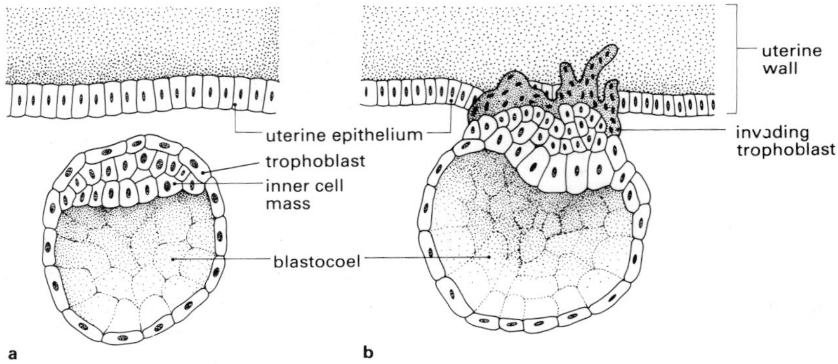

Figure 178
a A human blastocyst before implantation.
b Implantation of the blastocyst.
Based on Austin, C. R. and Short, R. V., Reproduction in mammals, Book 2, 2nd edn, Cambridge University Press, 1982.

A week after fertilization the blastocyst reaches the uterus. The cells of the trophoblast destroy cells in the lining of the uterus, so that the whole blastocyst burrows its way into the uterine wall. This process is called *implantation* (*figure 178b*).

The trophoblast forms a structure called the chorion. This encloses the developing embryo in a liquid-filled sac and also provides, because it is in contact with the uterine wall, a means by which materials can be exchanged between the embryo and mother.

c *What materials would need to be exchanged?*

At a very early stage information passes from the blastocyst to the maternal tissues that implantation has occurred.

d *How might such information be passed?*

Soon a more intimate connection develops between the embryo and the uterine wall. The associated chorion and uterine wall form a structure called the *placenta*. The embryo produces a bag-like structure, the amnion, that grows out within the chorion and in which the embryo grows and develops. The bag is filled with amniotic fluid and provides a protective environment against mechanical shock. As the amnion enlarges it fuses with the chorion. The embryo develops a blood

Figure 179
Section through a human placenta. *Based on Austin, C. R. and Short, R. V., Reproduction in mammals, Book 2, 2nd edn, Cambridge University Press, 1982.*

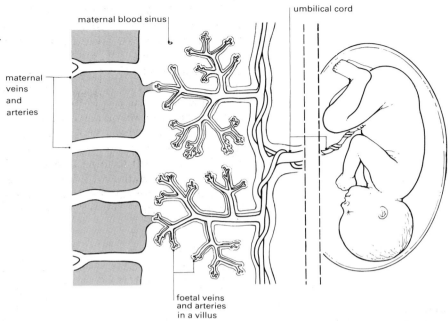

circulation to the placenta via the umbilical cord. The tissues of the uterine wall break down to varying degrees depending upon the mammalian group. In humans and other primates, the maternal tissues break down so that parts of the placenta are surrounded by pools (sinuses) of maternal blood (see *figure 179*).

An important point to appreciate is that, although the two circulations may come physically very close, there is never a direct connection between the embryonic and maternal blood systems.

e *In the human placenta what barriers remain between the embryonic and maternal blood circulations?*

f *Why would a direct connection between the two circulations be disadvantageous?*

The embryo develops in its carefully controlled environment until it is fully formed and it is then born into the harsher environment outside its mother.

Techniques are now available for studying the composition of amniotic fluid and of embryonic cells. Embryonic cells are sloughed off into the amniotic fluid and can be collected from samples of this. Alternatively, cells can be removed from the trophoblast. In the second case the procedure can be carried out as early as the tenth week of pregnancy. The cells can be subjected to cytological, histochemical, and biochemical tests. This procedure may be carried out when there is a fear that the embryo may have a genetic disorder such as Down syndrome, or a sex-linked disease such as muscular dystrophy or haemophilia. Analysis of the composition of the amniotic fluid can indicate the presence of open neural tube defects.

Summary

1. Organisms have a finite life-span and therefore must reproduce if the species is to survive (**22.1**).
2. The numbers of offspring that survive to maturity depend upon how well the genotype is adapted to the environment (**22.1**).
3. There are two major types of reproduction, sexual and asexual (**22.1**).
4. Asexual reproduction depends entirely upon mitotic cell division (**22.1**).
5. Sexual reproduction also involves mitosis, but in addition meiotic cell division and fertilization occur at some stage of the life cycle (**22.1**).
6. Sexually reproducing organisms possess different types of sexual life cycle depending upon the distribution of mitotic cell divisions in relation to fertilization and meiosis (**22.5**).
7. Fertilization involves the fusion of two gametes (**22.3** and **22.6**).
8. Sexual reproduction frequently enables genetic variation to be maintained in the population (**22.3**).
9. There are some life cycles, called subsexual, where fertilization or normal meiosis is omitted (**22.4**).
10. Gametogenesis immediately precedes fertilization (**22.6**).
11. In humans the process of gametogenesis is continuous in the adult male, but is partially completed at birth in the female (**22.6**).
12. In flowering plants, the pollen grain is the male gametophyte which produces male gametes. The female gametophyte develops within the tissues of the ovule (**22.6**).
13. In mammals, females are fertile at regular intervals, the periodicity being determined by the oestrous cycle (**22.6**).
14. The oestrous cycle (and the menstrual cycle in humans) is controlled by hormones (**22.6**).
15. For fertilization to occur in humans, the sperms must be introduced into the female genital tract close to the time of ovulation (**22.6**).
16. Normal fertilization involves the fusion of one sperm with one egg; a 'strategy' has evolved to prevent polyspermy (**22.6**).
17. The pollen grain requires a vector for transfer to the stigma (**22.6**).
18. In flowering plants, pollen grains germinate on compatible stigmas to form pollen tubes which grow towards the ovules (**22.6**).
19. Various 'strategies' have evolved to reduce interspecific hybridization. In animals these often involve behavioural responses. In plants, distance and specialized pollination mechanisms are important factors, as also are poorly understood physiological interactions (**22.6**).
20. More interspecific hybrids are found amongst flowering plants than amongst animals (**22.6**).
21. Some organisms have evolved breeding systems which favour conservation of a genotype rather than variation (**22.7**).
22. Strict conservation of a genotype is not advantageous in the long term as no environment is permanently stable. Too high a degree of variation can also be disadvantageous. Flowering plants reveal mechanisms that can favour one or the other, depending upon circumstances (**22.7**).
23. Plant embryos are protected with parental tissue (**22.8**).

24 The seed is a complex structure, often capable of surviving adverse conditions (**22.8**).
25 The embryo develops within the seed and is supplied with nutrients from the parent (**22.8**).
26 In egg-laying animals, the eggs are usually protected in some manner (**22.8**).
27 In many animals, the eggs are protected within their parents' bodies until a late stage of development (**22.8**).
28 In Eutheria ('placental mammals'), the embryo forms a complex organ, the placenta, through which exchange of materials with the mother takes place (**22.8**).

Suggestions for further reading

BARON, W. M. M. *Organization in plants*, 3rd edn. Edward Arnold, 1979. (Angiosperm life cycle.)

CLEGG, C. J. and COX, G. *Anatomy and activities of plants.* John Murray, 1978. (Development of the embryo sac, fertilization.)

GREEN, N. P. O., STOUT, G. W., and TAYLOR, D. J. *Biological science 2.* Cambridge University Press, 1985. (Reproduction and life cycles.)

GUILLEBAUD, J. *The pill.* Oxford University Press, 1983. (Straightforward popular account.)

HARRISON MATTHEWS, L. *The life of mammals*, volume 1. Weidenfeld & Nicholson, 1969. (Reproduction in mammals.)

HOLM, E. *The biology of flowers.* Penguin, 1974. (Very compact introduction to pollination mechanisms.)

KING, T. J. *Green plants and their allies.* Nelson, 1983. (Plant life cycles.)

MEEUSE, B. and MORRIS, S. *The sex life of flowers.* Faber & Faber, 1984. (Beautifully illustrated account of pollination mechanisms.)

ROBERTS, M. B. V. *Biology: a functional approach*, 4th edn. Nelson, 1986.

SIMPKINS, J. and WILLIAMS, J. I. *Advanced biology.* 2nd edn. Bell & Hyman, 1984. (Seed formation.)

YOUNG, J. Z. *The life of mammals: their anatomy and physiology.* 2nd edn. Oxford University Press, 1975. (Reproductive cycles in mammals.)

CHAPTER 23 THE NATURE OF DEVELOPMENT

23.1 The nature of development

The development of organisms exhibits both remarkable similarities and fundamental differences. In this chapter we are first going to consider some of the processes of animal development, and then to look at the major differences found in the development of animals and plants.

Almost all the animals we see about us come from a single cell, the fertilized egg.

a *What animals are exceptions to this rule?* Honey bee pg 28

How do the apparently similar and simple egg cells give rise to such diverse animals as worms, sea urchins, monkeys, and humans? In order to understand development we need to solve the problem of how the egg consistently gives rise to such different and complex organisms.

b *What is likely to be the major internal mechanism controlling the egg's development?* genes

During development, embryos acquire well defined organs such as eyes and limbs, all of which are made up of cells and their products. Thus the control of cell activities determines the form and nature of the various organs. While we wish, ultimately, to understand development in terms of molecular mechanisms, we do not yet have enough knowledge. However, if we could understand the changes and processes occurring at the cellular level then we could begin to consider their molecular basis. For example, cell movement and cell contractility play a very important role in moulding shape during development and considerable progress has been made in understanding the molecular basis of cell motility.

c *What will be the advantage to us of understanding the molecular basis of cell motility during development?*

Typical scheme of animal development

The process of development starts with fertilization and then proceeds through the following stages.

1 Cleavage. The egg divides a number of times. The pattern of early cleavage may follow a well ordered sequence, as in sea urchins (*figure 180a*), or a more variable one, as in mammals (*figure 180b*). In some animals, such as nematodes, annelid worms, and snails, the pattern of cleavage is remarkably constant and complex.

At the end of cleavage, the embryo typically consists of a few hundred to a few thousand cells arranged around a hollow interior; it is often called a *blastula*. Usually at this stage the cells all look alike. However, important changes have already taken place in the cells that make them different from one another.

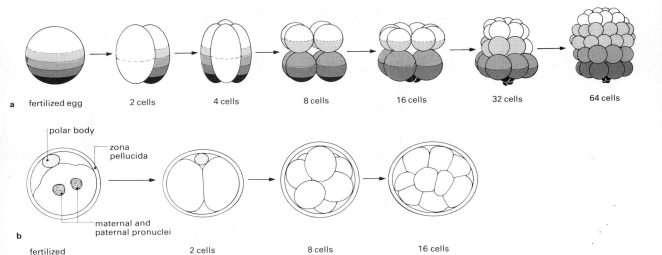

Figure 180
Diagrams to show **a** the ordered pattern of cleavage in a sea urchin embryo, and **b** the less precise pattern of cleavage in a mammalian embryo.
a *Adapted from Waddington, C. H.*, Principles of embryology, *Allen and Unwin, 1956.*
b *From Martin, G. R.*, 'Teratocarcinomas and mammalian embryogenesis', Science, **209**, *1980. Copyright © 1980 by A.A.A.S.*

2 Gastrulation is the next stage in higher animals. It involves considerable cell movements and changes in the spatial arrangement of the cells. Gastrulation establishes the three main body layers in appropriate positions for later development. On the outside is a cell layer, the *ectoderm*, that will give rise to the outer layer of the skin and to the nervous system. On the inside is another cell layer, the *endoderm*, that will form the lining of the gut. Between the ectoderm and the endoderm is the third layer, the *mesoderm*. The mesoderm gives rise to tissues such as muscle, cartilage, blood, and bone. Details of the way in which these three layers are established may vary from one group of animals to another.
3 Organogenesis follows gastrulation and is the process by which organs, such as the brain, eyes, and limbs, develop and the cells become obviously different.
4 Growth and increase in size occur when the main organs and structures have been laid down. This process is particularly clear in the development of vertebrates, and largely involves cell multiplication.

It is remarkable how, at least superficially, the early stages of development can resemble one another. In general, the early development of related organisms is similar; divergence occurs at later stages of development. If we look at Haeckel's drawings of the development of different vertebrates we can see that this is so (*figure 181*). Haeckel was, like many nineteenth century embryologists, interested in the relationship between embryonic forms and evolution. Up to stages well past the gastrula stage the basic body plan is very similar. Indeed, the embryo of humans passes through a phase in which structures similar to gill slits of more primitive animals are found.

Figure 181
Haeckel's drawings comparing the development of a fish, a salamander, a chick, and a human. In the early stages the basic forms are very similar. Note that the human embryo, at early stages, possesses structures similar to gill slits. *After Haeckel, E. The evolution of Man, London, 1879.*

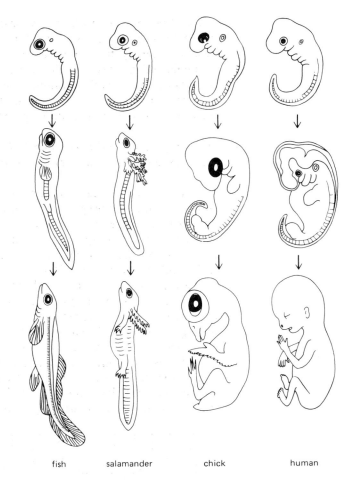

fish   salamander   chick   human

d *Why might this fact be of interest to evolutionary biologists?*

The main processes of development

Three main processes occur during the development of all animals from the egg. The first important process is the development of form, or *morphogenesis* (*morphe* is a Greek word meaning shape). This process is concerned with the physical forces which bring about change of form during development. In any body, whether living or not, for change in shape to occur, forces must be operating. In development these forces are provided by cells, often as tension brought about by cells contracting. As will be seen later, the neural tube, which will form the brain and spinal cord, starts off as a flat sheet, and contractile forces convert it into a tube.

The second process is *cell differentiation* – that is, the development of different kinds of specialized cells such as nerve cells, muscle cells, and red blood cells. These cells have very different structures and also contain different molecules. During the differentiation of red blood cells, for instance, the conjugated protein, haemoglobin, is made. This molecule is not made in any other kind of cell. Thus cell differentiation involves the synthesis of different molecules which characterize the different cell types. The most important amongst these molecules are proteins, particularly enzymes, and thus cell differentiation can be viewed as the process whereby the synthesis of particular proteins is initiated in particular cells. Since proteins are coded for in the DNA of the cell nucleus and the genetic information is the same in all embryonic cells, cell differentiation might be viewed as the activation of certain genes and the repression of others, as development proceeds.

e *How might a biologist set about finding experimental evidence that the genetic information is the same in all the cells of an embryo?*

The third process occurring in development is *pattern formation* – that is, the spatial organization of different types of cell. You can see at once that pattern formation differs from cellular differentiation by comparing your arm and leg. Both contain the same cell types – muscle, bone, skin, and so on, but their spatial arrangement is rather different. One way of looking at pattern formation is in terms of the problem of making a French flag. This consists of three stripes of blue, red, and white. Imagine a line of cells, each of which is capable of developing into a blue, red, or white cell. The pattern problem is to arrange that cells differentiate into a blue, a red, or a white cell in the right place, so that the line of cells form the striped pattern of the French flag, that is, one-third blue, one-third white, and one-third red (*figure 182*). In some ways pattern formation is a bit like painting, whereas morphogenesis is like clay modelling. The former involves assigning cells different states – like colours – while the latter involves moulding tissue.

Figure 182
The problem of how to generate the striped pattern of the French flag in a row of cells. The top line of cells can differentiate into blue, white, or red cells. The problem of pattern formation is to arrange that the first third of the cells differentiate into blue cells, the next third into white cells, and the last third into red cells (bottom line), to give the stripes of the French flag.

Cell activities

During development cells do a number of different things: they change the nature of the proteins that they make; they move from one place to another; and they exert contractile forces. Cells also change the stickiness of their surface and this, together with the forces that they exert, can play an important role in morphogenesis. They communicate with each other, thus allowing co-ordination of their activities, particularly in pattern formation. They also multiply, and this is associated with almost all developmental processes. This catalogue of cell activities is by no means complete but indicates some of the main activities of cells during development. It may seem surprising that the list is so small but it is one of the characteristic features of development that only a rather small number of cell activities may be involved. Often, what makes animals different from each other is how these activities are controlled in space and time. A useful comparison is with paper-folding (Origami). Although this only involves folding and unfolding and no differentiation, a remarkable number of different shapes can be built by changing the pattern and order of these activities.

One of the great challenges of research into development is to relate gene action to the structures that emerge during development. Genes essentially control the synthesis of proteins, and the causal sequence between the production of proteins and their effect on cell behaviour and finally on embryonic structure is a complex one and remains to be unravelled. However, as we begin, on the one hand, to understand cell activities in terms of gene action, and, on the other hand, development in terms of cell activities, we can begin to relate gene action to the forms that arise.

Understanding development is a central problem in biology. Not only is it of fundamental importance, but it is also the link between genetics and morphology. Development can be viewed as the process whereby genetic information is converted into body form. It is the conversion of the genotype into the phenotype. Studying development can help us to understand problems of great practical importance such as abnormalities, like spina bifida, that people are born with. These abnormalities are called congenital malformations or congenital defects.

f *What other examples of congenital defects have you come across? What do you know about their causes?*

SPINA BIFIDA
DOWN'S SYNDROME
SICKLE CELL ANAEMIA

23.2 Early development

As we have seen, in the early stages of development the fertilized egg undergoes a number of cell divisions, a process called cleavage.

a *How could you show that there was no overall growth of the embryo during cleavage?* Constant Mass

In animals the details of cleavage depend upon the amount of food reserve, or yolk, in the egg. In mammalian eggs, including human ones, and in eggs of other animals that also contain little yolk, the cleavage divisions divide the whole egg (*figure 180*). In contrast, in eggs that contain large quantities of stored yolk, such as those of birds and fishes, the cytoplasm streams to one pole of the egg and is there divided by cleavage divisions to give a cap of cells sitting on top of the yolk (*figure 183*).

> **Practical investigations.** *Practical guide 6*, investigation 23A, 'Tracing the early development of a nematode worm, *Rhabditis*', and investigation 23B, 'Embryonic development in shepherd's purse, *Capsella bursa-pastoris*'.

b *Most mammalian eggs contain little yolk, compared with the large*

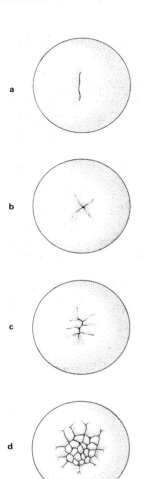

Figure 183
The cleavage divisions in the very yolky egg of a bird. The cytoplasm has streamed to one pole of the egg. Looking down on this pole, we can follow the series of cleavage divisions.
a The first cleavage divides the cytoplasmic cap only and does not involve division of the whole egg into two.
b The second cleavage divides the cytoplasmic cap into four.
c The third cleavage divides the cytoplasm into eight.
d Subsequent cleavages lead to the development of a cap of many small cells lying on top of the yolk.
Adapted from Saunders, J. W. Patterns and principles of animal development, Collier-Macmillan, 1970.

quantities of yolk in birds' eggs. What explanation can you give for this?

All the information required to make new individuals is encoded in the DNA of the fertilized egg. The question of how this information is used during development is a fundamental one. It can be investigated by taking cells from different stages of development and studying how much of an embryo each cell can produce.

The first cleavage division in mammalian eggs gives rise to two almost equally sized cells, or *blastomeres*. Each of these blastomeres contains all the information needed to make a new individual, as did the fertilized egg from which they arose. An experiment which shows this involves taking a mouse embryo at the 2-cell stage and destroying one of the blastomeres (*figure 184*). The remaining blastomere continues to develop and, when reimplanted into a foster mother, gives rise to a baby mouse.

Each blastomere at the 2-cell stage of human embryos can probably develop into a new individual too. The production of identical twins can

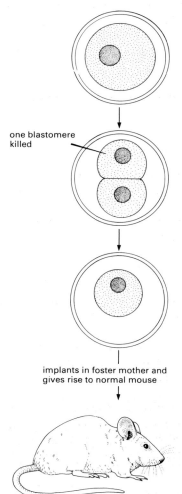

Figure 184
An experiment that shows that a normal mouse can develop from one of the blastomeres of a two-cell embryo.

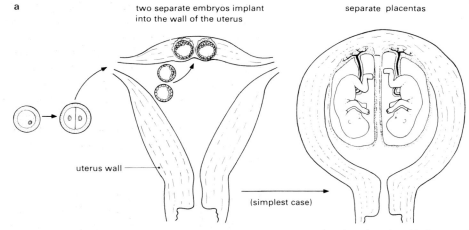
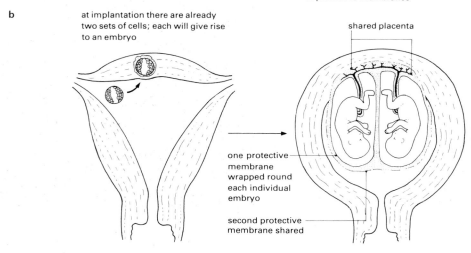
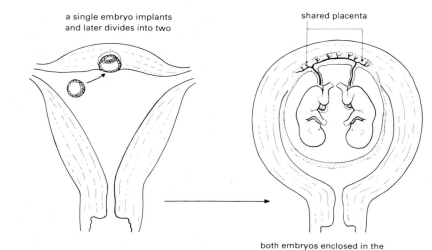

Figure 185
Three ways in which identical twins may arise.
a Splitting at the two-cell stage.
b Splitting later in development – this is the most common.
c Splitting still later in development.

sometimes be interpreted as being due to the separation of the blastomeres at the 2-cell stage. This is, however, the most rare type of identical twins. More frequently, identical twins appear to be formed by the splitting of the embryo at later stages of development (*figure 185*), since the embryos are wrapped in common protective membranes and may even share a common placenta. Indeed, in some cases, the embryo may not divide completely and 'Siamese' twins result (*figure 186*).

c *How could you detect whether human twins were the result of an embryo that had later split, or of the fertilization of two separate eggs?* impossible to prove scientifically

Figure 186
Eng and Chang, the original 'Siamese twins', after whom the term was coined.
Photograph, Mary Evans Picture Library.

The ability of daughter cells from later cleavage divisions to give rise to new individuals can be tested in mouse embryos. For technical reasons, the experiments do not involve implanting single cells at these later stages of development, but making mixtures of cells from embryos of different known genotypes, and then tracing the structures to which they give rise. It has been found that, up to the 8-cell stage, each individual cell has the capacity to participate in forming all parts of the embryo. However, between the 8- and 16-cell stages, cells in the embryo lose this capacity and become different from each other. Some of these cells can only give rise to the protective membranes that enclose the embryo and to the placenta.

STUDY ITEM

23.21 **Genes and cells**

A possible explanation for the differences between cells from the 16-cell stage mouse embryos, is that irreversible changes have taken place in their gene content, or alternatively in the genes able to function in that cell. This could be a general mechanism for establishing differences

Figure 187
A clone of *Xenopus laevis* produced by scientists at Cambridge.
Photograph from Gurdon, J. B., 'Egg cytoplasm and gene control in development', Proc. R. Soc. Lond. B., **198**, *1977.*

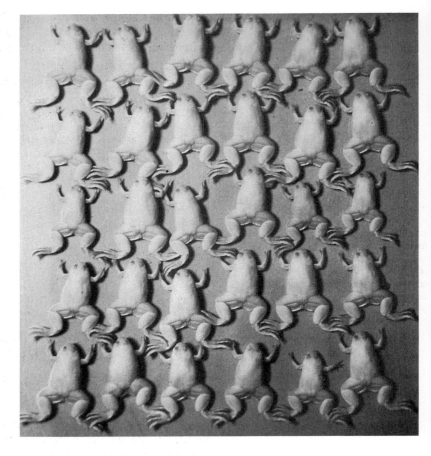

between cells and ultimately could result in specialization of cells. This possibility has been investigated using embryos of the toad *Xenopus laevis* (*figure 187*).

To answer the question of whether genes in specialized cells are irreversibly changed, nuclei of these cells were tested for their ability to support the development of an unfertilized egg. The nucleus of the egg

Figure 188
An experiment to test the ability of the nuclei of specialized cells to support development. In this example, the nuclei to be tested are taken from gut cells of a tadpole and injected into an enucleated frog egg.
After Gurdon, J. B., Gene expression during cell differentiation, Oxford University Press, 1973.

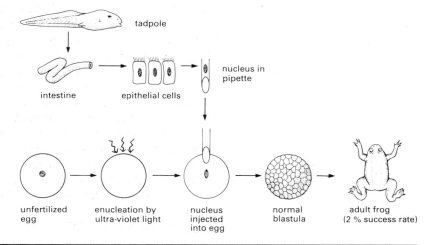

282 Inheritance and development

was destroyed by ultra-violet irradiation, and the nucleus of a differentiated cell was injected into the cytoplasm (*figure 188*). Table 37 gives the results of a series of such experiments.

		Per cent of total transfers reaching:			
		Tadpoles with functional muscle and nerve cells		Tadpoles with normal muscle, nerve, lens, heart, blood etc., cells	
Donor cells	Total transfers	1st transfers	1st and serial transfers	1st transfers	1st and serial transfers
Intestinal epithelial cells of feeding tadpoles	726	2½	20	1½	7
Cells grown from adult frog skin	3546	0.1	12	0.03	8
Blastula or gastrula endoderm	279	48	65	36	57

Table 37
The development and differentiation of embryos prepared by transplanting nuclei from specialized cells into enucleated unfertilized frog eggs.
Adapted from Gurdon, J. B. The control of gene expression in animal development, Clarendon Press, 1974.

a How do the nuclei from the different sources differ in their ability to support the development of the eggs?

b What do the results indicate about whether the genes of specialized cells are irreversibly changed?

The tadpoles that develop from such nuclear transplants will be identical to the individual which donated the tissue from which the nuclei were taken. This particular procedure of making identical copies of one individual is another example of cloning (section 22.2).

c What might be the advantages of this type of cloning in animals? What ethical problems could it pose?

Another possible explanation for the differences in the potential of individual cells from 16-cell stage mouse embryos is that, during cleavage, the cells inherit different regions of the cytoplasm of the egg. If there were differences in the cytoplasm, such that some regions contained substances that controlled the activities of the genes in the nuclei which came to lie within them, this could lead to differences in the behaviour of the cells. For the mouse egg, there is no evidence for the localization of specific constituents in the cytoplasm that could direct gene activity in this way. However, in sea urchin embryos, in which individual cells lose their ability to form complete individuals at about the same stage as the mouse (about the third cleavage division), the partitioning of the egg cytoplasm appears to be responsible for the differences that emerge between the cells.

For the mouse embryo, the position of a cell within the embryo appears to decide its future fate. By the 16-cell stage, some cells are clearly on the outside of the embryo and flatten against each other to enclose the remaining cells inside. The environment of the inside cells is then different from that of the outside cells. The cells on the outside of the embryo will go on to form parts of the placenta and the membranes that enclose and protect the embryo in the womb. The embryo proper will develop from cells on the inside.

STUDY ITEM

23.22 How do cells become different?

We can show that the future of the cell depends on its position within the 16-cell stage mouse embryo by means of experiments in which cells from different embryos are arranged in different positions in the formation of *chimerae*. These are animals that develop from embryos produced by combining genetically different cells from more than one embryo.

Genetic differences, such as coat colour, are used as markers. Thus one can take the cells of an 8-cell stage mouse embryo of a white strain and arrange them around the outside of an embryo at the 8–16 cell stage from a mouse strain with brown coat colour (*figure 189*). The composite embryo is then implanted into a foster mother.

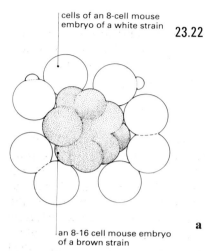

Figure 189
A diagram of an experiment to test the idea that the position of a cell in the mammalian embryo decides its fate.
Based on Hillman, N., Sherman, M. I. and Graham, C. F. 'The effect of spatial arrangement on cell determination during mouse development', Journal of Embryology and Experimental Morphology, **28**, *1972*.

a *What result would you expect if the position of a cell does determine whether it participates in forming the embryo proper or the placenta and membranes?*

b *What coat colour would the baby mouse possess?*

Another experimental system in which one might be able to answer the same types of questions about the mechanisms involved in making cells different is the development of teratomas. Teratomas are tumours that can arise spontaneously in the ovary or from cells of the testis. They contain undifferentiated cells that proliferate and give rise to many different cell types, such as muscle, bone, teeth, and so on arranged chaotically. One way in which we might find out more about how the

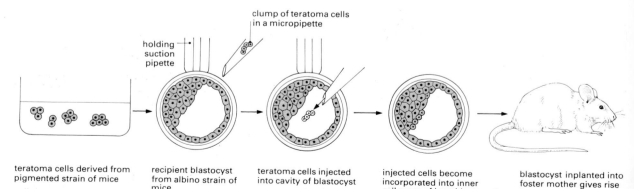

Figure 190
The experiment in which the behaviour of teratoma cells is tested in a normal environment.
After Alberts, B., Bray, D., Lewis, J., Raff, M., Roberts, K., and Watson, J. D., Molecular biology of the cell, *Garland, 1983.*

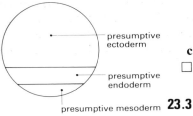

Figure 191
A fate map of a sea urchin embryo.
Based on Wolpert, L., The development of pattern and form in animals, Oxford Biology Readers 51, Oxford University Press, 1974.

control of differentiation in teratomas has broken down is to place some of the cells in a normal embryonic environment (*figure 190*).

c ☐ *What conclusions can be drawn about the behaviour of teratoma cells from this experiment?* Respond to cues in a normal embryonic environment + thus participate in forming normal structures — problems of cancer.

23.3 Development of the body plan

At the blastula stage, or even earlier, it is possible to mark cells and follow them during development in order to see what structures they form. One way of marking the cells is to colour them with a stain, such as Nile blue sulphate, which does not damage them.

a *Why do you think such stains are referred to as 'vital stains'?*

By marking the cells of the blastula in this way it is possible to construct a fate map. Fate maps show some remarkable features. If we examine that of the sea urchin (*figure 191*) we see that the cells which will give rise to three main layers of the body – ectoderm, mesoderm, and endoderm – are all, at this stage, on the outside of the embryo. So the future endoderm, which will give rise to the gut, an internal structure, is at this stage still on the outside. Similarly positioned is the future mesoderm, which is also an internal tissue. These areas are called *presumptive regions*, because the cells in each region will be expected, under normal conditions, to give rise to certain tissues in the later embryo.

It is the process of gastrulation that brings the regions into their proper positions. Gastrulation in sea urchin embryos involves the movement to the inside of the future mesoderm and endoderm. Because sea urchin embryos are transparent, it has been possible, using time-lapse cinemicrophotography, to follow the cellular activities during gastrulation (*figure 192*). The first event is the detachment from the *vegetal pole* (the region of the embryo where the cells contain most food

Figure 192
Diagrams, based on time-lapse film, of the cellular activities involved in gastrulation of a sea urchin embryo.
From Wolpert, L. and Gustafson, T., 'Cell movement and cell contact in sea urchin morphogenesis', *Endeavour*, **26**, 1967.

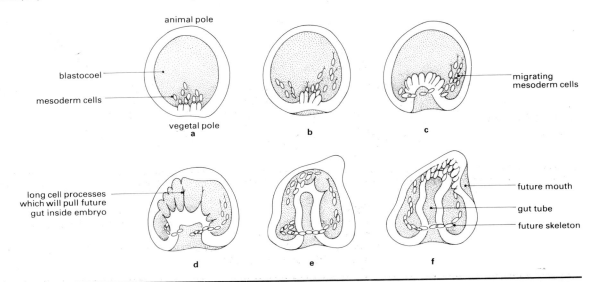

reserve), of about 40 cells, which will become mesoderm (*figure 192a*). They leave the wall and move to take up a quite specific pattern on the inner wall of the blastula (*figure 192a, b, c*). This pattern is determined by the pattern of adhesiveness on the inner wall. It is these cells that lay down the 'skeleton' of the sea urchin larva. Following their entry, the remaining cells at the vegetal pole, which are now presumptive gut cells, change their shape such that there is an inpushing (*figure 192b, c*). This invagination goes about one-third of the way across the inside of the blastula. The invagination is then pulled all the way across, by the cells at the tip sending out long processes which attach to the wall, contract, and so pull the future gut in (*figure 192d, e*). Eventually the tip of the gut fuses with the region of the future mouth (*figure 192f*). The gut is guided to this region by the pattern of adhesiveness in the wall.

In general, the main cell activities involved in gastrulation are changes in cell shape, adhesiveness, and cell migration.

We can also construct a fate map for the amphibian embryo (*figure 193*). In this case, there are particular problems in tracing the changes during gastrulation because the blastula wall is more than one cell thick and the movements of the individual cells cannot be directly observed.

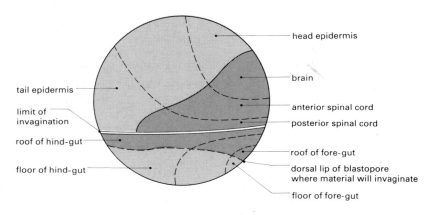

Figure 193
A fate map for the embryo of *Xenopus laevis*. At this stage, just before gastrulation, the embryo consists of two layers of cells surrounding a cavity. It is the fate of the cells in the outer layer that is shown in the diagram. The mesoderm comes from the inner layer of cells in the equatorial region of the embryo.
After Keller, R. E. 'Vital dye mapping of the gastrula and neurula of Xenopus laevis', Developmental biology, **42**, 1975.*

Practical investigation. *Practical guide 6*, **investigation 23C 'Morphogenesis in amphibians'.**

23.4 Tissue interactions and organogenesis

There are a large number of interactions during development, where one tissue influences the development of another. This is particularly prominent in the development of the ectoderm. One of the features of gastrulation in vertebrates is that the mesoderm comes to lie beneath the part of the ectoderm that will give rise to the neural plate (the neural plate folds over to form the neural tube, which in turn develops into the brain and spinal cord – see *Study guide I*, section 14.1). There is a very important interaction between these two tissues: the mesoderm influences the overlying ectoderm to form the neural plate. This process is known as *induction*. It is not yet known how the mesoderm induces the ectoderm to form neural tissues, but the effect is easily demonstrated. If a small piece of mesoderm from a gastrula is placed in the region of the future gut, a small neural tube will be induced in the overlying ectoderm. At later stages the local character of the ectoderm is also dependent on the nature of the underlying mesoderm. For example in birds, if the mesoderm from a feather-forming region is combined with ectoderm which would normally form, say, the cornea of the eye, feathers are formed. A dramatic example of such interactions can be found in the development of teeth. The enamel of teeth is secreted by ectoderm, whereas the rest of the tooth is of mesodermal origin. If the ectoderm that normally forms the epidermis of the skin of the back of a mouse is combined with the mesodermal tooth germ, the back skin ectoderm will be induced to secrete enamel.

23.5 Growth

The basic organization of most organs is laid down when they are very small. For example, when the humerus is laid down it is probably not more than about half a millimetre long. It must thus grow about 1000-fold to reach its adult size. This growth is brought about mainly by cell multiplication, but the secretion of extracellular materials such as cartilage and bone matrix is also very important. The main structure involved in the growth of the long bones like the humerus, radius and ulna, femur, and so on, is the *growth plate*. The growth plate is a cartilaginous structure, a few millimetres thick, that is found near both ends of the long bones. It is here that the lengthening of the bone occurs. Bone matrix itself is a hard extracellular material that cannot bring about increase in length. By contrast the cells in the growth plate divide

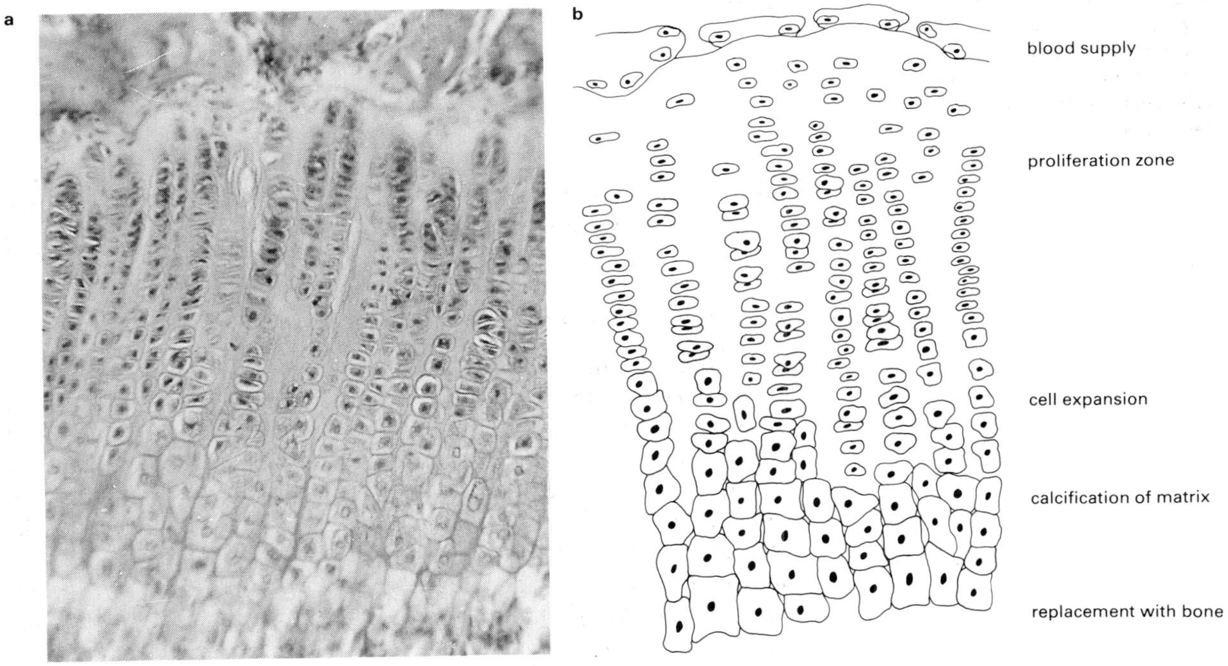

Figure 194
a A photomicrograph showing the formation of bone in a growth plate.
b A diagram showing the zones of cartilage tissue in the growth plate.
Photograph, Dr N. F. Kember.

blood supply

proliferation zone

cell expansion

calcification of matrix

replacement with bone

and get larger and it is this that causes the increase in length. The growth plate can be divided into three main regions: a proliferation zone where the cells are multiplying, a zone where the cells do not divide but get bigger, and a zone where the cartilage cells are broken down and replaced by bone. There is thus a continual traffic of cells from the region of proliferation to the region where the cells are broken down and replaced by bone (*figure 194*). How does the length of the muscle adjust to the growth in length of the bones? The muscles and tendons are inserted on different bones and as the bones get longer the tension they exert on the muscles makes them grow longer. Muscle cells do not divide; the increase in the length of the muscle is not due to cell division. Muscles grow by the increase in length of the muscle fibres. In this way the length of the bones and muscles is co-ordinated.

STUDY ITEM

23.51 Measuring the growth of humans

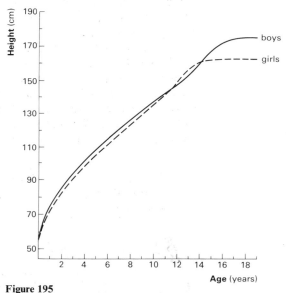

Figure 195
Typical growth curves for English boys and girls.
From Tanner, J. M., Whitehouse, R. H., and Takaishi, M., 'Standards from birth to maturity for height, weight, height velocity, and weight velocity: British children, 1965. Part 1', Archives of disease in childhood, **41***, 1966.*

Typical growth curves for English boys and girls are plotted on the graph in *figure 195*. The description 'typical' is important because growth curves vary considerably between different individuals even though the overall form of the curve is similar. While the overall form of the curves is clear, the most important features are better illustrated by constructing growth velocity curves, one for each sex. To construct a growth velocity curve using the data in *figure 195*, first draw up a table for each sex, giving age and increase in height in the previous year. When you have completed your table, plot the increase for each year against age. This is a velocity curve.

a *What differences are there in growth rate changes between boys and girls?* adolescent spurt earlier in girls

b *What do you think determines the different rates of growth?* Cartilaginous growth of cartilage under hormonal control

☐ **c** *Why do we stop growing?* no answer but fortunate we do unlike lobster, carp.

> Practical investigation. *Practical guide 6*, investigation 23D, 'Growth and development in the fore-limb of mice.

23.6 Fate maps and determination

It is possible to mark on a fate map of the amphibian blastula (*figure 196a*) the presumptive eye region and we can use this to illustrate some important concepts. Fate maps only indicate what structures different parts of the blastula would normally form. While a fate map reliably predicts what will happen during development, it does not show what different regions are capable of forming if interference takes place during development.

It has already been shown (section 23.2) that in the early development of the mouse embryo the fate of cells is not fixed but is determined by their position. Is the same true of the eye at the blastula stage of amphibian development? It is possible at this stage to graft the presumptive eye tissue to the belly region of an embryo and see what it will develop into (*figure 196c, d*). It does not develop into an eye but forms structures that are consistent with its new position, that is

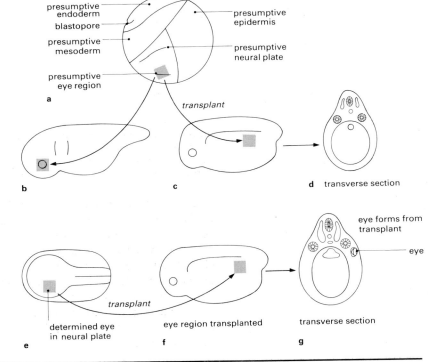

Figure 196
Diagram to show how the fate of a group of cells can be demonstrated, and a test for their determination.
After Wolpert, L. The development of pattern and form in animals, *Oxford Biology Readers 51*, Oxford University Press, 1974.

Chapter 23 The nature of development

structures associated with the gut. Thus the presumptive eye tissue has a much greater potential for development than is indicated by its presumptive fate. This is generally true for the whole of the blastula. However, with time the situation changes and shortly after gastrulation the potential of the eye tissue becomes restricted and the tissue is said to be *determined*. If, at this stage, the tissue is grafted to another site, it will retain its character and develop as an eye even if it is grafted to the flank (figure 196e–g). Determination occurs before any obvious structural changes take place and is thought to reflect chemical changes in the nucleus which commit the cells to certain pathways of differentiation. Different tissues and structures become determined at different times.

Some embryos, particularly those of vertebrates and sea urchins, have a considerable ability to regulate their development so as to form a normal embryo when parts are removed or placed in new positions. We have already seen this in relation to the formation of twins, when a presumptive half embryo can give rise to a whole embryo. So it is with the eye. Presumptive eye tissue, long before it is determined, can form other structures when placed in new positions. Even if the whole of the presumptive eye tissue is removed at an early stage, a normal pair of eyes will develop. The adjacent tissue which moves in to close the wound now forms the eyes. However, with time, as more and more determination occurs, the capacity for regulation becomes less and less.

23.7 Morphogenesis

Two important cellular activities in development, as we have just seen in gastrulation, are changes in shape and motility. These same cellular activities are called into play at later stages of development. Thus changes in cell shape lead to the formation of the neural tube, and the migration of neural crest cells is involved in the development of many different cell types including pigment cells and spinal ganglion cells.

Neural tube morphogenesis

At the end of gastrulation, following cell–cell interactions, the ectodermal cells in the part of the sheet overlying the dorsal mesoderm are determined to give rise to the nervous system. This flat sheet of cells rolls up to form a tube. This is accomplished by changes in the shape of the cells in the sheet. The process can best be visualized by considering the row of cells in a transverse section taken across the neural plate (figure 197). First the cells become elongated to form a row of columnar cells. Then the upper faces of the cells – those that abut the external medium – become shorter. As the lower surfaces of the cells remain the same length the cells become wedge-shaped. This results in the curving of the sheet, ultimately to give a tube. The shortening of the upper faces of the cells is thought to be due to the contraction of the bundles of microfilaments (made up of a contractile protein very similar to actin which is found in muscle fibres) that run just below these faces. The contracting filaments constrict the upper surfaces of the cells and act like a purse-string to draw the sheet into a curved tube.

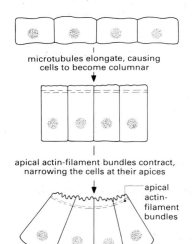

Figure 197
The diagram shows the changes in shape of the cells of the neural plate which lead to curving of the sheet of cells and ultimately to the formation of the neural tube.
After Alberts, B., Bray, D., Lewis, J., Raff, M., Roberts, K., *and* Watson, J. D., Molecular biology of the cell, *Garland, 1983.*

The edges of the curved sheet fuse to form the neural tube. Sometimes the neural tube may fail to close, leading to types of malformations that are included in the set of congenital malformations known as *spina bifida*.

STUDY ITEM

23.71 **The formation of the neural crest**

Neural tube fusion and the closing of the sheet of ectoderm over it are accompanied by the pinching off of a population of ectodermal cells into the body on either side of the neural tube (*figure 198*). These ectodermal cells are the *neural crest cells* and will migrate to give rise to a range of different cell types, including the pigment cells of the skin and the nerve cells of the autonomic system. In the head, the neural crest cells give rise to the skeleton. Thus the jaw bones, for example, are made up of cells of neural crest origin.

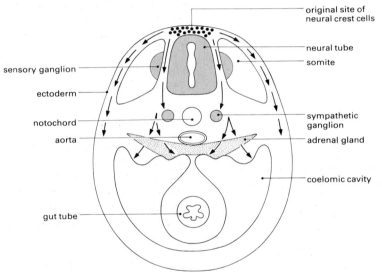

Figure 198
A section through a chick embryo showing the origin of the neural crest cells and the main pathways of their migration in the trunk.
After, Alberts, B., Bray, D., Lewis, J., Raff, M., Roberts, K., and Watson, J. D., Molecular biology of the cell, Garland, 1983.

a *Suggest two possible mechanisms that could account for the formation of tissues from the neural crest cells in their proper positions.*

A clear example of an experiment to test how neural crest cells form tissues in their proper positions is to graft forebrain neural crest cells, before migration, in place of hind-brain neural crest cells (*figure 199*). The forebrain neural crest cells will thus migrate into the territory usually occupied by hind-brain crest cells. Neural crest cells that migrate from the forebrain do not, in normal circumstances, give rise to any ganglia, whereas hind-brain crest cells give rise to the cranial ganglia.

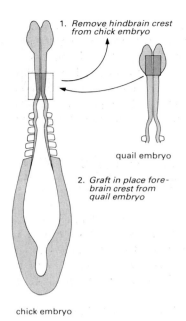

Figure 199
Transplanting neural crest cells from quail to chick embryo to investigate their differentiation.

Thus, this experiment provides a way of finding out the control mechanism for the differentiation of neural crest cells.

The fate of the transplanted cells can be traced. One way is to use quail forebrain neural crest to replace the hind-brain neural crest of a chicken embryo, because quail cells can be distinguished from chick cells by the staining properties of their nuclei. The result of this experiment is that the cranial ganglia which develop in the chick embryos following this transplant of the quail neural crest cells are made up of quail cells.

b *What does this result suggest about the determination of neural crest cells?*

c *What result would be expected from the reciprocal experiment in which quail hind-brain neural crest replaces chick forebrain neural crest cells?*

Although this result is typical of many other experiments in which neural crests from different regions are exchanged, for example neural crest from different regions in the trunk, more recent experiments suggest that the situation may not be so straightforward. There is now some evidence that, in certain cases, neural crest cells may be determined before they start migration.

23.8 Pattern formation in the development of limbs

The limbs develop from small swellings of the body wall, the limb buds, which consist of a mass of apparently homogeneous cells covered with ectoderm. This mass of cells gives rise to the specialized tissues that make up the limb.

The limb buds of mammals and other vertebrates such as birds are remarkably similar in appearance. Indeed, most experimental work has been carried out using chicken embryos since it is possible to carry out surgical procedures on the limb buds while the embryo lies inside the egg. The operations are carried out through a small hole cut into the egg shell. The embryo can then be left to develop further and the effects on the limbs assessed. Such operations cannot be carried out on mammalian embryos since methods of culturing these embryos outside the mother throughout the equivalent stages of development have not yet been perfected. However, the same basic mechanisms are involved in the development of the limbs of both chickens and mammals.

As cells multiply in the early buds, the buds elongate and grow out of the body wall. The successive parts of the limb are laid down in sequence as the bud elongates. Starting in the proximal part of the developing fore-limb (the part nearest to the body), cells begin to differentiate into the cells that will make up the specialized tissues of the upper arm, such as the humerus. Some cells in the core of the elongated bud differentiate into the cartilage cells that make up the limb skeleton of the young embryo. Later the cartilaginous skeleton is replaced by bone. Other cells differentiate into the specialized cells that make up tissues such as tendons, and a separate population of cells differentiates into muscle cells. This population of cells migrates into the region of the limb before the outgrowth even begins. They originate in the *somites*, which are segmented blocks of mesoderm that develop down the trunk of the body on either side of the neural tube. The succeeding parts of the limb, the lower arm and finally the hand, arise from the zone of undifferentiated cells that remain at the tip of the limb bud throughout outgrowth. The tissues of the lower arm are formed after those of the upper arm and are followed in sequence by those of the hand in which the digits are formed last of all.

> **Practical investigation.** *Practical guide 6*, **investigation 23D**, **'Growth and development in the fore-limb of mice'.**

The laying down of this sequence of structures depends on a thickened rim of ectodermal cells – the *apical ectodermal ridge* – at the tip of the developing limb. If the apical ectodermal ridge is cut away from a developing chick wing, further elongation of the bud is inhibited and the wing that develops is truncated. The level of truncation depends on the time at which the ridge is removed (*figure 200*).

By grafting the apical ectodermal ridge from a mammalian limb bud (this has been done using mouse limb buds) in place of the apical

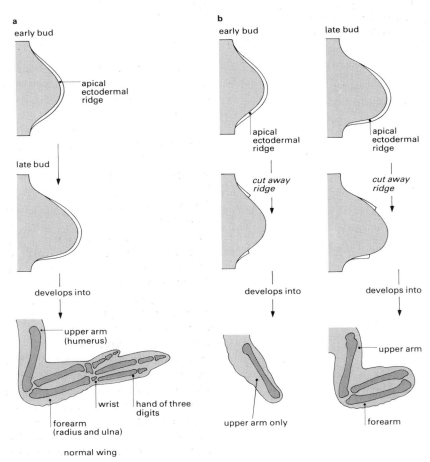

Figure 200
The laying down of the structures of the chick wing depends on the apical ectodermal ridge.
a Normal development of the wing.
b The effects of cutting away the ridge at different stages of development.
Partly based on Alberts, B., Bray, D., Lewis, J., Raff, M., Roberts, K., and Watson, J. D., Molecular biology of the cell, Garland, 1983.

ectodermal ridge of a chick wing bud, it can be shown that the ridge has the same function in both chick and mammal. The apical ectodermal ridge of the mouse limb bud allows outgrowth of the wing bud and a normal wing results. Such experiments give insights into the basis of some limb deformities.

a *Give a possible explanation for the basis of congenital malformations in which limbs are missing or truncated.*

The development of the limb involves more than just the differentiation of cells to make up its specialized tissues. These specialized tissues must also be arranged in the correct positions. Consider, for example, the spatial arrangement of cartilage cells that make up the early skeleton: there is a single rod in the upper arm, two rods in the lower arm, and a series of jointed rods that make up the digits. The generation of this ordered arrangement of specialized cell types is a problem of pattern formation. This problem can be thought about in terms of positional information. In these terms, the cells will be 'informed' of their position within the mass of cells of the limb bud and then, using this information, will differentiate to participate in forming the structure appropriate to that position. Thus the control of the spatial pattern of cellular differentiation depends on the cues that define position and their interpretation.

The position of a cell within the bud can be defined in relation to the three axes of the limb, the proximo–distal axis running from the shoulder to the tips of the fingers, the antero–posterior axis running at right angles across the limb (in the hand from thumb to little finger), and the remaining dorso–ventral axis (*figure 201*). For the first two of these axes, there are some clues about how the positions of cells are defined.

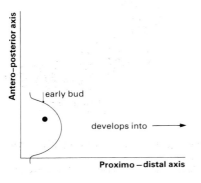

Figure 201
In the early chick wing bud the position of a cell can be specified in relation to three axes. The diagram shows two of these axes; the dorso–ventral axis runs perpendicular to these, *i.e.* into the paper.
Photograph, Dr Cheryll Tickle.

STUDY ITEM

23.81 How do cells 'know' where they are?

The way in which cells are 'informed' of their position along the proximo–distal axis appears to depend on a timing mechanism. Under the influence of the apical ridge, the sequence of structures along the limb is laid down. Proximal structures develop early, while distal ones develop later. One possibility is that the laying down of each set of structures depends on interactions between cells in successive structures. Cells that have formed, say, the lower arm, would inform adjacent cells to form digits. This possibility can be investigated experimentally in the developing chick wing. The undifferentiated tip of an old bud can be grafted in place of the tip of a young bud (*figure 202*). If cells in the undifferentiated tip depend on proximal cells for information about their position, then the grafted bud should develop to give a normal pattern of structures along the proximo–distal axis. However, the limb that develops has a defect – the lower arm is missing. This result suggests that the cells in the tip of a bud do not depend on information about position from structures already laid down. Instead, such information must be assigned by some mechanism intrinsic to the tip.

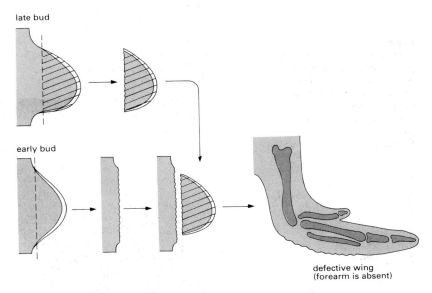

Figure 202
The undifferentiated tip of an old chick wing bud is grafted in place of the tip of a young bud.
After Alberts, B., Bray, D., Lewis, J., Raff, M., Roberts, K., and Watson, J. D., Molecular biology of the cell, Garland, 1983.

a *What is the reciprocal experiment?*

b *Predict results of this experiment that would be consistent with the idea that the tip develops independently.*

The following hypothesis has been put forward for the mechanism intrinsic to the tip that informs cells of their position along the proximo–distal axis. The apical ectodermal ridge maintains a zone of undifferentiated cells at the tip of the limb bud. Cells spill out of this zone, as the bud grows out, and are informed of their position by the length of time they were in the zone. Cells that leave the zone early form proximal structures while cells that leave late form distal ones (*figure 203*).

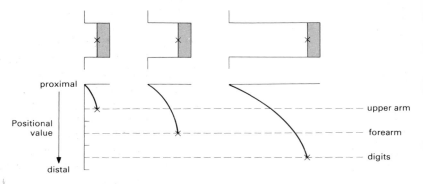

Figure 203
A schematic diagram showing the positional values of cells along the proximo–distal axis of a limb, for cells leaving the undifferentiated limb bud tip at three successive times.

c *According to this hypothesis, what would happen if most of the cells in the undifferentiated zone at the tip of the limb bud were killed by a toxic drug or radiation?*

d *If the surviving cells could continue to proliferate, what would the pattern of the limb be like?*

Figure 204
A child with restricted limb development, the result of the mother taking the drug Thalidomide during pregnancy.
Photograph, Times Newspapers Ltd.

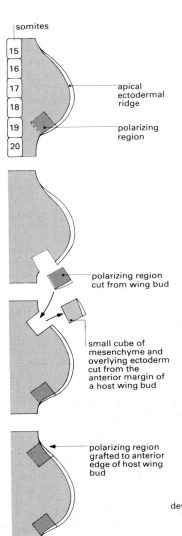

Such studies as these are potentially of great practical importance. For instance, in the 1960s a new drug appeared on the market called Thalidomide. Children born to mothers who were given this drug during early pregnancy commonly had congenitally malformed limbs (*figure 204*). The way in which Thalidomide acts to cause such malformations is still not clear, but research into limb development may find an answer.

Cells are informed of their position along the antero–posterior axis by a different mechanism. A small group of cells at the posterior margin of the limb bud, the *polarizing region* can affect the pattern of structures that develop across the antero–posterior axis of the limb. This can be shown experimentally by grafting the polarizing region from one chick wing bud to an anterior position in a second bud (*figure 205*). This second bud now has polarizing tissue at both posterior and anterior margins and develops into a symmetrical limb with a mirror image duplication of structures across the antero–posterior axis. Thus, in the hand, instead of the three digits of the chick wing, 234 (reading from the anterior to the posterior), there are now six digits in mirror image symmetry, 432234.

Mammalian limb buds such as those of mice and even humans have been shown to have tissue with polarizing activity at their posterior margins too. Thus grafts of this tissue from mouse or human limb buds to anterior positions in chick wing buds lead to the development of additional digits. These additional digits are chick digits, illustrating that the polarizing region is a signalling region and that the duplicate structures develop from the host bud tissue under the influence of the graft.

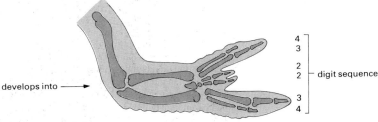

Figure 205
The polarizing region from one wing bud is grafted to the anterior margin of another wing bud. Compare the resulting digit pattern with the normal pattern shown in *figure 200*.
After Alberts, B., Bray, D., Lewis, J., Raff, M., Roberts, K., and Watson, J. D., Molecular biology of the cell, *Garland, 1983.*

STUDY ITEM

23.82 **How does the polarizing region exert its effect?**

It is not known how the polarizing region brings about a new pattern of cellular differentiation. However, many experiments have investigated its action. For example, by grafting additional polarizing regions to different positions along the antero–posterior axis one can investigate the effect of the amount of tissue between graft and host polarizing regions on the pattern of cellular differentiation. The data from such an experiment are shown in table 38.

Digits	14/15	15	15/16	16	16/17	17	17/18	18	18/19	19	19/20
Stage 18											
234		6									
2234		1									
32234		1	3								
432234			4	3							
43234			3	5							
4334					4	5	2	1			
434						1	4	4	1		
44											
4								1	2		
None									2	4	
Stage 20											
234	7	3	2	2							
2234		1	2		1						
32234		1		1							
432234		1	2	2	1						
43234		1		3	2						
4334					3	4*					
434						1	5*				
44											
4									2		
None							2	2	8	11	2	1

Position of graft with respect to somites

The data refer to the number of cases obtained.
* In some of these cases the identification of the anterior digit 4 was equivocal.

Table 38
The formation of digits between a grafted polarizing region and the host polarizing region when the polarizing region is placed in successive positions along the antero–posterior axis.
From Nature, **245**, 1975, 200.

a How is the number of digits affected as the amount of tissue between host and grafted polarizing region is reduced?

b Study the pattern of digits formed as the grafted polarizing region is placed in successively more posterior positions. What comparisons can you make of the effects on the formation of digit 2 and digit 4?

The following hypothesis has been put forward to account for these results. The hypothesis proposes that the polarizing region produces a chemical, a *morphogen*, that diffuses across the limb bud, thus setting up a concentration gradient. The concentration of the morphogen at any point would inform cells of their distance from the polarizing region and hence their position across the antero–posterior axis. Cells close to the polarizing region would experience a high concentration of morphogen and form posterior structures such as digit 4; cells a bit further away would be exposed to a lower concentration of morphogen and form digit 3, and cells still farther away would experience an even lower concentration and form anterior structures such as digit 2. This hypothesis is illustrated in *figure 206*.

c Using **figure 206**, make sketches to show how this hypothesis could account for the data in table 38.

Figure 206
This graph illustrates the hypothesis that the polarizing region produces a morphogen that diffuses across the limb bud and sets up a concentration gradient. The concentration of morphogen at each position would determine which digits form.

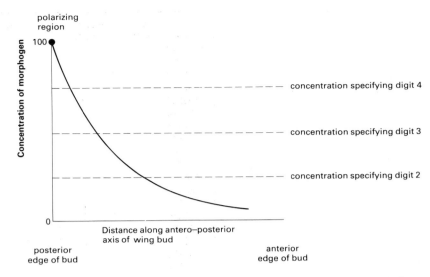

The wing illustrated in *figure 207* developed following a graft containing a very small number of polarizing region cells that was placed at the anterior margin of the bud.

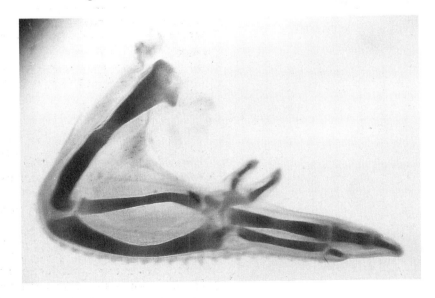

Figure 207
A wing that developed following a graft of a small number of polarizing region cells.
Photograph, Dr Cheryll Tickle.

    d *Identify the digits that have formed.*

    e *According to the hypothesis that polarizing region tissue produces a diffusible morphogen, how could you account for this result?*

        The positional cues from the polarizing region, and the time of exit of cells from the zone of undifferentiated cells at the tip of the limb, lead to the differentiation of the cells not only of the parts of the skeleton, but also of other connective tissues, such as tendons. In addition, the pattern of the connective tissue sheaths that surround the muscles is established in the same way. Thus the pattern of muscles in limbs that develop following grafts of polarizing region tissue is duplicated. However, the sheaths are filled by the muscle cells that arise from the population of cells that invade the limb area from the somites at an early stage.

There are other tissues as well that must migrate into the developing limb to take up their proper positions. Thus nerve cells in the part of the spinal cord that lies opposite the developing limbs put out long processes that invade the limb to link up with muscles and establish the normal pattern of motor innervation. The way in which the correct connections are made is an important problem in understanding the development of the nervous system as a whole. The development of the innervation of the limb is a good system in which to study this problem.

STUDY ITEM

23.83 **Establishing connections**

One idea about how proper connections are established in the nervous system is that nerves actively 'seek out' their appropriate target muscles. A second possibility is that nerves are passively 'guided' to their correct targets by either the maintenance of an ordered array of nerve outgrowth, or the timing of outgrowth. These ideas can be tested in the development of the innervation of the chick leg. At early stages, before the limb buds are formed, short segments of the spinal cord, from which the axons that will innervate the leg will later grow out, are cut out, rotated through 180° and replaced (*figure 208*). Thus when the leg buds develop, the nerve axons will grow out into leg tissue that they would not normally encounter. For example, nerves from the posterior end of the spinal cord segment, which after the operation are in an anterior position, will grow out into the anterior part of the developing leg rather than the posterior part that they would normally innervate.

a *What would be the predicted result if the normal pattern of innervation arises by nerves 'seeking out' their appropriate target muscles?*

b *What other operations could be carried out to perturb the relationship of the nerves and their target muscles, which would provide evidence to distinguish between the two possibilities?*

23.9 **The pattern of plant and animal development compared**

Plants and animals both begin their development as single, fertilized cells. The mechanisms involved in subsequent development to the adult organisms are similar, and the general scheme of development is basically the same. In animals, as we have seen, there is generally a phase of embryonic development, followed by a phase of maturation and growth, leading, in turn, to the adult capable of reproduction. Plants also show these phases, but they are less obvious. This is because plants possess an entirely unique tissue called the *meristem*. In meristems, new cells are continually produced by cell division and some of these cells then undergo differentiation to form different cell types arranged in a

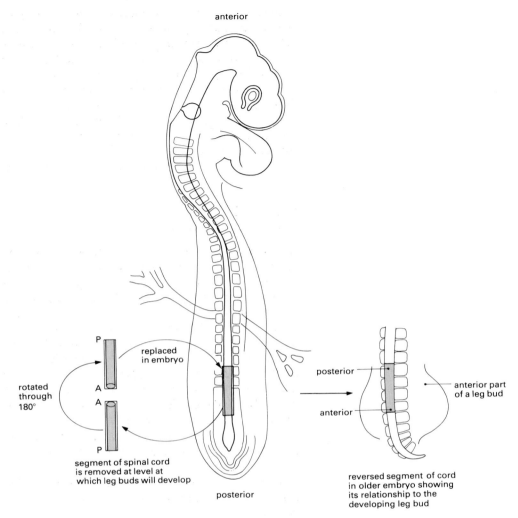

Figure 208
An experiment to test how proper nerve connections are made in the developing leg.

pattern to form tissues and organs. In animals this process of cell differentiation is usually associated with development. However, in plants cells differentiation and pattern formation take place even in adults.

a *There are examples in mammals of cells which differentiate throughout life. Name one example.*

Since every 'adult' plant possesses many meristems at any one time, they contain both fully differentiated and totipotent cells. Thus, although the higher plants start life as single cells and then embark on a phase of development, called embryogenesis, which ultimately finishes with the formation of the dry seed, the germinating seedling has root and stem meristems which effectively continue the process of embryogenesis.

STUDY ITEM

23.91 Growth of a root

The tip of a root contains a meristem. It can be used for analysing the process of growth.

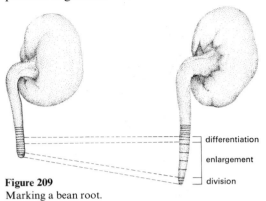

Figure 209
Marking a bean root.

Figure 209a shows the external appearance of the tip of a bean root marked with Indian ink. Each mark is about 2 mm from the next. In *b* the same root is shown after several days of development. Notice that some of the marks are no longer the same distance apart.

a *What two alternative hypotheses could account for this result?*

To investigate the change further two parts of the root were cut out. One part, cut exactly at adjacent ink marks, was taken from near the tip of the elongated root. A second part, again exactly between adjacent ink marks, was taken from the part of the root which showed the greatest amount of growth.

b *For which of your hypotheses would you expect to find all the cells in the two sections the same size?*

c *For which would you expect to find the same number of cells in the two sections?*

The photomicrographs *b*, *c*, and *d* in *figure 210* show longitudinal sections of typical cells from three different regions of an actively growing root. Each photograph is taken at the same magnification so that the field dimensions are standard. Count the number of cells in one longitudinal file of cells in each photograph.

d *Do your results agree with your predictions?*

e *Can either of your hypotheses be rejected as false? If so, which?*

f *When you have decided how you can best account for the elongation of the roots, examine photomicrograph a in figure 210, showing a longitudinal section of a young root. Where is cell division most likely to be occurring in it?*

Figure 210
a A longitudinal section of a broad bean root tip ($\times 23.5$).
b Cells from 2 mm behind the root tip ($\times 189$).
c Cells from 1–2 mm behind the root tip ($\times 189$).
d Cells from just behind the root tip ($\times 189$).
Photographs, Philip Harris Biological Ltd.

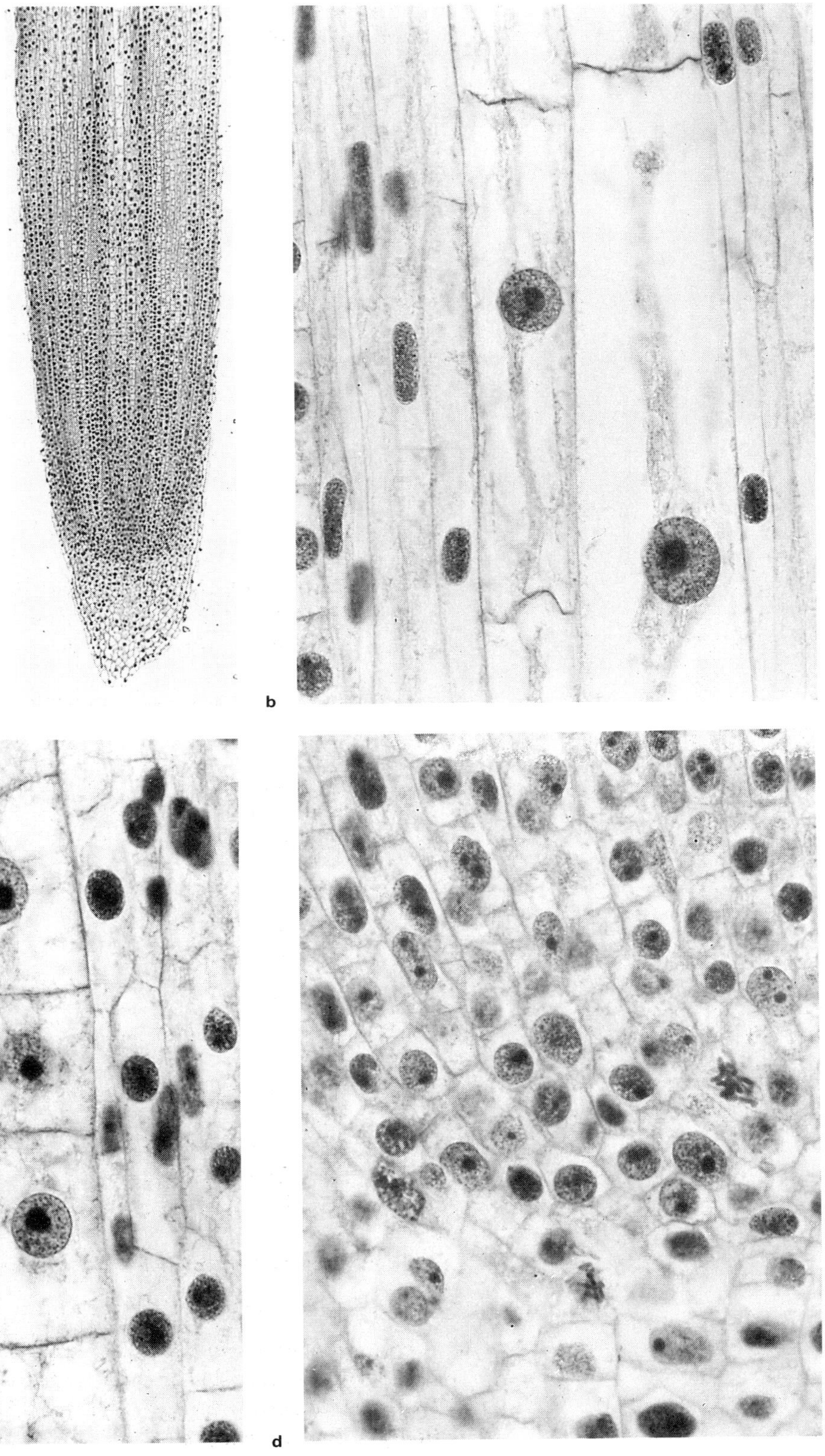

Chapter 23 The nature of development

Again you can check your prediction against the three other photomicrographs, **b**, **c**, and **d**, because if cells are dividing you would expect something to be happening to the nuclei before cell division.

g *Can you see any sign of this process in photomicrographs* **b**, **c**, *or* **d**? *If so, which?*

Finally, examine the cytoplasmic inclusions and size and shape of the cells in the photographs.

h *What other processes appear to take place in the cells during development?*

Meristems

There are three major types of meristem in most higher plants. These are the primary root and shoot meristems and the cambium. In each of these, cells continue the process of division and the daughter cells undergo differentiation. They represent, therefore, the continuation of the embryonic phase in the adult plant.

The root meristem produces cells which differentiate into new root components, and cells which will, in turn, initiate new meristems from which branch roots grow out.

In the shoot, the meristem is more complex than that of the root, but the same basic patterns of growth can be recognized (*figure 211*). The shoot meristem gives rise to separate growth centres which differentiate to give initiating centres of leaves and buds. At a later stage of plant growth the main shoot meristem may completely alter its character and start to produce cells that differentiate to form the reproductive organs, the flowers.

The third type of meristem is found in both stem and roots of many higher plants. This meristem, the cambium, produces cells which differentiate into the components of roots and shoots, and enables these structures to increase in girth.

Developmental aspects of the evolution of plant form

As injury to a single cell is generally fatal, unicellular algae either remained motile or evolved remarkable regenerative powers as in *Acetabularia* (see Study item 15.12). Alternatively, they evolved into multicellular organisms. Some cells could then be lost without effective loss of the whole organism.

When the mechanism of tip or apical growth evolved it pushed part of the organism towards the light. Thus at an early stage in evolution plants were probably organisms differentiated into a hold-fast, or root, and a thallus, or stem, which could photosynthesize. In time, the area of tip growth evolved into the meristem.

A major step forward was the evolution of meristems capable of replicating themselves. Thus, as we have seen, a shoot meristem produces many new daughter buds or meristems, each itself capable of growth.

Figure 211
Photomicrograph of a longitudinal section through a shoot, showing the meristem (×31).
Photograph, Biophoto Associates.

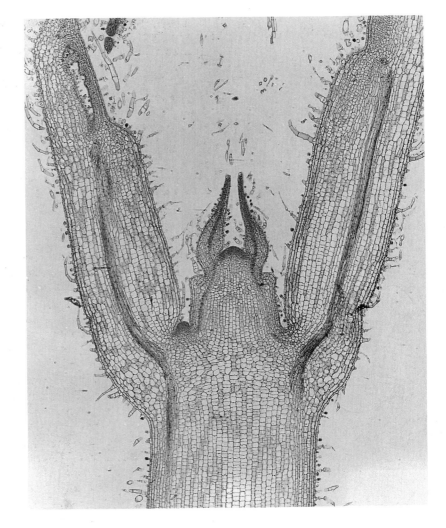

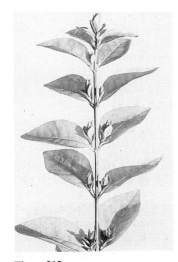

Figure 212
Photograph of privet, *Ligustrum ovalifolium*, illustrating the modular growth pattern of plants.
Photograph, Biophoto Associates.

Growth and development could thus continue in the adult plant at a large number of different parts of the plant body. If some were eaten or damaged, growth could continue at others until reproduction occurred. Additionally, the photosynthetic area could be increased by growth at a number of points and thus in turn provide materials to increase the number of reproductive tissues.

Higher plants can be thought of as modular organisms: they are made up of the same basic module structure repeated many times. In the shoot, the module is the leaf with its axillary bud or meristem, and, below ground, the root meristem. Growth takes place primarily by the addition of new 'modules' to the pre-existing body. Such a construction is entirely different from, say, a mammal in which all the various individual parts, such as the liver or kidneys, are specialized to perform very different functions. Animal construction can be described as unitary in contrast with the modular construction of plants (*figure 212*).

Although the evolution of flowering plants has resulted in the development of organs such as roots, stems, and leaves, these are not as

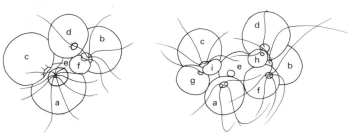

Figure 213
Daughter meristems of duckweed breaking off from the parent plant.
Photograph, Dr C. C. Smart and Dr A. J. Trewavas, Department of Botany, University of Edinburgh.

highly specialized as the organs found in a mammal's body. If one special function is located in only one tissue or organ, such as the mammalian heart, then even slight damage might kill the whole organism. In other words, specialization leads to vulnerability. Organisms subject to predator damage would not survive with excessive tissue specialization.

Several important consequences follow. In the vegetative plant, many tissues overlap in the functions they perform. For example, the stem and root as well as the leaves can be photosynthetic, and aerial roots can prop up the stem. Plants do not possess a vulnerable central, co-ordinating tissue, such as the nervous system. But this means that the individual growing, and indeed non-growing, parts of a plant must be very independent in their growth and their behaviour towards a local stimulus. This can be seen in its most extreme form in the duckweeds, in which the individual daughter meristems break off (*figure 213*) from the parent plant and live an independent existence.

The majority of cells in a higher plant remain unspecialized and capable of further development. It is relatively easy to stimulate many of them to undergo mitosis, forming callus (*figure 214*), and from this to regenerate or reorganize new meristems. Such a regeneration has been carried out even from single cultured cells (see sections 22.2 and 23.2).

> **Practical investigation.** *Practical guide 6*, **investigation 23E, 'Growing new plants from old'.**

Development in plants is adapted to adjust to the constant variation in the environment. In bad conditions a variety of ways have evolved to make meristems dormant, for instance in the seed or the dormant bud (see section 25.2). When conditions improve the rate of growth and

Figure 214
A carrot plant being regenerated from a small portion of carrot tissue.
Photograph, Roy DeCarava.

development of individual meristems is increased. If conditions improve still further more meristems break their dormancy and start to grow and produce new branches. The surroundings act as developmental signals to the plant, which grows and develops in response to them.

Adaptation to the environment can be very striking. Consider, for example, the difference between a dark-grown and a light-grown shoot; or a plant that will only flower when it has grown several weeks at low temperatures; or the massive production of new root meristems by roots growing in layers of soil rich in minerals. Thus, the patterns of growth and development are shaped by the place and surroundings in which the plant lives, as is the structure of its body. In a sense, every single plant contains within its body a developmental history of the environment it has experienced.

The concept of plastic development

'Plastic development' is probably the best term to characterize development in plants. It encompasses adapting to the environment, easy regeneration and limited tissue and organ specialization, presence of many growing points, and the degree of independence in behaviour of each of these. These strategies evolved in plants as adaptations which enabled them to be successful stationary organisms and to tolerate predation. Many of these plastic properties are shown by meristems, and the evolution of this novel tissue with its characteristic self-replication was the key to the success of plants.

STUDY ITEM

23.92 Differences between plants and animals

Why do plants show this pattern of on-going development even in the adult? The answer probably lies in a fundamental distinction which arose one or more billion years ago concerning the mode of nutrition.

One theory suggests that the earliest eukaryotic cells were all plant-like (autotrophs). They contained chloroplasts which carried out photosynthesis and they possessed a cell wall. Some cells, it is thought, may have dispensed with all their chloroplasts and instead lived by eating other cells (heterotrophs), and evolved into animals. Thus the earliest plants lived by capturing the energy of sunlight, whilst the earliest animals lived, in turn, by eating plants – and some ate each other.

a   *Why is locomotion not essential for autotrophs?*

b   *What plants, or motile cells of plants, have you met that show locomotion?*

c   *How can the angiosperms survive without a motile gamete?*

d   *Why would this be an advantage?*

e   *What are the disadvantages to an organism, such as an angiosperm, of being sessile?*

We left animal cells at a stage when they had lost their chloroplasts and cell wall and become heterotrophic feeders. The requirements both to sense the position of food and to move towards it, meant improving both the sensory systems and the systems involving movement. Once animals evolved which could eat other organisms, the strategies were to sense more quickly and move faster. In turn, a greater variety of surroundings would be experienced. Animals evolved with a distribution of labour between various parts of the body, with specialization into sensory organs and organs concerned with movement. A nervous system evolved which co-ordinated these activities and speeded up the response.

The price paid for this specialization was extreme vulnerability to injury. So adaptations also evolved in animals to reduce or avoid predation. However, the more specialized animals became, the more necessary it was to avoid predators; this in turn involved greater specialization and in turn greater vulnerability, and so the cycle continued. Perhaps this accounts for the explosive evolution of animals in the last five hundred million years.

Some details of animal embryogenesis are different from that of plants. The cells, as we have seen, can move around and thus need a complex system of sensing other cells on their surface. This is not a feature of plant embryogenesis. In animals the need for specialization involves the development of many obviously different cell types, each carrying out one special function. The production of such cells involves 'decisions' in embryogenesis which seem to be irreversible. Once a cell is committed to a pathway of development it remains relatively fixed in that pathway. We assume that this fixing occurs because of a very rapid 'switching-off' of very large numbers of genes. A consequence of this is that regeneration does not readily occur. However, remember that, as we have discussed (see section 23.2), the genetic information is still present in the nuclei, even if not used.

Compare the rather weak ability of animals to regenerate with the very easy regeneration of meristems, or of whole plants. Consider again the relative lack of cell specialization in higher plants. Nearly all plant cells have the same cytoplasmic structure, except shoot cells which have many chloroplasts.

b *How many different types of cells have you met in a mammal and in an angiosperm? Make a list of them.*

c *What examples of regeneration can you find amongst animals?*

d *What parts of a plant are capable of regeneration? List specific examples.*

Although there are basic similarities in the development of plants and animals, the environment has a much more important effect upon the development of plants than that of animals. Since animal and plant types probably separated in evolution when they were both single cells, it is not surprising that such differences are found. However, in practice, the degree of difference is tempered by the use of the same building block, the cell, with its limited range of activities.

e *Much of our knowledge of development has come from the study of a relatively small range of organisms, such as the sea urchin, frog, and mouse. How far do you think that it is reasonable to generalize from the conclusions from such work?*

Summary

1 Development begins with fertilization (**23.1**).
2 In the majority of animals development proceeds in three main phases: cleavage, when the zygote divides a number of items to form a ball of cells; gastrulation, which involves cell movements, giving rise to the three main body layers of ectoderm, mesoderm, and endoderm; organogenesis, when recognizable organs develop (**23.1**).
3 The early stages of development are followed by a period of growth before the animal becomes adult (**23.1**).
4 Development involves three main processes: morphogenesis, the development of the form of the animal; differentiation, the specialization of cells for particular functions; pattern formation, the organization of different types of cells in appropriate spatial arrangements (**23.1**).
5 As an animal embryo develops, its cells rapidly lose their totipotency and as a result are restricted in their development capacities (**23.2**).
6 It is possible to clone organisms by the transfer of cell nuclei (**23.2**).
7 Induction is an interaction, during development, between two tissues. The formation of the neural plate is an example of this process (**23.4**).
8 The growth plate is a structure involved in the growth of the long bones of the skeleton (**23.5**).
9 Fate maps can be constructed for embryos to indicate the structures that certain parts of the embryo would normally form (**23.3** and **23.6**).
10 Experimental embryology enables us to discover the mechanisms that control development (**23.8**).
11 In the flowering plants development follows the same general principles as in animals, but proceeds on a modular basis (**23.9**).
12 Plants, even when adult, possess meristems, a totipotent tissue (**23.9**).
13 Dormant meristems are a strategy adopted by plants to reduce the effects of damaging predation (**23.9**).
14 It is possible to regenerate complete plants from small parts, or even from single somatic cells (**23.9**).
15 Plants show a plastic form of development, which can be seen as an adaptation to being sessile organisms subject to predation. Compared with animals their development is more directly affected by environmental factors (**23.9**).
16 Animals are normally motile organisms, and have evolved more specialized, but potentially more vulnerable organ systems than plants (**23.9**).

Suggestions for further reading

BUTCHER, D. N. and INGRAM, D. S. Studies in Biology No. 65, *Plant tissue culture.* Edward Arnold, 1976. (Details of the various techniques available.)

CLOWES, F. A. L. Carolina Biology Readers No. 23, *Morphogenesis of the shoot apex.* Carolina Biological Supply Company, distributed by Packard Publishing Ltd, 1972. (Apical meristem and its functioning.)

EDWARD, R. G. Carolina Biology Readers No. 89, *Test tube babies.* Carolina Biological Supply Company, distributed by Packard Publishing Ltd, 1981. (Useful account, including the medical technology involved.)

GURDON, J. B. 'Transplanted nuclei and cell differentiation'. *Scientific American*, **219**(6), 1968, 24–35.

GURDON, J. B. Carolina Biology Readers No. 27, *Gene expression during cell differentiation.* 2nd edn. Carolina Biological Supply Company, distributed by Packard Publishing Ltd, 1970. (Nuclear transplantation, constancy of genome in development, etc.)

ILLMENSEE, K. and STEVENS, L. C. 'Teratomas and chimeras'. *Scientific American*, **240**(4), 1979, 86–98. (Early development.)

NEWTH, D. R. Studies in Biology No. 24, *Animal growth and development.* Edward Arnold, 1970. (A concise overview.)

SHARP, J. A. Studies in Biology No. 82, *An introduction to animal tissue culture.* Edward Arnold, 1977. (Generally concerned with the methods of tissue culture, rather than the potential of the technique.)

SHEPARD, J. F. 'The regeneration of potato plants from leaf cell protoplasts'. *Scientific American*, **246**(5), 1982, 112–121. (Tissue culture using cell protoplasts.)

WOLPERT, L. Carolina Biology Readers No. 51. *Development of pattern and form in animals.* 2nd edn. Carolina Biological Supply Company, distributed by Packard Publishing Ltd, 1977. (Useful introduction to pattern formation.)

WOLPERT, L. 'Pattern formation in biological development'. *Scientific American*, **239**(4), 1978, 124–137.

CHAPTER 24 CONTROL AND INTEGRATION THROUGH THE INTERNAL ENVIRONMENT

24.1 Endocrine communication and control in animals

> **Practical investigation.** *Practical Guide 6*, investigation 24A, 'The microscopic structure of endocrine glands'.

Animals possess means of communication between their cells to enable them to co-ordinate their responses to stimuli from the external and internal environment. This may involve chemicals already present in the environment; for example, the accumulation of carbon dioxide controls ventilation movements in insects and mammals (see *Study guide I*, Chapter 1 and 2). In all multicellular animals, however, the nervous and endocrine systems allow the development of sophisticated levels of co-ordination and integration. The role of the nervous system has been discussed in *Study guide I*, Chapter 11. Here, we shall examine in detail the role of the endocrine system, a field of study which we call *endocrinology*.

What is endocrinology?

Endocrinology is the branch of physiology that is concerned with the study of hormones, the glands that produce them, and the target organs that respond to them. What then is a 'hormone'? The classical definition states that hormones are chemicals produced by ductless glands and released into the blood, which carries them to other parts of the body where they bring about specific regulatory effects.

Nerves, hormones and neurohormones

The nervous system is always working and it is a discrete and localized system of rapid control. In contrast the endocrine system often functions intermittently and its actions are more diffuse and usually more prolonged. Most important, though, is the way the two systems are linked to regulate bodily functions. Most endocrine cells are controlled at a distance, by secretions either from specialized nerve cells, called *neurosecretory cells*, or from other endocrine glands. Neurosecretory cells are neurones which also possess glandular activity. They are not unusual in being secretory, as all nerve cells release chemical messengers at their synapses. In conventional neurones, however, these transmitters are short-lived and travel only about 20 nm across the synaptic cleft. The chemical messengers or *neurohormones* released from neurosecretory cells are often persistent and may act on much more distant receptors by entering the general blood circulation.

Neurohormones are often released into the blood at particular sites

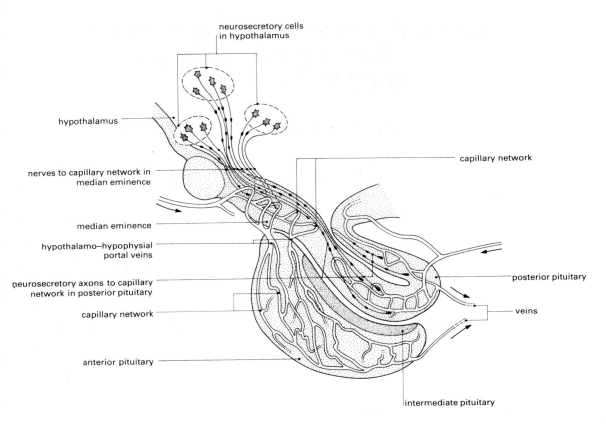

Figure 215
The anatomical relationship between the pituitary and the hypothalamus in a mammal. Neurosecretory cells in the hypothalamus send axons to the median eminence and the posterior pituitary. The axons terminate adjacent to capillaries in these two structures. Neurosecretory hormones produced in the hypothalamus are released into the blood. The pituitary portal system consists of a localized network of capillaries in the median eminence, taking blood via the portal veins to a second capillary network in the anterior pituitary. The hypothalamic releasing hormones which control the hormonal activity of the anterior pituitary thus reach the anterior pituitary by this portal blood supply.
Reprinted with permission of Macmillan Publishing Company from Hormonal control in vertebrates *by B. E. Frye. Copyright © 1967 by B. E. Frye.*

where neurosecretory axons are concentrated; in many of these, the accumulated endings of neurosecretory axons form specialized structures where release occurs. These structures are associated closely with the circulatory system and are termed *neurohaemal organs*; examples in the invertebrates are the insect *corpora cardica* (see *figure 229*) and, in the vertebrates, the posterior pituitary (*figure 215*). Neurosecretory cells form a third level of control: the *neurocrine system*.

The posterior pituitary of vertebrates contains axon terminals of neurosecretory cells whose cell bodies lie within the hypothalamus and produce peptides that function as true circulating hormones. Other neurosecretory cells in the hypothalamus have axon terminals in the median eminence, where they discharge hormones (hypothalamic hormones or releasing factors) into the blood in the hypothalamo–hypophysial portal system to control the synthesis and release of anterior pituitary hormones.

a *What features of neurosecretory cells make them more suitable than conventional neurones for controlling endocrine glands?*

The study of endocrinology

A classical experimental approach in endocrinology is to remove a suspected endocrine gland and test whether this results in a complex of symptoms (a syndrome) which are alleviated or reversed after subsequent replacement by re-implantation of the presumed endocrine gland.

It may be possible to correlate variations in the histological, ultrastructural, or histochemical features of suspected endocrine cells with certain physiological processes or events. Such correlations do not, however, indicate the exact nature of any relationship between a hormone and a physiological process or response. It must be determined whether it is a *primary* effect (a direct response of a target tissue to that hormone), or whether it is a subsequent or *secondary* effect consequent on an earlier event either in that target tissue or some other. For example, brain hormone stimulates moulting in insects (a secondary effect) but its primary action is to cause α-ecdysone production in the prothoracic glands (see page 329). In addition, we should realize that in living organisms many hormones are present in the blood simultaneously so that tissues are rarely exposed to one hormone only, but rather to a changing pattern of many. Often, hormones interact with each other so that one hormone antagonizes or potentiates the action of another. When two hormones act together to produce an effect they are said to be *synergists* (see page 340). Indeed the action of two hormones in combination may cause a response which neither hormone produces on its own.

The structure of endocrine cells

Endocrine cells show characteristics in their ultrastructure which relate to the type of hormone being produced.

STUDY ITEM

24.11 **The ultrastructure of endocrine cells**

A generalized protein- (or peptide-) producing endocrine cell is illustrated in *figure 216*. The major characteristics are the rough endoplasmic reticulum, a prominent Golgi complex, and large numbers of membrane-bound storage or secretory vesicles. The hormone is synthesized in the rough endoplasmic reticulum and packed into secretory granules in the Golgi complex.

Steroid-synthesizing cells (see *figure 217*) are quite different from those cells that produce protein.

a *What differences are apparent in the structure of these two types of cell and in the storage of the hormone?*

Neurosecretory cells and conventional neurones, have many features in common, including a similar morphology and the ability to conduct electricity. (See *figure 218*.)

b *In what ways do they resemble other protein-synthesizing cells?*

Neurosecretory cells are basically of two types: those that secrete peptides such as insect brain hormone and those that secrete biologically

Figure 216
Diagram of the ultrastructure of a typical protein-secreting cell.
After Fawcett, D. W. et al., Rec. Prog. Horm. Res. **25**, *1969, pp. 315–380.*

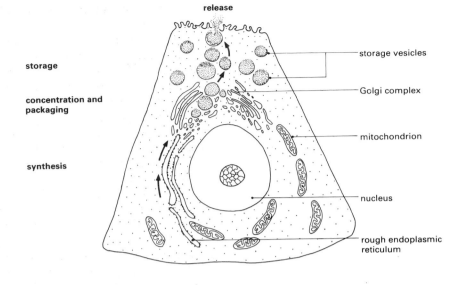

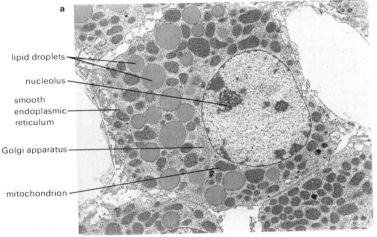

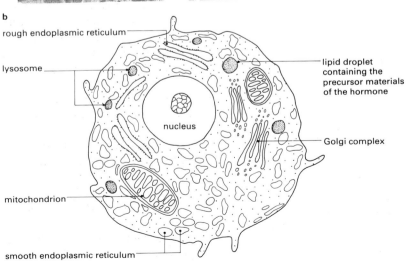

Figure 217
a Electronmicrograph illustrating the typical ultrastructure of a steroid-secreting cell. The cytoplasm contains many lipid droplets and there are numerous mitochondria. There is a prolific system of smooth endoplasmic reticulum which forms a dense tubular network. (× 4000).
b Diagram of the ultrastructure of a steroid-secreting cell.
Photograph, J. H. Kugler; the diagram in **b** *is after* Research in reproduction, **3**, *No. 5, International Planned Parenthood Association.*

Chapter 24 The internal environment   315

active amines such as adrenaline. Both store the hormones in membrane-bound vesicles.

c   *In what features does a neurosecretory cell differ from a conventional neurone?*

24.2 Hormone synthesis and release

The method of synthesis and release of hormones also varies with the type of hormone being produced.

Protein and peptide hormones

Protein hormones are synthesized initially as larger inactive molecules which are broken down subsequently to the smaller, active hormone. Protein hormone synthesis is well illustrated in the production of insulin in the β-cells of the islets of Langerhans of the pancreas. A precursor proinsulin molecule is synthesized in the endoplasmic reticulum and then broken down enzymatically to form proinsulin, which is packaged into vesicles in the Golgi region and converted into insulin.

The mechanisms of release of the hormone from the vesicles into the blood are not understood fully. One widely proposed mechanism is that of *exocytosis* (a kind of reverse pinocytosis) whereby hormone-containing vesicles approach the plasma membrane and the membranes of the vesicle and the cell fuse to release hormone into the extracellular fluid (see *figures 216* and *218*).

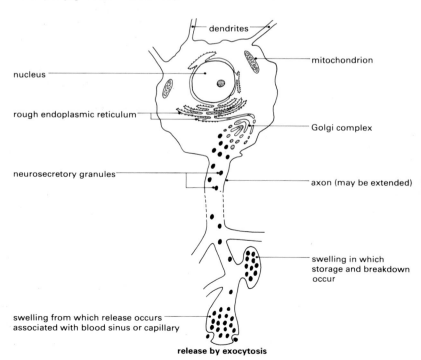

Figure 218
Simplified diagram of a neurosecretory cell. Neurosecretory material is synthesized in the cell body, packaged into granules, and transported along the axon to terminal swellings where it is released by exocytosis into the extracellular space. (Not drawn to scale.)

Steroid hormones

There is a common sequence of chemical reactions which lead to the synthesis of vertebrate steroid hormones. Cholesterol, which is always the precursor molecule, may be synthesized fresh from relatively simple molecules. Tissues that form steroids also obtain cholesterol directly, either by uptake from the blood or by hydrolysis of the cholesterol esters stored in cytoplasmic lipid droplets. Insects are unable to synthesize cholesterol, and therefore require a dietary supply of this steroid from which to synthesize α-ecdysone.

Tyrosine-based hormones

Thyronines. The thyroid gland gathers inorganic iodide from the blood and oxidizes it to iodine. This then replaces hydroxyl side-groups on the tyrosine residues of a large protein, thyroglobulin, to form thyronines (see *figure 219*). The thyronines formed are mainly thyroxine, T_4 (tetra-iodothyronine) and T_3 (tri-iodothyronine). Thyroglobulin does not accumulate in the thyroid cells but is secreted into the extracellular space of the thyroid follicles and stored (see *figure 220*).

When the thyroid gland is stimulated to release thyroid hormones, the thyroglobulin is taken back into the follicle cells and broken down to liberate the thyronines, which are then released into the blood.

Catecholamines. In mammals, the catecholamines adrenaline and noradrenaline are formed in the adrenal glands; noradrenaline is synthesized also as a chemical transmitter at sympathetic nerve endings and in the central nervous system. These hormones are synthesized from the amino acid tyrosine by the pathways shown in *figure 221*. In the medulla of the adrenal gland, two types of cells exist, one producing adrenaline and the other noradrenaline. The synthesized catecholamines are stored attached to a carrier protein called chromagranin, which is

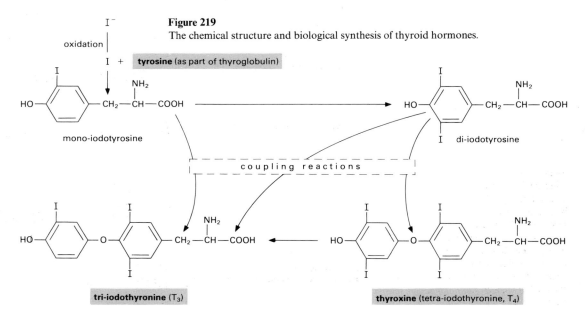

Figure 219
The chemical structure and biological synthesis of thyroid hormones.

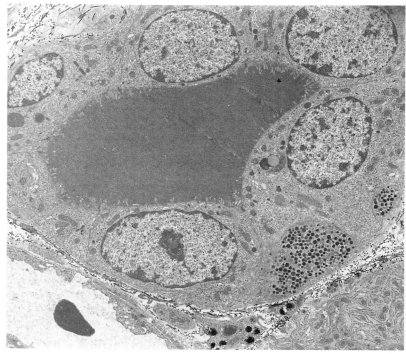

Figure 220
An electronmicrograph showing thyroglobulin stored in a thyroid follicle of a rat (× 4760).
Photograph, J. H. Kugler.

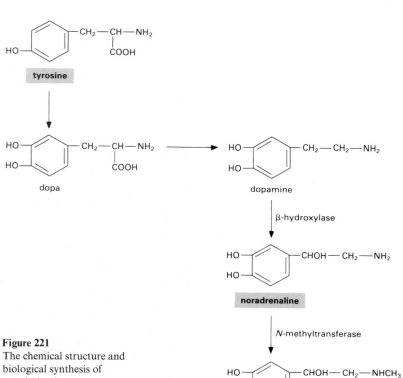

Figure 221
The chemical structure and biological synthesis of catecholamines.
After Bentley, P. J., Comparative vertebrate endocrinology, *Cambridge University Press, 1976.*

located in membrane-bound vesicles. Similar vesicles are found also in nerve endings where noradrenaline is synthesized.

Prostaglandins

A series of related 20-carbon unsaturated fatty acids, known as prostaglandins, are important components of many cellular processes. These substances were first found in fresh human semen. They attracted interest because they had a wide range of physiological and pharmacological effects, ranging from causing smooth muscle contraction, vasoconstriction, vasodilatation, and platelet aggregation, to the ability to mimic some hormones while blocking the effects of others.

Prostaglandins are synthesized by most tissues of the body, but their precise physiological role in many cases is not yet known; they may function as intracellular messengers, as local hormones, or as true circulating hormones. Certain anti-inflammatory, fever-reducing, and analgesic drugs such as Aspirin are thought to work because they suppress prostaglandin synthesis.

> **Practical investigation.** *Practical guide 6*, investigation 24A, 'The microscopic structure of endocrine glands'.

Control of hormone release

Animals must integrate the activities of their nervous and endocrine systems so as to maintain a constant internal environment in the face of changing external conditions. The control of hormone release is thus complex and can involve nervous, endocrine, and metabolic interactions. In general, however, we know very much less about the ways in which external stimuli influence hormone release than we do about the mechanisms by which the internal stimuli act. Of course, the internal stimuli may reflect external conditions; for example, excess heat and desiccation both tend to decrease the water potential of the blood.

If hormones are to regulate the function of their target cells satisfactorily, it is essential that the endocrine tissues should receive a constant supply of up-to-date information about the state of the target tissue. In other words, they operate as *closed loop* systems: the release of hormone into the blood initiates some action in the target tissue which influences further release of hormone. Usually this *feedback* of information is negative, in that it stops further release and tends to bring the system towards a steady state. Such closed loop control mechanisms vary in complexity (see *figure 222*). For example, the concentration of a hormone in the blood may influence its own secretion; this is an ultra-short (direct) negative feedback loop. Or the hormone in the blood, or a metabolite under the control of the hormone, may influence a higher centre in the central nervous system (or another part of the endocrine system) to modulate its own release directly; this is a short negative feedback loop. Finally, the target organ response may influence the

Figure 222
The control of hormone secretion in vertebrates by closed loop systems.
a Ultra-short negative feedback.
b Short negative feedback.
c Long negative feedback.

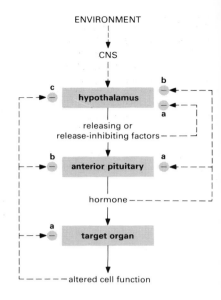

release of hormone by acting at a higher centre; this is a long negative feedback loop.

In vertebrates the higher centre in the central nervous system is usually the hypothalamus, which produces releasing or release-inhibiting factors to regulate the secretory activity of the anterior pituitary.

In *positive feedback*, the effect elicited by a hormone so alters the rate of release of that hormone that the net effect is greater; this type of feedback rapidly becomes unstable, however. Whereas negative feedback works to maintain some preferred steady state (homeostasis), positive feedback enables a discrete event to be achieved rapidly once initiated. Sexual behaviour, ovulation, and parturition are good examples of nervous and endocrine co-ordination that rely on positive feedback.

How does the arrival of the stimulus at the cell membrane of an endocrine cell initiate release of hormone?

Nervous stimuli, and some other chemical and hormonal stimuli, cause depolarization of the plasma membrane so that its permeability to inorganic ions is altered. In particular, the rate of Ca^{2+} influx increases, and intracellular Ca^{2+} pools may also be released, so that the concentration of Ca^{2+} in the *cytosol* (the fluid part of the cytoplasm excluding organelles and other solids) rises markedly. It is thought that this rise in intracellular Ca^{2+} stimulates movement of hormone vesicles towards the plasma membrane, and their fusion with it to bring about hormone release.

Modifications of this basic mechanism occur in different endocrine glands; for example, cyclic-3,5-adenosine monophosphate (cAMP) may be involved. Insulin release provides a useful specific example. The mechanisms by which secretion is linked to the presence of those stimuli which effect release are indicated in *figure 223*. Insulin release from the pancreatic β-cells is stimulated by glucose and also by glucagon.

Figure 223
Stimulus–secretion coupling in pancreatic β-cells. Exocytosis of the insulin-containing granules is stimulated by Ca^{2+} entering the cells during the action of glucose. Glucagon raises the level of cAMP in the cells, causing the release of Ca^{2+} from intracellular reservoirs and thus potentiating the action of glucose.
After Berridge, M. J., Advances in cyclic nucleotide research. **6**, 1975, pp. 1–98.

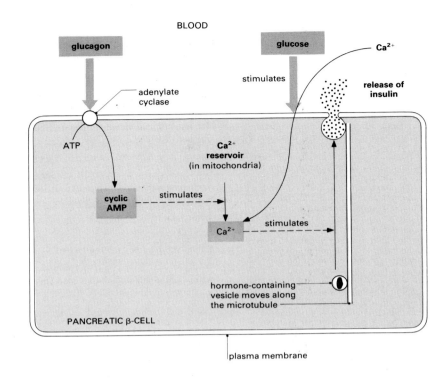

Glucose causes membrane depolarization and influx of extracellular Ca^{2+}; the consequent increase in intracellular Ca^{2+} leads to movement of the hormone-containing vesicles to the cell membrane and increased insulin release by exocytosis. This action of glucose is independent of cAMP. Glucagon, however, acts via an increase in the activity of an enzyme, adenylate cyclase. The increased levels of cAMP produced from ATP by this enzyme cause release of Ca^{2+} from the mitochondria. Glucose and glucagon thus both cause insulin release, but by different mechanisms (although both involve Ca^{2+}; see *figure 223*). The relative roles of cAMP and Ca^{2+} vary considerably from one tissue to another.

24.3 Dose–response relationships

Responses to injected hormones change with the amount of hormone applied (the dose). This is a function of the interaction of hormones with receptors in their target cells (see page 324). When hormones have several different actions, the dose–response curves for the various effects may lie far apart; therefore in the animal different actions will be evoked according to the concentration or *titre* of hormone in the blood.

STUDY ITEM

24.31 Dose–response curves to hormones

In mammals, insulin has four major metabolic effects

1 It inhibits the breakdown of muscle proteins, which provides amino

acids that can be used either to synthesize other proteins elsewhere, or to synthesize carbohydrate in the liver.

2 It inhibits the release of fatty acids from stores in adipose tissue.
3 It stimulates the entry of blood glucose into muscle and adipose tissue.
4 It stimulates glycogen formation in the liver.

The individual dose–response curves for each of these effects of insulin can be determined experimentally, and are shown in *figure 224*. The curves are almost completely separate. This means that when these effects are studied experimentally, the doses of insulin which bring about the four responses are quite different. The most sensitive response is the inhibition of the breakdown of muscle proteins, which is inhibited at quite low doses of insulin. The least sensitive is the stimulation of glycogen synthesis in the liver, which requires high doses of insulin. The ranges of the amount of insulin (the titres) found in the peripheral blood and in the hepatic portal vein of humans during different states are also indicated in *figure 224* by the dark horizontal bars.

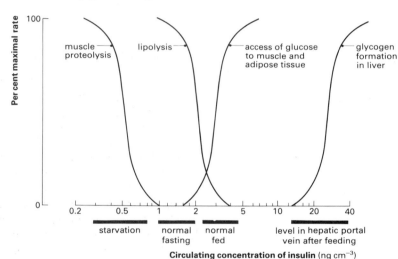

Figure 224
Dose–response curves for four important effects of insulin (note the logarithmic scale).

a   *From figure 224 suggest which of the effects of insulin are operating in different tissues under the following conditions:*
  1 shortly after a normal meal;
  2 three hours after a meal (called normal fasting);
  3 after two days of continuous fasting or starvation.

b   *Suggest reasons why the level of insulin in the hepatic portal vein is so much higher than in the general circulation.*

c   *What other hormones may be involved (in this case, released) to ensure survival during prolonged fasting?*

Variations in the response of the target tissue

The response of a target tissue to a particular concentration of hormone may be controlled by factors such as the stage of somatic and/or sexual development, the time of day (or month or year), and the previous endocrine experience. Later in the chapter we shall see that hormone sensitivity in amphibia and insects varies in a precise manner.

24.4 The fate of the hormone in the blood

Once a hormone enters the blood it may be taken up rapidly by its target organ, it may be inactivated by enzymes present in the blood or other tissues, or it may be excreted.

Many hormones are found in blood not as 'free' molecules, but 'bound' to specific proteins; this is especially true of steroid hormones, and also of those of the thyroid gland. The main function of binding proteins is probably to add specificity to the action of the hormone by regulating the distribution of hormone between target tissue and non-target tissue. Hormone will only be released from one binding protein to another (for example, the receptor) when the latter has a higher affinity for the hormone.

Activation and inactivation of hormones

Some hormones are modified in the blood or in other tissues to form active or more highly active derivatives. Little is known of the factors which regulate the nature or rate of these conversions, but they are important processes in the determination of hormonal activity.

In any system of chemical communication, mechanisms must exist to remove the chemical signal; it cannot be allowed to persist indefinitely or its target organ would behave like a motor car with no brakes. In fact, the rates of removal of hormones from the blood are remarkably varied (from hormone to hormone, animal to animal in different physiological or developmental states, and from species to species). Generally, hormones that influence developmental events have longer half-lives than those that exert rapid metabolic or physiological effects. Thus, for example, the half-life of thyroxine is about a week in humans, whereas that of insulin is from three to five minutes.

24.5 Mechanisms of hormone action

Nervous co-ordination is usually precise; for example, motor nerves stimulate discrete locations. Hormonal messages in contrast are broadcast widely in the blood in a manner which is essentially non-directional. In spite of this, endocrine coordination can also be directed precisely. This capability is a function of the target tissues; whilst blood-borne hormone molecules are carried to all cells, the 'message' is ignored in all except those equipped to receive it by virtue of their possession of specific receptors for the hormone molecules. How does the interaction of the hormone with its receptor lead to an appropriate response of the target tissue?

Protein and peptide hormones

In 1958 Professor E. W. Sutherland and his colleagues identified an important intermediate in the mechanism of action of adrenaline. This compound is cAMP, which we have already mentioned briefly. In the liver target cells, cAMP elicits the effect of adrenaline by activating the enzyme glycogen phosphorylase, which catalyses the breakdown of glycogen. Cyclic AMP is envisaged as the *second messenger*, responsible for the mediation of the intracellular actions of adrenaline (the 'first messenger').

The effects of many peptide hormones (as well as the responses to catecholamines such as adrenaline) are mediated by stimulation of the enzyme adenylate cyclase in the target cell. This enzyme catalyses the production of cAMP from ATP (see *figure 225*). The hormone (or 'agonist') binds reversibly to its receptor on the plasma membrane and forms a complex which is essential for its biological effect. As we have said, the specificity of hormone action depends on the distribution of receptors in the target tissues. Thus glucagon, for example, stimulates the breakdown of glycogen in liver but not in muscle, since muscle lacks the requisite glucagon receptor.

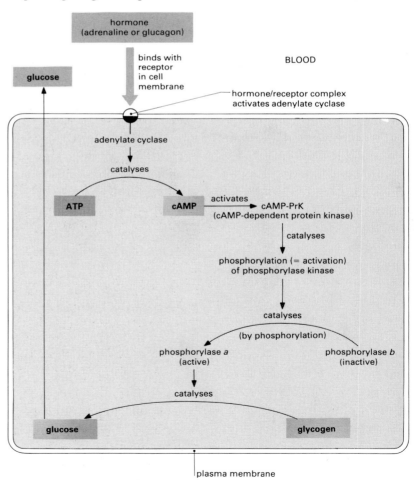

Figure 225
The mechanism of action of peptides and catecholamines in the control of metabolism. Glucagon and adrenaline bind to receptors in the cell membrane, and activate the enzyme adenylate cyclase; in this diagram, the receptor (upper semicircle) and the enzyme (lower closed semicircle) are shown as a single complex. Stimulation of adenylate cyclase causes the synthesis of cAMP, which mediates the release of glucose by activating the enzyme phosphorylase which catalyses the formation of glucose from glycogen. Increased levels of Ca^{2+} activate an enzyme which destroys cAMP.
After Berridge, M. J., Advances in Cyclic nucleotide research, **6**, *1975, pp. 1–98.*

How does an increase in the intracellular concentration of cAMP bring about the diverse functions of all those hormones which are thought to act by stimulating its production? Cyclic AMP binds to an enzyme, a protein kinase – known in full as cAMP-dependent protein kinase (cAMP-PrK) – and in doing so activates it. There are other effects of cAMP within cells (we have mentioned already its effects on Ca^{2+} regulation), but most attention has been focused on its activation of this enzyme. Many cellular proteins, including enzymes, undergo reversible phosphorylations and are physiological substrates for cAMP-PrK.

The particular importance of enzyme phosphorylation is that it converts one form of the enzyme to another; we talk of the enzymes being switched 'on' or 'off'. This is inaccurate, but is a convenient shorthand.

The role of cAMP as a second messenger is best known in relation to the effects of adrenaline on the breakdown of glycogen in muscle. Here, the cAMP-PrK phosphorylates a second protein kinase, phosphorylase *b* kinase, which catalyses another phosphorylation, that of glycogen phosphorylase *b*, converting it from this inactive form to an active '*a*' form (see *figure 225*).

Steroid hormones

Most steroid hormones appear to work by initiating the synthesis of specific proteins; thus, certain steroids (such as sex steroids in chicks and ecdysone in insects) induce the production of specific mRNA molecules (coded for specific proteins) in their target cells. This brief statement of their mechanism of action is deceptively simple. Steroid hormones which initiate growth and differentiation in their target tissues obviously act in a more complicated way, but the biochemical analysis of the numerous and intricate processes involved in development is very difficult. Nevertheless, the production of specific proteins serves as a useful starting point for the analysis of steroid hormone action.

What is the evidence that steroids act on the nucleus at the level of genetic transcription? If we inject ecdysone into larval insects and examine tissues such as the Malpighian tubules or the salivary glands, we can detect very rapid changes in the appearance of the chromosomes within the cells. These cells have what are called 'giant' or *polytene* chromosomes, which are visible under a light microscope (see page 35). The changes induced by ecdysone have been studied largely in the salivary gland chromosomes and involve the pattern of swellings or 'puffs' which can be seen along their length. These puffs are known to be sites of messenger RNA production and can be viewed therefore as sites of active genes (since transcription is occurring). We know that these puffs appear and disappear along different chromosomes in regular patterns, both in response to ecdysone injections and during normal development.

Support for the concept that steroids act at the nuclear level to control differential gene expression is obtained from observations that radioactively labelled steroids become localized in target cell nuclei. Thus we know that steroid hormones enter their target cells; indeed, they enter non-target tissues too. How, then, can steroid hormones be specific in their actions?

Figure 226
A tentative model for steroid hormone action.

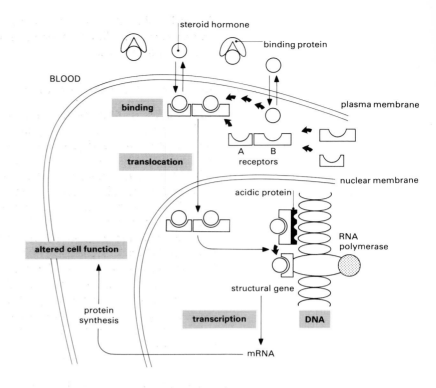

Steroids, because of their lipid nature, diffuse readily across the plasma membrane from the blood and enter most cells. In target cells, however, specific receptors bind the steroids and, in so doing, initiate the expression of the hormone's biological activity. Non-target cells lack such receptors.

One theory of steroid hormone action suggests that binding to the cytoplasmic receptor is followed by movement of the hormone–receptor complex into the nucleus, where it binds to some component of the chromatin and initiates transcription (see *figure 226*).

Thyroid hormones

Thyroid hormones affect growth, development, and metabolic activity in most cells. They stimulate oxygen consumption and increase RNA synthesis; they also have morphogenetic effects (that is, they bring about the arranging of cells to give form to organs or organisms of which they are part) which are especially prominent in some lower vertebrates. A further action, and one that is more specific, is the maintenance of normal growth hormone content in the pituitary; in the absence of thyroid hormone, the pituitary content of growth hormone declines rapidly.

Despite its considerable medical importance, the details of thyroid hormone action are still poorly understood. Broadly, it appears that thyroid hormone (TH) influences gene activity at the level of transcription, but it also stimulates protein synthesis at the level of translation in the cytosol. It may also affect the metabolic machinery of the cells directly by stimulating the oxidative metabolism of mitochondria. Binding sites for TH have been found in both the nucleus and the mitochondria.

24.6 Hormones in the control of growth and development

The life cycle of an organism consists of progressive changes which begin at fertilization and end shortly after death; these changes consist of periods of growth and replication during which a single cell, the egg, becomes an adult with millions of cells. These changes are known collectively as *development* (see Chapter 23).

In many animals, the larval or immediate post-embryonic stages are quite different in body form (and often in life style) from the adult; we call the change from one form to another *metamorphosis*. The intervention of metamorphosis in an animal's development allows the exploitation of different habitats and often involves dramatic differences in feeding, locomotion, and behaviour. The tadpole and the frog, the caterpillar and the butterfly, are familiar examples. An understanding of the nature of these changes and the means by which they are controlled are obvious and exciting challenges to biologists. We shall examine in some detail two of the best-known examples: metamorphosis in insects and amphibians.

The control of insect metamorphosis

The term *growth* has several meanings. It may be an increase in cell numbers, or in cytoplasmic or nuclear mass or volume, or in some constituent of protoplasm (such as protein or DNA). Thus an animal may be judged to have grown in one sense but not in another, according to the particular parameters employed. Insect growth, for example, is

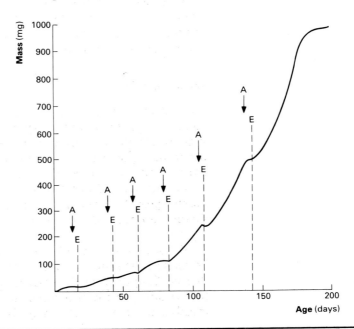

Figure 227
Growth in the stick insect, *Carausius morosus*, is a continuous process during each instar, the small discontinuities at each ecdysis being due to loss of the cast cuticle. E, ecdysis; A, apolysis.

Stage 1

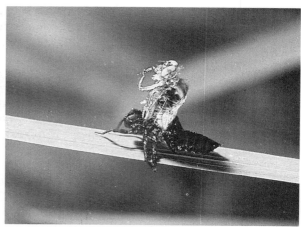
Stage 2

Stage 3

Stage 4

Stage 5

Figure 228
Libellula depressa moulting: five successive stages. *Photographs, Stephen Dalton/Natural History Photographic Agency.*

often described as 'discontinuous' because the semi-rigid exoskeleton, the cuticle, restricts any dramatic and visible changes in size to those occasions when the insect epidermis detaches itself from the old cuticle – a process known as *apolysis* – and grows a new cuticle, usually a bigger one. In terms of gain in mass (or of increasing DNA content) insect growth appears, however, to be almost continuous. As far as gain in mass is concerned the exceptions are those brief periods of reduced or zero growth when the old cuticle is discarded – a process known as *ecdysis*. This can be seen clearly in *figure 227* where the timing of apolysis and ecdysis are indicated; collectively, these two events constitute a moult.

Why do insects lose mass at ecdysis?

The first answer to this question is very simple; by definition, insects cast off the remains of the old cuticle when they moult. A second answer, however, requires careful examination of *figure 227*. Just before ecdysis insects stop gaining mass, or even begin to lose mass slightly. This is because just before apolysis they reduce, or stop completely, their feeding activity. Often, they undergo a distinct change in behaviour in order to find themselves somewhere safe from potential predators, because newly moulted insects are vulnerable until the new cuticle becomes hard (a process known as tanning). Perhaps, just as importantly, they find a position which makes moulting easier. You may imagine that it would be quite difficult to climb out of your clothes without using your hands! Not surprisingly, therefore, many insects seek places where they can hang upside down, so that they can make use of gravity to assist them in escaping from their old cuticle.

It is equally important that moulting is controlled carefully so that it occurs at a time when it is likely to be successful. That will be when the insect has sufficient reserves of energy with which to synthesize a new and larger cuticle (remember that feeding behaviour is severely curtailed during a moult), and when environmental conditions are appropriate. This control is achieved with the aid of hormones released in response to signals from the nervous system when the sensory input from the external and internal environment suggest that conditions are suitable.

STUDY ITEM

24.61 Which hormones are involved in insect metamorphosis?

The insect endocrine system is shown in a generalized scheme in *figure 229*. The brain is the source of a peptide hormone which causes the release of α-ecdysone, a steroid, from the prothoracic glands. α-ecdysone is converted immediately in the epidermis and elsewhere to β-ecdysone (*figure 230*), and it is this latter steroid which causes apolysis and therefore initiates moulting; β-ecdysone is thus a moulting hormone.

Because ecdysone causes insects to moult, and this allows growth in size to be expressed visibly, it has been described by some people as a growth hormone. Is this a reasonable conclusion?

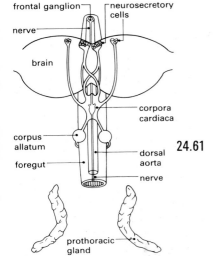

Figure 229
A diagram of the insect endocrine system.

Figure 230
The structures of α- and β-ecdysone.

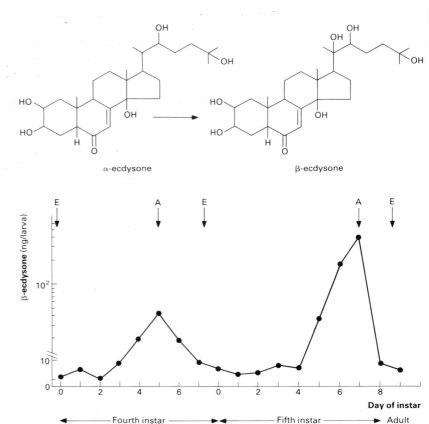

Figure 231
Titre of β-ecdysone in the locust. Ecdysone was extracted from whole fourth- and fifth-instar larvae, and from adults. The break in the y-axis represents a change in the scale from linear to logarithmic. E, ecdysis; A, apolysis. Note the coincidence of the peaks in β-ecdysone titre and the timing of apolysis.

a *Does the timing of the peak of β-ecdysone concentration (or titre) shown in figure 231 suggest it has a role as a growth hormone?*

We have seen that β-ecdysone allows visible expression of growth; it can also be said that it allows differentiation to proceed. In other words, when insects moult under the influence of β-ecdysone they can develop further towards the adult stage. But does this mean that β-ecdysone controls differentiation? There are two lines of evidence which argue against such a conclusion. First, in natural populations of termites, some of the workers undergo what are called 'stationary moults'. That is, they moult, but show no development either forwards or backwards.

b *Does this suggest that differentiation is an obligatory consequence of moulting?*

The second line of evidence comes from experimental studies in which adult insects are induced to moult. Adult winged insects do not normally moult, but they can be induced to moult under the influence of β-ecdysone. This can be achieved in two ways. The hormone may be injected, or the blood cavity of a final-stage larva which is producing its own β-ecdysone can be connected to the blood cavity of an adult by joining the animals together parabiotically. In either case the β-ecdysone causes the adult insect to moult. In both sorts of experiment, the new cuticle which is produced by the adult epidermal cells remains adult in character. If, however, the experiment is performed using younger larvae as the source of β-ecdysone, the epidermal cells of the moulting adult secrete new cuticle which has at least some larval characteristics.

c **What conclusions can be drawn from the results of these experiments?**

We know now that there is another hormone, juvenile hormone (JH), produced by a pair of small glands in the head called the corpora allata (see *figure 229*).

d **How could you demonstrate experimentally that JH is influencing differentiation?**

There are at least three ways in which we could imagine JH works: it could block adult development, it could maintain the existing stage, or it could actively promote larval characters. Let us examine these three possibilities.

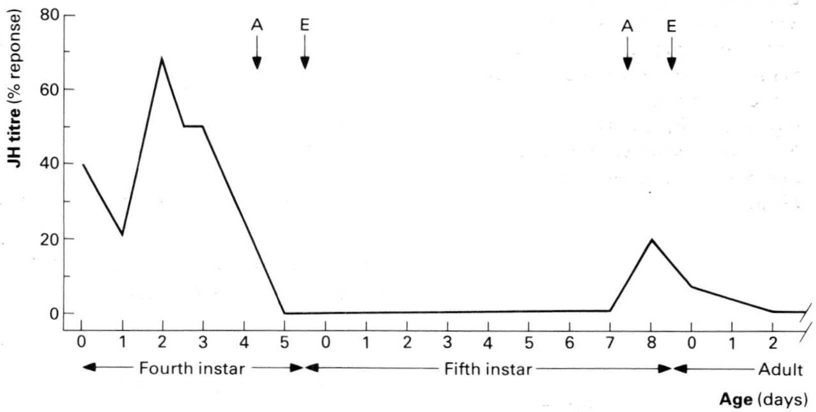

Figure 232
Titre of JH in the blood of the locust. Values are given in terms of the percentage of *Galleria* pupae showing a response when treated with an extract of blood from fourth- and fifth-instar larvae, and from adult locusts.
E, ecdysis; A, apolysis. Note that JH is absent for most of the fifth instar.

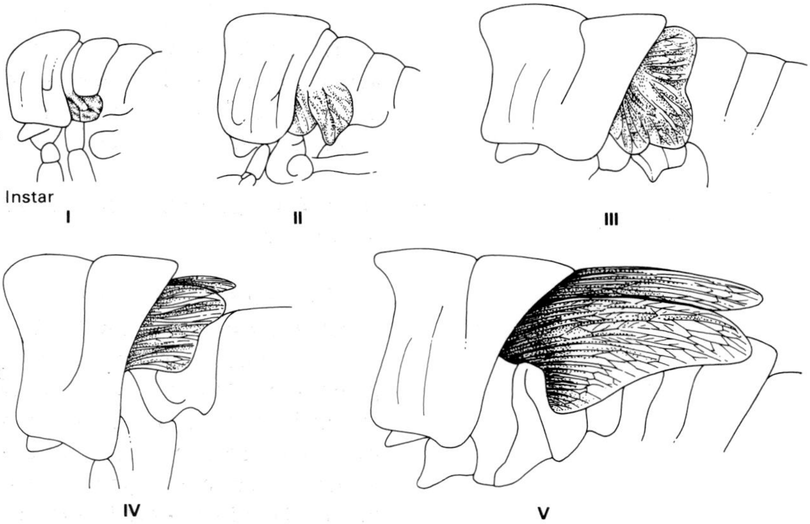

Figure 233
Changes in the shape and size of the wing pads in locust larvae. The general body form of locust hoppers may not appear to change much during the five larvae instars (**I** to **V**); there are, however, clear and progressive changes in the wing pads. These changes are controlled by the ordered decrease in the JH level.

Chapter 24 The internal environment

e *From the information shown in figure 232 do you consider JH stops development towards the adult? Give reasons for your answer.*

f *Does the evidence in figure 232 and figure 233 suggest that JH maintains the existing stage? Explain your answer.*

g *Is there any evidence from figure 233 and figure 234 to suggest that JH promotes larval characters?*

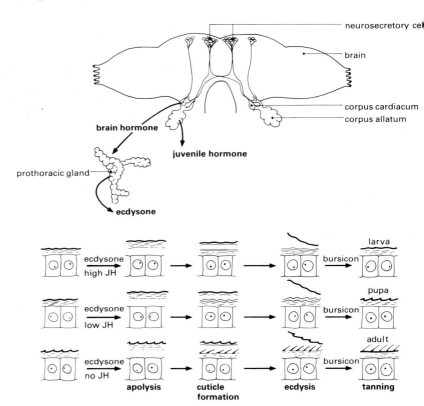

Figure 234
The insect endocrine system (above) and the effects of its hormones on growth and development (below). The hormone bursicon is secreted *after* ecdysis, and is responsible for the hardening or tanning of the cuticle.

What is the mechanism by which JH works?

Hemimetabolous insects, such as cockroaches, locusts, and bugs, have no pupal stage, and the adult tissues develop from the same cells (or their descendants) as those that form the larval body. Since all cells in an insect body have the same genotype, we must conclude that in hemimetabolous insects the larval cells carry a dual pattern – the visible larval pattern and an invisible adult one. In other words, the larval pattern, determined by one set of genes, is 'switched on' in the larvae whereas the adult pattern, determined by a separate set of genes, is 'switched off'. Correspondingly, in adults, the larval genotype is 'switched off', whereas the adult genotype is allowed phenotypic expression.

We can therefore describe the action of JH as being concerned with 'switching on' larval genes and 'switching off', adult genes, with a particular balance between the expression of the two extreme phenotypes being determined by the amount of JH present. We cannot as yet describe the exact mechanism by which JH switches genes on or off; we do, however, have some clues.

Earlier in this chapter we discussed the visible changes which take place in the appearance of the 'giant' chromosomes in response to ecdysone: these patterns change during development in such a way that we can recognize patterns of chromosome puffing which are characteristic of young larvae, and which are quite different from those seen in older insects. The significant finding here is that the particular puffing pattern depends upon the amount of JH, both during normal development and when injections of ecdysone with JH are administered. Experimental studies on the effects of ecdysone on polytene chromosomes of insects provided the first evidence that steroid hormones (like ecdysone) influence nuclear activity directly, but we now know a great deal more about how steroids affect gene activity (*figure 226*). In general, they stimulate specific messenger RNA production to alter cellular activity (see page 137).

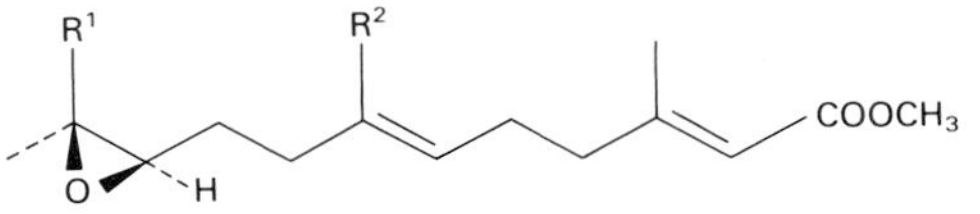

JH-I: $R^1 = R^2 = C_2H_5$
JH-II: $R^1 = C_2H_5, R^2 = CH_3$
JH-III: $R^1 = R^2 = CH_3$

Figure 235
Structures of insect juvenile hormones.
From Etkin, W. (eds Etkin, W. and Gilbert, L. I.), Metamorphosis, a problem in developmental biology, Appleton-Century Crofts, 1968.

Edcysone and juvenile hormone act together to control metamorphosis. This is an example of synergism between hormones. By acting together, they achieve an effect (in this case the control of metamorphosis) which neither hormone could bring about by itself.

STUDY ITEM

24.62 Insect metamorphosis

Holometabolous insects, such as flies, butterflies, moths, and beetles, undergo complete metamorphosis (that is, a pupal stage is present). In these insects adult tissues develop largely from the growth of small islands of tissue (*imaginal discs*) present even during development of the embryo in the egg. These imaginal discs grow during the larval stages, but take no part in the formation of the body of the larvae. In the pupal stage, however, the discs proliferate to form the adult tissues, the larval tissues being broken down. In hemimetabolous insects, cells which form larval tissues also form adult tissues. In the following passage evidence is given concerning both holometabolous and hemimetabolous insects. In answering the questions you should assume that the control of moulting and metamorphosis is similar in the two types of insect.

Part A. Hormones and moulting
Sir Vincent Wigglesworth at the University of Cambridge carried out most elegant but simple experiments with the blood-sucking bug (Hemiptera) *Rhodnius prolixus*. Some of the findings from his study are

given here. Normally, *Rhodnius* takes a blood meal only once during each larval stage.

Study the evidence given and use it to answer the questions.

Rhodnius larvae which are continuously starved survive for long periods but never moult. If a starved *Rhodnius* larva is allowed to take a blood meal, it moults a fixed time after the meal.

a *What conclusions can you make from this observation?*

If a starved *Rhodnius* larva is fed but then decapitated any time up to four days later, it does not moult. If, however, decapitation is delayed for longer than four days (this is called the critical period), the larva does moult.

b *What conclusions can be drawn from these observations about the importance of the head and the time of decapitation?*

If a starved *Rhodnius* larva is fed but the ventral nerve cord is cut at any time up to four days after the meal, the larva fails to moult.

c *What conclusions can be drawn from this experiment?*

Two *Rhodnius* larvae can be joined together by an interconnecting narrow glass tube, so that blood can mix between the animals. If two fed larvae are joined in this way they both moult. If one fed larva is joined to a starved larva, they both moult. If two decapitated larvae are joined but one of the larvae has been decapitated after the critical period, the pair moult; but if both are decapitated before the critical period, they do not moult.

d *What further conclusions can be made about the role of the head in moulting?*

Scientists have found that if a brain, or that part of a brain containing the neurosecretory cells, is taken from a fed *Rhodnius* larva and implanted into a larva decapitated before the critical period, the host larva moults.

e *What is the source of the control exerted by the brain over moulting?*

Another scientist, Professor C. M. Williams, performed experiments similar to those of Wigglesworth using silkmoths (holometabolous insects) instead of bugs. Silkworms are more or less continuous feeders (very unlike *Rhodnius*) but they still show a critical period, that is, a period during each larval stage, immediately after a moult, during which feeding is essential for moulting to occur. If silkworm larvae are decapitated after the critical period, they moult (as in *Rhodnius*).

If a ligature is placed between a silkworm's head and its thorax after the critical period the larva still moults, whereas if the ligature is placed before the critical period, it does not moult.

f *Is the importance of the head in moulting the same in silkworms as in Rhodnius?*

If a ligature is placed behind the thorax of a silkworm (to isolate the thorax and head from the abdomen), only the head and thorax moult regardless of whether the ligature is placed before or after the critical period.

g *What conclusion can be drawn concerning the possible role of the thorax in moulting?*

If a silkworm is decapitated before the critical period it does not moult. If, however, a brain from another suitable donor is implanted into the thorax of such a headless caterpillar, it moults. If the brain is implanted into the thorax of a headless caterpillar which has been ligated between the thorax and abdomen, only the thorax moults. If the brain is implanted into the abdomen of a headless caterpillar which has been ligated between the thorax and abdomen, there is no moult at all.

h *What is the role of the brain in moulting?*

A silkworm decapitated before the critical period will moult if prothoracic glands taken from a caterpillar fed for longer than the critical period are implanted into the larva; the glands will be ineffective, however, if they are taken from a caterpillar which has not fed for the critical period.

i *What is the role of the prothoracic glands?*

j *What is the relationship between the length of feeding (associated with the critical period) and the effectiveness of the prothoracic glands in promoting a moult?*

Part B. Hormones and metamorphosis
In the experiments in which Wigglesworth joined *Rhodnius* larvae together so that their blood systems were connected, he sometimes found that the rate of development of a larva was affected by its being joined to a larva of a different stage.

If a fourth-instar larva is decapitated before the critical period and joined to a fifth-instar larva (the final larval stage) which has been decapitated after the critical period, the pair moult. This is because they are under the influence from the fifth-instar larva, whose endocrine system was activated during the critical period when the head was present. But while the fifth-instar larva moults to a normal adult, the fourth-instar larva moults, not to a fifth-instar larva (as it would normally do), but to a small adult!

If, however, the timings of the larval decapitations are switched round – so that they moult this time under the influence of the endocrine system from the fourth-instar larva – an entirely different result is obtained. In this case, the fourth-instar larva moults to a normal fifth-instar larva, while the fifth-instar larva moults in a strange way so that it appears somewhat intermediate between a fifth-instar larva and an adult; it possesses some characters of both.

k   *What conclusions can be drawn from these experiments concerning the control of metamorphosis?*

Wigglesworth related his findings on metamorphosis to the hormonal activity of a gland called the corpus allatum (see *figure 229*), situated just behind the brain. He could perform decapitation experiments in two ways: either the brain was removed leaving behind the corpus allatum, or the brain and the corpus allatum were removed together. This latter procedure will be referred to as *allatectomy*, although strictly speaking it involves decapitation with removal of the brain as well as of the corpus allatum. Wigglesworth was able to show that allatectomy causes precocious metamorphosis. If a fourth-instar larva is joined to an allatectomized third-instar larva, when they moult the development of the fourth-instar larva is normal. The third-instar larva however moults almost to a fifth-stage larva, that is, its development is slightly accelerated. If the experiment is repeated using an allatectomized fourth-instar larva joined to a third-instar larva which retains its corpus allatum, development of the third-instar larva is normal but that of the fourth is retarded considerably.

l   *What conclusions can be made from this experiment? Relate your answer to your conclusions from the other experiments described in this section.*
☐

The control of amphibian metamorphosis

The familiar changes which occur during the transformation of the tadpoles of frogs and toads into adults are dramatic. The fully aquatic larva without lungs or limbs suddenly grows legs, crawls on to land, and becomes terrestrial. The general course of amphibian metamorphosis is well known and covered in numerous textbooks and will not be described in detail here. You should, however, familiarize yourself with anatomical and morphological changes associated with the three main phases of metamorphosis.

In *Rana pipiens*, the first phase of tadpole development lasts for about 20 days; we call this *premetamorphosis*. This is followed by *prometamorphosis* for a further 20 days, in which there are gradual changes (including growth of the hind limbs). This leads into the final 10 days of *metamorphic climax* when the transformation into a small adult frog takes place (see *figure 236*).

We can recognize similar questions here to those which were asked in relation to insect metamorphosis. What hormones control these events?

Figure 236
Pattern of metamorphosis in *Rana pipiens*. Data from one batch of normal animals raised at 23 °C shown by solid line. Comparable data for animals with thyroid removed (TX) and pituitary removed (HX) are shown by broken lines.
After Etkin, W., (eds W. Etkin and L. I. Gilbert) Metamorphosis, a problem in developmental biology, Appleton–Century Crofts, 1968.

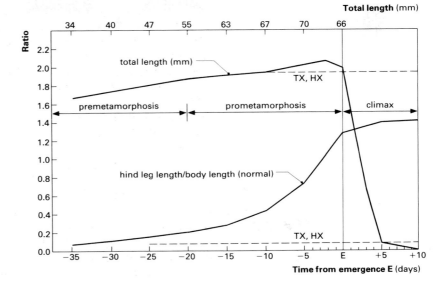

How do the hormones work? We know that at least two hormones are of major importance, thyroid hormone (TH) and prolactin (PRL).

Thyroid hormone and amphibian metamorphosis. Thyroid hormone (TH) was first implicated in amphibian metamorphosis many years ago when Gudernatsch fed extracts of thyroid glands to frog tadpoles and found that they underwent precocious metamorphosis. At that time (1912) this was a remarkable finding, because the general concept of the internal secretions, which we call hormones, was not yet established. We know now, of course, that the thyroid gland of vertebrates produces hormones (*figure 219*) which regulate energy production and influence growth. At that time the suggestion that TH controlled amphibian metamorphosis was a landmark in the development of our understanding of the role of hormones in controlling bodily processes.

In premetamorphic tadpoles there are low levels of TH in the blood, but at the onset of the prometamorphic stage the TH levels begin to increase. Why should this be so? Circumstantial evidence suggests that during the premetamorphic stage TH exerts an adult-type negative feedback control over the release of a hormone, *thyroid stimulating*

Figure 237
Changes in the control of TH release during tadpole development; + and − represent positive and negative effects respectively. TH, thyroid hormone; TRH, thyroid releasing hormone; TSH, thyroid stimulating hormone.

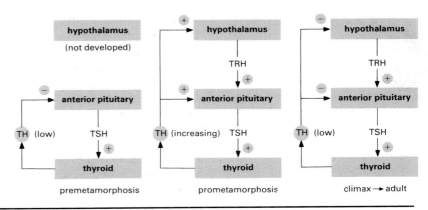

Chapter 24 The internal environment

hormone (TSH), from the pituitary gland (see *figure 237*). During prometamorphosis, however, the situation changes. At this time the pituitary gland and its associated structures are developing rapidly, and the ensuing release of a thyroid releasing hormone (TRH) from the hypothalamus stimulates TSH release from the pituitary. However, because TH itself may stimulate release of TRH (a positive feedback – see *figure 237*), the levels of TH begin to increase dramatically. This is the signal for the onset of metamorphosis. Increasing levels of TH stimulate further development of the pituitary gland, including that of the local blood vessels in the median eminence which carry hormones from the hypothalamus to the anterior pituitary (see *figure 215* and *figure 238*). One consequence of this is the release of a prolactin release-inhibiting hormone (PRL-IH) which is carried in these local blood vessels to the anterior pituitary and inhibits the release of the second hormone of major importance in controlling tadpole development, prolactin (PRL).

Prolactin and amphibian metamorphosis. In premetamorphic tadpoles, the PRL content of the blood is high. This promotes growth and inhibits metamorphosis. In this sense, the premetamorphic tadpole is stabilized; high levels of PRL and low levels of TH stimulate larval growth. However, as growth of the pituitary gland and its vascular connections with the median eminence proceeds, an inhibition of PRL release by PRL-IH develops so that, during prometamorphosis, PRL levels decline. Thus, changes in the levels of TH and PRL are mirror images of each other: TH increases while PRL decreases (see *figure 238*).

These dramatic changes are all consequent on the positive feedback influence of TH on the hypothalamus. This type of feedback, as we saw in section 24.2, characteristically becomes unstable, while negative feedback often underlies the maintenance of some preferred state (homeostasis). Positive feedback enables some discrete event (in this case metamorphosis) to be achieved rapidly once initiated. The changing patterns of the amounts of hormone initiate gross morphological changes. Specifically larval tissues like the tail, which have grown and been maintained under the influence of prolactin, are reabsorbed; adult tissues, such as limbs, differentiate. Once metamorphosis is complete, the control of TRH release in the new adult is by negative feedback, although the mechanism which brings about this change in the responsiveness of the hypothalamus is unknown. The appearance of this negative feedback loop however, restores a balance to the hormone levels in the adult (see *figure 238*).

How do thyroid hormone and prolactin work?

Precise answers to this question cannot be given, but we do know something of the broad effects of the two hormones and of their actions at the cellular level.

Thyroid hormone. A considerable understanding of TH action in the tadpole has been achieved by studies which involve incubating the isolated tadpole tail (*in vitro* experiments). Addition of TH brings about dramatic changes in the tail; with the exception of the epidermis, all the

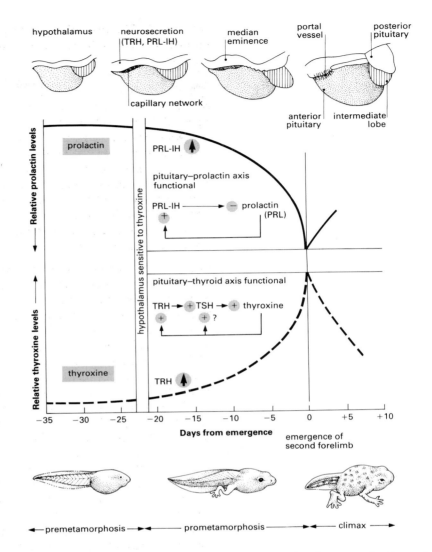

Figure 238
The role of hormones in the growth and development of the tadpole; + and − represent positive and negative effects respectively.

tissues begin to break down. The hormone stimulates production of mRNA which is coded for the synthesis of degradative enzymes, so that some hours after its addition, TH causes the regression of this typical larval tissue, the tail.

The action of TH is not always destructive (catabolic), however. For example, the epidermis in general responds to TH by growing and thickening – an adult characteristic. Different tissues and cells have different responses to the hormone. Indeed, the epidermis provides a splendid example of this. At the end of prometamorphosis, epidermis in the opercular region becomes thin, and forms what are called opercular windows through which the forelimbs emerge later on. If a pellet of cholesterol containing TH is implanted in this region during early prometamorphosis, the slow but prolonged release of TH from the pellet produces local thickening of most of the epidermis. In the area where the window is to form, however, the hormone evokes a thinning of the epidermis. Thus, epidermal cells which look identical, and which presumably carry the same genotype, give two completely different responses to the same hormone!

Prolactin. Prolactin is a polypeptide hormone containing about 190 amino acid residues. It is secreted from the anterior pituitary gland. It belongs to a family of structurally similar peptide hormones which includes growth hormone. Therefore it is perhaps not surprising that it has the effect in amphibian larvae of stimulating growth in larval tissues. Although most peptide hormones appear to act by stimulating cAMP production in their target cells, biologists are unsure how growth hormone or PRL act. Nevertheless, PRL stimulates protein synthesis by increasing mRNA synthesis in the nuclei of responsive cells. In the tail and the gill, for example, PRL stimulates connective tissue growth, especially the synthesis of collagen. Of course, the tail and the gills are characteristically larval tissues, but PRL may well have some influence on the growth of other tadpole tissues too.

Interactions of TH and PRL

It is clear that TH and PRL act in different directions. TH promotes development towards the adult, and PRL stimulates growth and retention of larval tissues. However, the hormones are antagonistic not only in their separate and opposite functions but also in their direct actions in the tissues. Thus injections of PRL will prevent TH-induced (metamorphic) changes and *vice versa*. The exact nature of their interaction at the cellular level is not known, but there is some evidence that PRl may be antimetamorphic in a way quite separate and distinct from the mechanism by which it stimulates growth (in larval tissues). Nevertheless, clearly PRL and TH control amphibian metamorphosis by a balanced synergism, similar to the system we have seen controlling insect metamorphosis. Hormones act together in different and often opposite ways to bring about a given process in an ordered manner. Metamorphosis is thus controlled via the complete pattern of the changing amounts and interactions of the two hormones.

General conclusions concerning the control of metamorphosis

In the two examples of metamorphosis which we have discussed, there are some common features. In each case hormones control the expression of genetic information in such a way that phenotypic expression varies with each developmental stage in a co-ordinated manner. Although both examples involve a hormone which can be described as antimetamorphic (JH and PRL), these hormones act in quite different ways and there do not appear to be important common features other than their general inhibitory effect. On the other hand, one conclusion we can draw from both examples concerns the general way in which all hormones are involved in metamorphic development. The different target cells or tissues are programmed to give precise individual responses. The hormones evoke these responses. Thus hormones signal that development should proceed, but the built-in programme of each cell determines what the response should be: in what form, and how far, that development should proceed.

24.7 Communication and control in plant growth

Farmers and gardeners in particular know that the weather affects the growth of plants. For example, plenty of sunshine, equable temperatures, and rain are needed to produce good plants. 'Good plants' in these terms usually means a greater quantity rather than a change in quality; favourable conditions may produce bigger and better cabbages but they don't turn into Brussels sprouts or radishes. To say that the features of different kinds of plants are due to their genetic make-up does not go far towards explaining how the genes achieve the characteristic shapes which we recognize. The differences in height, leaf shape and size must be achieved by precise co-ordination of the position and time of onset and duration of cell division and expansion. The growth of the root and the shoot are strongly interdependent and there must be co-ordination within a single plant of the growth of its different parts.

Sometimes this appears to be controlled by simple nutritional factors. This is illustrated in a study of the growth of radishes under high and low light intensity, coupled with either adequate or limited nitrate supply. Table 39 lists some results, which show that when light was limiting the shoots grew relatively better than the roots, but when nitrate was limiting the roots grew relatively better than the shoots. The simplest explanation would be that when a resource is limiting the part of the plant which has first access to the resource grows better than the part which has to manage on the 'left-overs' (nitrate enters via the roots.).

Experiment A. Restricted light

		Dry mass (mg)		$\dfrac{\text{Shoot}}{\text{root}}$
		Root	Shoot	
Low light	$\left(\dfrac{H}{2.5}\right)$*	20	80	4.0
High light	$\left(\dfrac{H}{1}\right)$	37	118	3.2

Experiment B. Restricted nitrate

		Dry mass (mg)		$\dfrac{\text{Shoot}}{\text{root}}$
		Root	Shoot	
Low nitrate	$\left(\dfrac{S}{10}\right)$†	260	188	0.72
High nitrate	$\left(\dfrac{S}{1}\right)$	357	400	1.12

Plants in Experiment A were very young seedlings, which had only their first true leaf emerging, and they were grown for 21 days in the experiment. In Experiment B the plants had four true leaves and were grown for 28 days under experimental conditions. By then they were producing good radishes which were classed as 'root' for this experiment.

* $\dfrac{H}{2.5} = 40\%$ of the full light intensity in the growth room

† $\dfrac{S}{10} = 10\%$ of the standard NO_3^- in a full nutrient solution.

Table 39
The effects on radishes of restricting either light or nitrate supply (class results; data are the means of ten plants per treatment).

These are responses to gross changes in major nutrients, however, as light intensity is directly related to sugar availability. Plants also respond to more subtle changes in their environment, such as uneven illumination, or changes in orientation to gravity, or seasonal changes in day length. All these factors are continually monitored by plants, and there is evidence that their responses may be mediated by chemical messengers.

Hormones as messengers in plants

By the late 1920s, the animal physiologists had discovered that there were specific chemical messengers which they called hormones, which were produced by endocrine glands and which were active at extremely low concentrations. Then, through the study of phototropism in cereal coleoptiles, scientists seemed to have discovered a similar hormone system in plants – except that plants did not have endocrine glands. This seemed to be a minor point compared to the obvious parallel with the concept of a specific, translocatable organic messenger which is active in extremely small quantities.

In 1934 indole-3-ethanoic (indole-3-acetic) acid (IAA) was identified as a naturally occurring growth substance or *auxin* which had a remarkably simple structure (see table 40) compared with animal hormones (see *figures 219, 221*, and *231*). There was evidence that this chemical travelled from the tip of a coleoptile towards its base and that its distribution could be affected by the stimulus of unilateral light. The stimulus and redistribution of the hormone were perceived and effected within the apical 0.5 mm or less, whereas the regulation of cell extension occurred 2 or 3 mm lower down the coleoptile (see *Study guide I*, Chapter 12). By use of this substance as the messenger between the site of perception and the site of response, the growth of the organ could be changed so that it became orientated into a position in which the seedling would be best able to absorb light for photosynthesis. Soon, the same substance was implicated as mediator of the gravitropic (geotropic) response in roots.

Subsequent work with recently discovered hormones indicates that the earlier work was based on an over-simplification of the mechanism. Nevertheless, the concept of specific messengers still stands. Table 40 summarizes the five groups of plant hormones discovered so far and indicates their main properties, as well as some commercial applications which have been developed.

As new plant growth regulators have been discovered, the basic idea that a hormone must act at a site distant from its origin has become less tenable. This is particularly so in the case of ethene (ethylene), which is a naturally produced gas. Ethene seems to be active in the tissues which produce it although, because of its mobility within a tissue via the intercellular space, it is not easy to pin-point a site of synthesis. Because there are many good examples which support the concept of a hormone as a chemical messenger, it would be premature to abandon it just yet. Some scientists would prefer to exclude ethene from the definition of a hormone; only further discoveries will help to resolve this dilemma.

Hormone group	Type substance and formula	Site(s) of synthesis	Some effects on growth and development
Auxins	Indole-3-acetic (-ethanoic) acid (IAA)	Dividing cells such as apical meristems, cambium, young leaves, seeds, and wound tissue	Promotes cell extension Suppresses outgrowth of axillary buds Promotes formation of lateral roots in stem cuttings, but inhibits their subsequent elongation Inhibits abscission of leaves, flowers, and fruits Promotes xylem differentiation from cambium
Gibberellins	Gibberellic acid (GA_3)	Probably as auxins but less clear; concentrations high in young leaves and apices, also in seeds	Promotes internode growth, especially in dwarf cultivars Promotes mobilization of seed reserves when embryo germinates Promotes fleshy development in fruits and, in conjunction with IAA, allows parthenocarpic fruit development Promotes bolting and flowering in long-day plants and biennials Breaks dormancy in cold-requiring seeds
Cytokinins	Kinetin	Probably as auxins; high concentrations in young fruits; may be significant upward transport from roots	Promotes cell division in callus cultures Encourages shoot formation in callus cultures Releases axillary buds from inhibition by IAA and favours their outgrowth Delays senescence in detached leaves Promotes germination in some dormant seeds
Abscisins	Abscisic acid (ABA)	Not clear; concentration in leaves rises under water stress and in senescing tissues	Imposes dormancy on non-dormant shoot meristems and seeds Promotes abscission of leaves, fruit, and flowers Causes stomatal closure
Ethene (ethylene)	$CH_2 = CH_2$	Not clear; concentration in some fruits rises dramatically at maturity; produced in developing seedling	Promotes fruit ripening Promotes senescence in leaves and flowers Produces downward curvature in shoots Promotes closure of plumular hook in seedlings Promotes internode extension in submerged aquatic stems

Table 40
The five groups of plant hormones discovered so far.

Metabolism of IAA

IAA enjoyed sole recognition as a plant hormone from the 1920s until the 1950s. All work in this field was focused on the compound, and more is known about its synthesis and breakdown than those of any other hormone. It will be taken as an example to illustrate the general ideas.

The indole nucleus is already present in the amino acid tryptophan and relatively minor modifications to the side-chain are needed to convert it into IAA. Oxidative deamination liberates the amino group and decarboxylation reduces the three-carbon side-chain to the two carbon ethanoate group. Enzymes are known which mediate these

Figure 239
The main pathways of synthesis and inactivation of IAA.

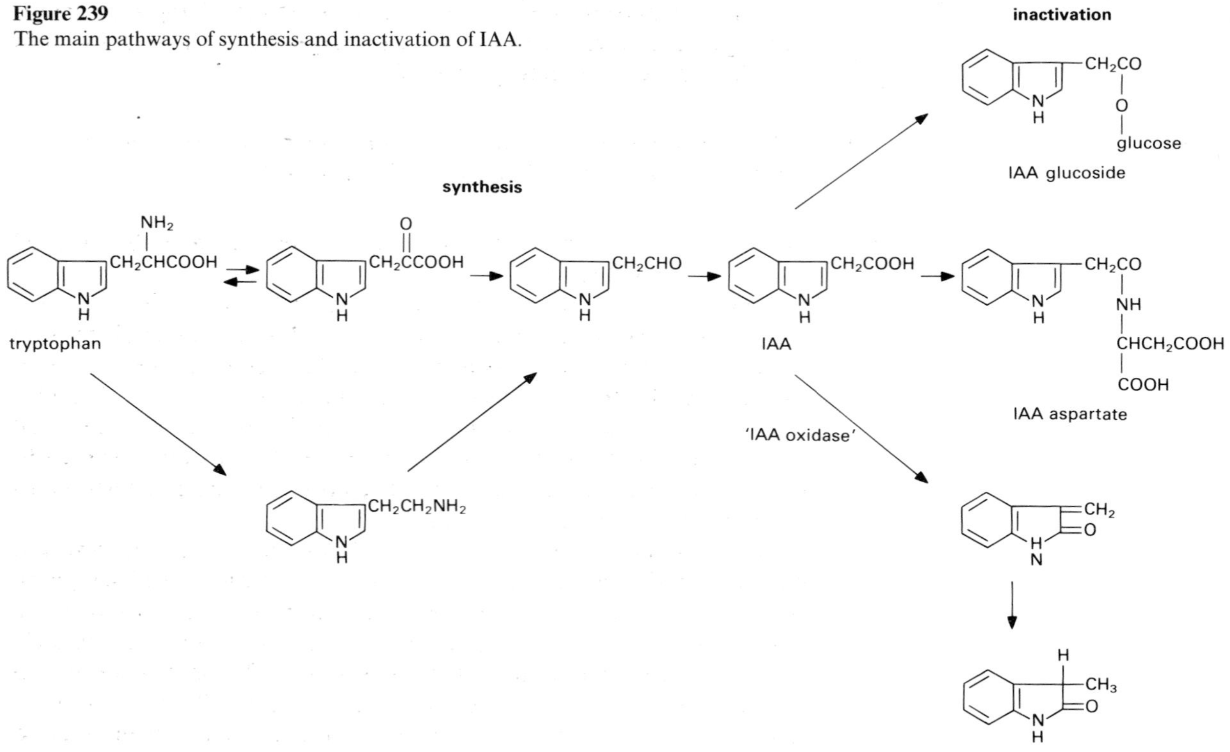

reactions and organisms adopt one or other of the two possible synthetic pathways (see *figure 239*).

IAA can form complexes with amino acids, and also conjugates with sugars such as glucose; in each case its auxin activity is removed. In these forms it could be stored in a reservoir of inactive product from which it might be released when needed.

A different type of inactivation is brought about by an enzyme called '*IAA oxidase*', which can be extracted from plant tissues. IAA oxidase will destroy IAA in the plant; it seems to be the most active in the older tissues and roots where IAA concentrations are at their lowest.

a   *Why is it necessary (apart from providing a means of storage) to have a system which inactivates or destroys a hormone within the plant?*

Mechanism of action of IAA in cell extension

If we take an elastic band and pull gently at each end it will stretch slightly, but as soon as we stop pulling it returns to its original size. This is elastic extension. It is this sort of extension which a cell wall undergoes when water enters the cell by osmosis. Tension is set up in the elastic walls as the cells become turgid.

If we gently pull a piece of Plasticine, however, it stretches – and becomes longer but thinner, and does not return to its original shape when released. This is plastic extension.

If the properties of the cell wall in a turgid cell changed so that it became a little more plastic and a little less elastic, then it would start to stretch as more water entered the cell.

b *Why does water enter the cell now, when prior to the increase in plasticity there was equilibrium at full turgor?*

When plastic extension ceases and only elasticity remains the enlarged cell will regain full turgor.

Elasticity is due to the strong cross-linking between the molecules which make up cell walls, such as cellulose and hemicellulose. If some of the cross-links were weakened then these long-chain molecules could slide past each other and we could observe plastic extension. The idea that IAA in some way modified the cross-linking between cell wall constituents has been around for a long time; the difficulty has been to identify the chemical nature of the bonds that change. The early 1980s saw two important advances in this area.

The first development concerns a knowledge of the nature and functioning of membrane-bound proton pumps which effectively pump H^+ from one side of a membrane to the other, using ATP to drive the pump. Proton pumps are important in chloroplasts and mitochondria where they are involved in the mechanism of ATP synthesis coupled to electron transport (see *Study guide I*, Chapters 5 and 7). All we need to recognize here is that if H^+ are moved from one side of a membrane to the other, then the side which receives them will become more acid (the pH will fall) and the donor side will become more alkaline. Energy will be required to drive the pump.

The second development is an increase in knowledge of the detailed chemistry of cell wall structure which suggests a sandwich-like arrangement of cellulose microfibrils alternating with hemicelluloses, and these are partially linked together by hydrogen bonding. These hydrogen bonds provide the cross-links that give rigidity and elasticity. Hydrogen ions in the vicinity of the hydrogen bonds would weaken them, so that the wall constituents could slide past each other and plastic extension of the wall results. The proton pump could provide the H^+, and IAA might control the proton pump rather like the starter controlling a car. How, or even whether, IAA controls the pump is not known.

Figure 240 illustrates this current concept that IAA promotes cell expansion by increasing wall plasticity.

One qualification needs to be made. When a piece of Plasticine is stretched it gets thinner; so would the cell wall. Careful observation records that the wall of a cell which has elongated is *not* thinner than it was when elongation began. Thus accompanying the plastic 'sliding' extension must be synthesis and incorporation of new wall material. This can be verified experimentally by demonstrating the incorporation of radioactively labelled cell wall precursors into the walls during extension.

Support of the view that hydrogen ions might be responsible for the plastic extension induced by auxins comes from the experimental

Figure 240
The possible role of IAA as a 'key' to switch on a proton pump in the plasma membrane which causes H^+ from the cytoplasm to move into the cell wall.

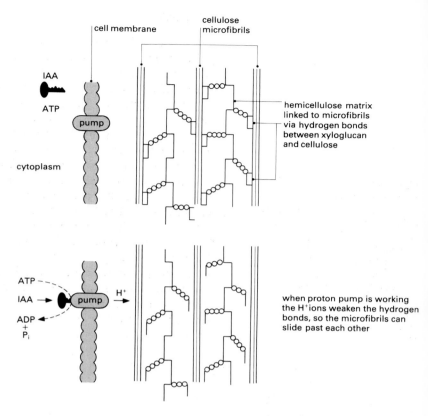

observation that placing untreated coleoptile segments into a weakly acid medium will itself favour cell extension and coleoptile elongation – though only for a short time. Chemical analyses of wall materials from coleoptiles treated either with IAA or with dilute acid shows that cross-linkages between cellulose polymers are weakened in the same manner by both treatments. The initiation of cell extension probably requires this change in cross-linking to allow for increased plasticity; but continued growth needs the synthesis of new wall materials which are added to the stretching wall to prevent it being thinned down. IAA-induced extension is accompanied by synthesis of new wall materials, as shown by incorporation of [^{14}C] glucose into cellulose of the wall, whereas acidic pH does not seem to stimulate wall synthesis. Research seems to be moving in the right direction but there is some way to go before we have a full explanation of how IAA mediates cell extension.

IAA also promotes root initiation in stem cuttings and inhibits lateral bud extension in shoots. At present, there is no obvious connection between the postulated mechanism for cell extension and these other processes. The transition from observations and manipulation at the physiological level to the explanation of what happens at the molecular level is always very difficult and slow; it requires different kinds of expertise. Plant biochemistry has been meagrely supported compared to human and microbial biochemistry – its time has still to come.

Gibberellins

This group of hormones derives its name from the fungus *Gibberella fujikuroi* (now called *Fusarium moniliforme*) which infects rice plants. The infected plants are easily recognized because they grow much taller than the uninfected ones and their leaves turn an unhealthy yellowish green. Japanese scientists were able to show that the fungus could be grown separately from the rice plants and that a culture filtrate, containing no live fungus, could produce similar symptoms in young rice plants. The Second World War interrupted research and the publication of results. It was not until the mid-1950s, when American and British scientists took up the search for the identity of this chemical which could stimulate elongation growth, that the substance was isolated, purified and chemically identified from large-scale cultures of the fungus. It turned out to be a much more complicated molecule than IAA and it was called gibberellic acid (table 40). The successful British team of chemists and biologists at Imperial Chemical Industries soon worked out a satisfactory method of producing the pure substance by extraction from the fungus. Then they showed that gibberellic acid was not confined to fungi but that it was a natural constituent of higher plants, seeds being a rich source. Furthermore, although the identical chemical to the gibberellic acid extracted from fungi (known as GA_3) was widely

Figure 241
Effects of the application of gibberellic acid in ethanol solution to Meteor pea seedlings, 26 days after treatment. From right to left: untreated seedling; control seedling (treated with ethanol only); seedlings treated with successively increasing doses of gibberellic acid.
Photograph by permission of Dr. P. W. Brian and H. G. Hemming, ICI plc.

distributed in plants, other substances of very similar composition could be extracted which had similar biological activity. Now over forty gibberellins are recognized, all with similar molecular structure but with various modifications to side-chains. All have biological activity and many can be interconverted to each other in the plant, but in some instances an individual gibberellin is effective whilst a closely related one is inactive. This suggests that in that particular case the organism does not have the facility to interconvert the two forms, perhaps because it has lost a particular enzyme during evolution.

As with all plant hormones the first method of assay depends upon the compound's specific effect on the growth of a suitable plant. Curvature or elongation of coleoptiles is used in auxin assays. Internode elongation in dwarf peas was used in the English assay for gibberellins at ICI, whilst Americans used a dwarf variety of maize.

> **Practical investigation.** *Practical guide 6*, investigation 24B, 'The effect of IAA on the growth of coleoptiles and radicles'.

It became apparent that gibberellins did not make all plants taller – only some, and these were all dwarf as compared with other known cultivars of the same plant. Quantities as small as one-millionth of a gram (1 µg) would completely transform the stature of dwarf Meteor peas (see *figure 241*).

Promotion of internode extension suggested that GA_3 might be an auxin, but when it was tested in the coleoptile assay it did not promote cell extension in that test. So gibberellic acid is *not* an auxin.

STUDY ITEM

24.71 Interaction between hormones

Experiments on the effect of IAA and GA_3 in the control of growth have been carried out using segments from young pea seedlings instead of coleoptile segments. IAA and GA_3 were applied independently and together, and the results of these experiments are shown in the graphs in *figure 242*.

a *From the graphs in* figure 242 *summarize the effect of IAA and GA_3 on elongation growth in pea internode segments.*

b *How is this similar to the way in which animal hormones act together?*

c *Suggest a possible reason why the two hormones act together to promote internode extension.*

d *How would you test the hypothesis suggested in c above?*

Other effects of gibberellins: promotion of flowering

One of the earliest discoveries was that application of gibberellins to plants could promote flowering. It soon became clear that this was not a

Figure 242
Effects of auxin (10 μg mm⁻³ IAA) and gibberellin (10 μg mm⁻³ GA₃) on elongation growth in excised pea internode segments.
Adapted from Brian, P. W. and Hemming, H. G., Ann. Bot., NS. 22, pp. 1–17, 1958.

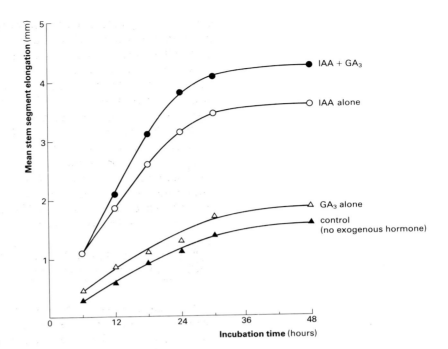

general phenomenon – GA₃ was not the flowering hormone for which physiologists had long been searching. It works only on those plants which require long days to promote flowering, such as henbane (*Hyoscyamus niger*) and spinach. When these plants are kept in short-day conditions (10 hours light and 14 hours dark, for example) they grow in a vegetative rosette. If the day length is then increased, say to 16 hours light and 8 hours dark, the plants 'bolt' and flower. The bolting is due to increase in internode lengths, so that a normal stem and leaves grow out from the centre of the rosette (*figure 243*).

If plants maintained in the short-day regime were sprayed with a very weak solution of GA₃, the result was very similar to that of the long-day treatment. If plants were kept in short-day conditions and then transferred to a long-day regime, analysis before and after the transfer showed that a large increase in gibberellin content took place soon after the change in day length was made.

c **What does this suggest about long days and the promotion of flowering?**

Similarly, externally applied gibberellin can substitute for the winter chilling which is required to induce flowering of biennials such as winter wheat, sugar beet, wallflowers, and Canterbury bells.

The requirement for chilling of the plant to promote flowering is called *vernalization* and should not be confused with stratification (which is chilling of seeds to break dormancy). Many biennial plants are used as food crops, because they produce storage organs such as the tap roots of carrots or the swollen stem bases of turnips, sugar beet, and swedes. They lay down these large food stores in their first season of

Figure 243
Henbane (*Hyoscyamus niger*) showing the rosette produced by short-day conditions and the flowering plant produced by increasing the day length.
Photograph, Dr J. W. Hannay.

growth and then, after winter chilling for several weeks, they commence flowering early the following season using the reserves of food to provide a good start for flowering (at least they would if we didn't harvest them for food first). If a carrot is sprayed with gibberellin during its first season then it doesn't form a swollen root but flowers straight away.

Control of mobilization of food reserves in seeds

Many 'seeds', such as cereal grains, have food reserves stored as endosperm; others, such as peanut and sunflower, store it in cotyledons. Some seeds require a period of *dormancy* before they will germinate. Normally, when a seed is given warm, moist conditions it will germinate and grow – but not if it is dormant. In some way, the state of dormancy must be communicated to the food reserves, otherwise as soon as the seed is moist and warm hydrolysis will start and the food reserves will be mobilized, even though the embryo is not going to grow until it has been released from dormancy. This would lead to disaster because the hydrolysis products, such as soluble sugars from starch, would leak out and provide a good substrate for fungal growth. Even if the embryo was not rotted, it might have no food reserves when it eventually germinated.

An experiment with barley grains illustrates how gibberellic acid appears to act as a signal from the embryo to indicate to the endosperm that it is now time to start mobilizing the insoluble reserves. Then they can be transported to, and used by, the germinating embryo.

> **Practical investigation.** *Practical guide 6*, **investigation 24C.** **'Stimulation of amylase production in germinating barley grains'.**

d *Why is it advantageous to store insoluble starch as a food reserve?*

It seems as though the embryo produces gibberellin which passes back to the endosperm, which in turn recognizes this as a signal to synthesize the hydrolytic enzymes necessary for starch mobilization. There are two reasons why we can say that fresh enzyme is produced, rather than that a previously stored enzyme reserve is activated. Firstly the gibberellin will not work if an inhibitor of protein synthesis is also present; secondly, if radioactive amino acids are provided along with the gibberellin radioactive amylase is produced (in the absence of the inhibitor of protein synthesis, of course).

e *Why should an inhibitor of protein synthesis affect amylase production?*

Commercial use is made of this effect of gibberellin on enzyme activity in barley. The first stage of ethanol (alcohol) production is the malting process, during which the starch is hydrolysed to soluble sugars (maltose and glucose) which can then be used by the yeast in fermentation. If the embryo of a barley grain is damaged, either during harvesting or during artificial drying, it will not germinate. If it doesn't germinate it won't produce gibberellin, so no amylase will be produced. Barley that has a high germination rate costs more than barley used for animal feed or for human food. The cheaper grain will serve perfectly well for malting, however, if it can be induced to produce amylase. This is achieved by spraying it with a weak gibberellin solution, after first scratching the outside of the grain, to break down its water repellency so that the gibberellin solution can penetrate. Not only does this help to ensure that all grains produce amylase; it also tends to speed up the malting process by ensuring that the enzyme is produced straight away – some grains might otherwise take several days longer than others to germinate so they would be slow to start malting. Although you may consider this to be an artificial process, it is only using a natural substance in one of its natural roles to manipulate the process to commercial advantage – an example of biotechnology.

Figure 244 illustrates the role of GA in the germination of a cereal grain.

Figure 244
Role of gibberellin in the germination of barley grains. The three grains have been halved and their embryos (which normally produce gibberellin) have been removed. The cut surfaces of the grains have been treated (from left to right) with plain water and with aqueous gibberellin solutions of concentration 1 and 100 parts per billion respectively. Two days after treatment, digestion of the starch-filled storage tissue of the gibberellin-treated grains has begun.
Photograph by Professor Joseph E. Varner, Department of Biology, Washington University.

Fruit growth

Some fruits are thought to be more desirable without their natural source of gibberellin production – the seeds. Citrus fruits such as orange, lemon and grapefruit are preferred with fewer 'pips', and seedless grapes make seedless raisins. Plant breeders have had considerable success in raising seedless varieties of fruit, only to find that when the seeds are missing, the fruit often develops poorly. In some cases application of GA_3 to the young fruit increases its final size – seedless grapes and citrus fruit are the commercial examples of this.

Figure 245
Seedless citrus fruits.
Audio-Visual Service Unit, King's College, Chelsea.

STUDY ITEM

24.72 Some effects of gibberellic acid on potato tubers

The following experiments which illustrate an effect of gibberellic acid were part of a larger investigation into the changes occurring in potato tubers during storage.

Mature potato tubers harvested in September were subjected to one of the following treatments:

1 buds removed (disbudded);
2 buds removed and wound plugged with lanolin paste which was renewed at monthly intervals (disbudded/lanolin);
3 buds removed and wound plugged with lanolin containing a fixed concentration of gibberellic acid which was renewed at monthly intervals (disbudded/lanolin/GA_3);
4 no treatment (intact).

The tubers were subsequently kept in an unheated store but protected from light and frost. Samples of tubers were taken and the mean concentration of reducing sugar was measured. The results are shown in *figure 246a* and *b*.

A further series of investigations was carried out on intact tubers in April. Cork borers were used to excise, with appropriate sterile

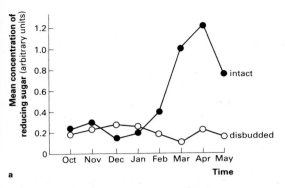
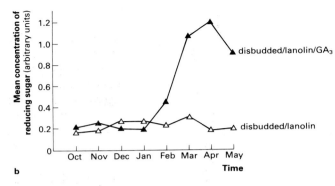

Figure 246
The effects of various treatments of potato tubers on the concentration of reducing sugar in their tissues.

techniques, cylinders of potato tissue each of which had a bud on it. The sterile cylinders with buds on were divided into two groups.

Group one cylinders were placed on moist filter paper in sterile Petri dishes and then treated in one of three ways as follows:

1 a control group;
2 a group with gibberellic acid solution applied to the buds;
3 a group with malonic acid solution – a Krebs (citric acid) cycle inhibitor – applied to the buds.

The Petri dishes were placed in darkness at 20 °C. Assays of reducing sugar content were made at intervals and the results are presented in *figure 247a*.

Group two was placed in sterile flasks. Half were treated with gibberellic acid solution and half were maintained as controls. All flasks were placed in darkness at 20 °C. A constant air flow was maintained in each flask and at intervals the carbon dioxide produced was measured in the effluent air from each flask. The mean results are shown in *figure 247b*.

a *How might bacterial and fungal contamination of the cultures and tissues have affected the results?*

b *How would the controls have been treated in groups one and two?*

c *Explain briefly how malonic acid interferes with the Krebs (citric acid) cycle.*

Figure 247
The effects of various treatments of potato tubers **a** on their reducing sugar content, **b** on the carbon dioxide they produce.

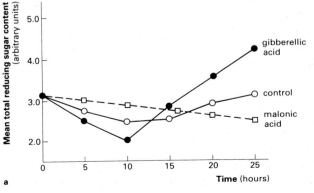
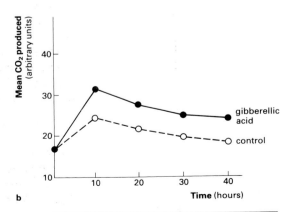

d *Relate the data in figure 246a to those in figure 246b.*

e *Account for the data obtained from the malonic acid treatment.*

f *How would you expect the mean carbon dioxide production to vary if cylinders with buds in group two had been treated with malonic acid solution? Use a graph to help explain your answer.* (J.M.B.)

Cytokinins

The discovery of cell-division promoters comes from work with plant tissue cultures in the mid-1950s. It was found that when small pieces of tobacco stem were cut out, using sterile techniques, and placed into a sterile medium containing mineral nutrients, sugar and certain vitamins then the excised tissue would enlarge.

Some of this excised tissue merely increased in mass, that is, the cells became large and bloated but did not divide. In other pieces it was clear that growth was occurring as a result of cell division and the original fragment could then be subdivided, placed into fresh medium, and the bits would continue to grow. All the pieces which grew well – and in which cells not only enlarged but also divided – contained bits of vascular tissue in the original explant. If the explant was entirely composed of parenchymatous pith then the callus did not divide. So it seemed that a cell-division factor ought to be sought.

An appropriate substance was first isolated not from extracts of tobacco vascular tissue but from autoclaved yeast nucleic acid. The substance was chemically identified as the purine derivative furfuryl-adenine (see *figure 248*), and it was named *kinetin*.

f *What connection is there between adenine and nucleic acids?*

It turned out that kinetin did not occur naturally in plants. Nevertheless, it had the important attribute that when added to parenchymatous pith tissue it stimulated cell division which did not occur otherwise. Other synthetic kinins have since been discovered, with chemical characteristics similar to those of kinetin (*figure 248*). Similar

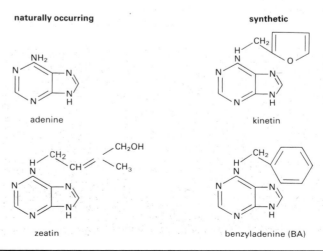

Figure 248
The molecular structure of some cytokinins.

molecules have since been extracted from plants. Especially rich sources seem to be the natural 'milks' which support growth of young embryos in fruits such as coconut, horse chestnut, or the immature maize grain at the stage when it is 'milky' rather than ripe and starchy. The cytokinin extracted from the immature maize grains is called zeatin and again it is an adenine derivative (see *figure 248*).

Whilst the theory that cytokinins are cell-division factors, which complement the cell-extension hormones like IAA, is an attractive one, it is a little too facile. The culture medium for the callus tissue also contained IAA. When IAA was omitted and kinetin added, no proliferation occurred (table 41). When IAA was added to the kinetin-containing medium cell division occurred – so IAA could have been called the cell-division factor too, if the observations had been made in that sequence.

		IAA	
		Omitted	Added
kinetin	Omitted	×	×
	Added	×	√

√ = good growth and cell division
× = no cell division

Table 41
Growth of tobacco pith callus in varying combinations of IAA and kinetin

Cytokinins and shoot development. Following on from the work with callus cultures it was found that with a few plants the callus could be induced to develop roots and/or shoots. Both IAA and kinetin were required; but when the concentration of IAA was high and that of kinetin relatively low, then only roots were formed. When kinetin was in fairly high concentration then shoots were formed and roots tended to be suppressed (*figure 249*).

Commercial developments from such plant tissue cultures include the vegetative propagation of orchids and other expensive flowering plants, the sales of which could cover the cost of the rather elaborate techniques.

Recently there has been great interest in vegetative propagation of

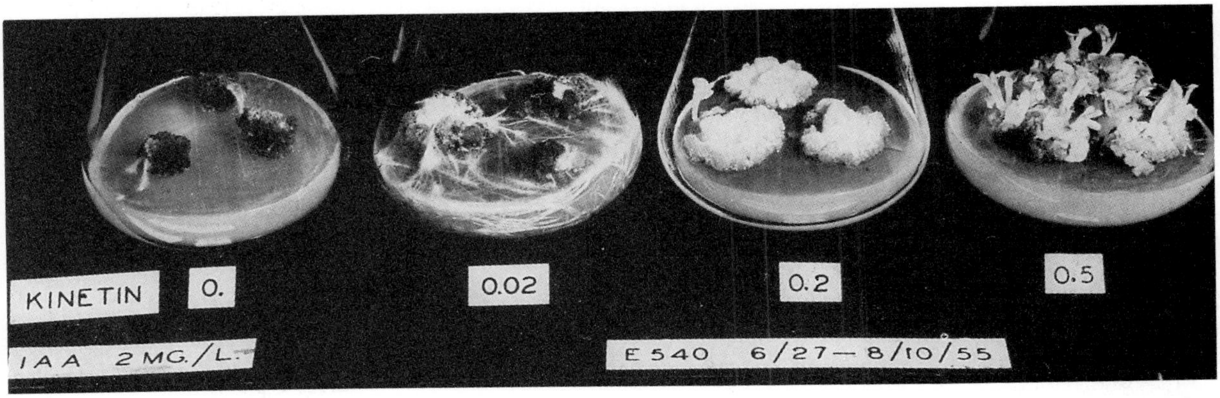

Figure 249
The effect of different concentrations of kinetin on growth and development of callus tissue from tobacco pith in the presence of 2 mg dm^{-3} IAA. With no kinetin, there is little growth. At low levels, roots develop. At an intermediate level, unorganized growth continues. At a higher level buds and shoots develop.
Photograph by F. Skoog and C. O. Miller, in Symp. Soc. Exp. Biol. **11**, *1957, pp. 118–31.*

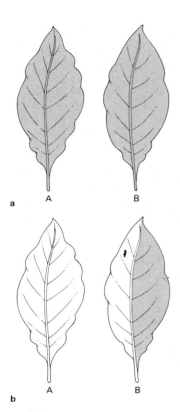

Figure 250
The effect of kinetin on excised tobacco leaves.
a The leaves were green and healthy when excised from the plant. The righthand half of leaf A was sprayed with water and that of B with kinetin solution. Both leaves were kept with their petioles in water.
b After a few days leaf A had gone completely chlorotic and yellow. The kinetin-sprayed half of leaf B stayed green and looked healthy.

certain monocotyledonous trees such as the oil palm. Oil palm is difficult to propagate vegetatively because there are no lateral shoots from which to make cuttings. On the other hand, some trees produce a lot more oil than others do. If clones could be made from the élite trees, then the whole plantation could be genetically uniform and all trees would have equally high productivity. Since productivity is likely to be controlled by a very large number of genes it is not easy to breed a consistently high-yielding strain with a stable genetic constitution. It is also a very slow process since the trees are fairly long-lived and take several years to reach peak production. In plants such as roses, chrysanthemums, and apples vegetative propagation is the only way to maintain the particular characteristics of colouring, scent, and flavour of a cultivar. Tissue culture can be useful when natural propagation is very difficult – but only a few plants have been successfully grown in tissue culture.

Cytokinins and senescence. When leaves are taken from a plant and their petioles are placed in water, or even in dilute nutrient solution containing all the essential inorganic elements, they usually go yellow and die. Moreover they die quicker than do the adjacent leaves left on the plant. This suggests that in their natural place on the plant they are receiving something which delays ageing (senescence).

If the excised leaf is provided with tiny amounts of a cytokinin it will stay green much longer. In fact, if only a small area is treated this will stay green, whilst the rest goes yellow and senescent (*figure 250*).

However, application of cytokinins to leaves on an intact plant has not yet produced any useful increase in the longevity of leaves. There is potential for improving crop yields if the natural senescence can be delayed so that the leaves continue to photosynthesize for a longer period.

There is a correlation between the 'shelf-life' of green vegetables and their cytokinin content. Certain varieties of Brussels sprouts tend to turn a rather unattractive yellowish-green a few days after they are harvested, whilst others stay green and look fresh for several days longer. Application of cytokinin to the varieties that yellow quickly can markedly delay this yellowing and increase their shelf-life. The varieties that have good shelf-life have relatively high natural contents of cytokinins, and plant breeders can select for this trait when developing new varieties, so that the use of external kinins can be avoided.

Abscisic acid

The hormone derives its name from the fact that it was discovered in an investigation of leaf abscission. If a factor which promotes leaf abscission can be used to cause leaf shedding just prior to harvest, then harvesting may be facilitated: the cotton boll, for example, is much easier to collect if it stands free of leaves. It was shown that a naturally occurring substance could be obtained from senescing leaves which, when applied to mature leaves, would cause them to abscise – hence the name given to the substance abscisin. Its chemical identity took some time to elucidate.

In the meantime, quite independently, work on shoot dormancy was directed towards extracting a dormancy-inducing substance which could cause the cessation of stem elongation and the formation of a resting bud in perennials, such as deciduous trees.

A substance was extracted from leaves of sycamore (*Acer pseudoplatanus*) which, when applied to young non-dormant sycamore plants, would induce resting bud formation and dormancy. This substance was called dormin. When abscisin and dormin were chemically identified they turned out to be identical. Because the name abscisin had been given first it took priority over dormin, and the substance became known as abscisic acid (ABA) (table 40).

Abscisic acid is not only involved in shoot dormancy; it also has a role in seed dormancy and experiments with lettuce seeds illustrate this quite well.

> **Practical investigation.** *Practical guide 6*, **investigation 24D, 'The effect of plant hormones on seed germination'.**

When dormancy is broken naturally, either in seeds or in shoots, it is usual to find a drop in the ABA content of the seed or bud. There is often an increase in gibberellin and cytokinins too, however, and these may interact to alleviate the effects of ABA (*figure 251*).

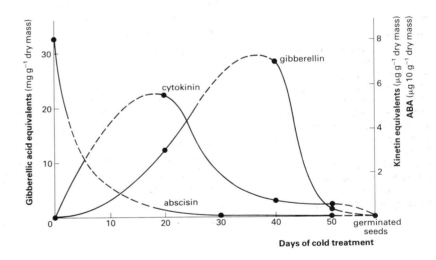

Figure 251
Changes in plant hormones in sugar maple seeds (*Acer saccharum*) during stratification at 5 °C.
Adapted from data of Webb, D. P., van Standen, J., and Wareing, P. F., J. Exp. Bot. **24**, *pp. 105–17, 1973.*

A further important role, which was only discovered when samples of extracted ABA became available for trials, was that ABA promotes stomatal closure. When leaves are placed under drought stress the ABA content rises quite quickly, and stomata close. If water is then restored the ABA content falls and the stomata reopen (*figure 252*).

Figure 252
The effect of osmotically induced water stress in tobacco plants: changes in leaf ABA content and in stomatal aperture during a cycle of stress and recovery.
After Boussiba, S. and Richmond, A. E., 'Abscisic acid and the after-effect of stress in tobacco plants', Planta, **129**, *1976.*

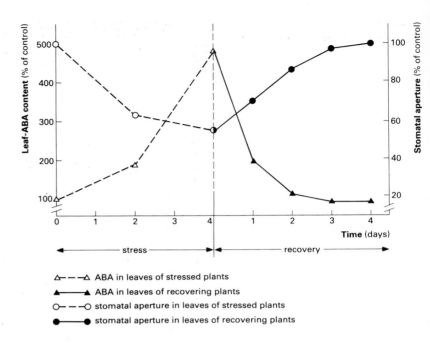

Ethene

The earliest recognition of the effects of ethene (ethylene) on plants was due to distorted growth in some pea seedlings exposed to a coal gas leak in a Russian laboratory around 1900. The substance responsible was shown to be ethene, and concentrations of pure ethene of as little as one part ethene in a million parts of air (1 p.p.m.) were found to induce a similar effect to that of coal gas. When town gas was installed into glasshouses for heating purposes in the 1920s in America, a great deal was written about the effects of ethene on plants grown in glasshouses where gas leaks occurred. It was recognized that some plants, such as young tomato plants and french marigolds, were extremely sensitive and produced distorted growth at concentrations of only 0.01 p.p.m. of ethene, whilst others such as ivy and ferns seemed to be unaffected The sensitive plants were useful for detecting leaking town gas because they responded to its ethene content before it reached a concentration that was easy to recognize by the characteristic sulphury town gas smell.

A major commercial role for ethene developed from work on fruit ripening carried out during the 1920s and 1930s. Imported fruit like bananas and apples had to be carried over long distances by sea, and during the journey they would become over-ripe and unattractive unless they were picked from the trees in a very under-ripe condition. Apples, moreover, were difficult to store, so in years with good harvests they had to be sold off cheaply in order to clear them quickly; if maturation could be delayed, it would mean apples could be stored. Because of the commercial importance of these problems, money was available for research on fruit ripening; results came fairly quickly, and have continued to yield useful applications.

In both apples and bananas the ripening process reaches a dramatic

Figure 253
Changes in apples during ripening.

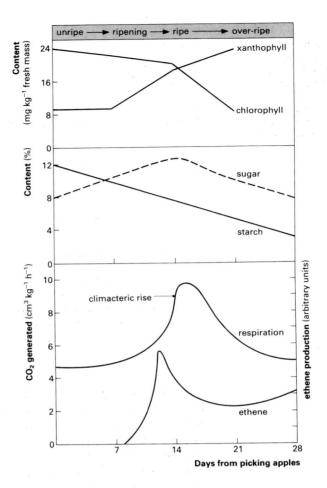

stage known as the climacteric, when several changes occur within a few days, notably a steep rise in respiration rate (*figure 253*). Ethene production by the stored fruit suddenly peaks at the onset of the climacteric, and the climacteric can be brought on earlier than it would normally occur by administering ethene externally to the fruit. If ethene production could be delayed or if an antidote to the effect of ethene could be found, perhaps the climacteric could be delayed. If it could, then maturation would also be delayed.

It has been shown that carbon dioxide can competitively inhibit the effects of ethene in many systems. Enriching the atmosphere with CO_2 to around 10 per cent coupled with reduction of oxygen to around 2 per cent plus cooling to 3–5 °C efficiently delays ripening.

Bananas can be picked green and kept green during shipment, and then ripened at their destination by placing them in air at room temperature and adding a trace of ethene to induce the climacteric and the desirable yellow coloration. Apples too can be stored in cool, carbon dioxide enriched atmospheres and released from store late in the season to fetch a good price.

The real stimulus to academic research on ethene was the advent of gas chromatography in the early 1960s. This made it easy to determine low concentrations of this gas in air, which previously had been possible but very tedious and difficult. As a result it soon became apparent that ethene was a natural product of all sorts of plant tissues, not just of certain kinds of fruit. It has a key role in natural leaf abscission, and any treatment that enhances ethene production in a leaf or petiole is likely to induce its abscission. It is believed that the herbicides 2,4-D and 2,4,5-T, which were used to defoliate forests in the wars in Korea and Vietnam, worked indirectly by enhancing ethene production. However, abscisic acid seems to affect abscission independently of enhancing ethene.

Since ethene is a gas it is not easy to administer outdoors but a compound called 'Ethephon' (*figure 254*) has been discovered which breaks down spontaneously at a pH higher than 4.1, liberating ethene. Ethephon is applied to plants in an acidic (stable) solution; when it enters the cytoplasm it encounters a pH which is usually much higher than 4 (often around 5 or 6), so ethene is liberated *in situ*. This compound is now used to enhance ripening of apples, improving their coloration and allowing some varieties to be marketed earlier in the season, when prices are higher. It is also used to enhance ripening in citrus fruits.

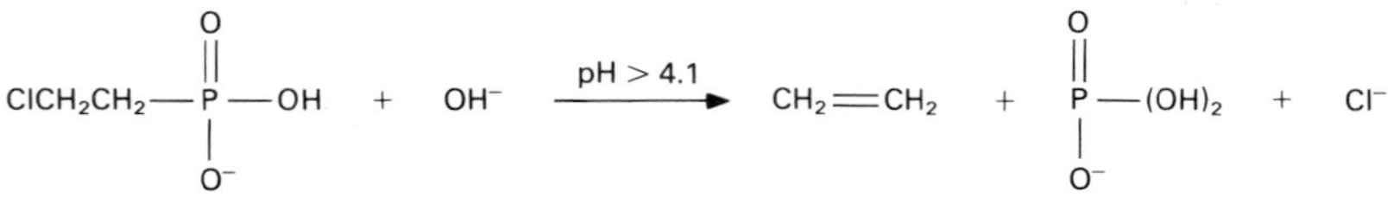

Figure 254
Breakdown of Ethephon to liberate ethene.

$$ClCH_2CH_2-\overset{\overset{O}{\|}}{\underset{\underset{O^-}{|}}{P}}-OH + OH^- \xrightarrow{pH > 4.1} CH_2=CH_2 + \overset{\overset{O}{\|}}{\underset{\underset{O^-}{|}}{P}}-(OH)_2 + Cl^-$$

(2-chloroethanephosphonic acid in ionized form (Ethephon or Ethrel))   (ethene)

Table 42 is a summary of synthetic growth regulators and their commercial uses.

24.8 The immune response

The immune system recognizes and eliminates from the body foreign material such as infectious micro-organisms and unrelated grafted tissues. There are two main forms of resistance to infection: *natural* or *non-specific immunity* and a more specific response known as *acquired* or *adaptive immunity*. A substance that provokes an adaptive response is called an antigen, and the soluble proteins that specifically combine with it to eliminate it from the body are called *antibodies*.

Natural or non-specific immunity

Our first defence against foreign material is skin, which acts as a physical barrier. Tears, as well as secretions from the nose, mouth and mucous membranes, contain enzymes such as lysozyme, which destroys the cell walls of some bacteria, and thus further impedes bacterial entry. If the pathogen circumvents these obstacles and enters the blood circulation it will be attacked by an enzymatic system of serum proteins known as *complement*. When the first component of complement is activated on the surface of the initiating agent it triggers the activation of several molecules of the second component in the sequence, each of which then acts on several molecules of the next component and so on until

Auxins 2,4-D (2,4-dichlorophenoxyethanoic acid) [structure: dichlorophenoxy-OCH₂COOH] NAA (naphthalenethanoic acid) [structure: naphthalene-CH₂COOH] IBA (indolebutanoic acid) [structure: indole-CH₂CH₂CH₂COOH]	*Antiauxins* TIBA (2,3,5-tri-iodobenzoic acid) [structure: tri-iodobenzene-COOH]	*Commercial uses* Promoting rooting of cuttings (IBA, NAA)* Controlling fruit drop (in apples): (a) inhibiting when crop small (IBA, NAA)* (b) promoting thinning when crop too heavy (TIBA) Certain weedkillers (2,4-D) *(Natural IAA is less suitable than synthetics because it is easily inactivated by the plant)
Gibberellins Synthetic gibberellins not available	*Anti-gibberellins* $(Cl^-)CH_3$ $CH_3-N^+-CH_2CH_2Cl$ CH_3 CCC CH_3 O $N-NH-CCH_2CH_2COOH$ CH_3 B9	Dwarfing cereals and pot plants (CCC and B9).
Kinins BA (benzyladenine) [structure: benzyladenine]	*Anti-kinins* Not used	Promoting shoot development in tissue culture (BA or zeatin)
Abscisin Synthetic abscisin not available	*Anti-abscisin* Not used	Abscisin is not yet used commercially: it is too expensive as a defoliator as compared with other substances which will produce a similar effect
Ethene From Ethephon $HO-\overset{O}{\underset{O}{\overset{\|}{P}}}-CH_2CH_2Cl$	*Anti-ethene* CO_2	Promoting ripening and colour in fruit (Ethephon) Delaying maturation in fruit and senescence in some cut flowers (CO_2) Facilitating latex flow in rubber trees (Ethephon)

Table 42
Synthetic growth regulators and their commercial uses.

thousands of molecules of the end components are actively generated. There are nine major components of complement. Activation of the last two components produces membrane damage resulting in lysis of the initiating agent. Certain bacteria can also induce complement activation by directly activating the third component, thus bypassing the first components of the sequence. Complement activation initiated by microbial infection is thus a very efficient immunological response. When antibodies combine with their target they often acquire the ability to activate complement, so that micro-organisms with antibodies adhering to their surfaces are frequently killed by the local activation of complement.

Phagocytosis is the main mechanism for removing particulate material, such as micro-organisms, from the blood and tissues. The most efficient phagocytic cells are the blood polymorphonuclear cells and the tissue macrophages. Both cell types contain a large armoury of lysosomal enzymes and have the potential to generate toxic oxygen radicals which can destroy micro-organisms or inhibit their growth. When the third and fifth components of complement are activated by bacteria, complement fragments are generated which induce an enhanced directional migration (chemotaxis) of the polymorphonuclear cells towards the source of the activation, and which also facilitates adherence of the phagocytic cells to the complement-coated micro-organisms. Thus complement not only itself produces direct lysis of bacteria but also augments the polymorphonuclear cells' response to these foreign invaders. Chemotaxis is probably the mechanism by which polymorphonuclear cells are attracted from the blood into infected tissue (see *figure 255*).

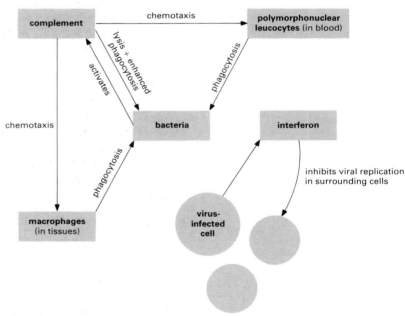

Figure 255
A summary of the processes involved in natural immunity.

If the infection is caused by a virus then a group of proteins, collectively termed *interferon*, is rapidly released from infected cells to inhibit the replication of the virus in neighbouring cells.

a   *Would you anticipate that individuals who are deficient in the last four components of complement would be more susceptible to microbial infection than people with a deficiency of the first component?*

Should natural immunity fail to resist an infectious micro-organism, the next host defence is that of acquired immunity.

Acquired or adaptive immunity

Memory is the essential component of the acquired response, as it implies recognition of foreign material as a result of previous contact. This is well illustrated by immunization against viruses such as those causing polio and measles. For this purpose, the virus in question is killed or treated so that it loses its infectious properties; this material will, when administered to an individual, stimulate the immune response to this virus, and is known as a *vaccine*. If the virulent organism infects the body at a later date, it will encounter a rapid and specific immunological response to its invasion because of the protective immunity acquired as a consequence of the immunization.

Lymphocytes are the main effector cells of acquired immunity. They are concerned with recognizing, processing, and responding to antigen. Like polymorphonuclear leucocytes and monocytes (the precursor cells of macrophages), they arise from the stem cells of the bone marrow. In embryonic development the yolk sac is the earliest source of these stem cells, which later migrate to the spleen and liver before permanently residing in the bone marrow. Acquired immunity consists of two different types of reaction, which are classified as *humoral* and *cell-mediated immunity*.

Humoral immunity. This form of immunity is mediated by antibodies whose function is to combine specifically with the antigens that initiate their production. By binding to viruses, bacteria, and other infectious micro-organisms these antibodies, which are secreted by a subpopulation of lymphocytes, enhance the removal of the pathogens by phagocytic cells. Humoral immunity in chickens was shown to be severely impaired if a small cloacal structure known as the bursa of Fabricius was removed shortly after hatching. Consequently, the bursa

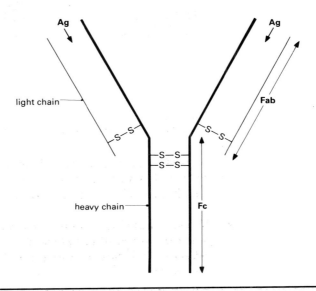

Figure 256
Schematic illustration of an antibody molecule. Each molecule contains four polypeptide chains: two heavy chains (relative molecular mass about 50 000) and two light chains (relative molecular mass about 22 000). The chains are held together by disulphide (—S—S—) bonds. Fab: fragment antigen binding. Fc: fragment crystallizable.

of Fabricius became recognized as a *primary lymphoid organ* responsible for producing the lymphocytes that are essential for antibody production. Such cells are called B (bursa-derived) lymphocytes. In humans the bursa equivalent has yet to be identified, though the bone marrow seems to be the strongest of the possible candidates.

Antibodies, often referred to as immunoglobulins, consist of three sections; two are identical (Fab: fragment antigen binding) and bind the antigen, whilst the third fragment (Fc: fragment crystallizable), which does not bind antigen, is responsible for the distribution of the antibody throughout the body and activating complement. The combination of antigen with antibody is called an *immune complex*. Often, these complexes activate complement and so attract polymorphonuclear cells which remove them by phagocytosis. To appreciate the complexity of antibody diversification it is well to remember that our immune system is programmed to produce probably millions of antibodies, each of which is directed against one specific antigen, and that each B cell makes only one kind of antibody. A B cell carries an antibody on its membrane which is believed to be identical to the antibody which the cell is committed to synthesizing. When an antigen comes into contact with its specific surface-bound antibody, which acts as a receptor, it triggers that B cell to differentiate and divide so that a large clone of B cells is assembled which can liberate specific antibodies in response to antigenic challenge.

STUDY ITEM

24.81 Response to an antigen

Most immunological experiments have been performed with 'inbred' strains of mice, that is, strains which have been closely mated for many generations. These animals are genetically very similar, thus ensuring that their immunological responsiveness is very similar and that cellular interchange with one another is possible without any adverse reactions. The experiment illustrated in *figure 257* was performed using such animals.

a *Why do you think there was an improved antibody response following the second contact with antigen?*

b *What is the benefit of such a response?*

c *What does the response shown in figure 257 tell you about the role of lymphocytes?*

Cell-mediated immunity (CMI). The observation that children with abnormalities of the thymus gland often had an impaired immune response implied that this organ had an important role in immunity. In the early 1960s, Miller investigated this possibility by performing experiments on mice. He found that thymectomy (removal of the thymus) at birth produced a reduction in the number of blood lymphocytes.

b *Offer an explanation for this observation.*

In another experiment adult mice were irradiated with X-rays so as to make their lymphocytes immunologically unresponsive by inhibiting their division. When bone marrow cells were injected into such animals, immune responsiveness was restored. But when the experiment was repeated using X-ray-irradiated mice that had been thymectomized at birth, the immune function was restored only after the administration of mature lymphocytes; injection of bone marrow cells was ineffective.

c *What conclusions would you draw from the results of this experiment?*

Lymphocytes, which acquire immunological competence in the thymus are known as T (thymus-derived) lymphocytes. At birth the thymus, which in humans lies just behind the sternum, is relatively large but it quickly atrophies with age. Although the thymus has the highest rate of cell production of any tissue of the body, most of these rapidly dividing cells (thymocytes) perish in the organ. The demonstration that several isolated thymic 'factors' enhance T cell maturation and

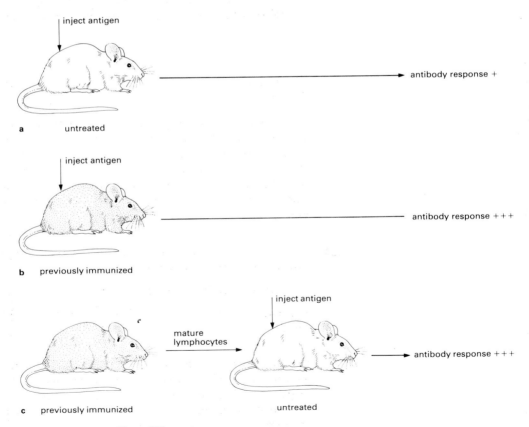

Figure 257
An experiment illustrating the antibody response of differently treated mice to antigen injection.

differentiation in the absence of antigen – although not as adequately as does the intact thymus – has cast doubt on earlier claims that the thymus is an endocrine gland.

Although most immune responses are of benefit to us in the fight against infections, occasionally the reaction induced is a harmful one. This is seen in those individuals who are sensitized to a particular antigen, in whom re-introduction of the antigen evokes a hypersensitivity reaction; that is, there is an excessive immunological response which results in tissue damage. The response appears in two forms, which are distinguished by the time that elapses before their onset.

In immediate hypersensitivity the response is of rapid onset and is antibody-mediated. It may be induced by a variety of antigens, such as grass pollen (which gives rise to hay fever) or certain drugs, or as a result of insect bites. The reactions are initiated by the antigens combining with antibodies which are already bound, via their Fc (fragment crystallizable) structures, to the surface of tissue cells which contain numerous cytoplasmic granules (mast cells). This interaction of antigen and antibody results in the extracellular discharge of potent chemical substances from the cytoplasmic granules, and it is these that induce the sneezing, itchy eyes and running nose characteristic of hay fever sufferers, and the 'wheezing' of asthmatic subjects.

Injection of soluble antigen into the skin of an immunized animal produces, after several hours, inflammation (heat, redness, pain, and swelling) at the site of the injection; this, because of its delayed appearance, is called a *delayed hypersensitivity* reaction. Histological examination of this inflammatory focus reveals a large number of infiltrating T lymphocytes and monocytes and very few polymorphonuclear cells.

Unlike acute inflammation, which may arise and subside within a matter of minutes or a few hours, the inflammation associated with delayed hypersensitivity is not apparent until after 6 or 8 hours. This is because of the time taken for sensitized T lymphocytes to respond and migrate to the antigen and synthesize non-antibody proteins (lymphokines) which maintain a CMI reaction. Such lymphokines recruit and hold other leucocytes to the inflammatory site. They also enhance the function of macrophages (phagocytosis) and stimulate other lymphocytes to generate more lymphokines. Inflammation continues until a peak level is reached between 24 and 48 hours after the start of the reaction. After this time the antigen is removed from the site and the inflammatory reaction subsides. Should the antigen persist, however, because it cannot be degraded (as, for example, in the case of asbestos) then macrophage colonization of that site will occur (granuloma).

d   *In the experiments shown in figure 258, state why the skin reaction in guinea pig B appeared before that of guinea pig A, and why no skin reaction was induced in guinea pig C.*

When killed lymphocytes were injected into an unsensitized animal no skin reactivity was observed following antigenic challenge, in

Figure 258
The skin reaction to antigen injection in guinea pigs following different treatments.

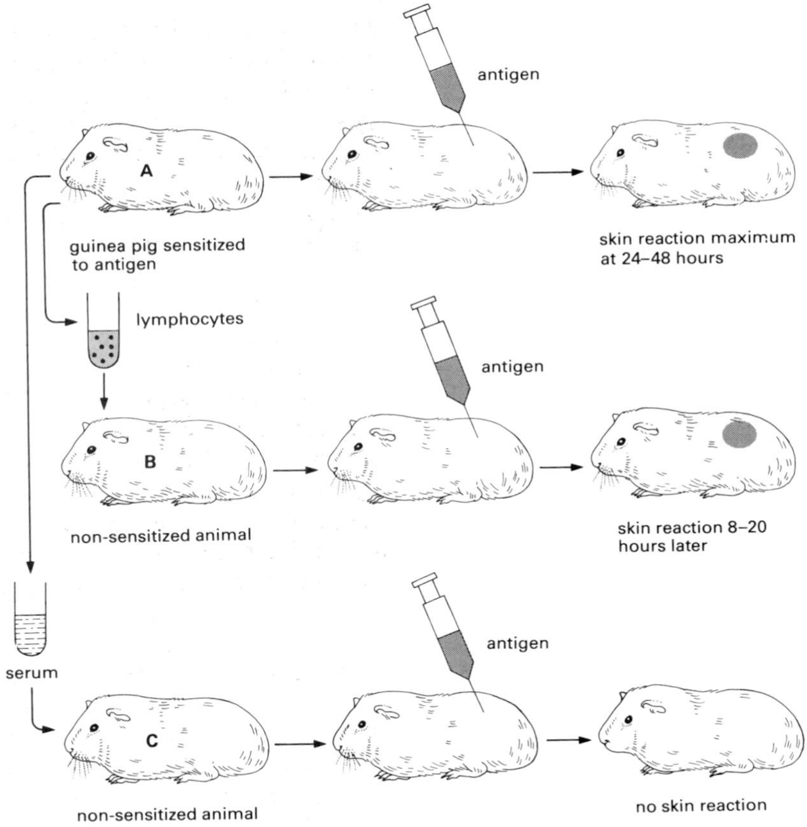

contrast to a positive skin reaction seen when a soluble extract from lysed T cells was transferred.

e **What additional information does this experiment give you regarding the mediation of delayed hypersensitivity?**

All reactions in which T cells participate are controlled by a region known as the major histocompatibility complex (MHC), which consists of a complex series of linked genes found in humans on chromosome 17. It is referred to as the human leucocyte antigen (HLA) system, because its gene products were originally identified on leucocytes. Within the MHC are several regions which are separable by genetic recombination, and which code for antigens present on all cells except red blood cells. The MHC exerts considerable genetic control over many immunological reactions.

MHC antigens on lymphocytes are identified either by the fact that target cells are lysed by antibodies in the presence of complement, or by the mixed lymphocyte reaction (MLR). The MLR is based on the observation that when lymphocytes from two genetically distinct individuals are mixed together (hence the term), there is an increase in DNA synthesis and in the size reached by cells before they undergo mitosis. This occurs as a result of the recognition of foreign HLA-antigens on each others' cells: the greater the difference between

Chapter 24 The internal environment

histocompatible antigens, the greater the response. These tests are routinely used for tissue typing.

Unlike bacteria, which can proliferate outside the host's cells, viruses need to replicate inside cells, thus making their elimination by antibodies very difficult. Fortunately, viruses induce new surface antigens on infected cells which, in conjunction with the cell's own antigens, are recognized as targets to be attacked by a subpopulation of T lymphocytes known as *cytotoxic* T lymphocytes and, under certain circumstances, possibly macrophages. Lysis of target cells by the cytotoxic T lymphocytes results in release of free virus particles which can be neutralized by antibody.

Organ transplantation is today becoming a common operation in some hospitals. Despite the use of the most up-to-date surgical techniques, many organs are rejected; that is, the recipient's immune response attacks and kills the foreign cells of the transplanted tissue. Such rejections are rarely seen between identical twins because the antigens of the donor and recipient cells are compatible with one another (that is, they have the same histocompatible antigens); the greater the incompatibility between host and recipient cell antigens, the greater the risk of rejection. The success of organ transplantation thus depends on the closeness of matching of the MHC antigens of the donor and the recipient. Graft rejection is predominantly mediated by lymphocytes. Almost all patients who have undergone transplantation, regardless of the degree of tissue matching, receive immunosuppressive drugs whose main mode of action is generally to inhibit dividing lymphocytes and so increase the chance of graft acceptance.

STUDY ITEM
24.82 Skin grafts

The following questions relate to the experiment illustrated in *figure 259*.

a *Why does the white mouse in (a) reject the skin graft from a black mouse (unrelated donor), yet if a newborn white mouse (b) is injected with mature lymphocytes from a black mouse it will later accept the skin graft from a black mouse?*

b *If a white mouse (b) has accepted a skin graft from a black mouse and is later injected with mature lymphocytes from a closely related white mouse, the graft is rejected. Explain why.*

c *Do you find it surprising that the young mouse in c which is injected with mature lymphocytes from a black mouse later rejects a skin graft from a black mouse?*

d *Offer an explanation why a mouse that has several months previously rejected a graft will expel a similar second graft more rapidly than the first.*

We have seen that lymphocytes consist of two distinct subpopulations: B cells, which mediate humoral immunity through the

Figure 259
Skin graft acceptance and rejection in mice.

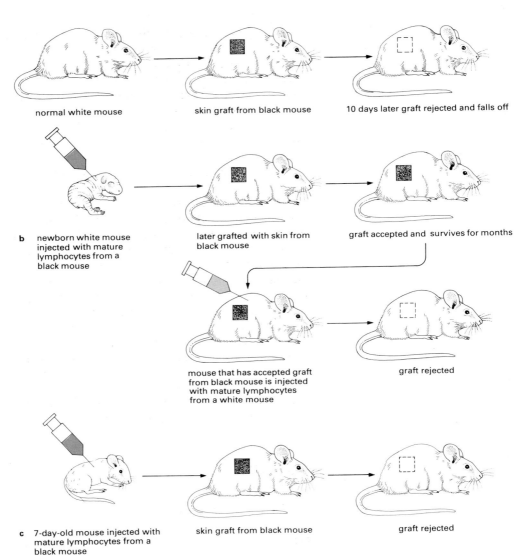

secretion of antibodies, and T cells, which do not secrete antibodies but mediate the classical reactions of CMI such as delayed hypersensitivity and graft rejection and which also limit intracellular viral infections. T and B cells arise from immature lymphocytes that have migrated from the bone marrow and undergone differentiation in the thymus and bursa (or its equivalent) respectively. These two organs, together with the yolk sac, liver, and bone marrow, constitute the primary lymphoid organs. Throughout the body are distributed a number of lymph nodes, which are small organs consisting of distinct areas of T and B cells. In conjunction with the spleen they ensure thorough processing and responsiveness to antigen, and are referred to as the *secondary lymphoid organs*. A small population of lymphocytes constantly recirculate between the blood and the lymph nodes; their access to these nodes from the blood vessels in peripheral tissues is via the afferent lymphatics, and their exit is by efferent lymphatics which converge to form the thoracic duct which returns the lymphocytes to the blood.

Although there appears to be a strict functional segregation between cell mediated immunity and humoral immunity, experiments such as those performed by Miller have shown that the antibody response of mice to certain antigens is impaired if the animals have the thymus removed at an early age.

Table 43 summarizes the results of a series of experiments designed to demonstrate the different antibody responses to sheep red blood cells (antigen) of mice which had received a transfer of lymphoid cells from either genetically similar normal mice or mice that had previously been immunized with sheep red blood cells. Before cell transfer the recipient animals were irradiated so as to render them incapable of eliciting an immune response.

Group of mice	Source of donor cells	Antibody response to sheep red blood cells
1	Normal bone marrow	weak
2	Normal spleen	3+
3	Immune spleen	4+
4	Normal or immune thymus	—
5	Thymus + bone marrow	3+

Table 43

f *What does this experiment tell you about the role of T cells in the humoral response?*

In addition to being present in the thymus, T cells in conjunction with B cells are located in the spleen. In another series of experiments (*figure 260*) lymphoid cells from the spleen of mice that had been rendered tolerant to sheep red blood cells were transferred to normal mice that were later challenged with sheep red blood cells.

g *Comment on the role of the splenic cells in modifying the humoral response of the recipient mice in* **figure 260.**

Let us call the effector T cells in table 43 type X and those in the figure type Y. In humans, a significant decrease in the ratio of X/Y (which is normally approximately 3:1) often reflects an immunological disorder. We also know that type Y cells contain cytotoxic T cells. A disease which illustrates an imbalance of the X/Y ratio, together with a low blood lymphocyte count, is the recently identified acquired immune deficiency syndrome (AIDS). Contraction of this disease induces such a marked immunodeficiency that the subject in question is unable to mount an immune response to the slightest infection. Consequently, mortality is high. The causative agent is a virus, HTLV III (human T-cell lymphotropic virus type III), which is mainly acquired through sexual contact, and occasionally through blood transfusions. Conventional drugs are of no benefit to AIDS sufferers.

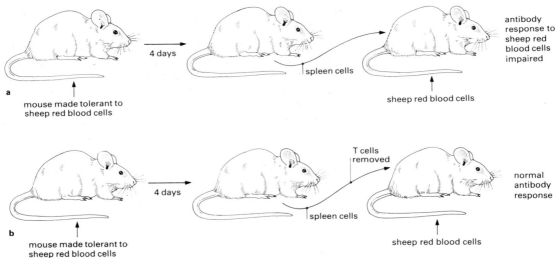

Figure 260
The role of spleen cells in the humoral response of mice to sheep red blood cells.

Immunization (vaccination)

At the end of the eighteenth century, a Gloucester physician named Edward Jenner noted that people who had previously contracted cowpox were resistant to smallpox. He performed the first successful immunization when, in an unethical experiment, he inoculated material from cowpox vesicles into an eight-year-old boy in order to afford him protection from smallpox.

The most efficient way of immunizing a subject against a pathogen is to administer an attenuated vaccine of the virus or bacterium. Attenuated vaccines are prepared by growing or treating the organisms in question in conditions in which they lose their virulence but not their capacity to stimulate a protective immune response. Louis Pasteur examined cultures of chicken cholera which had been left in the laboratory during a holiday, and found that they had lost their virulence and become attenuated. One of the best-known vaccines is the BCG vaccine, which protects against tuberculosis. In 1908 two French scientists, Calmette and Guérin found by chance that after thirteen years in a modified culture medium the bacillus *Myobacterium tuberculosis* became attenuated (its name derives from *Bacille de Calmette et Guérin*). This vaccine is a powerful activator of macrophages and natural killer cells (see page 372).

Certain bacteria release soluble proteins (exotoxins) which have a destructive action on local tissues, phagocytic cells, the central nervous system, and many other parts of the body. The most notable of these toxins, some of which are the most powerful poisons known, are those produced by diphtheria and tetanus bacilli. Fortunately, toxins can be detoxified by formaldehyde to form toxoids, which can be used for immunization.

Poliomyelitis is a good example of a viral infection from which recovery gives lifelong immunity. Continued immunity to other microbes may need occasional 'boosting' doses; this is the case with tetanus, for example. The route of administration of a vaccine is very important. For example, oral administration of a vaccine is the most suitable way of inducing local mucosal immunity to gut infection. Where attenuation is impracticable, as it is with pertussis (whooping cough) and rabies, inactivated vaccines are prepared from killed bacteria and viruses.

h *Why is the immunity acquired by inactivated vaccine not as efficient as that with the attenuated strain?*

i *For several months after birth measles and mumps vaccines are often ineffective in infants. Why do you think this is so?*

j *Children with hypogammaglobulinaemia, in whom antibodies are absent or nearly so, are immune to measles virus infection following recovery from the initial infection. Suggest how this non-antibody-dependent immunity is mediated.*

Tumour immunology

Development of malignant cells results in their acquisition of new surface antigens. By evading immunological recognition and elimination, malignant cells often proliferate into a tumour. The type of antigens present on a cancer (malignant) cell depends upon the inducing agent. Tumours induced by chemical carcinogens (cancer-causing agents) express unique surface antigens. This contrasts with virus-induced tumours whose surface antigens are characteristic of the virus. Virally induced tumours display strong surface antigens which elicit an efficient host immune response, while carcinogen-induced and spontaneously induced tumours elicit a poor response. Differentiation antigens, which are present on embryonic cells but absent from normal adult cells, are commonly found on tumours. Their appearance, which is probably due to re-expression of foetal genes, can serve as markers of certain tumours.

Several immune mechanisms induce tumour regression. Macrophages, particularly when activated by lymphokines released from T cells, inhibit the growth of some tumours whilst activation of complement by surface-bound anti-tumour cell antibodies can lead to cell lysis. Tumour cells may also be attacked by the cytotoxic T lymphocytes, which recognize virally infected cells, or by two other forms of cytotoxic lymphocytes which are referred to as 'natural killer' (NK) cells and 'killer' (K) cells. NK cells, unlike the cytotoxic T lymphocytes, do not require prior sensitization whilst K cells will attack only targets that are coated with antibody which the cell can recognize by having receptors for the Fc fragment of antibody (see *figure 256*) on its surface. K cells kill by a non-phagocytic cytotoxic action, and although they have been identified in the test tube the *in vivo* significance of these cells has still to be substantiated. The lineage of both types of killer cell in terms of T and B cells is uncertain.

Unfortunately, despite the plethora of immunological armoury, there is often an uncontrolled proliferation of tumour cells with fatal consequences. Three main possibilities are suggested to explain why tumours persist:

1 failure to be recognized by the host immune system;
2 inefficient immunological responsiveness in certain subjects;
3 the tumour itself inhibits the effector response of the immune system.

k *Why do you think tumour cells escape immune recognition? Release of antigens from the tumour surface may contribute to this phenomenon. Suggest why.*

l *In which group of individuals would you expect to observe an inefficient immunological response?*

m *In 3, it is probable that antibodies directed against the tumour augment its growth. Suggest how this could occur. Suggest any other mechanisms that might be involved.*

Though surgery, irradiation, and chemotherapy constitute the main methods at our disposal for the elimination and regression of tumours, an additional approach is the use of immunotherapy to boost the effector arm response of the immune system, which is often impaired in cancer patients. For example, it may be necessary to give the patient drugs which stimulate the proliferation of lymphocytes or which 'activate' macrophages so as to increase their anti-tumour effect. In this context it is of interest that BCG, which is known to activate macrophages and NK cells, is successful in the treatment of laboratory-induced tumours. Purification of the active components of BCG and possibly more potent immunostimulators of the immune response may be of value in future treatment. At the time of writing, clinical trials are being made of the action of a tumour necrosis factor (TNF), which is released from macrophages. In animal models of cancer this factor has been shown to destroy specifically tumour cells. The outcome of these trials is awaited with eager anticipation.

24.9 Resistance to disease in plants

The pathogenic fungi which cause plant diseases, such as rust in wheat and blight in potatoes, are responsible for serious crop losses. The resistance of some plants to such diseases is therefore of economic interest, and plant breeders have directed their efforts towards producing disease-resistant varieties, especially of food crops.

In the early twentieth century it was discovered that specific genes of the plant determine whether an attack by a particular parasite is possible. Later, a complementary gene system in the parasite was found to exist, which controls whether a parasitic relationship will develop after an infection. Close cooperation has therefore been needed between plant breeders and plant pathologists in an attempt to solve the problem of disease resistance in plants. The question that must be asked is: Is there a mechanism in plants similar to the immune system of animals?

In the 1940s K. O. Müller suggested that blight-resistant potato tubers produced substances which gave them protection against attack by the potato blight fungus, *Phytophthora infestans*. He called these substances *phytoalexins*. Phytoalexins exhibit an antibiotic effect and Müller proposed that it was not resistance itself that is inherited, but the ability of the plant to respond to fungal attack by producing phytoalexins.

Phytoalexins are not the only mechanism by which plants are protected against disease. However, it does appear that plants have a means of resisting disease by producing biochemicals which is biologically analogous to antibody production in animals. Much more is known about phytoalexins, but their further study is beyond the scope of this book and they continue to be the subject of current scientific research.

Summary

1 In addition to the nervous system multicellular animals possess an endocrine system, which is another important means of communication (**24.1**).

2 Endocrine glands produce chemical substances called hormones, which are released into the bloodstream and travel to specific target organs. The target organs respond to the hormones to bring about regulatory effects (**24.1**).

3 Nervous and endocrine systems are closely linked together to regulate body functions. Endocrine cells are usually controlled by secretions from neurosecretory cells or other endocrine glands. The neurosecretory system forms a third level of control – the neurocrine system (**24.1**).

4 Hormones often act together synergistically to produce a response which the individual hormones cannot produce on their own (**24.1**).

5 There are two principal types of hormone: protein hormones and steroid hormones. The cells which produce the two types are quite different; so are their methods of synthesis and release (**24.1**).

6 The control of hormone release is brought about by feedback of information through closed loop systems. This feedback can be either negative or positive (**24.2**).

7 The mechanism of hormone release seems to involve a rise in the level of intracellular calcium ions. The means by which this is brought about varies from one tissue to another and may involve changes in the intracellular concentration of cyclic AMP (**24.2**).

8 The responses of tissues to hormones are determined by the inbuilt programmes of the target cells, but the selection of particular parts of these programmes may be further controlled by the hormone titre (**24.3**).

9 The action of protein hormones is mediated through a second messenger, cyclic AMP. The hormone binds to a receptor site on the plasma membrane of the target cell, causing cyclic AMP to be formed

10. on the inside. Cyclic AMP initiates the response (**24.5**).
10. Steroid hormones act by a mechanism which involves the synthesis of specific proteins. They appear to act at the level of genetic transcription by inducing the production of specific mRNA molecules (**24.5**).
11. Hormones have an important role in the controlled development of animals. Metamorphosis in insects and amphibians illustrates the interaction of hormones to bring about a required, controlled effect (**24.6**).
12. Chemicals which provide a means of communication and control are also found in plants (**24.7**).
13. Plant hormones act as messengers between the specialized regions which perceive and monitor incoming signals from the environment – such as the quality of illumination – and the regions which respond either by changes in growth, such as tropisms, or by changes in development, such as flowering (**24.7**).
14. The auxin IAA promotes cell extension by increasing cell wall plasticity. This involves the movement of H^+ across the plasma membrane which weakens the hydrogen bonds in the cell wall and allows the wall constituents to slide past each other to give plastic extension (**24.7**).
15. Several plant hormones (or their artificial substitutes) are used to coerce plants to develop at a time or in a manner which is favoured by commerce (**24.7**).
16. Vertebrate animals possess an immune system which recognizes and eliminates from the body foreign material (**24.8**).
17. There are two main forms of resistance to infection: natural or non-specific immunity, and acquired or adaptive immunity which is more specific (**24.8**).
18. Natural immunity involves the activation of an enzyme system known as complement (**24.8**).
19. There are two kinds of acquired immunity: the humoral response which is mediated by antibodies, and the cellular response which is mediated by cells that attack the foreign material directly (**24.8**).
20. Lymphocytes are involved in both kinds of immunity. B lymphocytes initiate antibody production, while T lymphocytes initiate the cellular response (**24.8**).
21. Recognition and memory are two characteristics of the immune response and form the basis of immunization (**24.8**).
22. Tumours are the result of cells evading immunological recognition. Several immune responses induce tumour regression and this forms an additional approach to the treatment of cancer (**24.8**).
23. Mechanisms which give protection against disease also exist in plants (**24.9**).

Suggestions for further reading

BLOOM, F. 'Neuropeptides'. *Scientific American*, **243** (2), October, 1981.
COOPER, M. D. and LAWTON, A. R. 'The development of the immune system'. *Scientific American*, **231** (5), 1974. Offprint No. 1306.

EBLING, J. and HIGHNAM, K. C. Studies in biology No. 19. *Chemical communication.* Edward Arnold, 1969.

EDELMAN, G. M. 'The structure and function of antibodies'. *Scientific American,* **223** (4), 1970. Offprint No. 1185.

GOWANS, J. L. Carolina Biology Readers No. 87, *Cellular immunology.* Carolina Biological Supply Company, distributed by Packard Publishing Ltd, 1977.

GUILLEMIN, R. and BURGUS, R. 'The hormones of the hypothalamus'. *Scientific American,* **227** (5), 1972. Offprint No. 1260.

JERNE, M. K. 'The immune system'. *Scientific American,* **229** (1), 1973. Offprint No. 1276.

LUCKWILL, L. C. Studies in biology No. 129. *Growth regulations in crop production.* Edward Arnold, 1981.

PASTAN, I. 'Cyclic AMP'. *Scientific American,* **227** (2), 1972. Offprint No. 1256.

PIKE, J. E. 'Prostaglandins'. *Scientific American,* **225** (5) 1971. Offprint No. 1235.

PORTER, R. R. Carolina Biology Readers No. 85, *Chemical aspects of immunology.* Carolina Biological Supply Company, distributed by Packard Publishing Ltd, 1976.

STROBEL, G. A. 'A mechanism of disease resistance in plants'. *Scientific American,* **232** (1), 1975. Offprint No. 1313.

CHAPTER 25 DEVELOPMENT AND THE EXTERNAL ENVIRONMENT

25.1 Introduction

Earlier chapters have been concerned with the control of the development of organisms through the internal environment. This includes the mechanisms by which genes control biochemical systems and the ways in which genes combine to influence the orderly sequence of the events that result in growth and development. When differentiation was considered we saw that cells assume their respective roles in a precise and orderly way. Each differentiating tissue cell possesses its own inner control mechanism, but during its development the cell is part of a larger whole and throughout the period of differentiation it must therefore respond to controlling factors which are external to it. Finally, we considered the important role of hormones in the controlled development of animals and plants. Thus many processes together make up the internal environment of an organism and result in a balanced development.

But an animal or plant develops to a certain extent in response to the external environment as well. Certain aspects of the physical environment – temperature, light, availability of food, and living space are all examples – can affect both the rate and pattern of development. In this chapter we shall look at only a few aspects of the physical environment and they should be considered as examples with a wide application.

> **Practical investigation.** *Practical guide 6*, investigation 25A, 'The effect of temperature on root growth'.

25.2 The external environment in relation to growth and development

The ungerminated seed is usually a resistant phase in the life cycle of a plant capable of withstanding a level of stress from drought and/or temperature which would kill the vegetative plant. On the other hand, the seedling is a particularly vulnerable stage, so regulation of the time of germination can increase its chances of survival. In temperate climates like that of Britain, winter frost is a major threat to seeds germinating in autumn: few seedlings can survive it. Many of the native species which flower in late summer have a *dormancy* system, that is, their seeds will not germinate until they have been subjected to low temperatures for several weeks. The seeds thus remain dormant during the period of threat, and can monitor the passage of winter as a result of this sensitivity to low temperatures. A domestic refrigerator with temperatures around 2 to 5 °C can be used to 'cheat' such seeds into germinating.

STUDY ITEM

25.21 Control of the time of seed germination

The wild crab apple and the cultivated varieties of apples derived from it have this chilling requirement – which seems appropriate when you think of the season at which apple seeds are shed. Apple breeders often choose to break dormancy by artificially chilling the seeds in a refrigerator so that they can be germinated at any time of year. Table 44 shows some typical results of an investigation into which temperatures to use to break dormancy in apple seeds.

Temperature (°C)	Germination (%)
−1	0
+1	32
2	84
4	89
8	60
10	15
14	0

Table 44
Effect of low temperatures on breaking dormancy of apple seeds stored in moist sand for three months at each temperature and then set to germinate at 20 °C.

a *Why do you think that a subzero temperature was ineffective?*

b *If the seeds were kept dry at these temperatures, dormancy remained intact. Does this reinforce your answer to a?*

c *At 3 °C it took nine weeks chilling to get 50 per cent germination. Do you think there is any advantage in a requirement for such prolonged chilling?*

This process of breaking seed dormancy by chilling is known as *stratification*, a name derived from the practice of alternately layering damp sand and seeds in an earthenware plant pot and then burying it in the soil in late autumn to get the natural winter chilling. In the spring, the seeds can be reclaimed and will germinate.

The nature of the chemical changes brought about by chilling, which release the seed from dormancy, are not thoroughly understood but it seems probable that during the cold period there is a decline in abscisic acid, a hormone frequently involved in imposing dormancy (see page 350). There is also evidence that when the seed temperature is raised to allow germination, there is a marked stimulation of gibberellin synthesis in the stratified seeds as compared to non-chilled controls (see page 350). Sometimes dormancy of this kind can be overcome simply by germinating the seeds in a Petri dish in the presence of gibberellin.

The experiments with lettuce seeds (see *Practical guide 6*, investigation 11B, 'Effects of light on the germination of lettuce seeds') demonstrate a role for light in overcoming dormancy. They illustrate a mechanism for sensing certain wavelengths of light by which the seed is sensitive to being overshadowed by foliage. The increased proportion of far-red light that it receives is detected by the pigment phytochrome (see page 389) and results in its remaining dormant. Had it germinated, the seedling would inevitably have perished, since it would have received only light that had previously had its photosynthetic capacity absorbed by the overshadowing foliage.

STUDY ITEM

25.22 Controlling the quality of growth of the germinated seedling

Once the seed has germinated the seedling will often be in darkness, covered by soil or debris. Many seeds are tiny and have only meagre food reserves to meet the seedling's requirements for its growth into light, when photosynthesis can first augment these reserves and then take over food production. Growth will take the form of maximum elongation resulting in large thin-walled cells supported by turgor. Water loss will be minimal because of the dark, damp conditions, so minimal vascular tissue will be sufficient for water conduction. The growing point should be protected from abrasion by soil particles, because damage to the apex could prevent cell division and thus stop the seedling's growth. *Figure 261* shows two seedlings of white mustard, one grown in darkness and the other in the light.

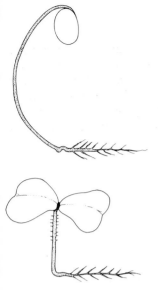

Figure 261
Two seedlings of white mustard which were sown in Petri dishes at the same time. The one above was grown in continuous darkness and the one below in natural lighting.

a *How does growth in darkness achieve protection of the shoot apex?*

b *What is the evidence that the cells are more elongated in the seedling grown in darkness? What further information do you need to confirm this?*

c *In what ways is the light-grown seedling adapted for life above ground?*

It is now known that it is the pigment phytochrome which is sensitive to the difference between light and darkness.

d *How could you test that phytochrome is responsible for the changed morphology of a seedling as a response to light?*

Quite a brief exposure to dim light will trigger the changes in morphology, although it may not be enough to bring about chlorophyll formation in the cotyledons. A slightly longer exposure to bright light will initiate that change as well.
    A seedling is therefore sensitive to the change in its environment as it grows out of the soil. Light-grown seedlings do have better developed vascular tissue than those grown in the dark and dark-grown seedlings have larger, thinner walled cells than light-grown ones have.

e *How do these two facts relate to the survival of seedlings?*

Up to a certain point, we normally expect an increase in temperature to increase the rate at which most metabolic processes take place. The effect of temperature on the growth of roots is investigated in *Practical guide 6*, investigation 11A, 'The effect of temperature on root growth'.

We could also expect the rate of development of some animals to increase with increasing temperature: in insects, for example, the length of the life cycle and the growth rate during the various stages in the life history are related to the external temperature. Another example is considered in the next Study item.

STUDY ITEM
25.23 Hatching trout eggs

Brown and rainbow trout are raised in England and Wales at commercial and river authority trout hatcheries. Only a few of the hatcheries raise fish primarily for food. The rest sell to angling clubs for restocking rivers.

Trout spawn in autumn and winter. The hen (female) trout scrapes a shallow trough in the stream bed and her eggs are fertilized externally. The parents do not look after the eggs or the young. A three-year-old (or older) female fish will produce about 750 eggs for every 0.5 kg of body mass. For about the first six weeks of their life the young are nourished from their own yolk sac.

The data in table 45 show how the incubation period for the development of trout eggs varies with temperature.

Temperature (°C)	Incubation period (days)
+2	205
+5	82
+10	41

Table 45
The influence of temperature on the incubation period of trout eggs.

a *From these data, make an equation to relate incubation period and temperature as exactly as possible.*

b *From your general biological knowledge, or from your equation, determine what you could expect the incubation period to be at $-1\,°C$, $+3\,°C$, $+8\,°C$, and $+15\,°C$.*

c *Can you explain how trout eggs fertilized throughout late winter and early spring tend to hatch at about the same time?*

d *What effect, if any, might this mass hatching have on the trout population?*

25.3 Light and plant growth

> Practical investigation. *Practical guide 6*, investigation 25B, 'Effects of light on the germination of lettuce seeds'.

The primary role of light in plant growth and development is to provide energy for photosynthesis. For effective photosynthesis the light needs to be of a fairly high intensity – most plants do not grow well in the living room unless they are placed on a window-sill. Light which is quite good to read by may be of very limited use for photosynthesis. The illumination also needs to be prolonged – two or three hours a day will not be sufficient. In other words, photosynthesis is a major energy transducer requiring a large energy input (see *Study guide I*, Chapter 7).

It follows that if light is to be used by a plant there must be a light-absorbing molecule to collect it; chlorophyll and other accessory pigments act as light collectors for photosynthesis. Plants do not absorb all colours of light and we already know that chlorophyll has absorption peaks in the red and blue regions of the spectrum but that it reflects most of the green light.

STUDY ITEM

25.31 The significance of action spectra and absorption spectra

If the amount of photosynthesis achieved by equal energies of different wavelengths of light is measured, an *action spectrum* for photosynthesis can be produced (*figure 262*).

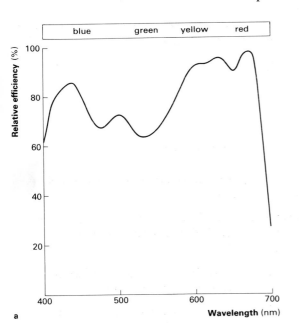
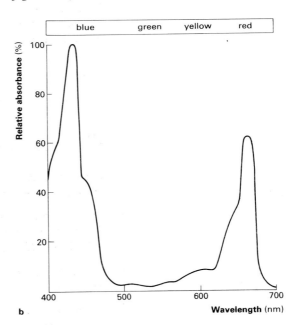

Figure 262
a The action spectrum of photosynthesis in bean leaves.
b The absorption spectrum of chlorophylls in ether.

a *How can the amount of photosynthesis be measured? What units would it have?*

If chlorophyll is extracted into an organic solvent its absorption spectrum (*figure 262b*) can be obtained by using a spectrophotometer. As expected this shows peaks in the red and blue regions but little absorption in the green and yellow regions.

b *Suggest why more photosynthesis results from light in the green region than you might predict from the absorption spectrum of chlorophyll.*

c *Do you think the absorption spectrum of a crude extract of chloroplasts would be a good match to the action spectrum of photosynthesis?*

The point that is being made is an important one. The action spectrum of the process gives a good clue to the kind of pigment molecule(s) involved in light absorption. The absorption spectrum of the pigment will be virtually identical to the action spectrum.

Photomorphogenesis

The effects of light on plant growth such as phototropism (see *Study guide I*, Chapter 12), breaking of seed dormancy, control of shoot development from darkness into light, and photoperiodism – the control of development by day length – are known as *photomorphogenic* effects. In these cases, unlike photosynthesis, only a short exposure to light of low intensity is effective. Photoperiodism might be different because it seems that day length is important (although we shall see later that it is really *night* length that is critical). Nevertheless, under experimental conditions at least, brief exposures to dim light can control the photoperiodic response of flowering.

These photomorphogenic responses are analogous to sensory systems as they monitor environmental information and plants grow and develop in a way which favours the survival and success of the species.

STUDY ITEM

25.32 Action spectra and photomorphogenic responses

Action spectra of various responses are illustrated in *figure 263a*.

a *How many different pigments do you think are involved? Could any of them be chlorophyll?*

b *The absorption spectra of four plant pigments (A, B, C, and D) are shown in* **figure 263b**. *Select the pigment which you believe best fits the action spectrum for each of the different responses shown in figure 263a.*

c *What colour pigment would you expect to be involved in a process for which you obtained the action spectrum shown in* **figure 263c**?

Photoperiodism

This is the ability of organisms to measure changing day length and to make an appropriate physiological adjustment. Its ecological significance is that it results in organisms being dormant during unfavourable conditions such as frost in temperate winters or summer heat and drought. In favourable conditions the organisms are active.

The migration of birds such as swallows is an obvious example of a seasonal effect regulated by day length. Higher plants cannot migrate

Figure 263
a The action spectra of various photomorphogenic responses.
b The absorption spectra of some plant pigments.
c The action spectrum of another pigment.

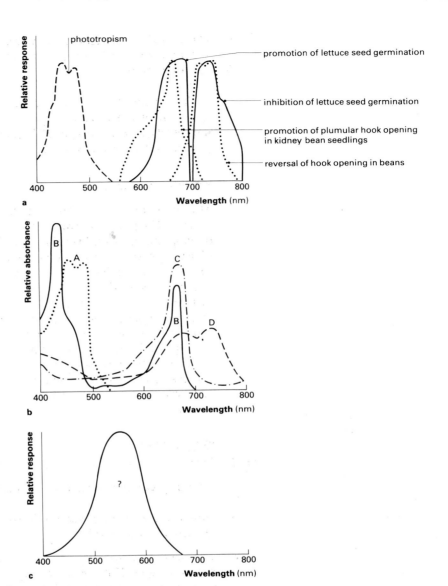

away from unfavourable conditions, however; they must survive the threat rather than avoid it. But the growth of plants to produce resistant phases such as seeds or resting buds is quite slow; there must therefore be an efficient 'early warning' system. Changes in temperature can be unreliable; for example in a British October warm sunny days (the Indian summer) are often followed by keen frosts at night. Day length, on the other hand, is always reliable.

a *On what dates are day and night of equal duration?*

b *Predict what changes in day length might induce a native British tree to go dormant.*

c *Suggest what day length might induce a perennial plant from a hot climate to go dormant.*

Chapter 25 The external environment 383

When photoperiodism induces a resistant phase to withstand a climatic threat its role is in a sense a negative one. Its more positive aspects appear when the threat has passed, in the termination of dormancy and the organization of a reproductive period which will leave plenty of time for seed production.

Many animals become sexually responsive in the springtime. For example, birds achieve their egg-laying, hatching, and the independence of offspring by midsummer, during which time food is abundant. Similarly, the development of testes in some male animals is photoperiodically controlled and is induced by short days and promoted by lengthening days.

In plants the onset of flowering is often under photoperiodic control (table 46) and in perennials, including many British deciduous trees such as birch and sycamore, the termination of winter dormancy may depend upon the lengthening days of spring.

Short-day plants	Long-day plants	Day-neutral plants
Monocotyledons:		
Rice	Barley	Annual meadow grass
(*Oryza sativa*)	(*Hordeum vulgare*)	(*Poa annua*)
	Oats	Maize
	(*Avena sativa*)	(*Zea mays*)
	Ryegrass	
	(*Lolium* spp.)*	
	Wheat	
	(*Triticum aestivum*)	
Dicotyledons:		
Red goosefoot	Clover	French bean
(*Chenopodium rubrum*)*	(*Trifolium pratense*)	(*Phaseolus vulgaris*)
Chrysanthemum	Henbane	Groundsel
(*Chrysanthemum* spp.)	(*Hyoscyamus niger*)*	(*Senecio vulgaris*)
Kalanchoe	Mustard	Shepherd's purse
(*Kalanchoe blossfeldiana*)	(*Sinapis alba*)	(*Capsella bursa-pastoris*)
Japanese morning glory	Petunia	Tomato
(*Pharbitis nil*)*	(*Petunia* spp.)	(*Lycopersicon esculentum*)
Poinsettia	Radish	
(*Euphorbia pulchernima*)	(*Raphanus sativus*)	
Cocklebur	Spinach	
(*Xanthium* spp.)*	(*Spinacea oleracea*)*	

* Plants often used for experimentation.

Table 46
Some examples of the response to day length for flowering

In Britain, however, many plants are unresponsive to day length, and flower after making a certain amount of vegetative growth: examples include peas, tomatoes, and of course all the ephemeral weeds such as shepherd's purse and groundsel. Those native plants in which flowering is photoperiodically controlled all tend to respond to the long days of late spring and summer, thus allowing time for seed production before the arrival of inclement weather. Some common garden plants such as dahlias and chrysanthemums flower in response to the shortening days of late summer, but these plants are not native to Britain. They originate from countries with warm winters, where growth is less threatened by frost than it would be by drought.

Indirect flowering. You may already have thought of the spring-flowering plants such as *Forsythia*, apple, bluebells, and daffodils. In all these cases flowering is not quite what it appears to be however. By 'flowering', the plant physiologist really means the flower *initiation* process. Usually the flower initials then proceed straight away to develop into the layman's 'flowers', but in some plants a period of bud dormancy intervenes between the initiation phase and the subsequent flowering. This is true of all the spring-flowering plants mentioned above, and it can easily be demonstrated by dissecting a flowering 'spur' on an apple or pear tree in autumn – the flowers are already there and recognizable, but their development has been arrested. A daffodil has formed next year's flower by the time the leaves have withered in July – it was initiated in the long days of summer, not in the short days of spring as it might appear. In both cases the flower bud dormancy requires winter chilling to break the resting stage. In *Forsythia* and apple the flowers then come out before the leaves; bulbs also have their flowers standing well clear of the leaves, which are only partly grown at the time when the flowers are out. This indirect flowering is not particularly common and occurs in perennials such as bulbs and trees where food stores have been laid down during the previous year. This ensures a good start for the flower before leaves are present to carry out photosynthesis.

To return to the general consideration of the mechanism of photoperiodic control of flowering: it is fairly obvious that two components are needed, a timing mechanism and a light (or darkness) detector.

The clock. The mechanism of the timing system is not understood at the molecular level, though it is the subject of a great deal of research. All organisms from humans to algae seem to have a biological clock, though it may not exist in prokaryotes as there is no evidence for it in bacteria.

d   *Can you think of any phenomena in humans which are connected with the biological clock?*

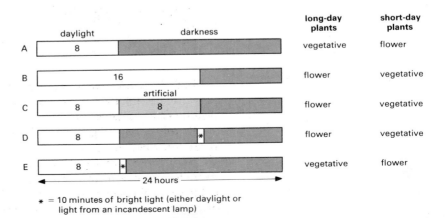

Figure 264
Diagram illustrating the flowering response of typical long-day plants and typical short-day plants when grown on specific regimes.

Chapter 25 The external environment   385

Figure 265
The effect of day length on the flowering of *Pharbitis nil*. The central plant is flowering after 5–6 weeks on a short-day regime. The plants on either side are of the same age, but have been kept on long-day regimes.
Photograph, Dr J. W. Hannay.

represents a series of differing day length regimes, together with the typical flowering responses of long-day and short-day plants (see also *figure 265*). A and B represent short-day (eight hours) and long-day (sixteen hours) treatments respectively. A comparison of regimes B and C shows that artificial light is as effective as daylight, and this is true even if the eight hours of artificial light in C is from a 40 watt lamp set a metre or so away from the plant.

e *What does this suggest about the importance of photosynthesis in the photoperiodic response?*

Treatments D and E show the effect of a break of ten minutes bright light during the sixteen-hour dark period. Plants which receive a light break around the middle of the sixteen-hour period behave oppositely to the plants in A, for which the dark period was continuous; a light break at the beginning of the dark period has no such effect. This is true for both long-day and short-day plants. From this it seems as though darkness is more important than light.

f *Suggest an experiment to check that the continuity of the light phase is not very critical.*

The treatments represented in D and E can be taken a stage further. After a standard light period of eight hours groups of plants have the following sixteen-hour dark period interrupted with one break of ten minutes bright light. Light breaks are given at hourly intervals

Figure 266
Effect on flowering of ten minutes of light given at different times during a period of darkness. Each point plotted on the graphs represents the flowering response of a group of plants which were all given the ten minute light period at the same time.
a Eight hours of light followed by sixteen hours of darkness.
b Eight hours of light followed by forty hours of darkness.

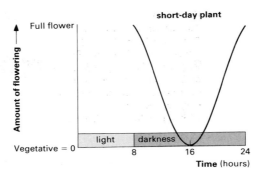
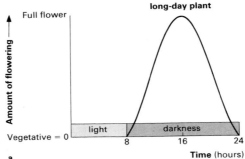

a

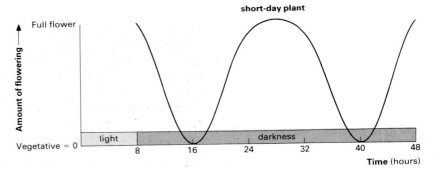
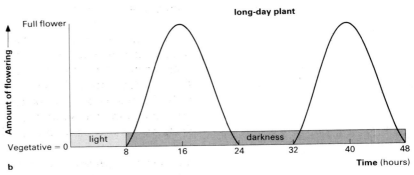

b

throughout the sixteen-hour period but to a different group of plants each time: one group receives the light break after one hour, another after two hours and so on. The results of such an experiment can be plotted as a graph (*figure 266a*).

The results of these treatments suggest that perhaps something goes on in the dark which is sensitive to interruption by light but not uniformly so: the nearer the light break is to the beginning or end of the dark period, the less is its effect.

If this treatment is modified so that an eight-hour light period is followed by a forty-hour dark period with interruptions of one ten-minute light break given as before, the results are as shown in *figure 266b*.

It is as though some timing mechanism is interacting to modify the effect of a standard ten-minute light break.

g *Can you deduce whether the timing mechanism acts like an egg-timer or like an oscillator (such as a pendulum)?*

Such a timing mechanism is called a *biological clock*. It seems as though the biological clock of a plant in this experiment is not simply following the light and dark of the environment but runs under its own rhythm even in continuous darkness. This is called an *endogenous* rhythm and it appears to be making an approximately twenty-four-hour period, reaching maxima and minima about twelve hours apart.

These results fit in with a multitude of experiments on biological clocks in other organisms. It is the way in which the clock is driven which remains a puzzle.

The light detector. We saw earlier in the chapter that if an organism uses light there must be a pigment to absorb it. Once it has been established that a ten-minute light break in the middle of the dark period can completely reverse the response from flowering to non-flowering in a short-day plant (and vice versa for a long-day plant), it is then a simple step to find out what pigment is responding to light.

h *How would you proceed to do this?*

In fact it was found that if 'red' light (wavelength 660 nm) was given during the dark period which would normally induce flowering in a short-day plant, no flowering occurred; giving far-red light did not have this effect, however. Furthermore, if far-red light was given after red light, the plant still flowered. If red and far-red light treatments were alternated, whether the plant flowered or not depended on which wavelength of light it received last.

In answer to question **a** relating to *figure 263a*, you may have suggested a pigment which exists in two forms, one absorbing red light and the other far-red light. This is exactly what phytochrome does.

i *What colour would you expect the pigment phytochrome to be?*

j *If ten minutes of red light was equivalent to ten minutes of white light, how would you check that phytochrome was the receptor pigment?*

There are two forms of pigment, R-phytochrome (P_R), which absorbs strongly in the red region of the spectrum at about 660 nm (620–670 nm), and FR-phytochrome (P_{FR}), which absorbs in the far-red region of the spectrum, around 735 nm (710–750 nm). (You may find these compounds referred to in the literature by the more precise shorthand names of P_{660} and P_{735} respectively.) When R-phytochrome is illuminated with red or white light it changes to FR-phytochrome. In the dark or in far-red light, FR-phytochrome changes back to R-phytochrome, which thus accumulates in the plant during darkness.

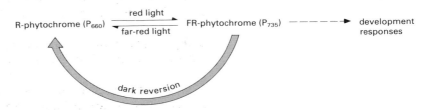

Figure 267
Phytochrome conversion.

The conversion process is thus reversible by red and far-red light and can be repeated many times *in vitro*. In darkness the metabolic conversion of FR-phytochrome to R-phytochrome can take many hours and is slowed down by low temperatures. It does not occur at all in the absence of oxygen, nor does it occur when the phytochrome has been extracted and purified, which suggests that it is brought about by an enzyme.

STUDY ITEM

25.33 **Control of flowering**

In a short-day plant, such as chrysanthemum, flowering involves the apex changing from a leaf producer of unlimited growth to a flower producer whose further growth is terminated by the flower. If the apex is where the flowering occurs, this may also be where the response to day length is perceived.

Figure 268 shows the outcome of an experiment to determine whether the apex perceives the photoperiod.

The apex of the plant A was given a short-day regime while the rest of the plant was given long days. This was achieved by placing a light-tight cover over the apex for sixteen hours each day, thus shortening its day to eight hours, whilst the rest of the plant received sixteen hours light and only eight hours darkness each day (that is, a long day). Plant B had the converse treatment: the apex had long days and the rest of the plant short days.

Plant B flowered whereas A remained vegetative.

a   *What do you conclude about the site of perception of the photoperiodic stimulus? How might this fit in with the idea of a flowering hormone?*

Before we become too enthusiastic about the idea of a flowering hormone it would be helpful to know whether the messenger from leaf to apex was needed in large or small quantities. It is impossible to say with certainty until we know what the messenger is – but we can get a clue. *Figure 269* illustrates the answer.

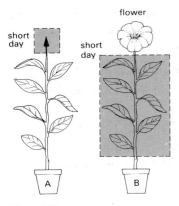

Figure 268
An experiment to determine whether the apex perceives the photoperiod.

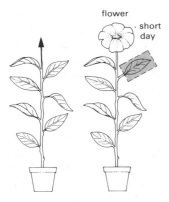

Figure 269
An experiment which demonstrates that a short-day plant will flower even if only one of its leaves is darkened each day.

If two short-day plants are grown in long days, neither will flower; but if only one leaf on one of them is darkened each day, so that that single leaf has short days, then that plant will flower. Thus only one of many leaves is sufficient to produce an effective signal; this indicates that only small quantities of the 'messenger' may be needed.

The name 'florigen' was invented for this hypothetical flowering hormone which travels between the leaf and the apex, but at the time of writing it has not been extracted and identified. Nevertheless there is further strong support for the evidence for a transmissible message. In some cases, a single leaf can be taken from a plant in flower and grafted on to another plant of the same species which is kept vegetative by being maintained under the non-flowering photoperiod (*figure 270*). This receptor plant will then flower even though it has never been exposed to the short-day induction period – neither before nor after grafting.

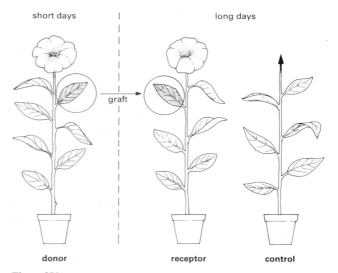

Figure 270
The effect of grafting a single leaf from a flowering short-day plant on to a plant of the same species kept under the non-flowering photoperiod (long days).

b *What does this tell us about the production of 'florigen' in relation to photoperiod being received by the leaf?*

The analogy has been made between infection with a virus disease (which can be transmitted by grafted leaves because an infected leaf continues to produce virus) and catching the 'flowering disease'. It seems that 'florigen' production continues once it has been induced, even though the leaf is now in the 'wrong' photoperiod. It is as though induction involves the unmasking of a gene which is held inactive until it receives the appropriate signal from the leaf. The idea is probably a good one but supporting evidence is very meagre at present.

One interesting observation is that it is not possible to 'catch' some florigen in an agar block, as auxins can be 'caught' in experiments on phototropism (*Study guide I*, Chapter 12). In the leaf-grafting experiment, if an aqueous gap intervenes between the base of the petiole of the leaf to be grafted and the stem of the receptor plant, no transmission of florigen is observed and the receptor remains vegetative. Only when a good cell–cell graft union is established will florigen be transmitted. On this basis, florigen is not likely to be IAA; in fact, none of the known plant hormones will induce a short-day plant to flower when it is maintained on long-day photoperiods. For whatever reason, 'florigen' remains a convenient shorthand for a will o' the wisp which has so far defied all attempts to track it down.

Notice that most long-day plants tend to stay in the rosette form of growth if held in short-day conditions (see *figure 243*). The grasses and cereals do the same sort of thing – they stay short and 'grassy' until the inflorescence forms, then up shoots the stalk with the 'ears' of grain on top. Application of gibberellin will produce this bolting response and then the appearance of flowers. Gibberellin is not 'florigen', however, because it does not induce flowering in short-day plants kept on long days.

In short-day plants it seems as though florigen production is limited by something other than gibberellin; whereas in long-day plants, addition of gibberellin is sufficient to allow 'florigen' production in any day length. It is known that induction by long days is accompanied by synthesis of gibberellins by the plant.

Very occasionally it is possible to find two plants which have opposing photoperiodic requirements (one long-day, one short-day), which are closely related and which can be intergrafted. The classical example is the graft between tobacco and henbane (*Hyoscyamus*), which belong to different genera in the family Solanaceae. Tobacco can be a short-day plant and *Hyoscyamus* is a long-day plant. If a tobacco plant is maintained in long days it won't flower, whilst under this regime the *Hyoscyamus* will flower. If a flowering shoot from the *Hyoscyamus* is grafted to the non-flowering tobacco, both plants still being maintained under long-day conditions, then the tobacco plant will start flowering.

c   *Can you suggest the appropriate conditions, and result, of the reciprocal experiment in which a flowering shoot of tobacco is grafted to non-flowering* **Hyoscyamus**?

d   *What do you conclude about the flower-inducing hormone in long-day versus short-day plants?*

Even though florigen defies identification, the indirect evidence that such a substance exists is fairly compelling. In short-day plants, none of the known hormones will induce flowering when applied externally so it seems likely that an unknown substance(s) will eventually be discovered. Many people have suggested that florigen may be a balance between several of the known hormones – and the effect of gibberellins on long-day plants could be viewed as support of this. If the imbalance in long-day plants is due to lack of gibberellin, however, then its external application soon puts things right. No equivalent is known for short-day plants, so it seems doubtful if it is simply a particular balance between known hormones which induces flowering – unless the particular substance never gets to the effective site when it is applied externally, which is possible but seems unlikely. The search for florigen must continue. It would be a very useful substance to have in a bottle; armed with that, we could switch on flowering at will.

The practical uses of photoperiodic control of flowering are production of flowers out of season – for instance, delaying flower production in chrysanthemums from October or November until around Christmas-time, when the price will be at its highest. This is achieved very inexpensively by using a one-hour light break during the middle of the dark period to maintain long days beyond their natural seasonal period, so that flower induction by the natural short days is delayed.

Conversely, the dwarf potted chrysanthemums which seem to be available all year round have to be 'cheated' into flowering during the long days of summer by artificially shading them in black polythene 'tunnels' which can be drawn over the greenhouse benches from 5 p.m. until 9 a.m. to give inductive long nights. The plants are also dwarfed by use of an anti-gibberellin, to keep them an attractive size for the pots.

Vegetative effects of photoperiod include deciduous leaf fall in autumn, which is a response to shortening days, and bud break in spring and bulb formation in onions, which are responses to lengthening days. Production of stem tubers in main-crop potatoes and Jerusalem artichokes and of root tubers in dahlias are all responses to shortening days.

STUDY ITEM

25.34 **The economics of growing a glasshouse crop using a controlled environment**

In order to achieve the high yields of agricultural and horticultural produce needed to meet our food requirements, it is necessary to modify or control the environment of the food plants while they are growing. Spacing the plants an optimum distance apart, keeping down competition from weeds and pests, watering, and the addition of fertilizers are all widely practised methods of control and modification. When a crop is grown in a glasshouse it becomes feasible to control the temperature and composition of the atmosphere as well.

Many of the large commercial growers concentrate on growing one crop only – this is called *monoculture*. For example, tomatoes may be grown as the only crop throughout the year although some growers use lettuce or flowers such as chrysanthemums as an alternative crop in autumn and winter.

If the environment is controlled, young tomato plants can be planted into a glasshouse in November. They will then start to bear fruit for picking in March and will continue to produce tomatoes until late October or November. By this time an individual plant could be up to seven metres long and may have produced ten kilograms of fruit. This is a very high yield of harvested product – that is, the part for which the crop is grown. Table 47 shows the harvest yields of some agricultural crops and glasshouse tomatoes.

	Harvest mass (tonnes ha^{-1})	Dry mass (approx.) (tonnes ha^{-1})
Wheat	4.4	3.7
Potato	24	2.4
Sugar beet	30	4.5
Tomato (monocrop)	270	14.9

Table 47
The national average yields of wheat, potatoes, sugar beet, and glasshouse tomatoes.

A comparison of the dry masses shows, for instance, that the long-season tomato crop produces more than four times as much useful product as a wheat crop.

a   *Why is it more useful to compare dry masses rather than fresh harvest mass?*

The best farmers may produce more than twice the national average wheat yields, but there is very little variation in yield amongst monocrop tomato growers.

b   *Suggest why there is less variation in the yields of tomatoes from different growers than there is in the wheat yields of different farmers.*

c   *Suggest two major factors which contribute to the higher yield of tomatoes as compared to wheat.*

Control of the atmosphere. Enrichment of the atmosphere with extra carbon dioxide is a technique that has been in use only since the mid-1960s and is a method only available for glasshouse crops (see table 48).

CO_2 concentration	Yield relative to untreated crop
unenriched (0.03% CO_2)	100
0.06% CO_2	120
0.09% CO_2	136
0.12% CO_2	139

Table 48
Tomato yields to the beginning of July from plants which received various levels of carbon dioxide enrichment from the time they were planted out in early January until mid-May.

d   *Why does the addition of carbon dioxide to the atmosphere in which plants are growing give an increased yield?*

For extra carbon dioxide to be effective the greenhouse must be well sealed, and the glass must be kept clean and have minimal interruption from glazing bars.

e   *What are the reasons for these measures?*

The idea of carbon dioxide enrichment of the atmosphere is not new; many investigators tried it in the 1930s. For each one who found it beneficial to plant growth there were many who found it caused injury to the plants. Nowadays the carbon dioxide used is produced by burning North Sea gas (mainly methane, CH_4), propane (C_2H_6), butane (C_4H_{10}), or low-sulphur kerosene. In the 1930s the extra carbon dioxide came, in many cases, from industrial flue gases.

f *Suggest why some of the early workers found injury to plants when they supplied extra carbon dioxide.*

Control of temperature. In order for tomato plants to benefit from extra carbon dioxide they must be grown at controlled temperatures. A minimum night temperature of 16 °C is usually recommended, together with a minimum day temperature of 20 °C with the ventilators opening for cooling when the temperature reaches around 22 or 25 °C. Carbon dioxide enrichment of the atmosphere in glasshouses is usually carried out only from November to mid-May.

g *Why is the supply of carbon dioxide stopped during the summer months?*

h *At what time during a twenty-four-hour period should the extra carbon dioxide be supplied for it to be beneficial?*

Control of water and nutrients. Water and nutrients must also be controlled in order to avoid stress in the plants, which could limit their growth. Many growers use peat bags and supply controlled amounts of water and nutrients through small tubes. Some growers use soil as the growing medium, supplying water and fertilizers in a similar way as for the peat bags, but the disadvantage of soil is the build-up of pests and diseases in it which means it must be sterilized in some way each season.

A method which is becoming increasingly used dispenses with a solid medium altogether and uses only a nutrient solution. The principle is not new and solution culture (hydroponics) has been used commercially for flower crops such as carnations for many years. The modification used for tomatoes is called Nutrient Film Technique (N.F.T.) (*figures 271 and 272*). The tomato plants are raised in peat blocks until they are about 20 cm tall and have several well-developed leaves. They are then set out in their final positions about a metre apart. The usual density is about 22 000 plants per hectare (2.2 plants per square metre).

The roots of the plant simply grow out into the nutrient solution in the polythene trough. The support surface for the polythene is gently sloped to provide a continuous gradient (about 1 in 75) down which the solution will flow forming a 'film' only about one millimetre deep.

i *Why is the maintenance of a thin 'film' an important consideration?*

Each plant is individually supported since it may grow to a length of six or seven metres. It is convenient to have a flexible system which allows the lower part of the plant to trail along the ground once it has borne

Figure 271
Growing tomatoes by the nutrient film technique.
a A schematic layout of the technique, showing just four rows.
b The polythene trough at the feed-pipe end.
c A method for training plants to facilitate fruit picking. A, a young plant; B, after several weeks' further growth.

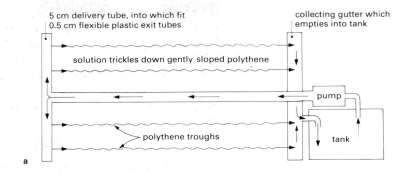

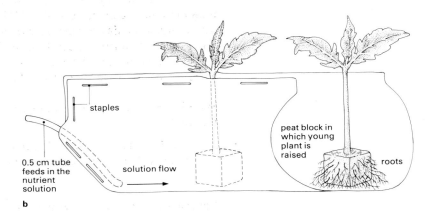

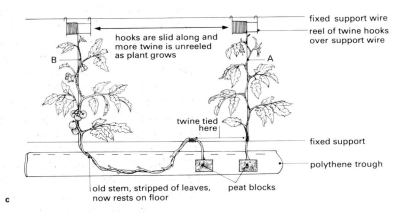

fruit and its leaves have been removed. In this way the fruit continues to ripen at a convenient height for picking.

The nutrient solution contains all the elements required for optimum plant growth. The composition and pH of the nutrient solution can be monitored and adjusted automatically.

j *List the elements that would be needed in such a nutrient solution.*

Control of light. Whilst it is not difficult to add extra lighting to a greenhouse its cost is very high, so supplementary lighting is used

Figure 272
Tomatoes being grown under the nutrient film technique, showing the supply tubes for nutrient solution.
Photograph, Glasshouse Crops Research Institute, Littlehampton, Sussex.

sparingly. Its use is usually restricted to the first three or four weeks after seed germination when the seedlings in their individual small pots or peat blocks can be packed closely together, and several hundred seedlings can be accommodated under a few fluorescent lamps (or equivalent light sources). With extra carbon dioxide as well, a much sturdier plant is produced more quickly than it could be grown without supplementary illumination, thus giving it a good start.

The variety of tomato seed used. So far we have considered the effects of controlling the environment, but the variety of seed used will also have an important effect on yield. A commercial monocrop seed variety costs around £42 per 10 g, compared with £9 per 10 g for a domestic variety.

k *Suggest reasons for the large difference between the cost of seed for a commercial variety and that for a domestic variety.*

About 7 g of seed will produce 1000 plants, which are planted at around 22 000 plants per hectare; the seed will therefore cost approximately £650 per hectare. As other input costs could be around £180 000 per hectare, money spent on seed is an important investment.

Environmental control gives extra yield but at what cost? The fact that commercial growers are using modern techniques for growing monocrop tomatoes implies that the cost of the environmental modification is repaid by the extra yield. The balance between profit and loss is very fine, however. The data which follow give a reasonable representation of the situation in the early 1980s in England and Wales.

Production costs: £ ha^{-1}
Plants, growing medium, CO_2, sprays etc. 25 000
Labour 33 000
Fuel for heating 38 000
  96 000

Marketing costs:
Packing, transport, commission etc. 39 000
  £135 000

Crop value:
This is the most difficult item to evaluate because it varies considerably from year to year, depending on sunshine to give good yields early in the season when prices are high. These prices are in turn affected by the prices charged by overseas competitors. Table 49 illustrates the interaction between price and production.

Month	Yield (tonnes ha^{-1} month^{-1})	Price (£ tonne^{-1})	Crop value (£ month^{-1})
February	2	1 300	2 600
March	12	1 250	15 000
April	34	1 220	41 480
May	47	830	39 010
June	52	590	30 680
July	46	480	22 080
August	43	380	16 340
September	24	440	10 560
October	18	420	7 560
Total yield: 278	Average price per tonne: £677		Total crop price: £185 310

Table 49
A typical tomato-grower's crop yields and values during the early 1980s.

Notice that the monocrop grower has harvested over 50 per cent of the total yield by the end of June and that its value is almost 70 per cent of the total crop value. The aim is to maximize the yield in February, March, and April when prices are highest and to this end there is a tendency to be planting out in November and December instead of January. A few growers even plant out in October, incurring extra heating costs to get the early yield. Of course, the price in November would be high too if the harvesting of good-quality fruit could be extended to later in the season.

You may already have decided that the margin of around £50 000 ha^{-1} is an excellent profit, but you would be wrong. This figure takes no account of the overheads of the business such as offices, telephones, depreciation of equipment, interest on overdrafts, interest on mortgage, and so on. In the 1981 season it seemed likely that well over 50 per cent of growers would sustain an overall loss rather than a profit. There is some evidence that the total area of heated glasshouses for tomato production has fallen since 1980. To build one hectare of heated glass and equip it to modern standards costs around £500 000, so that at an interest rate of 12 per cent per annum, repaying the interest alone would cost £60 000 ha^{-1} every year; it is therefore clear why virtually no new heated glass for tomatoes was built in 1981–82. If older glass becomes decrepit it is unlikely to be replaced at present.

Nevertheless the most efficient producers will survive and there have recently been clear indications of substantial reductions in the use of heating and labour to save costs, whilst yields meanwhile were being maintained.

l *The cost of producing carbon dioxide for enrichment is about £3800 ha^{-1}. Can you estimate its net benefit in crop value, assuming it improves overall yield by 30 per cent?*

m *In fact, it is likely to raise yields up to the end of June by 40 per cent and thereafter make little difference to the yield from July to October. Does this make much difference to the net value of the crop? Can you explain why its effects are minimal after June?*

This is just one example in which the application of science makes a contribution to an increased production of a good-quality crop at lower prices.

25.4 The study of ontogeny

In *Study guide I*, Chapter 13 the behavioural development (ontogeny) of animals was mentioned briefly. It is now appropriate to consider in a little more detail changes in behaviour which are due to the animal interacting with its environment.

The behaviour of animals, and of humans, changes during their lifetime. For example, we acquire certain skills, such as speech or riding a bicycle, only after reaching a certain age (though this differs a great deal from person to person). In animals, abilities such as being able to fly or to find food often first appear quite late in their development. The question that must be asked is whether behaviour patterns appear during life simply because an animal has reached a certain stage of maturity or whether they appear when they do because the animal has had to learn how to perform them. For example, pigeons, like most birds, begin to fly just after they have left the nest. Before they fly, they indulge in much wing-flapping. Is this flapping necessary for the development of flight? Are young, flapping pigeons learning how to use their wings?

To answer these questions, an experiment was conducted by Grohmann in 1939 in which young pigeons had their wings confined in cardboard tubes. These were then removed at the age at which unconfined pigeons began to fly properly. Grohmann found that the confined birds were able to fly just as well as those that had been free to flap their wings. These results suggest that, at least in pigeons, acquiring the ability to fly does not involve learning but is a feature of a particular stage in their development. Another example concerns dogs. A male puppy urinates in the same squatting position as an adult female, but when it becomes adult it cocks its leg. This is not a result of learning but of reaching sexual maturity. A puppy injected with male sex hormones promptly starts to cock its leg.

In contrast, many other things that domestic dogs do – such as walking to heel or sitting or fetching something when ordered to do so – are only done by animals that have been specially trained to do them. Clearly, such behaviour patterns are learned.

There are thus two kinds of influence on the development of behaviour:

1 Changes in behaviour that are due solely to increasing age or *maturation*. Such changes may occur when they do because an animal reaches a certain age or because it has reached a certain degree of muscular, neurological, or sexual development.
2 Changes in behaviour that are due to the animal interacting with its environment, or *learning*.

Let us now consider an example in which these two influences can be differentiated.

Pecking in chicks

Young chicks begin to peck at small objects soon after they have emerged from the egg. At first their pecks tend to be very inaccurate, but with time they become very good at picking up small objects. At the same time they rapidly become able to discriminate between small objects that are not food, such as stones, and those that are, such as seeds. It is essential that they develop this ability by the third or fourth day of life, by which time all their yolk is used up.

Consider first the improvement in the accuracy of pecking. This could come about through learning: a chick that misses an object could remember to aim a little to the left or to the right next time. Alternatively, it might be a maturation effect due to improving co-ordination between eyes and beak as a chick gets older.

a *How would you design an experiment to distinguish between these possibilities?*

b *Are there any other possible explanations you can think of for the improved accuracy of pecking?*

The alternative hypotheses proposed above were tested by Ekhardt Hess in an elegant experiment in which chicks were required to peck at a nail-head which was embedded in soft clay, so that it was possible to see exactly where the chicks' pecks fell when they pecked the nail. Hess fitted chicks with prismatic goggles that distorted their vision in such a way that they always pecked to one side of a target object. In control chicks without goggles, as expected, pecking accuracy improves with age (*figure 273a*). In the chicks with goggles, pecking became more accurate in the sense that pecks became less scattered, but not in the sense that they became directed closer to the target (*figure 273b*). If learning had been involved, we would expect the chicks to have compensated for the effect of the goggles so that they did hit the nail. Since they did not do so, we must conclude that their ability to aim their pecks more nearly to previous pecks is due to maturation, not to learning. In fact, if chicks are kept in goggles for a long time they will eventually learn to compensate for their effect, but this does not occur within the time scale investigated by Hess.

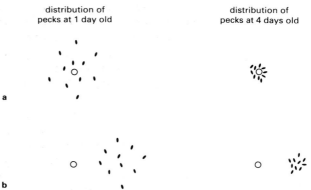

Figure 273
The distribution of pecks directed at a nail-head by 1-day-old and 4-day-old chicks, **a** without, and **b** with prismatic goggles.

The ability to discriminate between food and non-food objects has been studied by Hogan, whose experiments have shown that this is not present in newly hatched chicks, which are as likely to peck at sand as they are at food. If they eat some food, then within an hour they peck more at both food and sand. After many hours of pecking, however, they will peck more at food than at sand. Evidence that this is due to learning comes from the observation that chicks that are fed by having food pushed into their mouths are unable to discriminate between food and sand. It appears that chicks learn that pecking at food alleviates their hunger, whereas pecking at sand does not. Chicks do not, however, totally lose the habit of pecking at non-food objects. Throughout their lives they ingest a small number of stones, which are taken into the gizzard where they are necessary for the grinding up of food.

For many animals, there is a wide range of food objects in their environment, but often they show preferences for certain foods over others. In some species such preferences are acquired through learning, aspects of food objects such as their appearance or odour being associated with their beneficial nutritive consequences. Likewise, animals can learn to avoid foods that make them ill, a fact which makes it very difficult for pest control officers to eliminate rats by putting down poison. In some species, however, preferences may exist even before food has been experienced. One species of snake has been shown to have such a preference for worms, another species for crickets. In another snake, by contrast, a single feeding experience with a fish leads to the development of a feeding preference for fish.

The learning of a preference for a particular kind of food need not involve direct experience of that food. There is evidence that young rats that have just been weaned prefer the same diet that their mothers have eaten. It appears that the taste of the mother's food is passed to her young through her milk.

The ontogeny of bird song

One of the best examples of behaviour whose development has been studied in detail is bird song. In many small birds, males produce songs during the breeding season that serve either or both of two functions: attracting females and defending a territory against intrusion by other males. Bird song tends to be highly species-specific. With a few exceptions, each species has a distinctive song, a fact which makes it possible for ornithologists to recognize species on the basis of song. The songs of different species vary greatly in complexity; some are short and simple, others are long and highly elaborate. Some bird vocalizations, such as the cooing of doves and the crowing of cockerels, apparently do not have to be learned. In all song bird (passerine) species studied so far, however, some degree of learning is involved in the development of singing. Bird song has proved to be an ideal subject on which to study the interaction of inherited and environmental factors during development.

The scientific study of bird song development was pioneered by W. H. Thorpe. One of his major contributions was to perfect the use of the sound spectrograph, an instrument that converts patterns of sound into visual pictures called sonograms (*figure 274*). Sonograms mean that the detection of differences between songs can be done objectively and in great detail, and is not dependent on the ears and subjective assessment of observers.

Song birds that are hand-reared from soon after hatching will start to sing at the beginning of the following spring, but typically their songs bear only a superficial resemblance to the normal song of their species (*figure 275*). Clearly, normal song development requires some kind of experience, of which they are deprived in captivity. If hand-reared birds

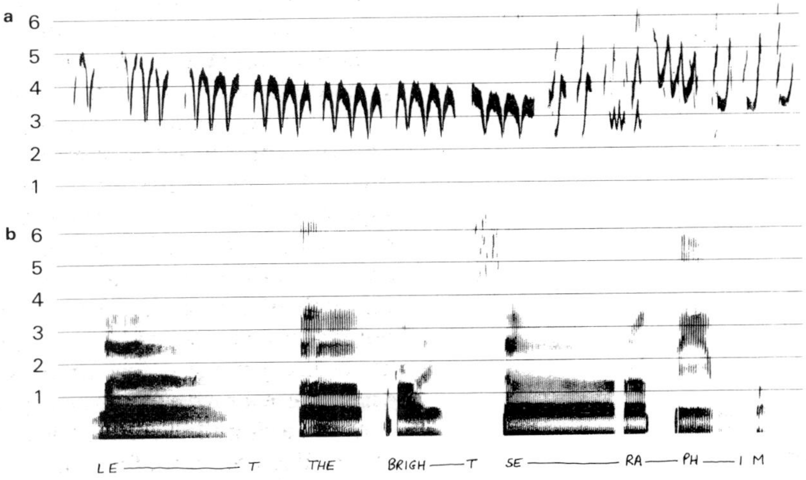

Figure 274
Sonograms.
a The song of the redwing (*Turdus iliacus*).
b An unaccompanied human voice singing the aria 'Let the bright seraphim' from Handel's *Samson*. The vertical axis shows sound frequency in kHz, the horizontal axis indicates time.
From Halliday, T. R. and Slater, P. J. B., Animal behaviour, Volume I, Causes and effects, Blackwell Scientific Publications, 1983.

Figure 275
Typical songs of a male swamp sparrow and a male song sparrow, as normally produced (above) and as developed by males kept in auditory isolation in early life (bottom).

Figures 275–277 are from Halliday, T. R. and Slater, P. J. B., Animal behaviour, *Volume 3, Genes, development and learning, Blackwell Scientific Publications, 1983.*

Figure 276
a and **c** Sonograms of two chaffinch songs that were played, as recordings, to young male chaffinches.
b and **d** The songs that they subsequently produced themselves. Note that the 'pupil' songs are very accurate copies of the 'tutor' songs except for a few features, such as the final flourish in **c** and **d**.

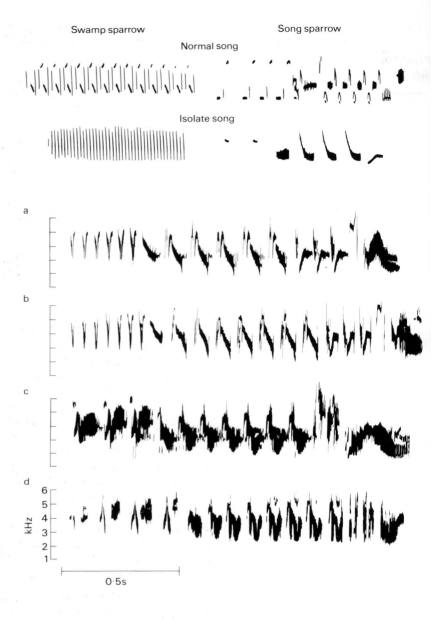

are allowed to listen at frequent intervals to a tape-recording of song during early life, they will produce a very accurate copy of that song in their first spring (*figure 276*). Thus it is clear that they must hear song in order to be able to sing correctly. Most birds will not learn *any* song that they hear, however, but will accurately copy only a recorded song that belongs to their own species; for example, P. Marler found that young white-crowned sparrows would learn their own species song but not those of song sparrows. Most birds are thus in some way predisposed to learn the song appropriate to their own species.

Marler also found that, for songs to be learned, they must be heard during the first 100 days of a young bird's life. He called this the

Figure 277
Marler's general model of bird song development.

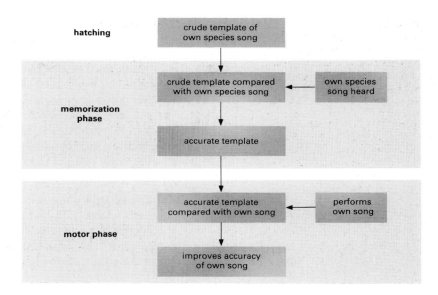

memorization phase and suggested that it represents a sensitive period. It is striking that the memorization phase is over long before a young bird starts to sing, when it is nearly a year old. Thus the behaviour (singing) and the learning process that leads to it are separated by many months.

Marler has proposed a model of how bird song develops (*figure 277*). At hatching, a young bird has a crude template of its own species song. During the memorization phase, this template is made more accurate through feedback from the song of adult males of the same species. If this feedback is prevented by social isolation, an accurate template is not developed. The following spring, the male starts to sing; this is called the *motor phase*. It hears its own song, compares it with the template it developed several months earlier, and perfects it. If feedback is prevented it is unable to produce an accurate song.

The scheme shown in *figure 277* is a very general one and does not consider the marked variations in song development between species. These variations occur in three aspects of bird song: the time at which memorization ends, the number of different songs that individuals will learn, and the accuracy with which songs are copied.

Timing of memorization. In the white-crowned sparrow, males do not learn songs after they are about 100 days old; thus memorization ends long before singing begins. In the chaffinch, males are still learning songs that they hear early in the year in which they themselves start to sing, but they stop learning new songs by the time they are fully engaged in singing. Indigo buntings continue to learn songs during their first breeding season and, as a result, are able to copy the songs of males who hold territories next to their own. Some birds, such as canaries, budgerigars, mynah birds, and parrots continue to learn new songs and other sounds throughout their lives.

Size of vocal repertoire. Whereas many birds produce a single distinctive song which makes them easy for ornithologists to recognize, others may develop a very large repertoire of different sounds. For example, marsh warblers migrate from Europe to East Africa during their first autumn and winter, when there is no adult marsh warbler song for them to copy. Instead, they learn elements of the songs of other species that they hear as they move south. As a result, a young male starting to sing in his first breeding season performs a song that contains, on average, elements of the songs of seventy-six other species. Because individual marsh warblers vary widely in terms of the species they hear during migration, and the order in which they hear them, each develops its own distinctive song. Clearly, the initial crude template with which marsh warblers are born is much less narrowly defined than that of a species such as the white-crowned sparrow.

Accuracy of copying. Young males of some song bird species copy the songs of adult males with almost perfect accuracy, while those of other species produce songs that contain novel features. Compare, for example, the 'tutor' and 'pupil' chaffinch songs in *figure 276*. Over many generations, the accumulation of such variations within a geographical area leads to the development of local 'dialects'. For example, P. J. B. Slater has found that chaffinches recorded in Sussex produce songs that are very different from those of Scottish birds, though both share the basic features of chaffinch songs. Furthermore, there is a population of chaffinches in New Zealand, where this species does not naturally occur, that is descended from about sixty birds collected in Sussex in 1860, and these sing songs that are very different from those of present-day Sussex chaffinches. The fact that there is variation in the songs of a single species, both between individuals and between populations, raises the possibility that bird song conveys much more information than was once supposed. It could be that birds can recognize, not only the species, but also the individual identity and home locality of a singing bird on the basis of its song.

STUDY ITEM

25.41 Mother–infant relationships

> **Practical investigation.** *Practical guide 6*, **investigation 25C, 'Early environment and later behaviour of mice'.**

In this section, we have seen how important early experiences can be in the development of behaviour. This has obvious relevance to human behaviour. Clearly, if one believes that experience in childhood affects adult behaviour, it is difficult to perform experiments on humans, for ethical reasons. However, a great deal of work has been carried out with non-human primates, such as rhesus monkeys (*Macaca mulatta*), particularly into the effects of disturbance of the mother–infant relationship on adult behaviour.

a *Do you think that it is legitimate to extrapolate conclusions from studies of non-human primates to humans?*

b *Do you think it is any more ethical to interfere with the relationship between a mother rhesus monkey and her infant than it is to do the same kind of experiment with humans?*

Work on mother–infant relationships in primates was pioneered by the American psychologist H. F. Harlow. The way that he came to start his work is, in itself, very interesting. In Harlow's own words:

' Our investigations of the emotional development of our subjects grew out of the effort to produce and maintain a colony of sturdy, disease-free, young animals for use in various research programmes. By separating them from their mothers a few hours after birth and placing them in a more fully controlled regimen of nurture and physical care we were able both to achieve a higher rate of survival and to remove the animals for testing without maternal protests. Only later did we realize that our monkeys were emotionally disturbed as well as sturdy and disease-free. Some of our researches are therefore retrospective. Others are in part exploratory, representing attempts to set up new experimental situations or to find new techniques for measurement. Most are incomplete because investigations of social and behavioural development are long-term. In a sense, they can never end, because the problems of one generation must be traced into the next.'
(*From Harlow, H. F. and Harlow, M. K.*, 'Social deprivation in monkeys'. Scientific American, **207** (5), *1962. Copyright © by Scientific American, Inc. All rights reserved.*)

c *What implications does this passage have for those concerned with the care of laboratory and domestic animals?*

d *What do you suppose is the meaning of the statement: '…investigations of social and behavioural development are long-term. In a sense, they can never end, because the problems of one generation must be traced into the next'?*

The emotional development of infant monkeys. For a long time, scientists thought that affection and consequent emotional stability in infants was generated by the satisfaction of feeding. Professor H. F. Harlow and his associates were impressed by the apparent need for some form of bodily contact by young rhesus monkeys and investigated the topic further.

' …we have sought to compare the importance of nursing and all associated activities with that of simple bodily contact in engendering the infant monkey's attachment to its mother. For this purpose we contrived two surrogate mother monkeys. One is a bare welded-wire cylindrical form surmounted by a wooden head with a crude face. In the other the welded wire is cushioned by a sheathing of terry cloth. We placed eight newborn monkeys in individual cages, each with equal access to a cloth and a wire mother (*figure 278*). Four of the infants received their milk from one mother and four from the other, the milk being furnished in each case by a nursing bottle, with its nipple protruding from the mother's "breast".

Figure 278
Cloth and wire 'mothers' used to test the preferences of infant rhesus monkeys. This infant is clinging to the soft cloth 'mother' even though its feeding bottle is attached to the wire 'mother' in the background.

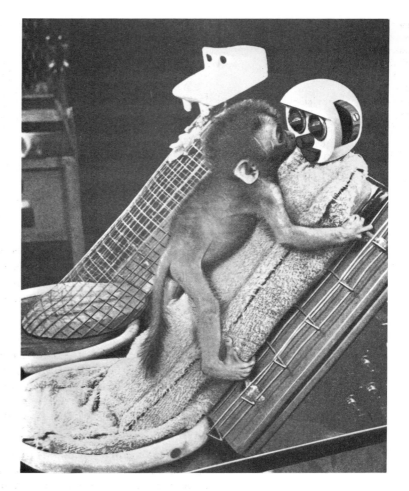

' The two mothers quickly proved to be physiologically equivalent. The monkeys in the two groups drank the same amount of milk and gained mass at the same rate. But the two mothers proved to be by no means psychologically equivalent. Records made automatically showed that both groups of infants spent far more time climbing and clinging on their cloth-covered mothers than they did on their wire mothers. During the infants' first fourteen days of life, the floors of the cages were warmed by an electric heating pad, but most of the infants left the pad as soon as they could climb on the unheated cloth mother. Moreover, as the monkeys grew older, they tended to spend an increasing amount of time clinging and cuddling on her pliant terry cloth surface. Those that secured their nourishment from the wire mother showed no tendency to spend more time on her than feeding required, contradicting the idea that affection is a response that is learned or derived in association with the reduction of hunger or thirst (*figure 279*).

' These results attest the importance – possibly the overwhelming importance – of bodily contact and the immediate comfort it supplies in forming the infant's attachment for its mother..

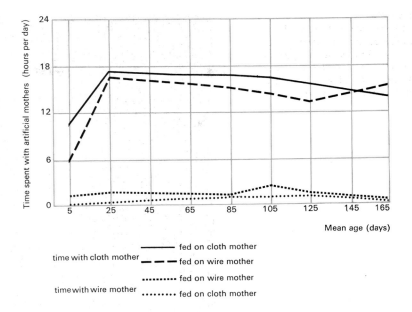

Figure 279
The preferences of infant rhesus monkeys given a choice of cloth and wire mother surrogates, one of which provided food.
Figures 278 and 279 are from Harlow, H. F., 'Love in infant monkeys', Scient. Am. **200**, *6, pp. 68–74, 1959. Copyright © 1959 by Scientific American Inc. All rights reserved.*

' When a child is taken to a strange place, he usually remains composed and happy so long as his mother is nearby. If the mother gets out of sight, however, the child is often seized with fear and distress. We developed the same response in our infant monkeys when we exposed them to a room that was far larger than the cages to which they were accustomed.

' In the room we had placed a number of unfamiliar objects such as a small artificial tree, a crumpled piece of paper, a folded gauze diaper, a wooden block and a doorknob. If the cloth mother was in the room, the infant would rush wildly to her, climb upon her, rub against her and cling to her tightly... its fear then sharply diminished or vanished. The infant would begin to climb over the mother's body and to explore and manipulate her face. Soon it would leave the mother to investigate the new world, and the unfamiliar objects would become playthings...

' If the cloth mother was absent, however, the infants would rush across the test room and throw themselves face-down on the floor, clutching their heads and bodies and screaming their distress. Records kept by two independent observers – scoring for such "fear indices" as crying, crouching, rocking and thumb- and toe-sucking – showed that the emotionality scores of the infants nearly tripled. But no quantitative measurement can convey the contrast between the positive, outgoing activities in the presence of the cloth mother and the stereotyped, withdrawn and disturbed behaviour in the motherless situation.

' The bare wire mother provided no more reassurance in this "open field" test than no mother at all.'

(*From Harlow, H. F., 'Love in infant monkeys', Scientific American* **200** *(6), 1959. Copyright © by Scientific American, Inc. All rights reserved.*)

The effect of isolation. In the following quotation from the same work the authors compare the social behaviour of monkeys with cloth 'mothers' and those reared in complete isolation, and then go on to describe further investigations.

'In fright-inducing situations the infants showed that they derived a strong sense of security from the presence of their cloth mothers. Even after two years of separation they exhibit a persistent attachment to the effigies.

'In almost all other respects, however, the behaviour of these monkeys at ages ranging from three to five years is indistinguishable from that of monkeys raised in bare wire cages with no source of contact comfort other than a gauze diaper pad. They are without question socially and sexually aberrant. No normal sex behaviour has been observed in the living cages of any of the animals that have been housed with a companion of the opposite sex. In exposure to monkeys from the breeding colony not one male and only one female has shown normal mating behaviour and only four females have been successfully impregnated. Compared with the cage-raised monkeys, the surrogate-raised animals seem to be less aggressive, whether toward themselves or other monkeys. But they are also younger on the average, and their better dispositions can be attributed to their lesser age.'

e   *To what extent are the cloth mothers adequate substitutes for real mothers?*

f   *Professor Harlow once wrote that 'the close behavioural resemblance of our disturbed infants to disturbed human beings gives us the confidence that we are working with significant variables...'. Is this statement justified?*

g   *In what respects are the conclusions you would draw from the results of the investigations adequate to explain the way psychological development of rhesus monkeys takes place in the wild?*

h   *What psychological factors other than those mentioned in the quotations may affect the development of young rhesus monkeys? Through further reading find out to what extent your answer is borne out by research findings that followed those mentioned here.*

Harlow's work showed that maternal deprivation caused gross behavioural abnormalities in young monkeys, both during the period of isolation and when they were returned to the company of other monkeys. Formerly isolated monkeys (referred to as isolates) frequently fought with other monkeys, tending to initiate fights in ways that non-isolates did not. They also frequently bit themselves and developed stereotyped behaviour patterns, such as rocking. When they became mature, they showed abnormal sexual behaviour. Nevertheless, some female isolates became pregnant, but then proved to be very abnormal in their maternal behaviour, often maltreating and wounding their infants. Such effects showed dramatically the importance for normal development of the mother–infant relationship. But, Harlow's experiments, because they involved periods of separation lasting several months, revealed very little of what it is that goes on during the time an infant spends with its mother that is so important. Subsequent work, notably by R. A. Hinde, has sought to answer this question, both by carefully observing what goes on during the infant phase, and also by performing separation experiments of much shorter duration than Harlow's.

Hinde's work showed how the mother–infant relationship changes with time. An obvious change is that, whereas the newly born infant clings to its mother all the time, as it grows older it spends more and more time out of physical contact with her and gradually makes larger and larger movements away from her. Such changes could be explicable entirely in terms of the infant's maturation. As it grows and becomes stronger and better co-ordinated, it is likely to become more independent. Hinde's work showed, however, that rhesus monkey mothers also play an important role in promoting their infants' increasing independence. Some of Hinde's results are shown in *figure 280*.

Figure 280
Changes in behaviour between mother and infant rhesus monkeys with age.
a Percentage of time that infants spend off their mothers.
b Percentage of time that infants spend more than 60 cm from their mothers.
c Percentage of attempts by infants to cling to the ventral sides of their mothers that are rejected by the mothers.
d The relative contributions of mothers and infants in maintaining contact: the higher the index, the greater is the part played by the infant.
In each graph, the solid line shows median scores for several mother–infant pairs, and the broken lines show the range of variation between pairs.
From Halliday, T. R. and Slater, P. J. B. 'Genes, development and learning', Animal Behaviour, Volume 3, Blackwell Scientific Publications, 1983.

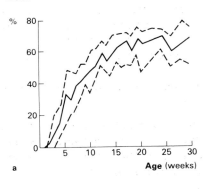
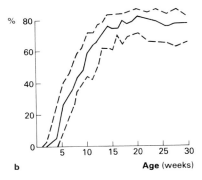
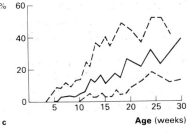
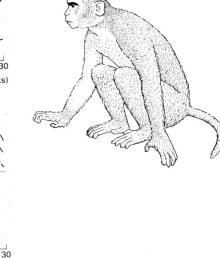

Figures 280a and *b* simply show how, as an infant grows, the amount of time it spends away from its mother increases, as does the distance it goes from her. An infant can come into contact with its mother in one of two ways: she may reach out and pull it towards her, or the infant may move on to her of its own accord. If the infant initiates a contact, the mother either may accept the contact or may reject it by pushing the infant away from her. *Figure 280c* shows how mothers' responses to contacts initiated by infants change over time.

i **In what way could the behaviour of mothers be responsible for the increasing independence of infants as they get older?**

Mother–infant contacts were used by Hinde to compute an index of which partner was more responsible for maintaining the relationship. This index is obtained by subtracting the percentage of contacts initiated by the mother from the percentage initiated by the infant: thus, the larger the value of the index, the greater the proportion of contacts that are infant-initiated. Changes in this index over time are shown in *figure 280d*.

j **How do the relative responsibilities of the mother and of the infant for maintaining their relationship change with time?**

k **What do you think is the significance of the variation in the contact index shown in figure 280d? What features of the behaviour of individual infants and mothers could account for this variation?**

Hinde has looked in detail at variations in the behaviour of many rhesus monkey mother–infant pairs and has identified certain types of mother. Some mothers reject a large proportion of attempts by their infants to climb on to them and are rarely responsible for establishing contact with their infants. These are referred to as 'rejecting' mothers. A few mothers rarely reject their infants and frequently initiate contacts with them; these are called 'possessive' mothers. Others reject many of their infants' approaches but initiate a high proportion of contacts: these are described as 'controlling' mothers.

Hinde also separated infants from their mothers for six days when the infants were between 21 and 32 weeks old. The mother was removed from the small social group in which she and her infant lived and the infant was left behind. The infants were clearly distressed during separation; they frequently gave distress calls and spent long periods being immobile. These symptoms were most marked in those infants whose mothers were in the 'rejecting' category.

Observational techniques have also been applied to human mother–infant relationships. For example, J. Dunn has followed children from the first ten days through to four years old to see if behavioural differences between very young babies are correlated with behavioural differences later on. She found, for example, that babies who suckled at the highest rate in the first ten days were smiled and looked at and affectionately talked to by their mothers more than babies that suckled at a low rate. At fourteen months old, the mothers of high-suckling babies were those who talked most to their children and who were most responsive to their vocalizations.

It is very important to remember that, in the mother–infant relationship, there are two individuals who are in continuous behavioural interaction. It is impossible, therefore, to attribute features of a relationship exclusively to the mother or to the infant. For example, the correlation between infant suckling rate and maternal smiling and talking observed at ten days old could arise either because mothers smile and talk in response to suckling, or because babies suckle in response to maternal smiling and talking. Alternatively, either or both could be responding to something that occurred even earlier in life. Long-term relationships such as those between mothers and infants are very complex, and it is extremely difficult to separate the roles played by the partners in their development.

Summary

1. Factors of the external environment (such as temperature and light), as well as the internal environment, have a part to play in the growth and development of animals and plants (**25.1** and **25.2**).
2. A short exposure to light of low intensity can have important effects on plant growth and development; these are called photomorphogenic effects (**25.3**).
3. If light is to be used by a plant there must be light-absorbing molecule to absorb it. The action spectrum of the process suggests the kind of pigment molecule involved. The absorption spectrum of the pigment will be almost identical to the action spectrum (**25.31** and **25.32**).
4. The ability of organisms to measure changing day length which is accompanied by physiological adjustment is called photoperiodism (**25.3**).
5. Flowering in plants is often under photoperiodic control (**25.3**).
6. The mechanism of photoperiodic control of flowering requires a timing mechanism – a biological clock – and a light detector which is the pigment phytochrome (**25.3**).
7. Phytochrome exists in two forms: one (P_{660}) absorbs in the red region of the spectrum, and the other (P_{735}) at the far-red region. P_{660} is converted to P_{735} in red light. In the dark or in far-red light the conversion is reversed and P_{660} accumulates (**25.3**).
8. The existence of a hormone (florigen) which induces flowering has been proposed but such a substance has not yet been identified (**25.33**).
9. There are also vegetative effects of photoperiodism.
10. The growth of crop plants can be improved by control of the external environment in such a way as to produce higher yields which help to meet food requirements (**25.34**).
11. The behaviour of animals changes during their lifetime and this may be because of increasing age (maturation) or the result of interaction with the environment (learning) (**25.4**).
12. Bird song is an example of behaviour the development of which has been studied objectively by the analysis of sonograms. A model of bird song development has been constructed (**25.4**).

13 The study of mother–infant relationships in primates has made an important contribution to an understanding of the effect of early experiences in the development of adult behaviour (**25.41**).

Suggestions for further reading

BLACK, M. Carolina Biology Readers No. 20, *Control processes in germination and dormancy.* Carolina Biological Supply Company, distributed by Packard Publishing Ltd, 1972. (An account of the control mechanisms that exist to regulate seed germination and which involve interactions between the seed and its environment.)

BRADY, J. Studies in Biology No. 104, *Biological clocks.* Edward Arnold, 1979. (An introduction to the whole subject of biological clocks.)

HENDRICKS, S. B. Carolina Biology Readers No. 109, *Phytochrome and plant growth.* Carolina Biological Supply Company, distributed by Packard Publishing Ltd, 1980. (An account of the discovery, isolation, and mode of action of phytochrome.)

HILLMAN, W. S. Carolina Biology Readers No. 107, *Photoperiodism in plants and animals.* Carolina Biological Supply Company, distributed by Packard Publishing Ltd, 1979. (A brief introduction to photoperiodism.)

HINDE, R. A. and HINDE, J. S. Carolina Biology Readers No. 63, *Instinct and intelligence.* 2nd edn. Carolina Biological Supply Company, distributed by Packard Publishing Ltd, 1980. (A discussion of some of the processes in the development of behaviour.)

KENDRICK, R. E. and FRANKLAND, B. Studies in Biology No. 68, *Phytochrome and plant growth.* Edward Arnold, 1978. (An interesting account of plant photomorphogenesis.)

LOFTS, B. Studies in Biology No. 25, *Animal photoperiodism.* Edward Arnold, 1970.

NAYLOR, A. W. 'The control of flowering'. *Scientific American*, 1952. Offprint No. 113 (An historical account of our knowledge of the mechanisms which make plants flower and their applications in agriculture.)

PALMER, J. D. Carolina Biology Readers No. 92, *Biological rhythms and living clocks.* Carolina Biological Supply Company, distributed by Packard Publishing Ltd, 1977.

SALISBURY, F. B. 'The flowering process'. *Scientific American*, 1958. Offprint No. 112. (This article includes accounts of experiments carried out in order to study the mechanism of flowering.)

VILLIERS, T. A. Studies in Biology No. 57, *Dormancy and the survival of plants.* Edward Arnold, 1975. (Dormancy is placed in an ecological context and entry into dormancy and how it is broken are described.)

PART FOUR ECOLOGY AND EVOLUTION

Note
Throughout this **Study guide** the end of a Study item is indicated by the symbol □.

CHAPTER 26 THE ORGANISM AND ITS ENVIRONMENT

26.1 Introduction to ecology

A living organism continually exchanges materials and energy with its environment. It meets members of its own and other species, chasing, killing, eating, being eaten, mating, decomposing, parasitizing, and digesting. The seething multitudes of organisms take part in countless interactions each day.

The job of an ecologist is to see patterns in this vibrant kaleidoscope, and to establish the principles which govern the lives of organisms. Why does each species occur in a limited geographical area, and why, within that area, is it confined to a particular habitat? Why are some species common and others rare? What factors regulate the population size of an organism? How does an individual species obtain its energy, and what role does it play in the pattern of energy flow and nutrient cycling in the community as a whole?

The answers to questions like these are not only valuable because they contribute to an understanding of how the World works, they are also relevant to the economic success of, for example, agriculture and forestry. Even in industrialized countries, humans depend on natural

Figure 281
The Earth from outer space. Organisms occupy a thin rind on the surface, which is 71 per cent ocean and 29 per cent land. The Earth as a whole is in energy balance. The energy received as short-wave solar radiation each year equals the energy lost to outer space as long-wave heat radiation. The global cycles of oxygen, carbon dioxide, nitrogen, and water are also balanced.
Photograph, National Remote Sensing Centre, R. A. E., Farnborough.

communities for food, oxygen, and fuel. Communities planted by humans, often composed of productive plants bred by humans, provide crops and wood on a large scale. Yet they are invaded by wild species, some beneficial, some pestilential. The more we understand the activities of species in nature, the more we can increase wood, crop, and fuel production, prevent outbreaks of pests, and limit pollution, soil degradation, and damage to human prospects.

Most interactions in nature are complex. A seemingly minor activity can have repercussions a long way away. For instance, local overgrazing in an area of low rainfall can result in sand being blown over a large area and begin the formation of a desert. There is a 'spider's web' of interactions which loosely binds together all the organisms in a community. Pull one strand, and distant parts of the web are affected. This is illustrated by the following example.

STUDY ITEM

26.11 The chestnut-headed oropendola bird

Chestnut-headed oropendolas live in colonies in the tropical forests of the Panama Canal Zone. Each colony has ten to a hundred nests, each built by one male and one female (*figure 282*).

The oropendola nests are parasitized by giant cowbirds, which act rather like cuckoos in Britain. When the female oropendolas are laying their eggs the cowbirds lay their own eggs in the same nests. A successful cowbird egg will hatch out first. The cowbird nestling ejects at least some of the young oropendolas from the nest and will be brought up by its foster parents as if it were one of their own offspring.

The interaction between the oropendolas and their cowbird parasites was investigated by Neal Smith, an American ecologist based at the Smithsonian Tropical Institution. In some areas the oropendolas were all 'discriminators' and in others they were all 'non-discriminators'. The discriminator birds tried hard to throw out any egg which was not their own. On the other hand, birds in the non-discriminator colonies were willing to accept foreign objects of all shapes and sizes in their nests, including cowbird eggs, without ejecting them.

The discriminator colonies were built near the nests of wasps or bees. The wasps and bees are abundant in the air in and around the discriminator colonies. None of the non-discriminator colonies was built near the nests of wasps or bees.

In some of the colonies of oropendola birds, large numbers of bot-flies were noticed. These bot-flies can lay their eggs inside the bodies of the oropendola birds, and their larvae feed on the internal organs. Fly paper was hung in both discriminator and non-discriminator colonies. It caught few bot-flies in discriminator colonies but very many flies in non-discriminator colonies.

a *Suggest why bot-flies did not occur in discriminator colonies. How could you test your hypothesis?*

In the non-discriminator colonies, in which bot-flies were abundant,

Figure 282
A female chestnut-headed oropendola bird (*Psarocolius wagleri*) at her nest (above). Below is a giant cowbird (*Scaphidura oryzivora*). In the inset, an oropendola egg (left) is compared with a mimic cowbird egg (centre) and a non-mimic cowbird egg (right).
From Ricklefs, R. E., Ecology, 2nd edition, Nelson, 1980.

the nestlings were often infected with bot-fly larvae, which sometimes killed or weakened them. In these colonies the number of nestlings with bot-flies seemed to be influenced by whether or not there were cowbirds in the nest (table 50).

	Number of oropendola nestlings in nests:	
	With cowbirds	*Without cowbirds*
With bot-fly	57	382
Without bot-fly	619	42

Table 50
The effects of cowbirds on bot-fly infestation in nests of the chestnut-headed oropendola bird in the Panama Canal Zone.
Adapted from Ricklefs, R. E., Ecology, 2nd edition, Nelson, 1980.

b *Apply the χ^2 test to the data in table 50, and work out the probability of obtaining these results by chance if the presence of bot-flies had been independent of the presence of cowbirds. Discuss these results and suggest a hypothesis to explain them.*

c *Explain why, in colonies built a long way from wasp or bee nests, it was advantageous to the oropendolas not to discriminate between their eggs and those of cowbirds.*

d *In non-discriminator colonies the cowbirds were ignored, but in discriminator colonies they were chased away by the oropendolas. Is this consistent with your hypothesis?*

Ecological terminology

Ecologists work at all levels, from the *individual* organism, explaining its distribution pattern in terms of its genetics and physiology, up to the *biosphere* – the volume of the soil, water, and atmosphere which contains organisms.

The basic ecological unit is the *species*, a group of individuals which resemble one another closely and can interbreed with other members of the same species to leave viable offspring. Each species is usually found only in a certain place, its *habitat* (considered in this chapter and Chapter 27). It plays a particular role in the community, known as its *niche*. For example, the habitat of a tawny owl (*Strix aluco*) is woodland and its feeding niche is that of a top carnivore, eating small rodents at night. The numbers of a species may remain stable, fluctuate randomly, or exhibit regular cycles (see Chapter 28).

Every species interacts with many of the other species which occur in the same limited area at the same time to form a *community*. The species composition of the community may alter in a predictable manner; in other words, the community may undergo *succession*. Different species interact in complex ways with one another and their environment. These interactions in total constitute an *ecosystem* (see Chapter 29). Ecosystem ecologists are particularly interested in *energy flow* and *nutrient cycling* in small areas of the landscape. They seek to compare different ecosystems by reducing them to simple patterns. The Earth's surface, of which 71 per cent is ocean and 29 per cent land, can be regarded as a series of *biomes*. A biome is a vegetation type, with a characteristic vegetation structure, animal life, pattern of energy flow, and pattern of nutrient cycling (*figure 283*).

Figure 283
An aerial view of the United Kingdom and Eire, taken from a satellite with an infra-red sensing camera. Several biome types are visible. These include the sea, and the freshwater lake Lough Neagh in Northern Ireland. Woodlands such as the New Forest and those near Thetford in Norfolk show up as black patches. Areas of heathland and bog, such as Dartmoor and Exmoor, show up white.
Photograph, National Remote Sensing Centre, R.A.E., Farnborough.

STUDY ITEM

26.12 Biomes; some examples of vegetation types which extend over large areas of the Earth's surface

Figures 284–287 show several types of biome. Look at the pictures and answer the questions on page 421.

Figure 284
Tropical rain forest vegetation in Queensland, Australia. Tropical rain forests occur in wet, humid climates near the equator, in four main areas – Amazonia, Zaire, and South East Asia and Australasia.
Photograph, J. Allan Cash.

Figure 285
Tundra in northern Iceland. It occurs at the most northerly and southerly latitudes where plant communities are found, and on mountain-tops at all latitudes. Grasses, sedges, plants in the heather family (Ericaceae), and mosses dominate large areas.
Photograph, J. Allan Cash.

420 Ecology and evolution

Figure 286
A boreal coniferous forest on the borders of Lake Bandak in Norway. Such forests occupy a band at about 60°N latitude which extends across Scotland, Scandinavia, Russia, and Canada. They are dominated by a few species of gymnospermous (cone-bearing) trees.
Photograph, J. Allan Cash.

Figure 287
Tropical savannah in Tsavo East Game Park in Kenya. Grass and scattered trees support large populations of many species of large mammalian herbivores.
Photograph, J. Allan Cash.

a *Where the boreal forest (figure 286) meets the tundra (figure 285) there is often a distinct boundary, known as the 'tree-line'. Discuss the factors which may prevent the trees from occurring in the tundra zone, and suggest why most of the tundra species do not grow beneath the trees.*

b *Discuss the various factors which cause the differences in tree frequency and species composition between tropical rain forest (figure 284) and tropical savannah (figure 287).*

Chapter 26 The organism and its environment 421

The study of biomes, ecosystems, or communities is known as *synecology* (literally, the study of organisms together in the household). Many ecologists, on the other hand, concentrate on individual species. This is *autecology*.

26.2 Making and testing hypotheses

Ecologists observe, measure, and count. They think up explanations for phenomena, and test them by experiment, either in the field or in the laboratory. The results are described in over a hundred thousand scientific research reports each year.

Many ecologists spend most of their research time observing or documenting natural phenomena in the field, or carrying out field experiments. Organisms in the field are influenced by a bewildering array of interacting environmental factors and other species. For this reason many ecologists do experiments in the laboratory, where this range of factors can be limited, and causal relationships more easily established. They also use the laboratory for chemical analysis, bomb calorimetry to determine the energy content of samples, computer modelling, and the analysis of data. The main advantage of laboratory experiments is control over variables.

STUDY ITEM

26.21 Determining a causal relationship (multiple choice)

a *If you observed a general agreement between the distribution of a species and a particular environmental factor, you might consider using one of the following methods to establish that there is a causal relationship between them. Which would be the best method to employ?*

A A controlled experiment in the natural environment with all the factors held unchanged except the one under investigation.
B An experiment with the species living in isolation from its natural environment under controlled conditions.
C Both (A) and (B).
D Neither (A) nor (B), but applying statistical methods to a large number of observations.

b *Give the reasons for your choice.* (*J.M.B.* – *modified*)

As an example of the difficulties in determining a causal relationship, consider the following. Field studies show that many lichen species do not grow near city centres, factories, or heavy traffic. The burning of fossil fuels creates carbon particles, carbon monoxide, sulphur dioxide, and oxides of nitrogen. Which is the culprit? This can be investigated in laboratory experiments in which the lichens are exposed to various doses of one pollutant at a time. In fact only sulphate(IV) ions are toxic in the laboratory at the concentrations which the lichens experience in the field. These ions are formed when sulphur dioxide dissolves in water. To lichens, sulphur dioxide is the most toxic component of industrial smoke (see Study item 29.52).

An experiment is an investigation prompted by curiosity. It must be designed to answer a clearly defined question. The essence of a successful experiment is that one factor at a time should be altered, so that the interpretation of the results is unequivocal. In the field, it is usually difficult to alter one factor only. Each species is exposed to a multitude of environmental factors which vary in space and time.

STUDY ITEM
26.22 The formation and testing of hypotheses

In deserts throughout the world, seeds are an important food source. They are collected and eaten by insects and rodents alike. The way in which an ecologist works can be illustrated by some experiments and observations carried out by Diane Davidson in deserts in the south-west of the United States during the mid 1970s. Why do some ant species occur in one place but not others? What determines how many individual ants there are?

Davidson first collected some seeds from both ants and rodents in the same desert habitat and measured the seed lengths. Ants and rodents overlapped in the sizes of seeds they were carrying. Were there plenty of seeds to go round? Or were ants and rodents competing with one another for a limited number of seeds of the same species?

This question was investigated by experiment. Davidson selected an area of homogeneous, level, desert scrub, and divided it into eight plots each 36 m diameter. In two plots, she trapped all the rodents and erected fences to keep out rodent invaders. In two more plots, ants were removed by repeated applications of insecticide. In two further plots, both rodents and ants were removed. The two remaining areas were left alone as unmanipulated controls. The results are shown in table 51.

	Control (both ants and rodents present)	Rodents removed	Ants removed	Ants and rodents removed
Ant colonies	318	543	—	
Rodent numbers	122		144	—
Seed density (relative to control)	1.0	1.0	1.0	5.5

Table 51
The responses of ants, rodents, and seed density to the removal of ants and rodents at Rodeo, New Mexico.
Data from Brown, J. H. and Davidson, D. W., 'Competition between seed-eating rodents and ants in desert ecosystems', Science, 196, 1977, p. 880.

a Discuss these results. Do you think that the ants and rodents are competing for seeds? State the evidence for your answer.

b Devise another experiment which might confirm your conclusion.

Davidson then wondered how it was that several different ant species could all live together at one site if they were all eating a limited supply of seeds. One might expect one species to be better at harvesting seeds than the others and to take over gradually. Perhaps different ant species prefer to collect different sizes of seeds? In order to test this hypothesis Davidson harvested seeds and seed fragments and presented them in the laboratory to each harvesting ant species in turn. Workers, and the seeds they were carrying, were then sampled and measured. The species overlapped considerably in the sizes of seeds that they took (*figure 288*).

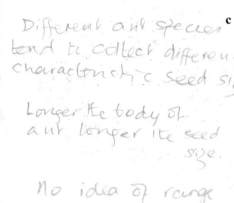

Figure 288
The relationship between worker body length and the sizes of seeds carried for eight species of seed-eating ants near Rodeo, New Mexico. Each point represents an average for a different species of ant.
Based on Davidson, D. W., 'Species diversity and community organization in desert seed-eating ants', Ecology, **58**, 1977, p. 711.

c *State two conclusions that can be drawn from the data in the graph.*

It seems from these results that several species prefer to collect the same sizes of seeds. Do they all find food in the same way?

In watching the ants Davidson realized that there were essentially two different foraging strategies, 'group' and 'individual'. In group-foraging species, the workers marched together in well-defined columns. Most of the searching and feeding took place in a restricted portion of the area surrounding the nest. When seeds were abundant group foragers had marked peaks of activity. Otherwise, when food was scarce, they entered a 'resting' state. Experiments showed that this foraging method was the most efficient when seed densities were high and the seeds were clumped. By contrast, in individual-foraging species, workers searched for and collected seeds independently of one another. As a result, all the area around the colony was continuously and simultaneously searched, even during spells of low seed density.

Davidson now examined the species composition of the ant fauna at five desert sites. When species of similar size exist together at a site they differ in foraging strategy, and when species of similar foraging strategy live together, they differ in size. Species which are similar in size and foraging strategy apparently do not live together.

Thus, as a result of these field experiments and observations the factors which determine the species composition of the ant fauna have been clarified. Each community of ants may not merely be a random mixture of species.

The purpose of much ecological research is to explain the distribution patterns and ecological preferences of organisms. Observations and experiments allow the most significant factors to be dissected out from the 'factor complex'.

> **Practical investigation.** *Practical guide 7*, **Investigation 26A, 'A qualitative study of a community'.**

26.3 Factors which influence the distribution patterns of organisms

Every organism is exposed to a multitude of environmental influences which vary in time and space (*figure 289*). These factors can be loosely classified into *physical* and *chemical* on the one hand, and *biotic* on the other. Each individual has a limited tolerance range to physical and chemical factors. Beyond these limits it will die. Physical and chemical factors therefore determine the geographical range within which a species can possibly occur. Many species, however, only occur in a small proportion of the possible range. Biotic factors, such as the uneven distribution of a food plant, predation, inefficient dispersal, or competition from other species, influence the presence or absence of each species on a small scale.

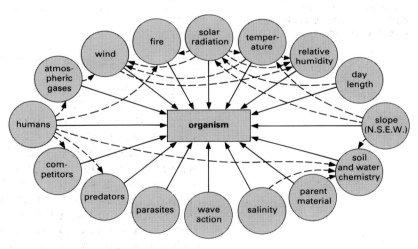

Figure 289
The factor complex which affects a plant or an animal species. All these factors can vary with time.
Based on King, T. J., Ecology, *Nelson, 1980.*

STUDY ITEM
26.31 Buttercups (experimental design)

a *It was observed that three species of* **Ranunculus** *(buttercups) were growing in a ridge and furrow grassland liable to flooding.* **R bulbosus** *occupied the ridge tops,* **R. acris** *the ridge slopes, and* **R. repens** *the bottom of the furrows. How would you test the hypothesis that this distribution of buttercups is determined by the amount of water in the soil?*
(J.M.B.)

It should be possible to define for every species the main factor limiting its distribution. If that critical factor is removed experimentally, the ecological or geographical range of the species should expand. Nettles (*Urtica dioica*), for example, do not grow well where the phosphate level in the soil is low. Cattle egesta contribute phosphate to soil. If the number of cattle dunging patches in an area increases, so may the nettle population (see *Practical guide 7*, investigation 26C).

26.4 Aquatic factors

The sea, lakes, and the streams and rivers flowing into them, constitute a range of habitats quite different from those in air or soil. This is probably why so few animals or plants can thrive in both water and air. The differences are as follows:

1 Organisms which live in water have no problems in conserving it.
2 Water has a high specific heat compared with air, and is therefore slower to heat up and cool down.
3 Water is 777 times denser than air, so that the organisms in it are better supported by the water.
4 Even large organisms are widely dispersed by ocean currents.
5 At the boundary between the water surface and the air, water exerts a considerable surface tension which for small organisms is more important than gravity. For some animals this surface film is their home.
6 Above all, the oxygen concentration is about thirty times less in water saturated with oxygen than in air. Organisms which live in water must have one of the following features: the capacity to respire anaerobically, large gills to absorb oxygen, the capacity to gulp oxygen during occasional visits to the surface, or the ability to take down large bubbles of gas which can be used for a long time.
7 Organisms in sea water are subject to salinity, tides, and wave motion.

STUDY ITEM

26.41 The influence of changes in river water

The graphs in *figure 290* show the changes in the per cent oxygen saturation, pH value, and temperature in river water during a period of twenty-four hours.

a *Explain why fluctuations in pH can be used as an indication of changes in the concentration of carbon dioxide dissolved in water.*

b *Describe and explain the changes in graphs A and B over a 24-hour period.*

c *What is the significance of the changes in graph C?*

d *In what way do the changes in graph C influence your interpretation of graphs A and B?*

Figure 290
Changes in the water of the river Lark during twenty-four hours: A, oxygen saturation, expressed as a percentage: B, pH: C, temperature.
Based on Butcher, R. W., Pentelow, F. T. K., and Woolley, J. W. A., 'The diurnal variation of the gaseous constituents of river waters', Biochemical Journal, **21**, 1927.

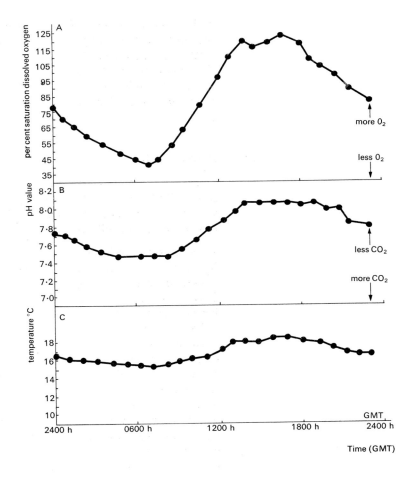

The restoration of the tidal Thames

The river Thames rises near Cirencester in Gloucestershire and flows 245 km to Teddington Weir, which is 31.5 km above London Bridge. The

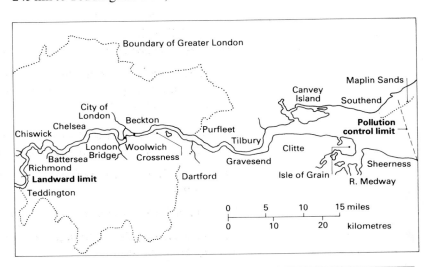

Figure 291
A map of the tidal region of the river Thames, administered by the Port of London Authority.
Based on Tribe, M. A., Eraut, M. R., and Snook, R. K., Basic Biology Course, Book 4, Ecological principles, *Cambridge University Press, 1975.*

Chapter 26 The organism and its environment 427

stretch of the Thames from Teddington to the sea is tidal, being inundated twice daily with sea water (*figure 291*). The effects of sewage pollution in the tidal stretch of the river since 1800 demonstrate the important role of oxygen levels in the water in influencing the patterns of fish.

Until 1815 London's sewage had been disposed of mainly onto neighbouring farmland, or taken by boat to be dumped on the Essex marshes. In 1800 there were about 200 000 houses in London, most with privies connected to private cesspools. In 1815 it became mandatory to connect cesspools to the sewers. This legislation had a disastrous effect on the Thames, for it caused the entire contents of the newly introduced water-closets to be discharged into the river in the centre of London.

The Thames through London became a fermenting sewer. Sewage solids covered the mud banks at low tide. When people saw an approaching paddle steamer stirring up the river, they fled in anticipation of the smell of hydrogen sulphide. In a letter to *The Times* of 9th July 1855, Michael Faraday recorded how pieces of card dropped into the river became invisible 2 cm below the surface. The climax came in 1858, 'the year of the great stench', when air temperatures reached 35 °C. The windows of the House of Commons were covered with disinfectant to counteract the smell. It was in that year that Parliament acted to reduce the nuisance.

Sewage in water is decomposed by bacteria, which take up oxygen from the water for their aerobic respiration. About 78 per cent of the oxygen which enters the water in the river dissolves into the water from the air above, but both the rate of supply, and the solubility of oxygen in water, are low. Many fish require a high oxygen content to survive – for salmon (*Salmo salar*), for instance, the water must be 35 per cent saturated with oxygen. When the oxygen content of water is below 5 per cent bacteria begin to produce gases such as ammonia, methane, and hydrogen sulphide (table 52). These are toxic to many organisms.

Substance	*Examples of bacteria involved*
N_2 or N_2O (denitrification)	*Pseudomonas* spp.
C_2H_4 (ethene)	
$Mn(II)$	*Metallogenium* spp.
$Fe(II)$	*Clostridium* spp.
H_2S (hydrogen sulphide)	*Desulphovibrio desulphuricans*
CH_4 (methane)	*Methanobacterium*
H_2 (hydrogen)	*Clostridium* spp.

Table 52
Ions and gases which appear in soil and water when oxygen is scarce. The ions and gases are arranged in order. At the top of the table are those which appear first in slightly deoxygenated water. Those at the bottom of the table appear when oxygen is very scarce.
Adapted from Etherington, J. R., Studies in Biology No. 144, Wetland ecology, *Edward Arnold, 1983.*

The tidal Thames was certainly rich in fish in the twelfth century, when records begin, and even as late as 1828 it was regarded as a good fishing river, with over 400 fishermen earning their full-time living from

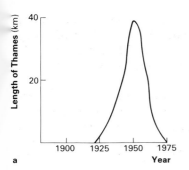

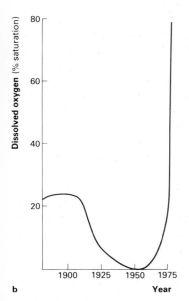

Figure 292
Oxygen levels in the river Thames.
a Length of the tidal Thames with dissolved oxygen less than five per cent of saturation (1883 to the present).
b Lowest point on the dissolved oxygen curve, expressed as a percentage of saturation (July to September each year from 1883 onwards).
Based on Wood, L. B., The restoration of the tidal Thames, Hilger, 1982.

the lower tideway. Records of salmon catches from Boulter's Lock, Taplow from 1794 to 1821 show, however, a decline from 1804 (39 caught per annum) to 1821 (3 caught per annum). In London itself fish were largely absent from the river from 1840 onwards.

The new sewer system designed by Sir Joseph Bazalgette in the 1860s led much of the sewage parallel to the river to chemical treatment works on the seaward side of London, from which most of it was discharged by pumping stations in a 3 km stretch of the Thames between Beckton and Crossness. When the system was completed, in the mid 1880s, the Thames through London was improved considerably. Data from the newly introduced monitoring scheme show that an average oxygen saturation of 18.5 per cent in 1885 had improved to 42.5 per cent in 1895. Bleak, dace and roach become plentiful even in the upper river, and flounders returned.

The changes which have occurred in the pollution of the Thames this century are summarized in *figure 292*. First, there was a marked deterioration. Although more efficient sewage treatment works were introduced, the population of London grew so rapidly that the extra volume of sewage washed up the river on every high tide deoxygenated the water. Synthetic detergents exacerbated the situation. They accumulated in large quantities in the water, lowering its surface tension and reducing by 16 per cent the rate at which oxygen could dissolve. In the worst years, 1950–1953, the oxygen level in the water was below 5 per cent along most of the tideway and 2 p.p.m. of hydrogen sulphide were detectable in the air. There were half a million tubificid worms (*Tubifex tubifex* and *Limnodrilus hoffmeisteri*) m^{-2}, characteristic of highly anaerobic water. The Thames was entirely devoid of fish over a stretch of at least 60 km from 1920 to 1964.

Enlarged treatment works, and improved methods of treatment which made the effluent innocuous, reduced bacterial activity and increased the oxygen levels in the water (*figure 292*). From 1964 onwards there was an increase in the numbers of fish species. This increase was documented by examining the fish caught in grilles in cooling water inlets for power stations (table 53). From 1976 onwards the numbers of fish increased rapidly, especially the populations of herring (*Clupea harengus*) and flounder (*Platichthys flesus*).

Year	1964	1965	1966	1967	1968	1969
Number of species	34	42	n.c.	51	56	63
Year	1970	1971	1972	1973	1974	1975
Number of species	71	82	86	92	96	97

Table 53
The number of fish species recorded in the Thames between Fulham and Gravesend from 1964 to 1975 inclusive.
From Wood, L. B., The restoration of the tidal Thames, Hilger, 1982.

Most of the main fish species prevalent in the North Sea now occur in the Thames as it passes through London. In 1983 the first two salmon were caught by Thames fishermen.

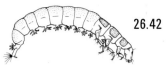

STUDY ITEM

26.42 The distribution patterns of some caddis fly nymphs in the river Usk in Wales

Many species of caddis fly spend the first months or years of their lives as predatory nymphs on the bottoms of streams or rivers. Some of these live in small tubes or 'cases' which they manufacture by secreting 'nets' around themselves and adding stones and other debris. Others build nets under stones. Six species of these caseless caddis fly nymphs (family Hydropsychidae) consistently occur in the same sequence from the upper to the lower reaches of British and French rivers. A study in the river Usk in Wales concentrated upon three of these species, *Diplectrona felix*, *Hydropsyche instabilis*, and *H. pellucidula*. In large unpolluted rivers their typical order of first occurrence from source to mouth is *D. felix*, *H. instabilis*, and *H. pellucidula*.

The distribution patterns of these species along the Usk are shown in *figure 294*. Some temperature data collected from the regions of the Usk where each of these species habitually occurs is shown in table 54. The graphs in *figure 293* show the relationships between respiratory rate and temperature for the three species.

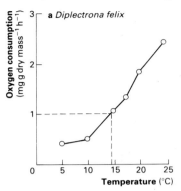

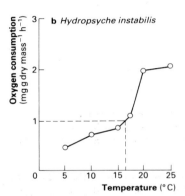

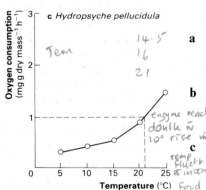

Species	Mean temperature of water in June–September (°C)	Maximum temperature	Range
Diplectrona felix	11.2	low	small
Hydropsyche instabilis	13.5	higher	large
Hydropsyche pellucidula	14.2	highest	small

Table 54
Temperature data for the river Usk for places where each of the three species of caddis-fly nymphs habitually occurs.
Data from Hildrew, A. G. and Edington, J. M., *Journal of animal ecology*, **48**, 1979, p. 557.

a Write down, from the respiration rate graph for each species, the temperature at which a respiratory rate of 1 mg O_2 g^{-1} dry mass per hour occurs. What does this result indicate?

b At what temperatures do the flattest parts on the graphs occur in 1 *Diplectrona felix* and 2 *Hydropsyche instabilis*? What is the importance of these flat portions of the graph?

c *Hydropsyche pellucidula* has a lower respiration rate at all temperatures than the other two species. Suggest what prevents it from surviving further towards the source of the river.

d Discuss critically whether temperature is likely to be a major factor which limits the distribution patterns of these species.

e Describe further experiments to test your hypotheses.

Figure 293
The relationship between respiration rate and temperature for the fifth instar larvae of three species of caseless caddis-flies, determined in laboratory experiments.
Based on Hildrew, A. G. and Edington, J. M., *Journal of Animal Ecology*, **48**, 1979, p. 557.

26.5 Soil factors

Soil is the medium in which plants grow. It is a mixture of six main components. Most of its mass consists of mineral particles, which range

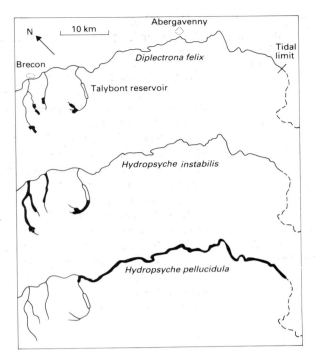

Figure 294
The distribution patterns of the larvae of caseless caddis flies (Trichoptera: Hydropsychidae) in tributaries and lower reaches of the river Usk in Wales.
Based on Hildrew, A. G. and Edington, J. M., Journal of Animal Ecology, **48**, *1979, p. 557.*

in size from large stones down to clay particles less than 2 µm in diameter. These particles are separated and held together by a sticky colloidal mass of decomposing organic matter, the humus. This holds water, which also fills up some of the space between the particles. The rest of the space is occupied by the soil atmosphere which, compared with the air above, is deficient in oxygen and rich in carbon dioxide. Nutrient ions occur, bonded to the charged clay particles and humus, and dissolved in the soil water. Plant roots constantly grow rapidly through the soil, searching for water and nutrients. Many other organisms live in the soil.

The soil is home for numerous species which take part in the disposal of detritus. They decompose excreta, dead bodies, and the discarded parts of plants. They include earthworms, mites, springtails, aphids, ants, fly larvae, ground beetles, spiders, millepedes, centipedes, and woodlice (table 55). These organisms are powered by energy which enters plant roots from their above-ground parts, and by the dead plant matter which falls onto the soil. Bacteria and fungi play a major role, breaking down dead cells and releasing nutrient ions into the soil. Many bacteria are involved in the nitrogen cycle. Most of them release ammonia from protein or help to convert it to nitrate(V), but some fix nitrogen gas into their own bodies, and others denitrify nitrate(V) to nitrogen gas.

The mineral particles in a soil have come from the breakdown of rock, either the rock on which the soil lies, or rock from elsewhere. Rocks

Figure 295
A scanning electronmicrograph of a soil mite (× 100).
Photograph, Dr J. M. Anderson, Department of Biological Sciences, University of Exeter.

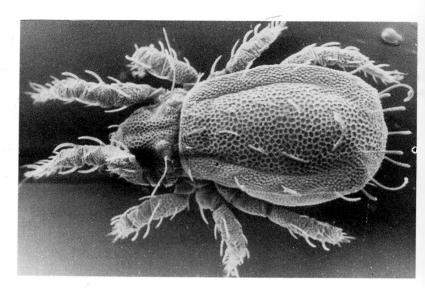

Depth (cm)	Collembola (springtails)	Hemiptera (bugs)	Acarina (mites)	Others	TOTAL
0–15	43.5	15.5	121.15	9.75	190
15–30	18.25	2.5	42.25	11.25	77
TOTAL	61.75	18.0	166.5	21.0	267

Table 55
The numbers of arthropods at different depths in the soil of a permanent grassland. The figures are in thousands per square metre.
From Salt, G., Hollick, F. S. J., Raw, F. and Brian, M. V., 'The arthropod population of pasture soil', Journal of animal ecology, **17**, *1948, p. 139.*

are broken down by the freezing and thawing of water in cracks, by sand-blasting in deserts, by expansion and contraction, and by erosion by water. Soil particles created by glaciers or river erosion can be carried by ice, water, or wind to accumulate as alluvial soils elsewhere.

The pressures exerted by plant roots, the water rich in carbon dioxide which they provide, and acids secreted by plants like lichens all etch rocks and help to form smaller particles. The plants provide humus. This allows decomposer organisms to establish themselves. Ultimately a diverse community of respiring soil organisms is established, supported mainly by the dead and decomposing parts of plants.

Humus tends to be a black, sticky colloidal mixture of organic compounds. It therefore holds water and glues soil particles together. The organic acids it contains are negatively-charged and bind positively-charged nutrient ions. Its organic compounds provide food for earthworms and soil bacteria, and a long-term source of soil nitrogen. Humus darkens the soil, warming it both because darker bodies reflect less solar radiation, and because the micro-organisms it contains give out heat when they respire. Some soils, such as peat consist almost entirely of organic matter.

The chemical composition and size of the soil particles are an important influence on the soil's chemical and physical properties. Sand is composed of large particles (2.0–0.02 mm), composed mainly of silicon dioxide, without charged surfaces. The pores are large and water is not held strongly by surface tension forces. Water therefore runs through the pores and out of the soil. As a consequence, sandy soils tend to lack water and to be well aerated. The nutrient ions within the soil solution are also washed out easily, especially since the sandy soils often contain little humus, to which the ions might otherwise be bound.

Clay particles, on the other hand, are tiny (less than 2 μm across) and consist of potassium aluminium silicates. Each particle consists of a multiple sandwich of flat plates, each with negative charges on its surface to which positively-charged ions can bond. Water is held very strongly by surface tension forces in the tiny pores between the clay particles, often so strongly that it cannot be extracted by plant roots under normal circumstances. Clay soils tend therefore to be waterlogged, but they usually have concentrations of nutrient ions adequate for plant growth.

Most fertile soils have a mixture of particle sizes. Such soils are known as loams. A well developed crumb structure is a desirable feature in a fertile agricultural loam. Clay, sand, and silt particles are held together by water, fungal hyphae, and sticky humus to form small lumps of soil which average about 4 mm in diameter. Such a crumb structure is ideal for seed germination. Water and nutrient ions are held by the clay and humus within the crumbs, and can be extracted by roots. Nevertheless, water can trickle down and air can pass through the relatively large gaps between the crumbs.

Nutrient ion uptake by plants

Plants must absorb enough ions from the soil, or the watery medium in which they grow, to provide their cells with enough atoms for ATP, DNA, proteins, phospholipids, chlorophyll, and enzyme cofactors. Otherwise their growth rates may be limited. Plant cells concentrate the ions, and heterotrophs depend on this concentrated source of ionic food to satisfy their own mineral nutrient requirements.

Certain elements are required by all green plants and some are constantly required in greater quantities than others. In the following list of essential elements, each symbol is followed by its concentration in plant material at levels considered adequate: N (1000 μmol g^{-1} dry mass), K (250), Ca (125), Mg (80), P (60), S (30), Cl (3), B (2), Fe (2), Mn (1), Zn (0.3), Cu (0.1), Mo (Molybdenum, 0.001). The first six elements are known as *macronutrients* and the rest as *micronutrients*. Some halophytes, that is to say, plants which grow in salty conditions such as glassworts (*Salicornia* spp.), also require sodium. Wheat (*Triticum durum*) and legumes require cobalt, and some plants (for example, a vetch, *Astragalus racemosus*) even require selenium.

Only roots that grow rapidly can exploit new sources of nutrients. A single maize (*Zea mays*) root tip can produce 17 500 new cells an hour, or 420 000 a day. During the development of a root tip meristem cell into a root parenchyma cell it expands by about thirty times. As a consequence, maize roots can extend by 50–60 mm a day. In a famous experiment, Howard C. Dittmer grew a single rye plant for four months in a container 30 × 30 × 56 cm. He found the striking rates of root growth listed in table 56.

Surface area of total root system (hairs included)	639 m²
Area of root hairs	402 m²
Length of total root system (hairs excluded)	622 km
Length of root hairs	10 620 km
Mean daily root growth rate (hairs excluded)	4.99 km
Mean daily growth rate of root hairs	89 km

Table 56
The extent and average daily growth rate of the root system of a four-month-old rye plant (*Secale cereale*), grown singly in a soil volume less than 0.06 m³.
Modified from Dittmer, H. J., American Journal of Botany **24**, *1934, pp. 417–20.*

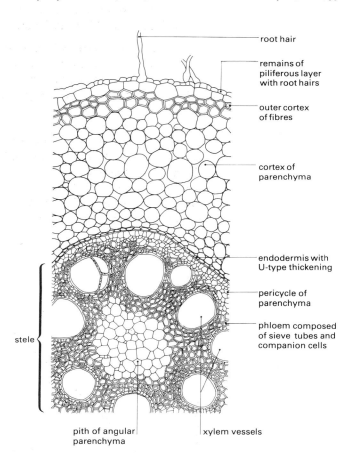

Figure 296
A cross-section of a monocotyledonous root in the absorbing zone. The soil solution saturates the walls of the parenchymatous cells in the cortex, and extends up to the endodermis. The cortex cells absorb ions from their cell walls across their cell membranes by active transport.
Based on Bracegirdle, B. and Miles, P. H., An atlas of plant structure, *Volume 2, Heinemann Educational Books, 1971.*

A cross-section of a root in the absorbing zone (*figure 296*) reveals that a cortex of parenchyma cells surrounds a central stele which contains the transporting tissues, the xylem and phloem (see also *Study guide I*, Chapters 3 and 8). The function of the cortex is probably to expose a large surface area of cell membrane to the soil solution (*figure 296*). The soil solution, that is, the water-carrying dissolved nutrient ions which surrounds the humus and the soil particles, does not merely extend up to the epidermis of the root. Provided that the soil is wet, the soil solution extends throughout the wet cell walls of the cortex, right up to the endodermis. Circumstantial evidence for this came from the immersion of thoroughly washed barley roots in solutions containing radioactive ^{32}phosphate. After a few minutes, the roots were blotted dry and placed in de-ionized water. The concentration of ^{32}phosphate attained in the water in equilibrium with the roots indicated that the radioactive solution had occupied about 29 per cent of the root space, that is, about the proportion of the root volume made up by wet cell walls.

a *Discuss the assumptions underlying this experiment.*

The ions are absorbed by active transport across the cell membranes. The evidence for this is as follows.

1 The ions are absorbed from a low concentration in the soil solution to a higher concentration within the cells. Transport against a diffusion gradient requires energy.
2 If roots are deprived of oxygen, they continue to take up water but not ions.
3 Cyanide inhibits ion uptake by roots. Cyanide inhibits aerobic respiration since it inhibits electron transport by cytochrome a_3 (cytochrome oxidase), the terminal cytochrome in the 'respiratory chain' in the mitochondrion.
4 Ion uptake is strongly temperature-dependent. Results like those illustrated in *figure 297* are obtained if washed roots are immersed in a solution of radioactive potassium ions.

b *Explain the results in* **figure 297**.

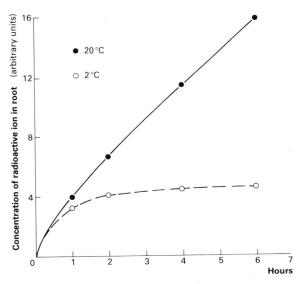

Figure 297
The rate of uptake of potassium ions by washed carrot root slices at two different temperatures.
Based on Sutcliffe, J. F., Mineral salts absorption in plants, Pergamon, 1962.

STUDY ITEM

26.51 Earthworms and soil fertility

'Earthworms, though in appearance a small and despicable link in the chain of nature, yet, if lost, would make a lamentable chasm. For, to say nothing of half the birds, and some quadrupeds, which are almost entirely supported by them, worms seem to be great promoters of vegetation, which would proceed but lamely without them; by boring, perforating and loosening the soil, and rendering it pervious to rains and the fibres of plants; by drawing stalks of leaves and twigs into it; and most of all, by throwing up such infinite numbers of lumps of earth called worm casts, which, being their excrement, is a fine manure for grain and grass.'

(*From Gilbert White*, The natural history of Selborne, *letter 77, 1777.*)

Earthworms, particularly the surface-casting species *Allolobophora longa* and *A. nocturna*, are very abundant in fertile neutral soils, although they hardly exist in soils of pH less than 4.5. Typical densities are 500 to 750 m^{-2} in old grassland and 7 to 10 m^{-2} in land cultivated for a long time.

As Gilbert White first indicated, earthworms may have dramatic effects on soil. For instance, the channels created by their burrowing allow roots to obtain oxygen and water to drain. Some species drag dead leaves into the soil and eat them. In beech forest, for example, about 90 per cent of the energy in dead leaves passes through worms. This speeds up the decay of organic detritus to humus. Earthworms digest dead leaves and humus, and release nutrient ions to the soil in their alkaline secretions and egesta. By eating soil from lower down the profile and 'casting' on the surface, they take nutrient-rich soil to the upper, leached layers and their casts produce soil crumbs. This accumulates as a stoneless layer of silt and clay on the surface, providing an ideal tilth for seed germination.

In some apple orchards at Wisbech, Cambridgeshire, so much copper(II) sulphate was sprayed on the trees as a fungicide that high levels of copper in the soil killed the earthworms and other soil fauna. Compared with unpolluted orchards nearby, polluted orchards had more leaf accumulation on the surface, more waterlogging, and a drastically reduced apple crop.

a *Explain these three effects on the basis of the normal activities of earthworms.*

An experiment was designed to test whether earthworms improved the yield of winter wheat drilled directly into uncultivated soil. The experiment was set up in 1977 on clay loam at Bosworth, Cambridgeshire. The field had been sown with direct-drilled winter wheat for the previous six years, and had a low earthworm population.

Twelve rectangular plots, each of 500 m^2, were arranged at random in the field in September 1977. Nine of them were fumigated intensively to kill the earthworms. The crop was drilled in October. On 10th November 1977 chopped straw was distributed evenly over the plots and earthworms were added to some of the plots as follows.

Treatment A (3 plots) Fumigated, then each inoculated with 300 deep-burrowing earthworms, *Lumbricus terrestris* and *Allolobophora longa*.

Treatment B (3 plots) Fumigated, then each inoculated with 300 shallow-burrowing earthworms, *Allolobophora chlorotica* and *A. caliginosa*.

Treatment C (3 plots) Fumigated, no earthworms added.

Treatment D (3 plots) Unfumigated, no earthworms added.

Some of the results appear in table 57.

	Treatment			
	A	B	C	D
Mean number of earthworm casts m^{-2}	165	73	11	21
Number of plants m^{-2} of row	12	3	2.5	1.75
Number of shoots per plant	7.7	14.5	19.8	20.3
Mean plant height (mm)	280	220	235	210
Mean dry mass root material (g m^{-2} of row)	13.3	6.4	3.6	3.4
Mean grain mass (g m^{-2})	8.6	7.0	6.2	6.2
Mean dry mass straw on surface (g m^{-2})	94	217	236	219

Table 57
The results of an experiment concerning the influence of added earthworms on the growth of winter wheat. The crop was sown in October 1977, after fumigation. The earthworms were added to treatments A and B in November. The results above were recorded in the following June.
Data from Edwards, C. A. and Lofty, J. R., 'Effects of earthworm inoculation upon the root growth of direct drilled cereals', Journal of Applied Ecology, **17**, *1980, p. 533.*

b *Discuss these results in detail. In what ways do the earthworms affect the growth of the winter wheat?*

In an experiment carried out in boxes in a greenhouse, the creation of artificial vertical holes in soil which lacked earthworms significantly increased the height of barley seedlings. When the artificial holes were in the top 7.5 cm of soil, growth was greater than when the vertical holes were 7.5–15.0 cm down. This in turn exceeded barley growth with holes 15.0–30.0 cm down, which nevertheless still exceeded growth in control boxes without any artificial vertical holes. These results suggest that the effect of earthworm burrows on cereal growth is partly physical. However, in the same experiment, growth of barley in boxes to which the earthworms *Lumbricus terrestris* and *Allolobophora longa* had been added was very much the best of all.

☐ **c** *Why was this?*

The Broadbalk wilderness

At Rothamsted Experimental Station in Hertfordshire, an area of 0.1 hectare was enclosed by a fence after the wheat harvest of 1882. It has not been cultivated since. It is now woodland, mainly of hawthorn, oak, ash,

Figure 298
The Broadbalk Wilderness at Rothamsted Experimental Station, Hertfordshire. It is now woodland, mainly of hawthorn (*Crataegus monogyna*), oak (*Quercus robur*), ash (*Fraxinus excelsior*), and sycamore (*Acer pseudoplatanus*). *Photograph, Rothamsted Experimental Station, Hertfordshire.*

and sycamore (*figure 298*). Some trees are over 18 m high. Over the period from 1881 to 1964 the top 68.5 cm of soil accumulated about 4.4 tonnes ha^{-1} of nitrogen and 50 tonnes ha^{-1} of organic carbon.

c *Explain in as much detail as you can how the soil has accumulated nitrogen and carbon in such quantities.*

Soil varies a great deal in depth and composition from place to place. To document these differences soil scientists examine soil profiles, which are vertical sections of the upper soil. Profiles often exhibit distinct layers, or horizons, which differ from one another in colour, stoniness, particle size, composition, and so on. The profile reflects the local history of the soil (*figures 299–301*).

For example, in many areas of acidic sandy soil the humus does not break down rapidly. The acid conditions eliminate earthworms and limit the growth of bacteria. As plants die, their remains accumulate on the surface and form a layer of humus. Water falling onto the surface picks up organic acids and washes them down through the profile. There they may combine with iron(III) and aluminium(III) ions. In many areas with high rainfall these ions are washed 10–30 cm down before they accumulate to form an iron pan, a hard layer of encrusted iron and aluminium oxides (*figure 299*). Above it is a layer bleached by the loss of its coloured salts. Below it is a coloured horizon which grades into the bedrock. Such a profile as a whole is called a podzol.

The characteristics of soil vary from place to place and affect the geographical and local patterns of plants and animals (table 58). The main limiting factors are:

1 salinity, which affects the water potential of the soil;
2 waterlogging, which influences the oxygen supply to the roots;
3 the concentrations of toxic ions;
4 deficiencies in essential ions.

Table 58
Some of the physical and chemical factors which may limit plant patterns in nature, in addition to those mentioned in the text. In each case competition from other species, predators, pathogens, and other factors may also be important.

Low temperature	during fruit ripening, prevents fruit development at northern end of U.K. range of broad-leaved lime (*Tilia cordata*).
Low frost tolerance	prevents many coastal plant species from occurring further inland.
High salinity	prevents some inland species from invading salt-marshes and sea-cliffs (the soil water potential is too low).
Low soil water	prevents many mesophytes from occurring in deserts because they have limited mechanisms for conserving water.
Low soil oxygen	limits species of ragwort (*Senecio jacobaea* and *Senecio squalidus*) to dry soils.
Iron(II) ions toxic	prevents dog's mercury, *Mercurialis perennis*, from invading wetter soils in Cambridgeshire woods.
Copper ions toxic	prevents many species from invading copper-rich soils in Zaire, but allows copper plants such as *Bechium homblei* to flourish.

Figure 299
A profile of a podzol, from the Jurassic sandstone of the North Yorkshire Moors. The natural vegetation was ling heather (*Calluna vulgaris*), but this site has been forested with conifers. Conifer litter lies on top. It is slow to break down in this acid soil which lacks earthworms.
After Etherington, J. R., Studies in Biology No. 98, Plant physiological ecology, *Edward Arnold, 1978.*

Chapter 26 The organism and its environment

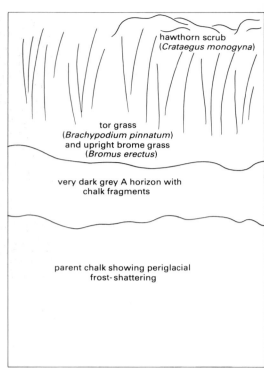

Figure 300
A rendzina soil from the chalk of the North Downs in Surrey. The parent rock has weathered to give a shallow soil with large chalk stones and small chalk pebbles. They have been mixed throughout the profile by past cultivation, frost-heaving, or the activities of soil organisms.
After Etherington, J. R., Studies in Biology No. 98, Plant physiological ecology, *Edward Arnold, 1978.*

> **Practical investigations.** *Practical guide 7*, **investigation 26B, 'The effects of different water regimes on plant growth', and investigation 26C, 'The relationship between nettle distribution and soil phosphate'.**

The preference of some species for basic or acidic soils

The pH amongst the wide diversity of soil types which exists in Britain varies from 3.1 to 10. Different plant species have different pH optima. Some, for example, ling heather (*Calluna vulgaris*) and Scots pine (*Pinus sylvestris*) hardly ever occur naturally on soils where the pH exceeds 6.0. Others, such as old man's beard or traveller's joy (*Clematis vitalba*), rock-rose (*Helianthemum chamaecistus*), and many orchid species are seldom found at pHs less than 6.5. The species which grow on acidic soils are called *acidophilous*, and those which grow mainly on soils of high pH are *basophilous* (see *figure 302*).

It is easiest to investigate the patterns of acidophilous and basophilous species in an area containing a mosaic of soil types, since

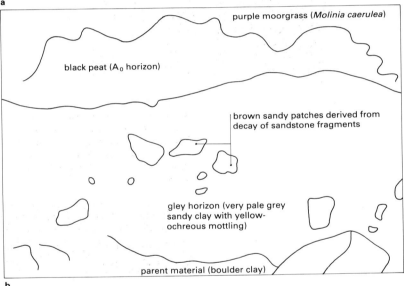

Figure 301
A peaty-gley soil on boulder clay. Above the parent clay is a waterlogged zone which is grey and mottled with grey and black. The mottling is not easily visible in a black-and-white photograph. The predominant grey colour is partly the result of the reduction of Fe(III) to Fe(II) by bacteria. The black spots in the gley horizon are manganese dioxide (MnO_2).
After Etherington, J. R., *Studies in Biology No. 98*, Plant physiological ecology, *Edward Arnold, 1978.*

one can be fairly certain within a small area that the soils of high and low pH are subject to the same regime of temperature, rainfall, evaporation, and so on. The results of such an investigation in grassland growing on shallow glacial drift overlying carboniferous limestone are shown in *figure 303*. Of the two species *Galium saxatile*, heath bedstraw, is an

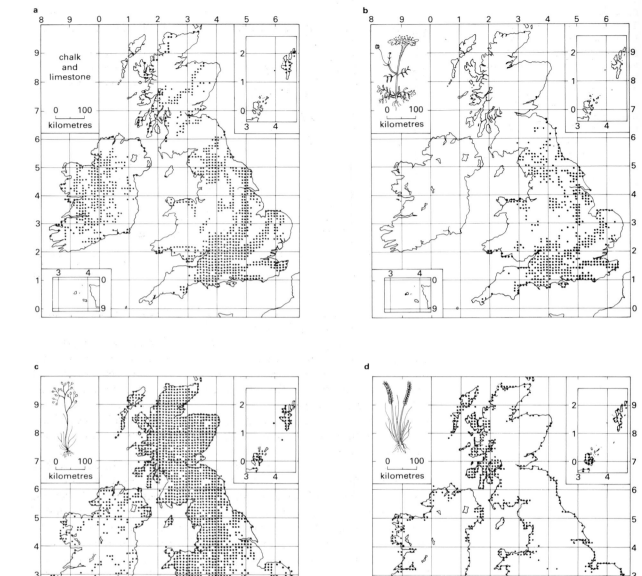

Figure 302
The distribution patterns in Great Britain of **a** chalk and limestone bedrock, **b** a basophile, small scabious (*Scabiosa columbaria*), **c** an acidophile, wavy hair grass *Deschampsia flexuosa*, and **d** sea arrow grass (*Triglochin maritima*). What factors may be important in influencing the distribution limits of the species?
Based on Perring, F. H. and Walters, S. M., Atlas of the British flora, 2nd edn, EP Publications, 1976.

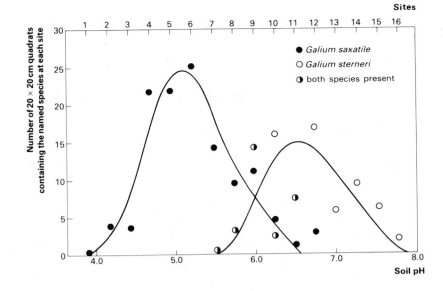

Figure 303
The distribution with soil pH of *Galium saxatile* and *G. sterneri* in grassland growing on shallow acidic drift overlying carboniferous limestone at Malham, Yorkshire. At each site the frequency of each species was assessed in a 1 × 1 m quadrat by noting its presence or absence in each of twenty-five 20 × 20 cm sub-quadrats.
Based on Etherington, J. R., Studies in Biology, No. 98, Plant physiological ecology, *Edward Arnold, 1978.*

acidophile, most abundant on soil of pH 5.0, whereas *G. sterneri* is more basophilous, with an optimum at pH 6.5.

Arthur Tansley carried out a well-known experiment in 1917 to investigate the relationship between these two species. He grew them singly and together, both on soil in which *G. saxatile* grew and on soil where *G. sterneri* grew. Both species grew on both soil types when alone, but when sown in the presence of the native species the alien species was suppressed by it. This indicates that the range of pH in which a species is found in nature may be limited by competition from plants which normally grow in different soil conditions.

Why do some species grow on acidic soils and others basic soils? Let us first consider acidophiles. Many acidophilous species suffer from 'lime chlorosis' when grown on basic soils. This is a yellowing of the leaves, caused by lack of chlorophyll. Iron is necessary as a cofactor for one of the enzymes in chlorophyll synthesis. The chlorosis can be corrected by spraying onto the leaves a fine spray of Fe-EDTA, a chelated solution of iron(III). Acidophilous species are probably iron deficient in basic soils because the hydrogen carbonate ion interferes with the uptake of iron, and because the solubility of iron(III) ions in the soil solution decreases markedly with increasing pH.

Basophilous plants, on the other hand, such as small scabious (*Scabiosa columbaria*) and squinancywort (*Asperula cynanchica*), may not occur in nature on acidic soils owing to aluminium(III) toxicity. Aluminium(III) ions show the same curve of increasing solubility with decreasing pH as iron(III) ions. Aluminium(III) ions, immobilized on the surfaces of the soil particles at higher pH, dissolve into the soil solution at a low pH. Seedlings of *Scabiosa columbaria* and the stemless thistle, *Cirsium acaule*, are killed in cultures by solutions containing more than 1 p.p.m. of aluminium(III) ions. The basophilous bog-rush. *Schoenus nigricans*, in Ireland occurs in those bogs where the Al(III) ion concentration is less than 1 p.p.m. and is absent from the others.

26.6 Air factors

The relative humidity and temperature of the atmosphere and the effects of wind, frost, and snow influence the distributions of animals and plants. Species most affected by drying out are those with gas exchange surfaces on the outsides of their bodies, such as woodlice and mosses. Woodlice take up oxygen into gills underneath. Mosses and liverworts often lack a waxy cuticle and their leaves are usually one cell thick.

> **Practical investigation.** *Practical guide* 7, **investigation 26D, 'The autecology of woodlice'.**

A complex of interacting environmental factors influences the geographical area and habitats in which a species is found. In the case of the stemless thistle, *Cirsium acaule*, the influence of climate on its geographical distribution has been investigated experimentally.

STUDY ITEM

26.61 The distribution pattern of the stemless thistle *(Cirsium acaule)*

The stemless thistle (*Cirsium acaule*) is a low-growing herb which occurs in short infertile grasslands on chalk and limestone throughout southern Britain (*figure 304*). It has a rosette of spiny leaves which produces one or two flowering heads on stalks 1–2 cm long. It reproduces both sexually, by insect-pollinated flowers, and asexually, by slow-creeping

Figure 304
The stemless thistle, *Cirsium acaule*. This plant occurs mainly in chalk and limestone grasslands where the vegetation is less than 20 cm tall.
Photograph, Dr M. C. F. Proctor, Department of Biological Sciences, University of Exeter.

Figure 305
The geographical distribution of the stemless thistle, *Cirsium acaule*, in the British Isles. Each dot represents at least one record in a 10 × 10 km square of the National Grid. The bold lines over the map show isotherms for mean day temperatures in August.
Based on Perring, F. H. and Walters, S. M., Atlas of the British flora, 2nd edn, EP Publications, 1976.

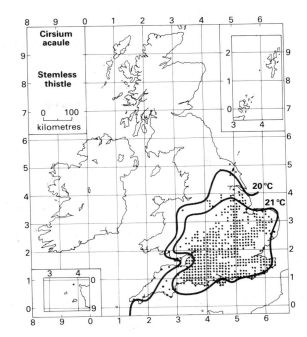

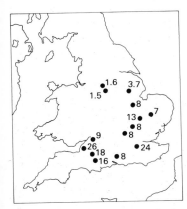

Figure 306
The mean numbers of ripe fruits per flowering head in samples of *Cirsium acaule* collected at fourteen sites in September 1963.
Based on Pigott, C. D., 'The response of plants to climate and climatic change', in Perring, F. (ed.), B.S.B.I. Conference Report 11, The flora of a changing Britain. Classey, 1970.

underground rhizomes which may divide. Each flowering head contains many florets – the species is a member of the family Compositae – and can potentially produce about 75 fruits, one per floret. Although the fruits are heavy they are attached to parachutes of hairs and may be dispersed by the wind. This thistle has a striking distribution pattern in Britain (*figure 305*). It hardly occurs north of a line from the Bristol Channel to the Humber. Examine the following statements and answer the questions beneath.

1 The number of fruits ripened per head varies across the range of the species (*figure 306*).
2 At the northern end of its range, in the Pennines, the species occurs only on south-facing slopes.
3 Stemless thistle plants flower in early August on the south coast, but in early September in the northern part of its British range.
4 In an experiment carried out in the Pennines by Donald Pigott of Lancaster University, artificially-pollinated flowering heads, enclosed in polythene bags, produced significantly more viable fruits than similar flowering heads left uncovered.
5 Several flowering heads were divided into two halves by a partition extending in a north-south direction. One side of the head, selected at random, was sprayed with water in a fine spray at 0800 hours each morning. The other side remained unsprayed. Many more viable fruits developed on the unsprayed side than on the sprayed side.

a *Account for each of these five observations and construct a hypothesis to explain them.*

b *Suggest some different experiments to test your ideas.*

Chapter 26 The organism and its environment 445

26.7 Zonations

Species often occur in zones of abundance which stretch across the landscape. A zonation exists when some of the species in an area are confined to well-defined belts. Examples include the zonation of species in concentric circles around a lake (see *figure 323*), the zonation of plant species from the top to the bottom of a salt marsh, the zonation across a sand dune system, and the zonation of species down a rocky shore. In the first three cases, the zones move with time. They probably represent different stages in a successional sequence. In other words, at any one place, one zone will develop into another. On a rocky shore, however, the zones do not move. The pattern is determined by a complex of factors which is difficult to investigate.

Zonation on rocky shores

A rocky shore may seem to us to be an inhospitable environment for organisms. Often, on a surface so smooth that it is difficult to cling to, the organisms are desiccated when the tide goes out, and pounded by saline water twice a day when the tide comes in. Nevertheless, thousands of species live in this habitat. On vertical cliffs, or where there is a constant slope, a zonation of species exists and is so marked, especially amongst the plants, that it is as if straight lines had been drawn along the rock. The zonation of animals is probably less obvious than the zonation of plants because animals are mobile and move up and down with the tide.

Table 59 lists the patterns of some of the species common around the rocky coasts of Britain. The factors causing this zonation are complex and have not been fully resolved for many species. For plants they probably include storm and hurricane damage, physical factors such as

Zone	Producers (green plants and algae)	Animals
Above upper shore	Lichens, e.g. *Verrucaria* Green algae in pools, e.g. *Ulva*, *Enteromorpha*, and *Cladophora*	*Littorina neritoides* (a periwinkle) *Ligia* (a crustacean) No insects
Upper shore	Brown algae, e.g. *Pelvetia canaliculata* (channel wrack) and *Fucus spiralis* (flat wrack)	Barnacles, e.g. *Balanus* spp., *Chthamalus*, and *Elminius* *Patella* (limpet) *Mytilus edulis* (mussel) Sea-anemones, e.g. *Actinia equina*
Middle shore	*Ascophyllum nodosum* (knotted wrack) *Polysiphonia* *Fucus vesiculosus* (bladder wrack)	*Patella* (limpet) *Gibbula* (a top shell) *Nucella* (dog-whelk) *Littorina littorea* (common periwinkle) Crabs, e.g. *Carcinus maenas*
Lower shore	*Fucus serratus* (saw wrack) Encrusting red algae, e.g. *Lithophyllum* Laminarians (kelps)	Diverse fauna including sponges, hydroids, bryozoans, and tunicates

Table 59
Some of the species frequently found at different levels on rocky shores around the coasts of Britain.

resistance to desiccation and temperature tolerance, and biotic factors, such as the ability to utilize light when shaded by other plants. The position of an animal species in the zonation probably depends on its temperature tolerance, desiccation tolerance, its ability to feed and breed in air as compared to water, the positions on the shore of its preferred food, and the distribution pattern of its predators.

Zonation has been investigated experimentally in the acorn barnacles *Chthamalus stellatus* and *Balanus balanoides*. In Britain *C. stellatus* occurs at the upper end of the shore, in a distinct band above the average level of high tides. Below this is a narrow region where both species occur. Below this, with its centre around mean tide level, there is a broad band of *B. balanoides*.

a *List the differences in environment between the* **C. stellatus** *zone and the* **B. balanoides** *zone.*

Barnacles are arthropod crustaceans which occur clamped to rock in very dense populations (*figure 307*). These populations are formed when their larvae settle on bare rock surfaces. They tend to settle where there is only moderate water movement in areas where other barnacles already occur.

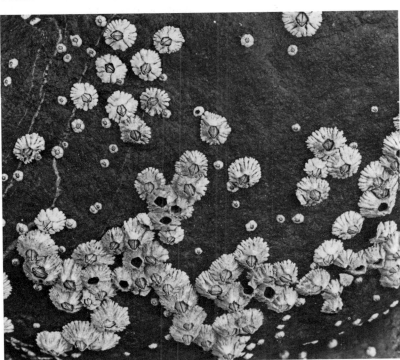

Figure 307
Barnacles on a rocky shore.
Photograph, D. P. Wilson/Eric and David Hosking.

b *In an experiment* **Balanus** *larvae were presented with slowly rotating slates, half of which had been thoroughly cleaned and the rest of which had been treated with* **Balanus** *extract. One hundred larvae settled on the clean slates, but 2197 established themselves on the treated ones. What conclusions can you draw?*

Chthamalus occurs higher up the shore than *Balanus* partly because *Chthamalus* is more tolerant of desiccation and high temperatures, a tolerance necessary if it is exposed for longer. *Chthamalus* continues to be active in water up to 39 °C (optimum, 32 °C), whereas *Balanus* is inactive above 32 °C (optimum, 20 °C). *Chthamalus* in particular can survive several days out of water in hot sun.

In a well known study at Millport, Isle of Cumbrae, Scotland, Joseph Connell was recording barnacle populations. In the spring of 1955 he sampled a freak area which had not been covered with sea water for several days. Between the censuses of February and May, *Balanus* one year old suffered a mortality of 92 per cent and those two years old, 51 per cent. In the same area, over the same period, the death of *Chthamalus* aged seven months was 62 per cent and those one and a half years old, 2 per cent. Those *Balanus* in the top quarter of the distribution range suffered by far the greatest mortality. The upper limits of both species seem to be determined by their desiccation tolerances.

Connell also investigated the lower limit of *Chthamalus*. He found stones with both species on them and placed these at different tidal levels. Half of each stone was left untouched. In the other half, all the *Chthamalus* individuals were freed from competition by chipping away any *Balanus* individuals in contact with them. The barnacle populations on each stone were then counted at intervals.

At the lower levels, *Chthamalus* slowly disappeared in the untreated areas. Some were undercut or lifted by growing *Balanus*, and others were smothered by the *Balanus* individuals which grew on top of them. At these levels, measurements showed that *Balanus* grew far faster than *Chthamalus*. The faster it grew, the faster the *Chthamalus* was eliminated. In virtually all the areas kept free of competing *Balanus*, however, *Chthamalus* grew far better. It would certainly survive well at the lower levels if competition was prevented.

Thus it seems that the upper limits of these two barnacle species are determined by their resistance to desiccation and heat (*figure 308*). The lower limit of *Chthamalus* is determined by the intensity of inter-specific competition for space with *Balanus balanoides*. Competition is one of the biotic factors which is discussed in the next chapter.

It is obvious from these examples that the geographical area within

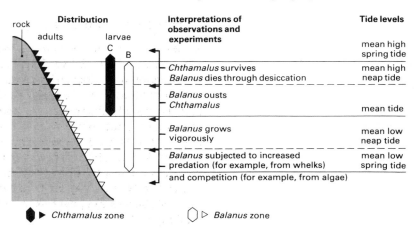

Figure 308
Factors which control the distribution patterns of two barnacle species, *Chthamalus stellatus* and *Balanus balanoides*, between the tide marks on a rocky shore near Millport, Scotland.
Based on Connell, J. H., 'The influence of interspecific competition on the distribution of the barnacle Chthamalus stellatus', Ecology, **42**, 1961, p. 710.

which a species could live is determined by its tolerance to air, water, and soil factors. Yet it is common for most species in nature to occur only in a small proportion of their potential geographical range. Many species must be limited by a dispersal barrier, such as the sea or a mountain range. Muntjac deer and ginko trees, for instance, both introduced from China, thrive in captivity in the British climate; presumably they were not native in Britain because they could not get here. Some species may be restricted in range by a competitor, or because their food plants only occur within a limited area, or because a parasite or a predator eliminates them elsewhere. All these biotic factors are discussed in the next chapter.

Summary

1. Organisms in biomes, communities, and populations are involved in a complex web of interactions with each other and their physical environments (**26.1**).
2. Ecologists make and test hypotheses. They make observations, suggest explanations, and as far as possible test their ideas by controlled experiments to try to expose causal relationships (**26.2**).
3. The geographical and local distribution patterns of organisms in nature are influenced by physical factors, such as light intensity and relative humidity, and by other organisms (**26.3**).
4. The main factors which influence the patterns of animal species in water, such as oxygen levels and temperature, are discussed with examples (**26.4**).
5. A brief account of soil formation and composition leads on to a description of nutrient ion uptake in plants. Earthworms exert a considerable influence on soil fertility; the chemistry of the soil greatly influences the distribution patterns of many plant species (**26.5**).
6. Climatic factors and their effects on the distribution patterns of organisms are illustrated with reference to the stemless thistle (**26.6**).
7. Finally, the zonation of barnacles on a rocky shore is used as an example of the interactions of physical and biotic factors in influencing species distribution (**26.7**).

Suggestions for further reading

ANDERSON, J. M. *Ecology for environmental sciences: biosphere, ecosystems and man.* Edward Arnold, 1981. (A pithy, concise introduction to ecology, with an applied emphasis.)

ATTENBOROUGH, D. *Life on Earth.* Collins, 1979. (A good bedside read, beautifully illustrated. It is mainly an introduction to the groups of organisms, but conveys plenty of ecological information.)

BREHART, R. N. Studies in Biology No. 139, *Ecology of rocky shores.* Edward Arnold, 1982. (It includes a description of factors influencing the zonation of species on rocky shores.)

COLLINVAUX, P. A. *Why big fierce animals are rare.* Penguin, 1981. (A series of short essays guaranteed to set even a sceptical student thinking.)

ETHERINGTON, J. R. Studies in Biology No. 154, *Wetland ecology*. Edward Arnold, 1983. (A masterly compression of a vast amount of information difficult to obtain elsewhere.)

JACKSON, R. M. and RAW, F. Studies in Biology No. 2, *Life in the soil*. Edward Arnold, 1967. (A concise introduction to soil animals, their activities, and importance.)

KING, T. J. *Ecology*, Nelson, 1980. (This book provides several simple examples of the effects of the environment on the distribution patterns of organisms.)

OWEN, D. F. *What is ecology?* 2nd edn. Oxford University Press, 1980. (A readable introduction to ecological principles, illustrated with examples from the author's own research.)

RICKLEFS, R. E. *The economy of nature*. Chiron Press, Portland, Oregon, 1976. (An elegant and fascinating introduction to ecology – the best available.)

RUSSELL, E. J. *The world of the soil*. 4th edn. Collins, 1967. (A classic introduction to soils, their fauna, and the relationships between soils and agricultural production.)

TOWNSEND, C. R. Studies in Biology No. 122, *The ecology of streams and rivers*. Edward Arnold, 1980. (A clear, straightforward description of the main factors influencing the distribution patterns of animals in inland water habitats.)

CHAPTER 27 ORGANISMS AND THEIR BIOTIC ENVIRONMENTS

Organisms are affected by others in many ways. Their physical and chemical environments are influenced by their neighbours and the conditions created by the activities of plants and animals in the past. Plants may be grazed. The diets of animals, decomposers, detritivores, and parasites consist of other organisms. Thus the distribution patterns of particular species on a geographical or a local scale may be determined by the presence or absence of others.

The effects of one organism on another are known as biotic factors. This chapter explores the biotic factors which seem to be important in nature, such as competition, predation, parasitism, commensalism, mutualism, and dispersal. It ends with a discussion of the complex interactions between species in succession and the climax. All these terms are now defined, in the order in which they are discussed in the text:

Competition Competition occurs when two individuals obtain resources which are in short supply, so that the growth rate of both individuals is decreased. It also occurs when one individual, whose growth rate is not affected, greatly reduces the supply of resources to another, which is suppressed and may die. When the two competing individuals belong to

Figure 309
Pattern of pine trees (*Pinus cembra*) above the tree line in the Gurglertal, Tyrol, Austria. The pine is confined to the ridges and completely absent from the intervening depressions. This may be because the nutcracker, *Nucifraga caryocatactes*, preferentially stores seeds as a source of winter food on the more accessible ridges; the snow there is less deep in winter than in the depressions.
Photograph, Professor Walter Tranquillini.

the same species intra-specific competition occurs. Competition is interspecific when the two individuals belong to different species.

Predation Predation occurs when a living organism, the predator, catches another organism, the prey, and eats it dead or alive. In this sense predators include herbivores, but not decomposers or detritivores.

Parasitism Parasitism occurs when one organism, the parasite, spends some time inside or on the surface of another organism, its host, from which it obtains some food. The parasite benefits from the association but the host is harmed, sometimes only slightly.

Commensalism In commensalism, individuals of two different species often live together. One partner benefits from the association, but the other neither benefits nor is harmed.

Mutualism In a mutualistic association, individuals of different species are habitually found together and both the species benefit from the association.

Dispersal The movement of individuals of a species to new sites. Some species are better dispersers than others. Many plants are dispersed by animals. Dispersal by humans has considerably affected the geographical patterns of many species.

Succession Succession is a gradual change in the species composition of a community with time. The direction of succession can often be accurately predicted from the changes which have occurred in similar communities in the past.

Climax In the climax state, changes in the species composition of a community occur only very slowly.

27.1 Competition

The growth rates and reproductive outputs of individuals in communities are often reduced by the presence of neighbours. This is known as competition. Competition occurs when two individuals obtain resources which are in short supply, so that the growth rates of both individuals are decreased. It also occurs when one individual, whose growth rate is not affected, greatly reduces the supply of resources to another, which is suppressed and may die.

What resources do species compete for? In order to survive, flourish, and leave descendants, plants require water, nutrient ions, light, and pollinators, and animals need food, mates, water, nest space, and shelter. Organisms may compete for any of these factors. In some deserts, for example, rodents and ants compete for seeds as food (Chapter 26, Study item 26.22). On rocky shores in Scotland, two barnacle species compete for space on rock surfaces (Chapter 26, section 26.7).

The mere fact that two adjacent organisms both require the same resource does not prove that they are competing for it. For instance, two individuals in a dense cereal crop might shade one another but their leaves might all be light-saturated. In that case they would not be competing for light. Competition would occur if their growth rates were limited by nitrate supply and each plant took up nitrate ions which the other might have used.

How can competition be demonstrated? It can sometimes be inferred from field data, but the only convincing method is to do experiments in the laboratory or the field. In the laboratory the potential competitors can be grown singly and in various relative proportions. In the field each of the competing species can be removed from, or added to, the community. The effects on the potential competitors can be measured. Examples of experiments such as these are in Study items 27.11, 27.12, 27.13, and 27.14.

What are the effects of competition on organisms, and what characteristics should we measure to demonstrate competition? Organisms exposed to competition increase in dry mass more slowly than if their competitors were absent. They may produce fewer or smaller offspring, or they may die. In fact there are two extremes to competition. In a 'scramble', the competing organisms are all about the same size and deprive one another of scarce resources. They all grow slowly; most of them survive but their reproductive output is diminished. This is what happens to competing larvae of the fruit fly *Drosophila*. The other extreme is a 'contest'. Some individuals obtain the lion's share of the resources. They grow rapidly. The rest become 'suppressed weaklings' and ultimately die.

The individuals which compete may be of the same species (intra-specific competition) or of different species (inter-specific competition). In general, two individuals of the same species and of the same age are more likely to require the same environmental resources than individuals of different species and the same age. Two closely related species which habitually occur together in nature often compete little. Their individuals may nest at different heights, eat slightly different sizes of food item, use different pollinators, or require water and nutrient ions at different times of year from different levels in the soil.

We may be seeing here the 'ghost of competition past'. Competition between species may have played a major role in evolution. Perhaps when two species occurred in the same community, those individuals of both species which were most similar were eliminated by competition. This process would produce two sets of individuals with markedly different phenotypes.

The next four study items demonstrate various aspects of competition in artificial and natural communities.

STUDY ITEM

27.11 The effect of density on the growth of monocultures

It is important for those who grow large areas of plants of a single species to understand the effects of competition on the final yield. In that way the crop, be it apples (*Malus* spp.) or maize (*Zea mays*) can be established at the density which in the long run will produce the optimum quantity and quality of produce.

Examine the results of an experiment in which maize plants were grown at a range of densities (*figure 310*). At the higher densities the total yield rises to a plateau. The total mass of the fruiting heads (cobs), however, reaches a peak and then begins to decline. Since this crop is

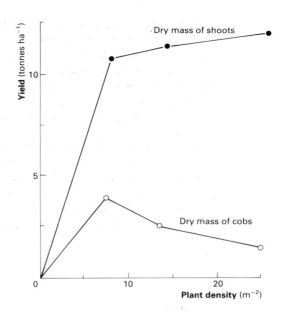

Figure 310
The effect of plant density on the dry masses of shoots and cobs in maize (*Zea mays*). Notice that the yield of cobs, which represent the main harvest, is at its maximum at only seven plants per square metre.
Adapted from Harper, J. L., 'Approaches to the study of plant competition', in Milthorpe, F. L. (ed.), Mechanisms of biological competition, *Symposium of the Society for Experimental Biology*, **15**, 1, 1961.

grown mainly for its fruiting heads the optimum sowing density is about 7.5 m^{-2}, a density at which the chances of attack by fungi are also reduced. At higher densities the increase in the number of plants per unit area is more than offset by the fewer seeds produced per plant.

With reference to figure 310, explain why, with increasing numbers of plants per unit area, the rate of dry matter production by each remaining plant is reduced.

The −3/2 power 'law'

The time at which a seedling emerges greatly influences its later response to increased density. The seedlings which emerge first are free of

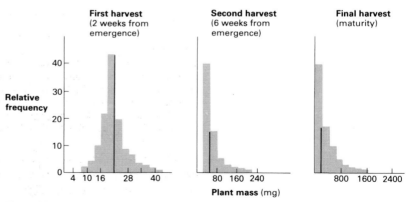

Figure 311
Changes with time in the frequency distributions of plant masses in populations of flax (*Linum usitatissimum*) at high density (3000 plants m^{-2}). The dark vertical lines represent the mean plant mass (mg) at each harvest.
Data from Obeid, M., Machin, D., Harper, J. L., 'Influence of density on plant to plant variations in fiber flax, Linum usitatissimum', *Science*, **7**, 471, 1967, copyright © 1967 by A.A.A.S., and Harper, J. L., *The population biology of plants*, Academic Press, 1977.

competition for light, water, and nutrient ions, and grow quickly. The latecomers have to compete on unequal terms with individuals which are continually expanding their volumes of roots and shoots. The results in *figure 311*, which shows the frequency distributions of mass with time of flax plants (*Linum usitatissimum*) at high density, confirm that in many plant populations the commonest individuals are the 'suppressed weaklings', many of which may die without flowering.

Imagine a dense population of a single species, such as that of flax (*figure 311*) in the early stages. At first the number of plants per unit area is high. As the individuals grow, increasing their average mass, some suppressed individuals will die, lowering the density of the remaining plants. Survivors increase their mean mass, shading the smaller survivors, which will grow slowly and may also die. After some time a few large plants remain, with a high mean mass, with access to all resources.

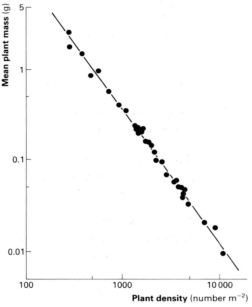

Figure 312
The relationship between the density of surviving plants and their mean dry masses in a single population of plantains (*Plantago* spp.). In a dense population (bottom right) individuals die. As they die, the mean plant mass of the remaining plants increases (towards top left). Notice that both scales are logarithmic.
From Yoda, K. et al., 'Self-thinning in overcrowded pure stands under cultivated and natural conditions', Journal of Biology, Osaka City University, **14**, 107, 1963.

The relationship between the mean plant mass and the density of the remaining plants can be expressed on a graph. When logarithmic scales are used, the relationship appears as a straight line (*figure 312*). The point which represents the newly-sown plants is in the bottom righthand corner of the graph, where the plants are numerous but do not weigh much. As time goes on, the population moves up the graph to the left, with the density declining and the mean masses of the survivors increasing.

The remarkable feature of graphs like these is that they nearly always have the same slope, $-3/2$. The relationship holds in at least eighty wild and cultivated populations, ranging from mosses to forest trees.

STUDY ITEM
27.12 Shepherd's purse (Experiment design)

a *Shepherd's purse,* Capsella bursa-pastoris, *is a common annual weed of disturbed ground. By means of a field-based investigation, how would you test the hypothesis that seed production in shepherd's purse is related to its population density?* (J.M.B.)

> Practical investigation. *Practical guide 7*, investigation 27A, 'Intra-specific competition in *Drosophila*'.

STUDY ITEM
27.13 Competition between two flour beetle species in a limited environment

Thomas Park's experiments with flour beetles demonstrated that an environmental shift can affect the outcome of competition. He reared mixtures of equal numbers of *Tribolium confusum* and *T. castaneum* at various different temperatures and relative humidities. He cultured them in glass vials containing sieved, homogeneous flour. Each vial was kept in the same environmental conditions throughout. Every two weeks the beetles and their eggs in each vial were sieved away from the flour, and placed in fresh flour. The experiment was done on the grand scale, with thousands of tubes.

Park found that each species could survive indefinitely on its own in any of the combinations of temperatures and relative humidities which he tried. Whenever the two species were reared together, however, one species ultimately became extinct in each tube, although extinction took more than a year in some tubes. Park's results are summarized in table 60. They illustrate the principle that the outcome of competition between two species depends on the environment in which competition occurs.

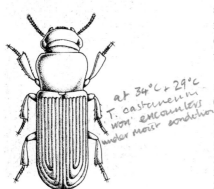

Figure 313
The flour beetle *Tribolium confusum* (actual length about 3 mm), which Thomas Park used in his competition experiments. *Tribolium castaneum* is very similar but its eyes are closer together and its antennae have a three-segmented club on the end.

Temperature (°C)	Relative humidity (%)	Climate	Percentage of experiments won by	
			T. confusum	*T. castaneum*
34	70	hot–moist	0	100
34	30	hot–dry	90	10
29	70	temperate–moist	14	86
29	30	temperate dry	87	13
24	70	cold–moist	71	29
24	30	cold–dry	100	0

Table 60
The outcome of competition between the two flour beetles *Tribolium confusum* and *T. castaneum* in homogeneous flour under various constant environmental conditions.
Adapted from Park, T. 'Experimental studies of interspecies competition. Temperature, humidity and competition in two species of Tribolium', *Physiological Zoology,* **27**, 177, 1954

a *Comment on these data.*

b *In what ways could the two species have interacted within the vials so* that one species was placed at a disadvantage compared with the other?

c If the temperatures and relative humidities within a set of vials containing both species had been altered every twelve hours between hot–moist and cold–dry, to simulate the changes between night and day, what might have been the outcome of the experiment and why?

> **Practical investigations.** *Practical guide 7*, investigation 27B, 'Inter-specific competition between two *Lemna* species', and investigation 27C, 'Inter-specific interaction between clover and grass'.

STUDY ITEM

27.14 Competition between two ant species

In Britain two ant species of the genus *Lasius*, the black ant *L. niger* and the yellow hill ant, *L. flavus*, are abundant in grasslands. *L. niger* nests

Figure 314
a Workers of the black ant, *Lasius niger*, feeding their larvae (segments obvious) and tending their worker pupae (segments not obvious) in their nest.
b Workers of the yellow hill ant, *Lasius flavus*, tending the pupae of queens. These queen pupae are much larger than the worker pupae in photograph **a**. A *L. flavus* colony can produce up to 500 queens a year. The number of queens produced seems to be an index of the vigour of the colony and is related to the area of the underground foraging territory of the workers around the nest.
Photographs, Dr A. J. Pontin, Department of Zoology, Royal Holloway and Bedford New College, University of London.

under superficial stones, and is black, with normal compound eyes. It forages both below and above ground. *L. flavus*, on the other hand, builds permanent grass-covered mounds. It is yellow-brown, with reduced eyes, and is entirely subterranean. These species often live together in the same grasslands, feeding mainly on root-living aphids and their secretions, but do they compete?

To test the competitive effect of one species on nests of its own and the other species, John Pontin carried out a transplantation experiment on a field called Aldehurst, at Englefield Green, Surrey, England. Pontin had already established that the production of queen ants by a nest was a sound indicator of the vigour of a colony and the area around the nest which it exploited. In fact the square root of the number of queens produced by a colony is proportional to the radius of its territory.

In 1966 Pontin placed cement slabs on each nest. The queens congregated in the warmest galleries beneath each slab and were easily collected and counted. By this means the foraging areas around all the *Lasius niger* and *L. flavus* nests at the site were mapped.

In March 1967 Pontin transplanted some of the *Lasius flavus* mounds. The mounds which he moved were dug out to a depth of 15 to 20 cm below the surrounding surface, placed intact on a sheet of galvanized iron, and slid into new positions. The ant colonies in these mounds all survived this rather drastic treatment!

Pontin knew from his map which species already 'owned' the territory into which each ant-hill had been transplanted. Over the next two years, he assessed the changes in queen production both near the spaces vacated by the transplanted colonies, and near their new positions (table 61).

	Effect on queen production by neighbouring nests of					
	L. niger			L. flavus		
Treatment	No. of nests	Total queen production		No. of nests	Total queen production	
		Number	Biomass		Number	Biomass
Removal of *L. flavus* nests producing 47 g of queens	2	+560	+14.5 g	5	+1378	+24
Addition of *L. flavus* nests producing 47 g of queens	2	−270	−7.0 g	7	−1093	−19

Table 61
Maximum changes in queen production following transplantation of six nests of *Lasius flavus*.
Data from Pontin, A. J., *Competition and coexistence of species*, Pitman, 1982.

a **From the data in the table, assess the effect of Lasius flavus colonies on 1 other L. flavus colonies and 2 L. niger colonies.**

b **What were two adjacent colonies competing for? Explain in detail the various ways in which the vigour of a colony might be reduced by another colony nearby.**

c **How would you expect the foraging behaviour of L. niger to change when a new colony of L. flavus appears nearby?**

d *What is the importance of the result that competition within one species may be more severe than competition between species?*

In fact, inter-specific competition may limit the geographical distributions of many plant species in nature. Plant species which are commonly confined to salt marshes, deserts, pond margins, or serpentine rock can often flourish in gardens on ordinary soils if competition is prevented. This is because these species are slow growing, and on average soils they cannot compete well with other species.

Competition and the properties of sun and shade plants

Many plant species grow beneath others and suffer permanently from shading by a neighbour. At the base of a dense leaf canopy, like that of beech (*Fagus sylvatica*), the lower leaves of beech are shaded just as much as those of a woodland herb. In both cases, however, the leaves survive under conditions where they receive less light, of altered spectral quality. Understory species are genetically programmed to withstand shade, whereas the production of 'sun' and 'shade' leaves by the same tree is an example of phenotypic plasticity in organs which are composed of cells of identical genotype.

The effect of light intensity on the net photosynthesis in sun and shade leaves of evergreen oak (*Quercus ilex*) and beech (*Fagus sylvatica*) is shown in *figure 315*. The sun leaves have a greater potential rate of net photosynthesis and can better exploit high light intensities. The shade leaves reach maximum net photosynthetic rate at much lower light intensities than sun leaves. If light intensity is reduced, the shade leaves will show a proportionately smaller reduction in net photosynthesis than the sun leaves.

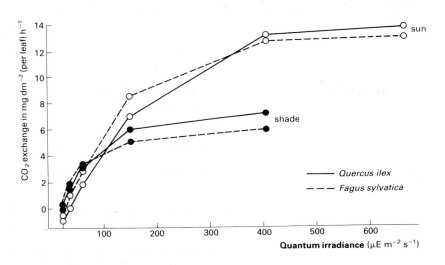

Figure 315
The photosynthetic response to light intensity of the sun and shade leaves of evergreen oak (*Quercus ilex*) and beech (*Fagus sylvatica*).
From Larcher, W., 'The effect of environmental and physiological variables on the carbon dioxide gas exchange of trees', *Photosynthetica*, **3**, 167, 1969.

A leaf is only worth while to the plant if it repays the plant's investment by a net production of photosynthetic products. A certain respiratory rate is essential to provide the energy for biochemical reactions within the leaf cells and phloem transport. A leaf will only be viable if its daily photosynthesis exceeds its daily respiration plus (in a C_3 plant) photorespiration (see *Study guide 1*, Chapter 7, 'Photosynthesis').

Thus of fundamental importance to any discussion of sun and shade leaves is the concept of *compensation point*. At compensation point the rate of photosynthesis of a leaf equals its rate of respiration over 24 hours, so that there is no net carbon dioxide or oxygen exchange. If a leaf is deprived of light or carbon dioxide, photosynthesis stops but respiration continues.

As the light intensity or carbon dioxide concentration around the leaf is slowly increased, there comes a point where the rate of photosynthesis equals the respiration rate. The light intensity at which this occurs (at a particular temperature and CO_2 concentration) is called the light compensation point. The carbon dioxide concentration at which this occurs (at a particular temperature and light intensity) is known as the carbon dioxide compensation point. In most cases the light and temperature compensation points are lower in shade leaves than sun leaves.

The main differences between sun and shade leaves are shown in table 62. In general these differences are also evident between different species which habitually grow in the sun and in the shade. Some species, therefore, grow better in shade than in full sunlight.

Characteristic	Sun leaves	Shade leaves
Light compensation point	high light intensity	low light intensity
Light saturation	high light intensity	low light intensity (about one third)
Respiration rate per dm² leaf surface	high	low
Leaf area/unit dry mass	small	large
Leaf thickness	thick	thin
Palisade cell layers	2–3	1
Cuticular thickness	thick	thin
Stomata per unit area	more	fewer
Chlorophyll content per unit area	low	high
Sensitivity of photosynthesis to temperature	high	low
Content of RuBP carboxylase enzyme per unit area	high (ten times more)	low
Ratio photosynthetic to non-photosynthetic cells	lower	higher

Table 62
Differences between sun and shade leaves on the same plant.

27.2 Interactions apart from competition

The next four interactions, unlike competition, involve some benefit to at least one of the participating organisms. They are predation (section 27.3), parasitism (27.4), commensalism (27.5), and mutualism (27.6).

In predation an animal feeds on a living species, which may be a plant or an animal. A parasite gains its energy by feeding from a larger organism, its host, at the host's expense. In commensalism two organisms habitually live in an association from which only one derives obvious benefit and the other neither benefits nor is harmed. Mutualisms benefit both the participating organisms. Not every association can be put in one of these categories with confidence. There are only four categories to describe hundreds of thousands of complex relationships, and it is often difficult to prove that an organism derives benefit from an association.

These associations can be distinguished by a code in which a 'plus' indicates that an organism benefits, a 'nought' shows that an organism neither benefits nor is harmed, and a 'minus' indicates that an organism is harmed by the association. On this basis predation is $++--$, parasitism is $+-$, commensalism is $+0$ and mutualism is $++$.

> **Practical investigations.** *Practical guide 7*, **investigation 27D, 'The holly leaf miner (*Phytomyza ilicis*) and its parasitic insects', and investigation 27E, 'A study of inter-specific interactions'.**

27.3 Predation

A predator is an animal which feeds on living species, either plant or animal. Predators are often stealthy and speedy, and have sophisticated senses of sight, smell, or hearing. Their heads are equipped with devices to bite the prey. On the other hand, potential prey are often equipped to avoid capture and death. Animal prey often find safe hiding places, or possess camouflage, mimicry, poisonous or foul-tasting chemicals, a sophisticated sense of smell, or the speed to run away. Plants may be protected by toxic chemicals such as cyanogenic glucosides, silica bodies, spines, stings, or sticky traps.

Predators may influence the geographical or local patterns of their prey, and the prey may determine where the predator occurs. When Klamath weed (*Hypericum perforatum*) was introduced into California from Britain it became a troublesome pest in agricultural crops. The beetle *Chrysolina quadrigemina*, which eats its shoots and leaves, was

Figure 316
A chameleon catching its prey.
Photograph, Frank Lane Picture Agency.

introduced to control it. The beetle reproduces rapidly on the weed in agricultural fields, but cannot complete its life-cycle on the shaded heads of *H. perforatum* in scrub. Therefore the weed is almost confined to scrub, and the beetle is only abundant where its food plant grows.

Tens of thousands of other insect species depend upon a single plant species for food. In these cases the distribution pattern of the food plant determines the local distribution pattern of the predator.

27.4 Parasitism

Parasites feed on their hosts, just as predators feed on their prey. Whilst parasites are smaller than their hosts, however, predators are usually larger than their prey. A parasite tends to be physically attached to its host, either inside or outside, for a significant proportion of its life-cycle, whereas encounters between predator and prey are brief! Should vampire bats, aphids on plants, and mosquitoes feeding on humans be called parasites or predators?

The list that follows gives some of the best known parasitic species. Many are of considerable economic importance. Some, the ectoparasites, live on the outside of their hosts. This is a precarious existence but at least they avoid the host's immune system. The majority are endoparasites, which live inside their hosts. Naturally the geographical and local distribution patterns of the parasites are related to those of their hosts.

Well known parasites
All viruses (*e.g.* measles, T_4 bacteriophages)
Many bacteria (*e.g.* tetanus, tuberculosis)
Many protozoans (*e.g.* sleeping sickness *Trypanosoma*, malaria *Plasmodium*).
Many fungi (*e.g.* dutch elm disease)
Many nematode worms (*e.g.* potato root eelworm)
Many flatworms (*e.g.* beef tapeworm, liver fluke)
Many insects (*e.g.* flies, fleas, hymenopterans)
Brood parasites (*e.g.* cuckoos, cowbirds)
Angiospermous parasites (*e.g.* dodder *Cuscuta* spp.)

As an example of the life history of a parasite, consider a protozoan endoparasite, the malarial parasite of humans. (See *figure 317*.)

STUDY ITEM

27.41 The ecology of malaria (*Plasmodium* spp.)

Human malaria ('bad air') is a disease caused by four parasitic protozoans of the genus *Plasmodium*, transmitted by about 60 species of mosquitoes of the genus *Anopheles*. It is a disease of the tropics and sub-tropics, particularly between 60° N and 40° S. Despite efforts at control which have eliminated it from 36 out of 143 countries, over 200 million people suffer from it, and it kills about 2 million people a year, about half of whom are in Africa. Children less than a year old are particularly susceptible.

Of the four main species, *Plasmodium falciparum*, which causes malignant tertian malaria, is the most deadly, responsible for more mortality in Africa than any other cause. We shall concentrate upon it. Every forty-eight hours, at night, there is a bout of fever characterized by sweating and rapid shivering. When as many as 25 per cent of the red blood cells in the body are infected, the patient may suffer from anaemia, spleen enlargement, blood clotting, and respiratory failure, and finally may die.

Figure 317
The life cycle of the malignant tertian malaria parasite, *Plasmodium falciparum*, in human and the mosquito. All stages are drawn to approximately the same scale. The other common malarial parasites of humans all have similar life cycles. In infections of *Plasmodium vivax*, which causes 'benign tertian malaria', confined to the sub-tropics, and *P. ovale*, bouts of fever occur every 48 hours. In 'quartan' malaria (*Plasmodium malariae*) the crises are 72 hours apart, but the infection may be chronic.
Adapted from Vickerman, K. and Cox, F. E. G., The Protozoa, John Murray, 1967.

a *Suggest two advantages to the parasite in promoting bouts of fever in the host which occur every forty-eight hours at night.*

In the malaria parasite's life-cycle (*figure 317*) there are three phases of reproduction, two in the human host and one in the mosquito. Mosquitoes are responsible for the transmission of the disease from one human to another. Only the female adult mosquitoes suck blood – the males feed on nectar, and the larvae and pupae feed on aquatic organisms and debris.

b *Name another parasite which has two alternative host species and which reproduces within each. In what ways does the possession of two alternative hosts benefit* **Plasmodium**?

c *Explain the advantages to* **Plasmodium** *of producing short-lived blood phases rather than circulating in the blood plasma.*

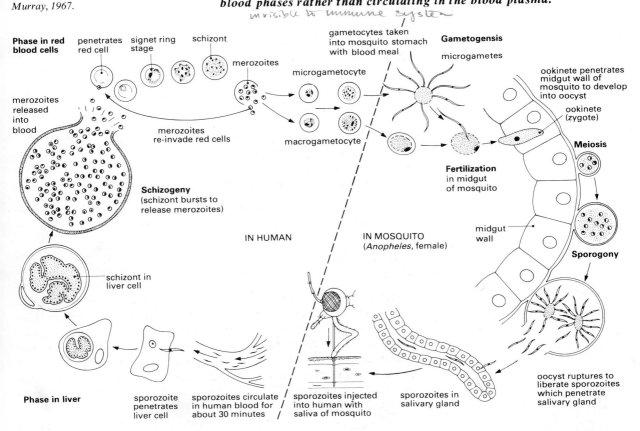

Many attempts have been made to eliminate malaria by annihilating the mosquitoes. The life cycle of the mosquito has four phases: egg, larva, pupa, adult. After one or two blood meals, a fertilized female lays several hundred eggs, singly, on water. At 30 °C the whole life-cycle can be completed in seven days.

d *Suggest some methods of eliminating mosquitoes.*

Besides the eradication of *Anopheles* mosquitoes, a second line of defence is to prevent the mosquitoes from feeding on humans. This is accomplished by placing mosquito netting over doors, windows, and beds, by wearing protective clothing, and by the application of insect repellent to exposed skin.

A third possibility is to try to kill the *Plasmodium* parasites before they can infect the liver, or to kill them in the bloodstream during a malarial attack. Quinine, from the bark of trees of the genus *Cinchona*, was first eaten by Indians in the Peruvian Andes and was the only anti-malarial drug available up to the Second World War. It kills the schizonts of *P. falciparum* which are developing inside the red blood cells. Of the many synthetic anti-malarial drugs, the most useful has been chloroquine (1945), which acts as a blood schizonticide. It is effective in bringing under control a raging fever and high temperature.

Unfortunately *Plasmodium* itself is becoming resistant to chloroquine. Nevertheless, some combinations of drugs, such as pyrimethamine and sulphonamide, can still be effective. The main promise now is of an anti-malarial vaccine, either against sporozoites or, more usefully, against the merozoites.

e *Why is a vaccine against merozoites likely to be more effective than a vaccine against sporozoites?*

27.5 Commensalism

In commensalism ('eating at the same table'), individuals of different species are habitually found together, and whilst one species benefits from the association the other neither benefits nor is harmed. Two examples will suffice.

The water shrimp *Gammarus* often has living on its surface the protozoan *Vorticella*. *Vorticella* is a filter-feeder and probably benefits because as the water shrimp moves around, the *Vorticella* encounters new sources of food. It may even eat particles of food or egesta which the water shrimp discards. It seems unlikely that the shrimp benefits from the *Vorticella* on its surface.

Similarly, trees are often festooned with individuals of other plants, which grow on their trunks and branches. These 'epiphytes' grow nearer the light than they would have done on the ground. Whether the epiphytes are lichens, or flowering plants such as orchids or bromeliads, it does not seem very likely that the trees are greatly affected by their presence.

27.6 Mutualism

In mutualistic interactions, both partners benefit from the association, as in ants and acacias, mycorrhiza, lichens, and cellulose-digesting bacteria and protozoa in the gut. A wide variety of mutualistic associations exists. On a gross scale, every living thing is mutualistic. Green plants produce food and oxygen which all organisms use. The other organisms 'feed' plants by producing carbon dioxide and releasing nutrient ions. In many mutualisms, such as bees and flowers, there is no prolonged contact between the mutualists. At the other extreme, however, there are pairs of species which only exist in mutualistic associations. The two members of the pair depend on one another for survival and their physiologies tend to be complementary.

Four well-known mutualistic associations

1 *Ants and ant acacias* (e.g. the bullhorn acacia, *Acacia cornuta*, in Costa Rica). The acacias have large spines which queen ants (e.g. of the genus *Pseudomyrmex*) hollow out to establish colonies. The ants eat nectar from nectaries on the petioles of the acacia, and protein and fats from the 'Beltian bodies' attached to the sides of the leaves. Experiments have shown that bullhorn acacias from which the ants have been removed are rapidly smothered by vines or attacked by herbivores. The ants chew off the meristems of vines which emerge near the base of the acacia, and attack caterpillars and birds.

Figure 318
Acacia sphaerocephala, often associated with mutualistic ants.
After Gösswald, K., Unsere Ameisen, Kosmos, Stuttgart, 1954.

2 *Mycorrhiza.* The rhizoids and roots of many land plants, especially where they grow on nutrient-poor soils, are habitually associated with specific species of fungi. Some of these fungi form sheaths which surround the rootlets; their feeding hyphae penetrate the cell walls of the roots. Others form vesicles, arbuscules or haustoria, different types of feeding hyphae in contact with the cytoplasm of the root cells. Although the fungi absorb large quantities of photosynthate from the plants with which they are associated, experiments have shown that they markedly promote the growth of their mutualistic partners. They greatly increase the rate of mineral salt uptake by roots (see *figure 319*).

3 *Lichens.* Each lichen is a slow-growing organism which is a combination of a fungus and an alga. Lichens may be leafy (foliose), branched (fruticose) or closely appressed to a rock surface (crustose). The lichen looks quite different from both the fungus and the alga when they are grown separately. The fungus, which may make up 95 per cent of

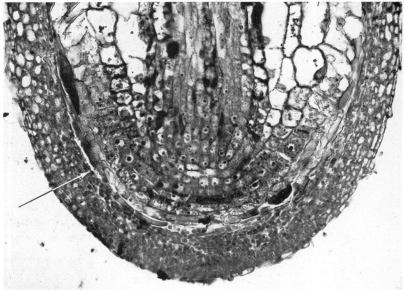

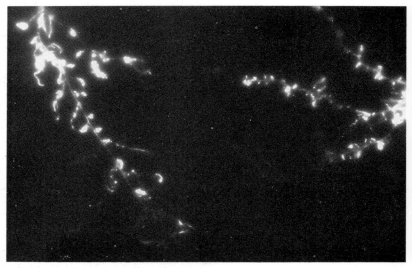

Figure 319
The mycorrhizal root-system of the beech, *Fagus sylvatica*.
a Except for the main axes, the whole system is covered by a sheath of fungal tissue.
b The longitudinal section through the apex of a mycorrhizal root shows the sheath of fungal tissue (arrowed). The outside of the sheath is connected in nature to the soil by hyphae, which have been broken off in preparing the section. From the inside of the sheath, hyphae penetrate the cell walls of the root cortex.
c An autoradiograph of part of an absorbing mycorrhizal root-system. The lighter the colour the greater the uptake of radioactive phosphate, ^{32}P. The parts of the system which are covered by a sheath of fungal tissue are more active in uptake than the main axes and uninfected lateral roots.
Photographs **a** *and* **c** *are by Dr J. L. Harley, Department of Forestry, University of Oxford, photograph* **b** *is by Dr F. A. L. Clowes, The Botany School, Oxford.*

the lichen dry mass, obtains organic compounds from the photosynthesis of the alga. How the alga benefits is not clearly established, but it receives more carbon dioxide, it can occupy a wider range of habitats, and can better withstand desiccation.

4 *Bacteria and protozoa in the guts of ruminants and termites.* The high densities of gut mutualists receive a constant supply of food and, in the case of ruminants, a continuously high temperature which is ideal for rapid metabolic activity. The ruminants and termites benefit because they can absorb the products of digestion of cellulose. The gut micro-organisms may also produce B-vitamins, and convert inorganic nitrogen into amino acids which are absorbed by the animal.

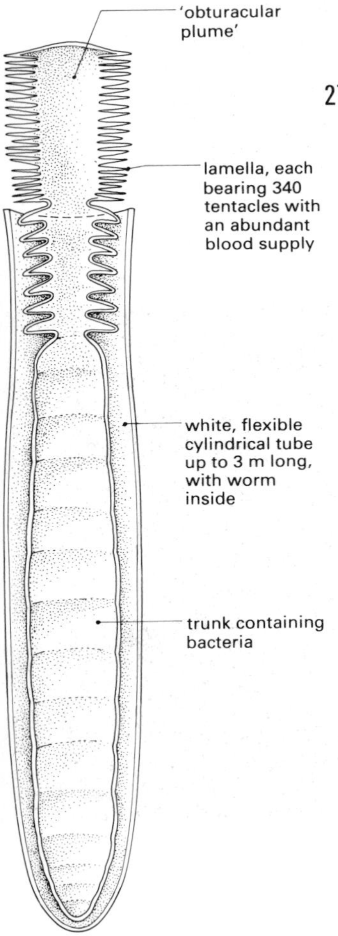

'obturacular plume'
lamella, each bearing 340 tentacles with an abundant blood supply
white, flexible cylindrical tube up to 3 m long, with worm inside
trunk containing bacteria

Figure 320
The Galapagos vent-worm, the annelid *Riftia pachyptila.* *Reconstructed from photographs in Jones, M. L.,* 'Riftia pachyptila *Jones: Observations on the Vestimentiferan worm from the Galápagos rift*', Science **209**, *4505, 333, 1981. Copyright © 1981 by A.A.A.S.*

STUDY ITEM

27.61 Equilibrated physiological interdependence in *Riftia pachyptila*, the Galápagos vent-worm

In the Pacific ocean, between the Galápagos Islands and mainland South America, is a rift valley about 2500 m below sea level known as the Galápagos rift. Running along the base of this chasm is a small ridge which rises 60 m above the floor of the valley. Along this ridge groups of animals occur sporadically. The animals occur in vent areas 30 to 100 m wide, in which warm water at 7 to 17 °C escapes from the tiny volcanic fissures in the rock. The unique communities are characterized by huge clams, blind crabs, mussels, fish and limpets, and, in particular, by large tube-worms, up to 3 m long, and 10 cm in circumference (*figure 320*).

At this depth none of the organisms can photosynthesize, because they live in complete darkness. The whole community seems to be supported by chemo-autotrophic bacteria, which oxidize hydrogen sulphide. This is present in the vent water at concentrations of up to 350 μmoles dm^{-3}. The energy from its oxidation is used by the bacteria to reduce carbon dioxide to organic compounds such as glucose.

Each worm lives within a vertical, white, blind-ending, flexible but extremely sturdy tube. Numerous worms, ranging in length from 1.5 m to 0.75 mm, were collected from Dandelion, Garden of Eden, and Rose garden geothermal vents, and their nutrition was investigated.

a ***Consider in turn each of the points below. State the inferences which you would draw from each piece of information. At the end, state a hypothesis to explain how*** **Riftia** ***gains its energy.***

1 The tube-worms are bright red. Hb. (not warning since dark)
2 They lack a mouth, gut, and anus! Cannot eat absorb small molec
3 At the top of the worm, protruding from the thick-walled tube and in contact with the water, is the 'obturacular plume', which consists of tentacles. There are 228 000 tentacles, in the type specimen, each with up to 200 pairs of vascularized projections. ↑ SA for absorbing
4 A related worm, *Siboglinum fiordicum*, was exposed to radioactively-labelled amino acids and carboxylic acids. It absorbed some of these compounds but the rates of uptake were totally inadequate to account for the metabolic rates of the worms. not enough in sea to maintain Hem

5 *Riftia*'s body cavity contains the sex organs and tissue known as the trophosome. The trophosome contains many blood capillaries. It consists of large numbers of closely-packed bacteria (10^9 cells per gram fresh mass).

6 In most specimens of *Riftia* there are sulphur crystals in the trophosome up to 100 μm across.

7 In the trophosome the concentrations of the enzymes ribulose biphosphate carboxylase and ribulose-5-phosphate kinase were comparable to those in fresh spinach leaves (*Study guide I*, Chapter 7). These enzymes are absent from muscle.

8 *Riftia*'s blood contains a sulphide-binding protein which appears to concentrate sulphide from the environment. The capacity of the blood for sulphide (1 to 3 mmol dm^{-3}) approaches that for oxygen (4 mmol dm^{-3}).

9 The obturacular plume tissue of *Riftia* is insensitive to sulphide poisoning compared with the gills of sharks, yellowfish, tuna, and rockfish, which do not inhabit vents. In aerobic organisms, sulphide poisoning usually affects the cytochrome c oxidase system (see *Study guide I*, Chapter 5).

b *The blood exhibits a small Bohr effect (**Study guide I, Chapter 4**) (a shift of the oxygen equilibrium curve to the right with increased CO_2 levels) compared with that of other worm species which inhabit permanent, well-ventilated burrows or tubes. What might be the significance of this?*

c *Compare the problems of gas transport by the blood in* **Riftia** *and a mammal.*

d *Is it justifiable to call* **Riftia** *the first example of an autotrophic animal at the base of a food chain?*

STUDY ITEM

27.62 Crops without chemicals

With the world's population currently at over 4.8 thousand million and increasing at an annual rate of about 2 per cent, improvements in our total crop production and distribution are essential if we are to avoid a catastrophe of widespread starvation. In the 1960s, the Green Revolution (Study item 28.52) provided the undernourished world with cereal strains of much improved yield, but these made heavy demands on the fertility of the soil. So far the only answer to this problem has been the use of expensive nitrogenous fertilizers. The fertilizers themselves have produced problems when their application on farm lands has led to the *eutrophication* of lakes and waterways. This is one of the reasons why agricultural botanists have been directing their attention for some time towards the prospect of developing cereal crops which can directly assimilate atmospheric nitrogen and so reduce the demand for nitrogenous fertilizers (see Chapter 21).

In nature, nitrogen fixation is an ability confined to a few genera of bacteria and cyanobacteria. Some associated with host plants while others are free-living. The activity of most of these nitrogen-fixing organisms is limited by oxygen. Members of the genus *Klebsiella* are *facultative anaerobes* and are found free-living, or in association with plants. Some bacteria, however, such as *Azotobacter*, survive only in aerobic conditions. Perhaps the best-known nitrogen-fixer is the bacterium *Rhizobium* which is found in the root nodules of leguminous plants. Legume crops have long been used in rotation with non-legumes to improve the fertility of the soil.

Rhizobium bacteria enter the legume roots through an infection thread which develops as an invagination of the cell wall of a root hair. The tip of the thread ruptures, the bacteria are released, and they invade cortical cells of the root which are stimulated into meristematic activity. The root nodules which develop may appear pink because of the synthesis of leghaemoglobin by the legume root tissue. This is the only form of haemoglobin known in the plant kingdom; like animal haemoglobin, it has the ability to bind oxygen. In nodules which do not produce leghaemoglobin, it has been shown that the *Rhizobium* bacteria do not fix nitrogen.

In the laboratory, it has proved possible to cultivate legume root tissue on permeable membranes. When these were applied to the surface of agar plates innoculated with *Rhizobium*, the bacteria were able to fix nitrogen in anaerobic conditions. The membrane-loaded plant culture was then removed, leaving behind some of the bacteria, and these continued to fix nitrogen for several hours. *Rhizobium* is normally considered to be an *obligate mutualist*, but when in anaerobic conditions and provided with a pentose sugar and a dicarboxylic acid – substances present in all plants – its powers of nitrogen fixation are unimpaired.

This accumulation of knowledge about nitrogen fixation is now being put to good use in the search for ways of producing nitrogen-fixing cereal food crops. Much effort has been expended in trying to persuade *Rhizobium* from legumes to invade the roots of wheat and other cereal plants, but with little success. Even if nitrogen-fixing cereal crops can be produced, however, there may be serious drawbacks. Nitrogen fixation consumes a great deal of energy (up to 24 molecules of ATP for each molecule of nitrogen fixed) and the presence of nitrogenase in many cereals would turn out to be something less than the economic miracle anticipated (see Chapter 21, 'The principles and applications of biotechnology').

a   Give the meaning of the following terms:
    1 *eutrophication*
    2 *facultative anaerobes*
    3 *obligate mutualist*.

b   Describe two ways in which nitrogen fixed by **Rhizobium** bacteria in leguminous plants may enter the soil.

c   1 Suggest a function of the leghaemoglobin produced in the root nodules of legumes.
    2 Explain how this benefits the **Rhizobium** bacteria.

d *State the sequence of events in* **Rhizobium** *which result in the synthesis of nitrogenase enzyme.*

e *Give a reason for the comment that 'the presence of nitrogenase in many cereals could turn out to be less than the economic miracle anticipated'.*

☐ (*J.M.B. – modified*)

27.7 Dispersal and other historical factors

One can explain the patterns of many animal species on the basis of their evolutionary history and of human influence. Marsupials, for example, are abundant in Australia, occur in small numbers in America, but are absent from Africa, Europe, and Asia. This is because, 200 million years ago, their ancestors only occupied the southern part of Gondwanaland, before the continents split apart.

Mountains, rivers, or the sea may restrict species to islands or continents. If the dispersal barrier were removed, the species could extend its range (table 63). New islands are formed by volcanic action. They are rapidly colonized by plants and animals, even mammals. Presumably mammals reach islands by accidental human dispersal, or by passive dispersal of pregnant females or pairs on drifting logs.

Rabbit (*Oryctolagus cuniculus*)	Europe to Britain in late 12th century	Bred for fur and meat. Escaped. Wild population expanded in 18th and 19th centuries. Major farm pest until myxomatosis in 1953–5. Only 1 per cent survived. Now expanding again. Similar population explosion in Australia.
Plague (*Pasteurella pestis*)	S.E. Asia to Britain 1348 (Trade)	Killed humans and black rats.
Prickly pear cactus (*Opuntia* spp.)	S. America to to Australia 1788	Infested 24 million ha grazing land by 1925. Moth *Cactoblastis* imported, achieved rapid control.
Oxford ragwort (*Senecio squalidus*)	Sicily to Oxford Botanic Garden 1794	Spread to London on railway ballast. Vigorous weed.
Canadian pondweed (*Elodea canandensis*)	Canada to British Isles 1836	Spread along canals and waterways and blocked many in the late 19th century. Has since declined.
Grey squirrel (*Sciurus caroliniensis*)	Eastern USA to Britain 1876	Kills beech; poisons used in squirrel control may kill other species.
European starling (*Sturnus vulgaris*)	Europe to New York 1890	Now present throughout USA and most of Canada.
Muntjac deer (*Muntaicus* sp.)	S.E. Asia to Britain 1900 (Park deer)	Graze crops and woodland plants.
Japweed (*Sargassum muticum*)	Pacific to Britain 1973	Competes with native seaweeds.

Table 63
Well-known instances of dispersal of species to new areas followed by rapid population explosion.
Data from King, T. J., Ecology, *Nelson, 1980, and Taylor, J. C. The introduction of exotic plant and animal species into Britain,* Biologist **26(5)**, *229, 1979.*

STUDY ITEM
27.71 Dispersal onto Surtsey

Figure 321
Surtsey, a newly formed island off the coast of Iceland, is still being colonized by plants and animals.
a *Rhacomitrium* moss is forming small mats on the lava.
b Sea-birds may be important agents by which the spores of mosses and the seeds of flowering plants are dispersed onto Surtsey from the mainland. Here you can see fulmar (*Fulmarus glacialis*), kittiwake (*Rissa tridactyla*), and black guillemot (*Cepphus grylle*) which rest on the ledges of the cliffs.
c Sea sandwort (*Honkenya peploides*) is one of only four C_4 plants in the British flora. The C_4 mechanism may allow it to grow in such a dry habitat as this. (See *Study guide 1*, Chapter 7, 'Photosynthesis', sections 7.6 and 7.7.)
d A general view of the terrain of Surtsey. The sea sandwort (*bottom left*) is now the most common of the species of higher plants on the island.
Photographs, Dr Sturla Fridriksson, Agricultural Research Institute, Reykjavik.

On 14 November 1963 a volcanic eruption on the ocean floor 37 km from the south coast of Iceland began to produce the island of Surtsey. This now consists of 155 ha of volcanic lava. Micro-organisms and a fly were found in 1964, the first higher plant appeared in 1965, mosses invaded in 1967, birds first nested in 1970 and soil nematodes were first found in 1971. Table 64 shows the number of moss species on the island:

	1967	1969	1971	1972	1980
Number of moss species	1	8	40	63	70

Table 64

a   Plot the data in table 64 on a graph.

b   Account for the shape of the curve.

c   Why did micro-organisms and birds occur only in very small numbers for the first seven years?

d   For each of these groups, suggest two ways in which they might have reached the islands: 1 spiders, 2 mosses, 3 protozoa, 4 higher plants.

e   Explain how the proximity of other islands might affect the invasion of species onto Surtsey.

f   Suggest three ways in which sea birds might affect the composition of the flora.

g   Speculate on the changes which will take place in the flora and fauna of Surtsey over the next fifty years.

In the intensively-cultivated landscape of northern Europe, humans have continually altered the distribution limits of plants and their associated animal species (table 63). Plants valuable for food or floral display have been imported, and herbs valuable in cooking and medicine have been grown in captivity. Marshes have been drained, grasslands ploughed and converted to produce grain crops, rivers dredged, woods cut down, and new habitats created such as roadside verges and spoil heaps. Herbicides have been used on a massive scale in agriculture. Thus the present distribution patterns of a species may be greatly influenced, just as their numbers may be, by human policies over the centuries.

Practical investigation. *Practical guide 7*, Investigation 27F, 'A grassland survey using sampling techniques'.

27.8 Successional changes

The vegetation and animal life at a particular place may slowly change. This is known as *succession*. Since many species are only abundant at a single stage in the successional sequence, the local distribution pattern of a species may change as conditions become unsuitable for it or it invades new sites. The general changes which are likely to occur during succession at a particular place, such as the replacement of shrubs by

Figure 322
Stages of secondary succession in the oak-hornbeam forest in southern Poland. From upper left to lower right, the succession has occurred for 0, 7, 15, 30, 95, and 150 years respectively.
Photographs, Z. Glowacinski, from Glowacinski, Z. and Jarvinen, O., 'Rate of secondary succession in forest bird communities', Ornis Scand., **6**, *33, 1975.*

Figure 323
An aerial view of the zonation of plant species around Sweat Mere, North Shropshire, (looking East) May 1960. The white area at the fringe of the centre of the lake is covered by floating aquatic plants. Around it in grey is the reed swamp. The sedge zone is visible on the right-hand side as a darker area outside the reed swamp. An extensive alder carr (woodland) surrounds the sedge zone.
Photograph, J. K. St. Joseph, Committee for Aerial Photography, University of Cambridge. Crown copyright.

long-lived trees, may be predictable on the basis of studies made in similar situations. Because of random factors, however, it is difficult to predict with certainty the relative abundances of the species in each phase (see section 27.9).

When a succession occurs on bare rock and causes the gradual development of soil and vegetation it is called a 'primary succession'. When it happens on soil which was formed a long time ago, has some seeds in it and has been cleared, burned, ploughed, or wind-thrown, a 'secondary succession' occurs (*figure 322*).

Successions often become apparent because of zonations. The species composition of the different zones up a salt marsh or around a lake is often quite distinct. Each zone seems to be at a different stage in a sequence. At any one place, the vegetation goes through a regular sequence of changes.

Zonation and succession
Around many lakes in Britain, such as Esthwaite Water in Lancashire, Crosmere in Shropshire, and some of the Norfolk Broads, the vegetation exhibits a series of distinct zones, one inside the other (*figure 323*). In this case the zonation in space reflects a succession in time. Long-term studies in permanent quadrats have shown that at any one place there is

474   Ecology and evolution

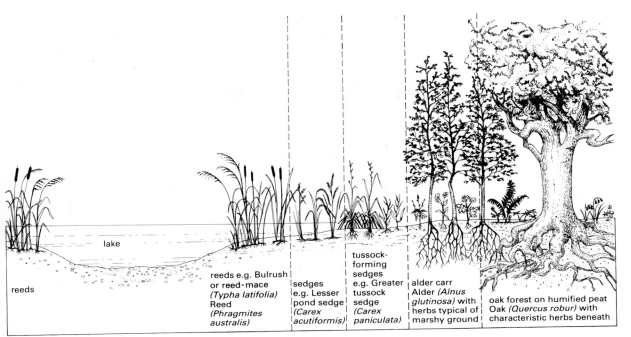

Figure 324
A diagram showing the hydrosere succession. Silt accumulates around the bases of the plants, which raise the level and create conditions suitable for the invasion of other species. Evidence for the oak forest stage is scanty. In some areas, *Sphagnum* bog may form instead.
After King, T. J., Ecology, *Nelson, 1980.*

predictable change in species composition. Furthermore, if the peat is excavated, plant remains from the previous zones in the succession can be found beneath the present vegetation at any point. The hypothetical sequence of species which occurs around nutrient-rich lakes is called a *hydrosere*, and might in some cases even turn a lake into *climax* forest on dry land, given several thousand years!

Suppose that the lake has an inflow stream, likely to carry silt, and an outflow stream. The margin of the lake is first colonized by reed, *Phragmites australis*, or bulrush (reed-mace), *Typha latifolia* (*figure 324*). Both species spread by underground rhizomes and can colonize water 60 cm deep. These species live with their leaves exposed to severe evaporation and their roots in anaerobic mud. In experiments, the roots and rhizomes of *Phragmites* have survived for over a week without oxygen. They were placed in water with an atmosphere of 85 per cent nitrogen, 10 per cent hydrogen and 5 per cent carbon dioxide, with a palladium catalyst present to remove any oxygen.

Silt and clay particles from the inflow stream accumulate around the stem bases and rhizomes of the reeds. As they do so, the level of silt around the edge of the lake gradually rises. Sedges (*Carex* spp.) then invade, and replace the reeds or bulrushes around the edge. By this time, however, the reeds or bulrushes have spread further towards the centre of the lake, migrating over a platform of silt accumulated by their underground organs.

As the silt level rises, the sedges around the lake margins are replaced by different species of tussock-forming sedge, such as *Carex paniculata*, the greater tussock sedge. Its leaves and rhizomes form dense hummocks up to 1.5 m high and 1 m in diameter. They accumulate still more silt. Their tops, sides, and the spaces between them are soon invaded by marshland herbs.

The area containing these sedge mounds is next invaded by alder (*Alnus glutinosa*), a tree able to grow in soil which is periodically waterlogged. Its roots are associated with nitrogen-fixing micro-organisms, which may help to increase the level of soil nitrogen. Alder forms a woodland around the margins of the lake (*figure 324*), with marshy vegetation beneath. This is known as 'alder carr'.

All the time, the vegetation zones have been marching towards the centre of the lake. Eventually they may meet in the middle, and if the level of silt continues to rise, each zone in turn will disappear at the centre. The result will be alder carr, oak forest, or *Sphagnum* bog, established over alluvial soil.

a   *If a cylinder of peat was taken out of the 'soil' beneath this 'climax' forest, what fragments of vegetation would you expect to obtain?*

A rather oversimplified view of the succession and the zonation has been presented here. The rate of succession differs considerably from lake to lake, since it depends on the volumes of the inflow and outflow streams, the area and depth of the lake and the silt, and the nutrient content of the water. Lake levels may rise for geomorphological reasons or because of human interference, reversing the successional trend. Recent surveys of peat analyses suggest that, as a result of this succession, *Sphagnum* bog is a more likely climax vegetation in many parts of Britain than oak forest. Although the rate of change is imperceptible, this zonation of species can be seen along canals and in lake margins throughout Britain.

STUDY ITEM

27.81   Changes in production and biomass during succession

Measurements of the changes in production and biomass on a site as succession proceeds have not been made for many communities because scientific ecology has not existed for long enough. Adjacent communities of different ages similar in species could be compared, but the measurements are complicated by the change in species composition with time. The best data available come from naturally-sown populations of the Norway spruce or Christmas tree (*Picea abies*) in Karelia, U.S.S.R. (62° N, 34° E). All these sites were on humus-iron podzols, soil pH 4.1 to 4.4, at altitudes of 80 to 200 m (table 65).

a   *Plot the data in table 65 on a graph.*

b   *Examine these results and state the main changes which are occurring with time during the development of a mature stand of Norway spruce. Pay some attention to the ratios of photosynthesizing to respiring to dead tissue.*

c   *At what age is the spruce growing fastest?*

d   *Calculate the ratio of production to biomass for each of the stands. Explain the trend.*

Stand age (year)	22	37	45	54	68	82	98	109	126	138
No. trees ha^{-1}	34800	13750	9420	4820	2336	1898	1319	1080	856	1087
Tree height (m)	2.6	6.8	8.8	11.1	14.2	17.1	19.6	20.0	22.6	22.8
Leaf area index (LAI): leaf area (m^2) per m^2 ground area	1.8	3.0	3.2	3.6	3.8	3.8	3.6	3.2	2.7	2.4
Dry mass living vegetation (t ha^{-1})	32	75	94	120	162	178	227	238	254	248
Total net production including litter and root production (t ha^{-1} year^{-1})	4.4	6.1	6.5	7.1	7.0	6.6	5.6	4.7	3.7	3.2
Net production litter (t ha^{-1} year^{-1})	1.7	2.5	2.55	3.15	2.3	2.2	2.9	2.7	2.4	2.3
Net production roots (kg ha^{-1} year^{-1})	520	480	520	530	500	420	250	150	60	20

Table 65
Changes in production and biomass with age in stands of *Picea abies*.
Data from Kazimirov, N. I. and Morozova, R. M., pp. 629 in Reichle, D. E. (ed.) Dynamic properties of forest ecosystems, *Cambridge University Press, 1970*.

 e *Calculate the ratio of litter fall production to total production for forests of each age. What do these data indicate about the trees as they mature?*

 f *How would the mass of detritus at the soil surface change during succession?*

27.9 Succession and the climax

The speed of succession, and the apparent sequence of dominant species, probably depends on several factors. The range of available plants depends on the rate at which their seeds invade the site. The species succeed in a particular order because of two mechanisms: facilitation and inhibition.

Some species create, during succession, conditions which are suitable for other species with different environmental preferences. This is known as facilitation. At the same time they may alter the micro-environment so that conditions become less favorable for other species ('inhibition'). These mechanisms depend on the fact that different species differ in their tolerances to a range of environmental factors, such as the shading which often increases during a successional sequence.

It is convenient to think of each succession as ending in a 'climax' state. At climax, the species composition of the community begins to fluctuate between narrow limits. It may remain in this state for thousands of years. The type of climax which develops is partly influenced by local site conditions. On acid soils in lowland southern Britain, for instance, a pine forest might develop, but on basic soils it might be a beech forest or a yew wood. Human interference during succession, such as the introduction of grazing animals, might produce a 'deflected climax', such as a heathland on acid soils or chalk grassland on basic soils. This would be maintained for as long as the grazing regime persisted.

The concept of the climax, however, has recently been challenged. The glaciation episodes over the last 1.5 million years have continually shifted the distribution ranges of plant species. There is good evidence for migrations of forest trees over the last 15 000 years. Many forests around the world are devastated periodically by fires or hurricanes and never reach the climax state. Tropical rain forests consist of a mosaic of areas in different stages of secondary succession, because of the shifting agriculture practised by Indians over the centuries.

In fact a 'climax' community such as a forest which contains a mixture of species is bound to consist of a mosaic of areas in different stages of secondary succession. When a dead tree falls, it creates a gap which is often colonized by pioneer species. A mini-succession takes place there. The forest as a whole never reaches the uniform species composition which the originators of the climax concept envisaged.

Perhaps the closest approach to the climax occurs in those few mixed forests in North America where devastations by fire and hurricanes are rare. In some of these forests there is good evidence amongst climax species for 'reciprocal replacement'. Imagine that species A establishes well under species B but not under its own canopy, while species B grows well under species A but not under its own canopy. The result will be a stable mosaic.

a   *Suggest some reasons why saplings of species A might be able to grow better under the canopy of trees of other species than of trees of species A. Remember that biotic factors such as those outlined in this chapter, not just physical and chemical factors, may be involved.*

b   *How could you test your hypotheses?*

Summary

1   The interactions between one species and another are known as biotic. They include competition, predation, parasitism, commensalism and mutualism, and often influence the local and geographical distribution patterns of species.
2   Individuals of the same or different species may compete, that is, obtain the same resources from the environment when the resources are in short supply. Competition may reduce growth rate, reduce reproductive rate, or cause death (**27.1**).
3   Many plant and animal species which live together in the same environment differ in ways which reduce competition between them. Shade plants, for example, survive beneath others by having leaves which have lower respiration rates, lower light compensation points, and optimum photosynthesis at lower temperatures (**27.1**).
4   Predators may influence the geographical and local patterns of their prey, and the prey may determine where the predator occurs (**27.3**).
5   The life-cycle of one of the malarial parasites (*Plasmodium falciparum*) is used to illustrate a parasite and the means by which it may be controlled (**27.4**).
6   In mutualisms, as in vent worms and legume root nodules, two different species habitually live together in a long-term association which provides mutual benefit (**27.6**).

7 The geographical patterns of species are strongly affected by their dispersal ability and by human activity (**27.7**).
8 On bare rock or in disturbed habitats succession takes place; this is a change in the relative abundances of the animal and plant species present at a site (**27.8**).
9 Successions may end in 'climax' communities in which the relative abundances of species fluctuate between relatively narrow limits (**27.9**).

Suggestions for further reading

HARLEY, J. L. Carolina Biology Readers No. 12, *Mycorrhizae*, Carolina Biological Supply Company, distributed by Packard Publishing Ltd, 1972.

HILL, T. A. Studies in Biology 79. *The biology of weeds.* Edward Arnold, 1977. (The chapter on 'What weeds do' summarizes the various ways in which weeds affect crops.)

HOLLOWAY, J. K. Weed control by insect. *Scientific American* **197**, 1957, 56. (Describes the effects of Klamath weed (*Hypericum perforatum*) as a weed in California, and its control by the introduction of Chrysolina beetles.)

HUTCHINGS, M. J. Ecology's law in search of a theory. *New Scientist* **98**, 1983, 765. (Clear account of the $-3/2$ power law.)

PHILLIPS, R. S. Studies in Biology 152. *Malaria.* Edward Arnold, 1983. (Excellent summary of the life cycles and control of these important parasites.)

POSTGATE, J. Studies in Biology 92. *Nitrogen Fixation.* Edward Arnold, 1978. (Elegant description of the process and its potential.)

SCOTT, J. C. Studies in Biology 16. *Plant symbiosis.* Edward Arnold, 1969. (Introduction to the main mutualisms, including lichens and mycorrhizas.)

VICKERMAN, K. and COX, F. E. G. *The Protozoa*, Murray, 1967. (Contains a brief, well-illustrated summary of the life-cycle of *Plasmodium* spp.)

WHEELER, B. E. J. Carolina Biology Readers No. 74, *The control of plant disease.* Carolina Biological Supply Company, distributed by Packard Publishing Ltd, 1975. (Includes a brief illustrated account of potato blight fungus and the ways of controlling it.)

CHAPTER 28 POPULATION DYNAMICS

28.1 How can population size be assessed?

Population dynamics is the study of changes in the population sizes of organisms from generation to generation. In order to study populations, ecologists need methods to estimate population sizes accurately. This can be rather difficult. Large numbers of small mobile animals, many hidden from view, cannot be counted directly. In the case of plants, it is easy to count individual shoots or rosettes. However, since shoots may be connected underground, the number of genetically different individuals is difficult to estimate.

The most appropriate technique to assess population size depends on whether or not distinct individuals exist in the population. An individual is a product of a zygote. Some organisms, such as humans, fruit-flies, and most other animals, develop from the zygote into a 'unitary' structure which is a distinct individual. Others, such as colonial animals and all higher plants, develop from the zygote by producing more and more identical units, known as 'modules'. In the coelenterate *Hydra*, for example, a module is a body with tentacles, and in many flowering plants it is a piece of stem with a leaf attached and a bud in the leaf axil.

Unitary organisms such as humans grow to a certain size and then stop growing. Individuals can move about, sensing their environments. Usually, separate male and female individuals can find one another. The reproductive output of an individual tends to reach a peak with time and then decline. When resources are in short supply, individuals may die.

On the other hand, modular organisms, like many higher plants, are potentially immortal. They can usually continue to grow throughout life by producing more and more modules. Modular organisms tend to be static and to lack sophisticated sensory mechanisms. Organisms like this often possess both male and female gamete-producing organs. The fertility of a modular organism may continue to increase throughout life, and the organism does not obviously 'grow old'. When resources are in short supply, module production may decrease, or module mortality may increase, but the modular organism may live on. Seen in this light, a clump of the bracken fern, which spreads horizontally underground by creeping rhizomes, is equivalent to a tree on its side.

Thus, whilst unitary organisms are easy to count, modular organisms often present problems. The buttercup rosettes scattered across a pasture might all be modules derived from the same colonizing seed. Genetically, they may all be parts of the same individual, but it is almost impossible to tell whether or not this is the case. We would, however, be justified in studying the dynamics of rosettes, leaves, or buds. In other modular species we might study the dynamics of branches or polyps.

Whatever sort of organism we are interested in, some method of *sampling* the population must be employed. It is usually impossibly time-consuming to attempt to count every individual or module. Sampling allows the average number of 'individuals' per unit area to be estimated. This estimate is more accurate (that is, closer to the real value) and more precise (that is, likely to be accurate within narrower limits) the larger the sample size employed.

To count modular organisms such as plants, a square or circular quadrat is often used. The number of modules of a species within each quadrat is counted, and the results are averaged. Quadrat size should be adjusted to the type of vegetation. An appropriate size is 10×10 m in a forest, 1×1 m in heathland vegetation or 0.5×0.5 m in species-rich grasslands. To avoid sampling bias, the quadrats should be placed at random on a pair of axes at right angles. Pacing is sufficient.

It is often more convenient to record per cent frequency or per cent cover rather than the number of 'plants' per unit area. The *per cent frequency* is the proportion of sampled quadrats in which a species occurs. To determine it accurately, a large number of quadrats must be surveyed. Even then, the resulting value is imprecise. The per cent frequency also depends on the size of quadrat which is used. The *per cent cover* is the proportion of the ground area covered by a perpendicular projection onto it of the aerial parts of the species. It can be estimated by eye for each species in each quadrat, or with a point quadrat frame.

A point quadrat frame consists of a series of needles which move vertically through a horizontal metal or wooden frame. Each pin is lowered in turn and the species touched by the tip are recorded. The per cent cover of a species is the proportion of pins which touch the species. Point quadrats are valuable when the vegetation is to be sampled at the same place at yearly intervals to assess the changes in composition. However, for simple vegetation comparisons, point quadrats are too time-consuming. It is more efficient to estimate the per cent covers by eye, and to average the results.

Counts in quadrats are also appropriate for some populations of unitary organisms, such as brittle starfish in a tidal creek, or earthworms in a field. Large organisms, such as wildebeest or caribou, in herds, can be photographed from the air and the counts can be made on the photographs. For small, highly mobile animals often hidden from view, techniques such as beating, sweeping, and pitfall-trapping are frequently employed.

In 'beating', the animals which live in leaves and branches are detached by knocking the vegetation with a stick. They fall onto a 'beating tray' from which they are rapidly collected. Beating is useful for identification, but because it is difficult to reproduce the technique exactly on different sampling occasions, is less valuable for assessing population changes. In 'sweep-netting', a fine-mesh net is swept through grassy vegetation a certain number of times in a standard way. All the organisms can be collected and the technique is fairly reproducible. Animals which move over the ground can be sampled by pitfall traps, which are yoghurt cartons or jam-jars sunk into the surface.

Another method employed widely for censusing small animals is the capture-mark-recapture technique. It is suitable for animals such as grasshoppers, woodlice, snails, ants, pond-skaters and small mammals. Since it is non-destructive, it can be repeated time and time again on the same population to assess how the numbers of individuals are changing. In its simplest form, a number of animals are captured, marked and released (C). After enough time has elapsed to allow these marked animals to mingle randomly with the rest of the population, a certain number of animals is recaptured (R). Some of these (M) will be marked and the rest will be unmarked. The population size is then CR/M.

> **Practical investigations.** *Practical guide 7*, investigation 28A, 'Sampling methods for small invertebrates', and investigation 28B, 'Capture–mark–recapture technique'.

STUDY ITEM

28.11 The capture–mark–recapture method (Short answer question)

A population of grasshoppers in a meadow was sampled by sweep netting on two successive occasions. All of the grasshoppers captured in the first sampling were marked by a spot of harmless quick-drying paint on the dorsal side of the thorax. Twenty of these marked grasshoppers were kept inside a net cage which was left out in the meadow. All the remaining marked individuals were released.

Two days later, the population of grasshoppers was sampled again in exactly the same way as before. Details of the animals captured are given in table 66.

Grasshoppers captured on Day 1	Marked grasshoppers released on Day 1	Marked grasshoppers captured on Day 3	Unmarked grasshoppers captured on Day 3
200	180	30	120

Table 66

a 1 Calculate the estimated population size for the grasshoppers in the meadow.
   2 Suggest a reason why some of the original marked grasshoppers were kept enclosed in a cage and not released.

When the size of a population of woodlice was estimated by using the marking-recapture method (involving overturning stones and sucking up individual woodlice with a 'pooter'), the population estimate was always much lower than with other methods of calculating population size.

b Explain how the marking-recapture method as practised here can introduce large errors in the estimation of population size.

c Suggest two further circumstances in which population estimates using marking-recapture methods might be inaccurate. *(J.M.B.)*

A refinement of this method is often used to estimate the population size of small mammals. A large number of baited animal traps are sited at random. Longworth traps, for example, consist of a hollow tube, square in cross-section, with a trap door at one end and a removable metal plate at the other. The individuals caught are marked and released. The traps are laid a second time. The proportion of marked individuals caught again is noted. The unmarked individuals are marked. The same procedure is repeated several times until almost all the population has been marked.

28.2 Population growth

The fluctuations in numbers of individuals or modules in a population are determined by the rates of birth, death, emigration and immigration. To state the relationship formally:

Number of organisms at time t + Births + Immigrants − Deaths − Emigrants
 = Number of organisms at time $t + 1$

Apart from the large-scale migrations of insect, bird, and mammal populations, many organisms move or are blown around the landscape so rapidly that population ecologists neglect immigration and emigration at their peril. In a patch of well defined habitat, separated from other habitats, immigration and emigration may sometimes be unimportant. In this case population fluctuations can be worked out in terms of birth rates and death rates.

Population ecologists therefore do not merely document changes in numbers. They work out the extent to which the changes in numbers are due to births, deaths, immigration, and emigration. Digging more deeply, they then try to discern what factors cause the birth rate, death rate, immigration rate, and emigration rate to vary from year to year. This is obviously a complex problem in the field, but important principles have emerged from studies in the laboratory on simple populations such as yeast and bacteria.

STUDY ITEM

28.21 The growth of a population of yeast

If we introduce some yeast cells into a sterile nutrient broth and take

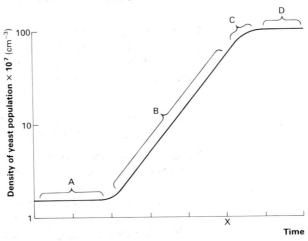

Figure 325
Increase with time in the numbers of yeast cells in broth. Twenty million cells per cm³ were introduced into the broth. The yeast cells were counted at intervals with a haemocytometer. Notice that the vertical axis is logarithmic, not linear.

samples at intervals, we can calculate the number of yeast cells per cm³ of medium. Typical changes in numbers of cells in such populations are shown on the graph in *figure 325*.

a *What phase in population growth does A in* **figure 325** *represent? What is happening to the cells at this stage?*
b *Describe the manner of increase in the number of cells at stage B.*
c *What differences would there be in the graph if the population growth had not been limited after time X?*
d *What is happening in the culture to give the shape of the curve at C?*
e *Name two factors which could be causing the shape of the curve at D, assuming that the temperature remains constant.*
f *What would be the shape of the graph if we used a sample of the culture, in its stationary phase, to inoculate a fresh flask of nutrient broth?*
g *Would you describe the possible interactions between the yeast cells at C as competition? For what resources could they be competing?*

Like yeast, many organisms invading new habitats undergo population increases with time which approximate to a S-shaped curve. The collared dove in Britain, pheasants on Protection Island off Michigan, and probably rabbits in Australia have exhibited similar patterns of population growth. In each case dispersal by humans has overcome a natural dispersal barrier. Without its usual predators and parasites, the number of individuals increases slowly at first (the 'lag' phase), then at an accelerating rate (the 'log' phase), and eventually does not increase at all (the stable phase).

> Practical investigation. *Practical guide 7*, investigation 28C, 'Population growth in *Chlorella*'.

STUDY ITEM
28.22 Population growth of bacteria

Bacteria reproduce in favourable conditions by binary fission. Each cell divides into two daughter cells. In studies of the population growth of bacteria it is frequently assumed that all the bacteria divide at regular intervals. Is this assumption justified? How can we determine the generation time? A simple mathematical model may help. Look at the diagrammatic model of bacterial population growth shown in *figure 326*. This is based on the assumptions of binary fission and a constant time interval between divisions.

With this model in mind, let us devise an equation which relates the number of bacteria in a culture (x) at any time (T) to the number of bacteria in the original innoculum (x_0) and the generation time (t).

At time zero, $x = x_0$. With each generation, the number of organisms doubles. Thus, after one generation $x = 2x_0$, after two generations

Figure 326
A diagrammatic model of population growth in bacteria.

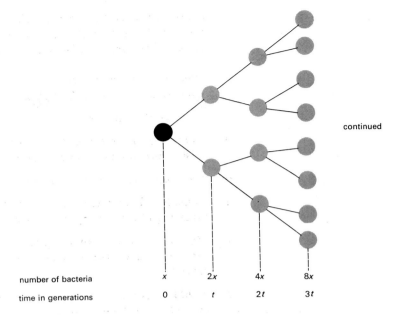

$x = 4x_0$, after three $x = 8x_0$. To generalize, $x = 2^n x_0$, where $n =$ number of generations elapsed. The number of generations which have passed at any time is T/t, where T is the time elapsed and t is the generation time. Thus $x = 2^{T/t} x_0$.

The generation time (t) may be especially interesting to a microbiologist. The following simple procedure shows how t may be determined.

Figure 327
Growth of a population of *Lactobacillus bulgaricus*. From Snell, E. E., Kitay, E., and Hoffogenson, E. Archives of Biochemistry and Biophysics **18**, 495, 1948.

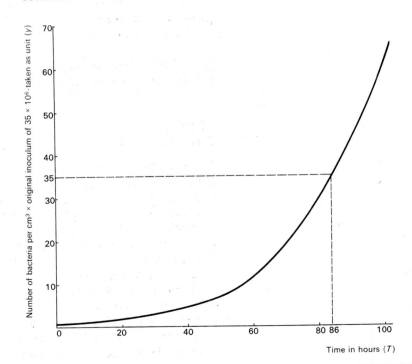

Chapter 28 Population dynamics 485

a *Plot a graph, on your own graph paper, of the equation $x = 2^n$ for values of $n = 0$ to $n = 5$. The symbol n represents the number of generations which have elapsed.*

b *Compare your graph with the graph in figure 327 which shows the observed growth rate of a population of the bacterium* **Lactobacillus bulgaricus**. *What conclusions do you draw from this comparison?*

c *For* **Lactobacillus bulgaricus**, *plot a graph of the $\log_{10}$ numbers of the bacterium against the time in minutes. Use for the horizontal axis the same linear scale as in figure 327. This graph is most simply constructed on log-linear graph paper. If the bacteria divide by binary fission, 1 do they divide at regular intervals? 2 what is the generation time?*

In nature, populations rarely rise like this because individuals die. Every population, at any phase in its growth, has an 'age structure'. This is the frequency distribution of ages in it. Populations of constant size may have totally different age-structures, depending on the pattern of death with age. The age of death may depend not only on death from old age, but also on the susceptibility of the organism at different stages in its life-history to predation and parasitism.

STUDY ITEM

28.23 Age structure of a fish population

The plaice (*Pleuronectes platessa*) is a shallow water flatfish which is an important commercial species in the North Sea. The plaice population has remained fairly stable, except in the war years 1914–1918 and 1939–1945, when fishing was reduced and stocks increased.

The females spawn when they are 5–7 years old, in mid-winter near the coasts. They can lay up to 350 000 fertile eggs each, but this is balanced by an equally high mortality. The eggs float in the plankton until hatching, and the larval plaice are carried about in the currents. At about two months of age they settle out in nursery areas off the sandy coasts of Holland, Denmark, and Germany. They stay there until they are 2–3 years old, when they begin to move away from the coast and towards the middle of the North Sea. They begin to be fished in abundance at 3–5 years old, at a length of 20–30 cm.

Table 67
Catch of plaice (*Pleuronectes platessa*) of various ages in the southern North Sea in three successive seasons. The numbers are catches per 100 hours of trawling.
Data from Gulland, J. A., Estimation of growth and mortality in commercial fish populations. *U.K. Ministry Agr. Fish, Fish Invest. Ser. 2*, **18**(9): 1, 1955.

Age of plaice	Catch 1950–1	Catch 1951–2	Catch 1952–3
2	39	91	142
3	929	559	999
4	2320	2576	1424
5	1722	2055	2828
6	389	982	1309
7	198	261	519
8	93	152	123
9	95	71	106
10	81	57	61
11	57	60	40
12 + older	94	87	99

The figures in table 67 show the catch of plaice, per 100 hours of trawling, in the southern North Sea over three seasons. Examine these figures and answer the questions below.

a How can the ages of fish be determined?

b How did the total catch differ from year to year?

c Is it justifiable to conclude from these data that the plaice were more abundant in some years than others?

d Concentrating on the age-classes 4–11 only, calculate for each catch figure the percentage of the year's catch which it represents.

e On the basis of these percentages, calculate the proportion of plaice which survived until the next year, for each of the age-classes surrounded by the rectangle.

f Examine the percentages which you have calculated in part e. How do the chances of survival of the fish differ 1 between one year and the next, 2 between ages?

g Does the age-distribution change significantly with time? What will be the age-distribution in ten years time?

h Discuss the changes which might occur in the age distribution if
1 fishing intensity increased so that twice as many fish were caught per unit time
2 fishing ceased
3 the mesh size of the nets was greatly reduced.

STUDY ITEM

28.24 Expectations of life of different animal species

Figures 328a, b, and c show the percentages of animals of different kinds which survive to a given age. The arrows indicate time of first reproduction. Notice that in each case the vertical axis is logarithmic, not linear.

Figure 328
Per cent mortality in animals of different ages.
a dall mountain sheep, lapwing, sardine
b night herons (*Nycticorax nycticorax*) in the zoo and in the wild
c human population of U.S.A. (1930) and London (seventeenth century).
Oxford Entrance, Additional Biology, Oxford University Press, 1982.

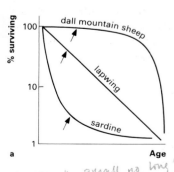
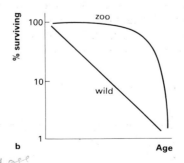
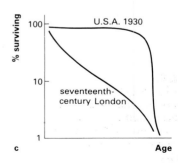

a State what each of the curves in figure 328 shows. Discuss the reasons for the differences between the curves.

b Comment on figures 328b and 328c in the light of the differences in the curves in figure 328a.

In a rapidly-growing population of a species in the log phase of population increase, there is a high proportion of young individuals and a small proportion of old ones. There is also considerable genetic variation amongst the individuals. In a declining population of the same species, old individuals predominate and genetic variation may be more limited.

28.3 Population limitation

Many species in nature have population sizes which fluctuate between narrow limits from one year to the next. Of course most species vary considerably in numbers throughout the year, since they produce all their offspring at the same time. Yet if a species is censused at the same time each year, its numbers will usually appear to be relatively constant. This is unexpected in view of the very high reproductive rates of many species.

Population sizes remain the same if the birth rate equals the death rate. The death rate depends on two main factors. Firstly, the genetic programming of the species sets an upper limit to the life span. Secondly, there are many factors which kill individuals, such as death from hypothermia, competition for food or water, the inability to find a territory, disease, and being eaten. These factors account for the deaths which occur between each stage in the life histories of most organisms.

STUDY ITEM

28.31 The population dynamics of the great tit (*Parus major*) in Marley Wood, Wytham Woods, near Oxford

More is known about the population dynamics of the great tit, *Parus major*, than any other organism. This small woodland bird is particularly suitable for detailed study (*figure 329*). Compared with many other species, it is abundant, nesting at four to six pairs per hectare in most European woodlands. Since it is a resident bird, its mortality can be studied outside the breeding season, impossible in a migratory bird. The young stay in the nest for three or four weeks after hatching, being fed by the parents, and this allows chick development and death to be studied in a manner impossible in those species whose young get up and walk away almost immediately. Most importantly, the birds normally nest in holes and prefer to nest in nest-boxes whenever they are provided. This allows the broods to be counted, manipulated, and easily observed time and time again. It was for these reasons that the ornithologist David Lack decided in 1946 to set up a long-term study on great tit population dynamics in Marley Wood, a mixed woodland not far from Oxford.

This investigation still continues. Every great tit in the wood, and many around it, is ringed with a colour code identifiable with binoculars. They all nest in nest-boxes provided by the management. The design of these boxes has evolved over the years and now limits predation of the brood by weasels. The chicks can be removed from the same nests at intervals and weighed. Mass can then be related to the risk of death. Some nest-boxes have had cameras built into the back. Whenever a

Figure 329
The great tit, *Parus major*. Its stout but pointed beak enables it to feed on a wide range of foods from strong-shelled hazel nuts to small aphids. It is the largest common tit.
Photograph, Dr M. S. Wood/Eric and David Hosking.

parent perches on the entrance hole with food in its beak for the youngsters, a flash camera is automatically triggered (*figure 330*). The photographs allow the food item to be identified, and since a timepiece is

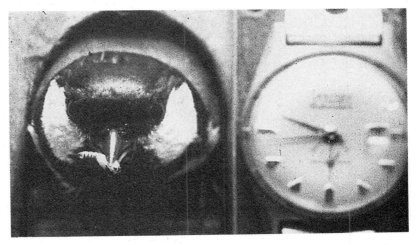

Figure 330
Three frames from a film taken with an automatically-triggered camera inserted in the back of a great tit nest-box. Great tits mainly eat insects in summer. Note how rapidly the visits were made.
By courtesy of Dr C. M. Perrins, Edward Grey Institute of Field Ornithology, Oxford.

also photographed, the frequency of the visits can be related to the food supply, the time of year, and the rates of development of individuals in the brood.

The fluctuations in the number of breeding pairs in Marley Wood over the first 38 years of the study are shown in *figure 331*. There were only seven pairs in the first year, probably because of food shortage in the very hard winter of 1947, and as many as 61 pairs in 1961. The numbers have often been between twenty and thirty pairs. What causes these fluctuations, and what causes the relative stability?

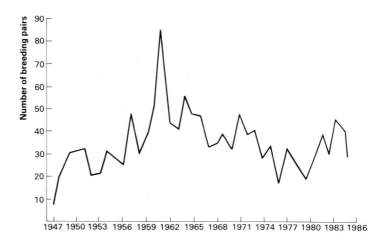

Figure 331
The numbers of pairs of great tits breeding in Marley Wood, Wytham Woods, Oxford, 1947–85.
Perrins, C. M. '*The great tit*, Parus major', Biologist **27(2)**, *73, 1980; subsequent data by courtesy of Dr Perrins.*

In general terms, the great tit is a perennial in which the probability that a breeding bird will survive to the next season is about 50 per cent (*figure 332*). Each breeding pair produces 8–10 eggs. Although there is some mortality of eggs and nestlings, most of the deaths occur after the fledglings have left the nest and it is this mortality which, as we shall see, largely determines the size of the breeding population next year.

The males establish territories by singing in January (see also *Study guide I*, Chapter 13, for a discussion of the recognition of one male by another). Territory size seems to be rather flexible in great tits. Males compete with one another for territories. In 1968, in a small wood on Wytham Hill, J. R. Krebs shot seven of the pairs in mid-March, when territories had been established. Five new pairs settled within six days. In a similar experiment the following year Krebs showed that the replacement birds had previously established territories in the inferior hedgerow habitats around the wood. Suitable nest sites are scarce in hedges and the broods only have a 20 per cent chance of survival.

The males attract females and mate. The date when the first eggs have been laid has varied from about 10 April to 10 May. This date tends to be earlier in warm than in cold years, probably because the female has to produce her own mass in eggs in twelve days. The warmer the weather, the more insect food there is to catch and the richer her diet. After about fourteen days, the eggs hatch and the adults begin to work hard to feed them. At about this time, there are plenty of caterpillars of the winter moth (*Operophtera brumata*) and the green tortrix (*Tortrix viridiana*) in the oak canopy and the parents may pay 800–1000 feeding visits per day

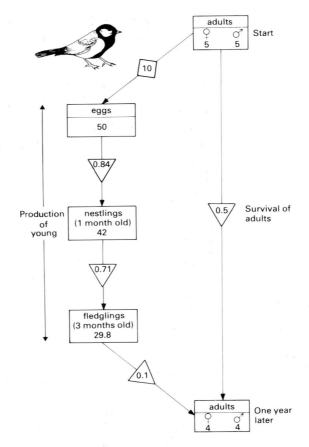

Figure 332
Diagrammatic life-table for the perennial great tit, *Parus major*. Population sizes are per hectare. The figures in triangles are the probabilities of survival.
Data from Perrins, C. M. 'Population fluctuation and clutch size in the great tit (Parus major L.)', Journal of Animal Ecology **34**, 601, 1965.

to the nest. Even so, they only take 1–2 per cent of the available caterpillars. The youngsters grow feathers and become fledglings after about two weeks. The parents feed them for another two weeks, during which they learn to fly and ultimately leave the nest for an independent existence.

During this phase in the life history, several factors have been shown to affect the numbers of birds which a nest ultimately produces.

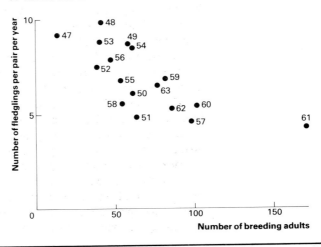

Figure 333
Number of young great tits raised per pair per year in Marley Wood in relation to the density of breeding pairs. Years are indicated by numbers against the points, *e.g.* 57 = 1957.
Lack, D. Population studies of birds, Oxford University Press, 1966.

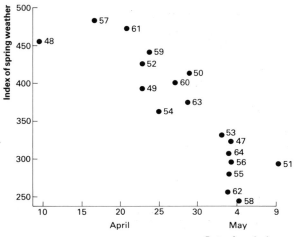

Figure 334
Average date of breeding of great tit in relation to spring weather. The index of spring weather was obtained by adding the average for the maximum and minimum temperatures each day from 1 March to 20 April. The date of laying is the average of the dates on which each female laid its first egg. Years are indicated by numbers against the points, e.g. 57 = 1957.
Perrins, C. M. 'Population fluctuation and clutch size in the great tit (Parus major L.)', Journal of Animal Ecology **34**, 601, 1965.

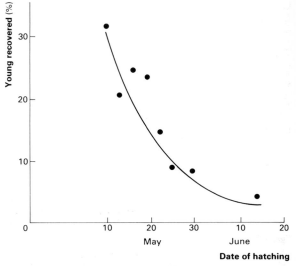

Figure 335
Post-fledging survival of young great tits in relation to their hatching date. Survival is measured by the percentage of the young hatched at a particular date which were subsequently recovered alive.
Perrins, C. M. 'The great tit, Parus major', Biologist **27(2)**, 73, 1980.

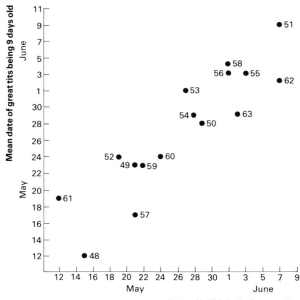

Figure 336
Relationship between date of great tits in Marley Wood being nine days old, and date by which half the winter moths (*Operophthera brumata*) pupated under oaks in Wytham Wood.
Lack, D. Population studies of birds, *Oxford University Press*, 1966.

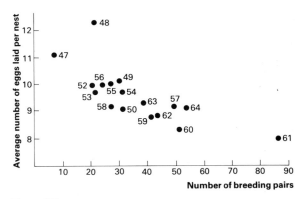

Figure 337
The relationship between the average number of eggs laid per nest, and the number of breeding pairs each year, for the great tit in Marley Wood. Years are indicated by numbers against the points, e.g. 57 = 1957.
Lack, D. Population studies of birds, *Oxford University Press*, 1966.

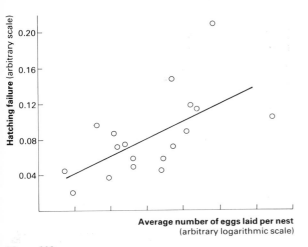

Figure 338
The relationship between the numbers of eggs per nest which did not hatch out, and the average number of eggs laid per nest in any one year, in the great tit. Each point represents data from one year.
From Krebs, J. R., 'Territory and breeding density in the great tit, Parus major L.', Ecology **52**, 2, 1971.

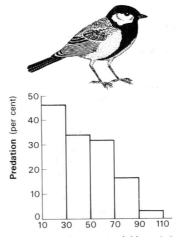

Figure 339
The influence of territory size on the risk of nest predation by weasels in the great tit.
After Krebs, J. R. 'Territory and breeding density in the great tit, Parus major L.', Ecology, **52**, 2, 1971.

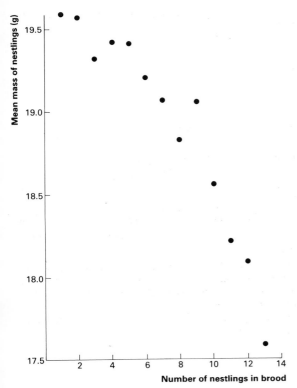

Figure 340
Mean mass of nestling great tits in relation to brood size.
From Perrins, C. M. 'Population fluctuation and clutch size in the great tit (Parus major L.)', Journal of Animal Ecology **34**, 601, 1965.

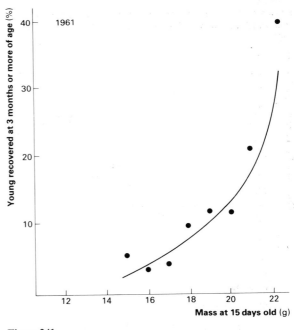

Figure 341
Post-fledging survival of young great tits in relation to their nestling masses on the 15th day after hatching. Survival was measured by the percentage of the young, of any given mass, which were subsequently recovered alive.
Perrins, C. M. 'The great tit, Parus major', Biologist **27(2)**, 73, 1980.

Chapter 28 Population dynamics 493

a *Study the nine graphs shown in figures 333–341.*
  1 *Summarize what each graph shows.*
  2 *Discuss the reasons for the shape of each graph.*
  3 *Define the conditions in which a nest might produce most offspring and those in which it might produce fewest offspring.*

Whilst weasels can only attack nestlings, sparrowhawks (*Accipiter nisus*) only take flying birds to feed their young. One pair brought eleven great tits to the nest in a single day. The rings from the legs of the great tits accumulate in their nests. Sparrowhawks probably eat as many as thirty-five per cent of the young birds.

Quantitative studies have made it clear that in Marley Wood the extent of death of the fledglings during the summer months largely determines the numbers of great tits available to breed the following year.

The birds which are heaviest when they leave the nest, and those which grow feathers earliest, have the most chance of being caught again in autumn. This suggests that light birds and those which fledge late stand more chance of dying in summer. Perhaps they die from starvation, or they stand more chance of being predated. Starvation and predation are also possible causes of winter disappearance.

b *Explain why great tits require more energy in summer than in winter.*

c *List the various factors which can 'regulate' the population size of great tits in Marley Wood, that is, factors which tend to bring the population size down when it is high, but which cause little mortality when the population is small.*

The deaths which occur between one stage in the life history and the next can in many cases be attributed to a single environmental factor. Two types of factor operate to kill the birds. One set of factors differs considerably in intensity from year to year. Some winters will be warm for instance, and there will be few deaths from hypothermia; other winters are cold for a long time, and there are few survivors. If this was the main cause of great tit deaths in winter, the numbers would be high after a warm winter and low after a cold winter. The population size would exhibit large fluctuations from year to year. Such factors are called density-independent, and seem to have a major influence on the population sizes of many insects.

d *Why are butterflies, aphids, and locusts common in some years and rare in others?*

On the other hand, density-dependent factors kill a higher proportion of the individuals when the population density is high than when it is low. For example, in the years before weasel-proof nest-boxes were introduced, weasels destroyed 21 per cent of great tit nests within 40 m of their neighbouring nest, but only 11 per cent of isolated nests (*figure 339*). When great tits were abundant, many of their nests were destroyed. When the tits were sparse, weasels did not greatly reduce their numbers. Density-dependent factors, such as weasel predation of great tits, prevent the population size from becoming too high or too low, and therefore reduce the amplitude of the fluctuations in numbers.

> **Practical investigation.** *Practical guide 7*, **investigation 28D**, '**Population regulation in the water flea, *Daphnia*'.**

STUDY ITEM
28.32 Density-dependent, density-independent, and inverse density-dependent factors

The graphs in *figure 342* show various hypothetical relationships between the proportion of the population dying and population size for species in nature. The first four parts of the question refer to unitary organisms, and death is considered to be a frequent consequence of competition for scarce resources. In modular organisms the consequence of competition for scarce resources may be a reduction in the rate of

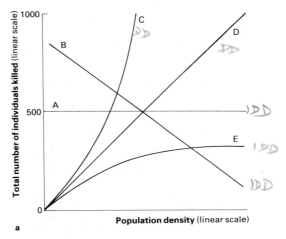

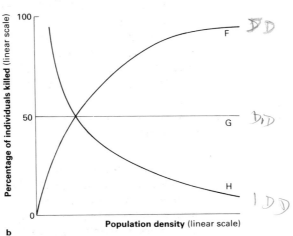

Figure 342
Hypothetical curves showing the effects of stress of different kinds on populations of organisms of different sizes.
a total number of individuals killed (linear scale) against population size (linear scale)
b per cent of individuals in population killed (linear scale) against population size.
Considerably modified from Solomon, M. E. Studies in Biology 18. Population dynamics, 2nd edition. Edward Arnold, 1976.

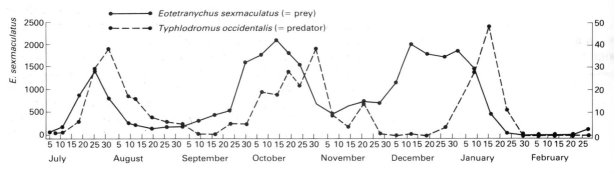

Figure 343
Population fluctuations between predator and prey in a laboratory experiment. The six-spotted mite *Eotetranychus sexmaculatus* eats oranges. *Typhlodromus occidentalis* eats it.
Huffaker, C. B. Experimental studies on predation: dispersion factors and predator-prey oscillations. Hilgardia, **27**, 343, 1958.

module or seed production, which is equivalent to mortality in the long run.

a State for each of the lines A–H whether it represents a density-dependent, density-independent or an inverse density-dependent relationship. (Inversely density-dependent factors have the opposite effect to density-dependent factors – they kill a larger proportion of the population when the population size is small than when it is large.)

b Explain why the shooting of a wood-pigeon population is likely to have only a short-term impact on the population size.

c Discuss for each of the following factors whether it is likely to be density-dependent, density-independent, or inversely density-dependent.

1 a large population of seedlings being grazed by a single slug;
2 death of birds from hypothermia in winter;
3 death of birds from hypothermia in winter if there is only a limited number of warm places where they can shelter;
4 death of birds from hypothermia in winter if their low body temperature depends on how much food they have collected in competition with other birds in the population;
5 the mortality of locust eggs in dry sand;
6 a pest population being attacked by an insect parasitoid capable of locating high concentrations of its prey.

d The graph in figure 343 illustrates cyclical fluctuations in population size of a predator and its prey, of a type which can often be achieved in laboratory cultures. Explain these cycles of abundance in terms of the density-dependent factors acting on the populations.

e Discuss the significance of density-dependent factors in the regulation of the population sizes of animal and plant species in nature.

f Imagine a population of plants, such as that of creeping buttercup (**Ranunculus repens**) in a meadow. The plants grow by producing stolons, overground runners which produce new stolons at their tips. Discuss the various factors which might keep its population size relatively constant from year to year.

Studies of the population regulation of pest species are particularly worthwhile. The results may suggest ways in which humans can intervene to lower the population size of the pest.

It seems that, in nature, many parasites and predators keep their host or prey populations down to 10^{-4} or 10^{-5} of the numbers which might exist if the parasite or predator were absent. When a pest expands in numbers it is often because its usual predator or parasite is absent. In 'biological control', the predator or the parasite of the organism is added. This often brings about a considerable reduction in population size (table 68).

1863	Transfer of caterpillars of South American *Dactylopius ceylonicus* to India to control prickly pear cactus, *Opuntia vulgaris*.
1889	Importation to Californian citrus groves from Australia of the predatory vedalia beetle *Rodiola cardinalis*, to control cottony cushion scale insect *Icerya purchasi*, a native of Australia.
1925	Introduction from Argentina to Australia of caterpillars of the moth *Cactoblastis cactorum* to control prickly pear cactus (*Opuntia* spp.), native of South America, which infested 24 million ha of Australian pasture, making 12 million ha useless.
1940	Transfer to California from Europe of the plant beetle *Chrysolina quadrigemina* to control the Klamath weed, *Hypericum perforatum*, introduced from Britain, in cereal crops.
1960s	Control in the southern United States of the cattle parasite, the screw-worm fly, *Callitroga* spp., by the large-scale release of sterilized irradiated male flies.
1970s	Use of a spray preparation containing the bacterium *Bacillus thuringiensis* to limit the spread of lepidopteran larvae on forest trees in the United States.

Table 68
Some famous instances of successful 'biological control' of pests.

Practical investigation. *Practical guide* 7, investigation 28E, 'The responses of a predator to changes in the number of its prey'.

STUDY ITEM

28.33 Pest control in glasshouses: biological control or pesticides?

The growth in glasshouses of cucumbers and tomatoes is a multi-million pound industry, worth about £82 million a year at 1980 prices (see Chapter 25). The warm moist conditions are, however, ideal for the growth and reproduction of insects, mites, fungi, nematodes, and other organisms which might reduce the quality or the productivity of the crop. Pesticides and biological control have both been used effectively to control these pests.

In the winter the heating is turned off and the soil is sterilized with methyl bromide, which kills nematodes and other pathogens. In the spring, the heating is switched on and young cucumbers and tomatoes are spaced out in the house. Occasional whitefly (*Trialeuroides vaporariorum*) enter with these plants, whilst two-spotted spider mites (*Tetranychus urticae*), awakened from hibernation, come down from the upper structure of the glasshouse.

The whiteflies can do considerable damage if not controlled. They increase rapidly because at greenhouse temperatures their life cycle

Figure 344
The life cycle of the greenhouse whitefly (*Trialuroides vaporariarum*) and its parasitoid *Encarsia formosa*. The female parasitoid lays one egg inside the 3rd instar larva. The whitefly then continues to develop, but at the pupal stage it turns black when parasitized, instead of remaining greenish–white. The adult *Encarsia* makes a round hole in the pupal case, and emerges.
Samways, M. J. *Studies in Biology 132. Biological control of pests and weeds*, Edward Arnold, 1981.

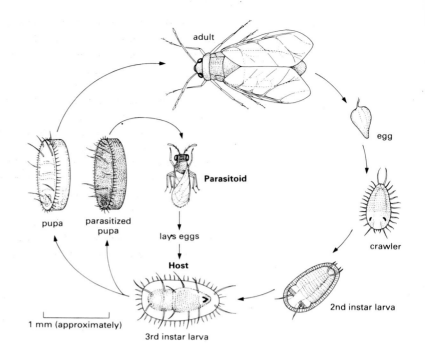

(*figure 344*) only takes ten or twelve days. They suck the phloem juices of the plants, may transmit viruses, and produce honeydew, their sticky egesta, which acts as a focus for infection by 'sooty mould' fungi.

The whitefly is parasitized by a tiny wasp, *Encarsia formosa*. This minute organism, 2–3 mm long, was introduced from South America and was first noticed in Britain by a market gardener at Cheshunt in 1926. *Encarsia* lays its eggs inside the third instar larvae or 'scales' of the whitefly, which are abundant on the lower surfaces of the leaves. Each female wasp can lay about fifty eggs, one per scale. Each egg hatches into a larva which eats the whitefly larva and then pupates. A single adult wasp emerges from the black pupal case by making a hole in the surface. *Encarsia* produces several generations of females during the summer, and breeds rapidly enough between 24 and 27 °C to exert control over the whitefly population.

In order to establish control of the whitefly the parasitic wasp must be present in sufficient numbers at the beginning of the growing season, in late April or May. The problem is that *Encarsia* released into the glasshouse cannot breed in the absence of the whitefly. Some growers release the wasps, bred for the purpose, at ten-day intervals in early May. Others may actually introduce whitefly scales in mid April, and then add a large number of parasitized whitefly scales two weeks later. This produces large numbers of *Encarsia*, which quickly cope with any subsequent whitefly invasion from outside. Control can be established most easily in sunny conditions. The grower knows that control is progressing well when about half the whitefly scales under the leaves have turned black.

The two-spotted spider mites damage cucumber leaves by sucking out the contents of the plant cells. Plants with damaged leaves grow poorly. The predatory mite *Phytoseiulus persimilis* is introduced by growers into glasshouses to control it. It encounters patches of the red spider mite and consumes eggs, young stages, and adults voraciously. When it has eliminated two-spotted spider mites in one area, it goes in search of others. The interaction between these two species is unstable. *Phytoseiulus* is such an efficient predator that it may eliminate its prey and become extinct itself. Reintroduction may be necessary.

Of course insecticidal sprays can be used to devastate any economically damaging infestation by whitefly or spider mite. In the 1950s and 1960s, for example, DDT was widely available and fairly effective. The pests have now become resistant to most of the commonly-used insecticides, and the biological control of the pests once again become popular.

a **What are the difficulties of integrating biological control with pesticide application in the same control programme?**

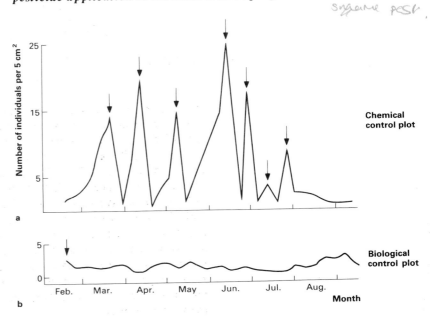

Figure 345
Control of the two-spotted spider mite *Tetranychus telarius* on glasshouse cucumbers in Finland by chemical control using dicofol (arrowed) and biological control employing the predatory mite *Phytoseiulus persimilis*.
After Markkula, M., Titanen, K., and Nieminen, M. Annales Agriculturae Fenniae **11**, 74, 1972.

b **The graphs in figure 345 show an attempt to control a spider mite under experimental conditions with a insecticide and b a predatory mite. Comment on the graphs.**

c **Compare the advantages and disadvantages of the biological control of pests (adding a parasite or a predator to control the pest) with the use of pesticides to control the pest, under the following headings:**
  1 speed of action
  2 specificity to pest
  3 toxic effects
  4 cost of programme
  5 host resistance to pesticide, parasite or predator
  6 effect on beneficial insects
  7 ability to kill a range of potential pests

d *Imagine that there was a widespread outbreak of the sugar-cane borer (Diatraea saccharalis, Lepidoptera), on sugar cane (Saccharum spp.). Draw a flow diagram to represent the measures which you might take to assess the scale of the problem and to control the pest.*

28.4 Population cycles

Many small mammals, and several insect species, have population sizes which fluctuate rhythmically with a cycle of four years or more. Consider the species of periodical cicadas (*Magicicada* spp.) in the northern United States. They become abundant every thirteenth or seventeenth year, depending on the species. In this case a genetically-determined 'biological clock' controls the rate of nymphal development of cicadas on underground tree roots, and the adults all hatch simultaneously. Presumably, other periodicities appeared in the past, and may have been selected against by predators which evolved similar life-cycles.

The most intensively studied cycles are those of lemmings, but the reasons for these regular cycles are still obscure.

Cyclical fluctuations in lemming populations (Lemmus spp.)
Several rodents, such as the meadow mouse (*Microtus agrestis*) in British woodlands, exhibit regular and marked cycles in numbers, with peaks three and a half years apart. These cycles have attracted much attention. The regular four-yearly peaks in lemming numbers in the arctic and sub-arctic have been investigated since Charles Elton first drew attention to them in 1924. Lemmings are not shot for fur, and so their cycles are not

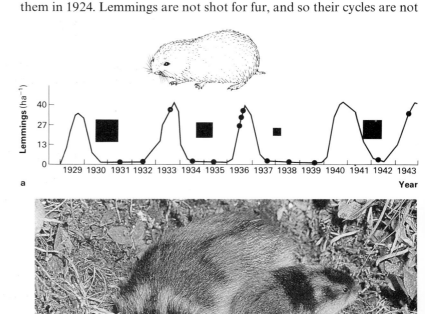

Figure 346
a Population density of the lemming *Dicrostonyx groenlandicus* at Churchill, Manitoba. The squares show the invasions of the snowy owl (*Nyctea scandica*).
b A lemming with young. *The diagram in **a** is from Shelford, V. E., 'The relation of snowy owl migration to the abundance of the collared lemming', Auk,* **62***, 592, 1945.*
*The photograph in **b** is by Eric Hosking.*

so well-documented as those for bears, lynx and arctic fox. Some of the scarce data appear in *figure 346*.

The amplitude of the cycles is striking. In a recent study of the brown lemming (*Lemmus trimucronatus*) in Barrow District, Alaska, for example, the population density at the peak over sixteen years was 612 times the lowest population density at the crashes.

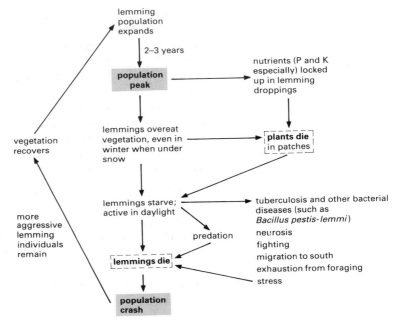

Figure 347
Flow diagram showing the main factors which have been implicated in the marked four-year cycles of abundance in lemmings (*Lemmus* spp.) in Arctic tundra.

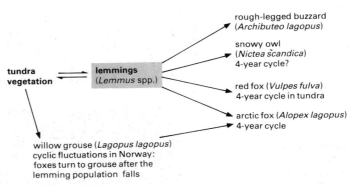

Figure 348
Some feeding relationships of lemmings in tundra.

The main factors which have been implicated in the cyclical fluctuation in lemming population size are shown in *figure 347*. Lemmings eat tundra vegetation of grass, mosses and herbs, and are in turn eaten by several predators, such as the arctic fox (*figure 348*). The predators, however, do not seem abundant enough to cause the population crashes of lemmings. They themselves exhibit four-year cycles of abundance. These may be fuelled by the high lemming population densities every four years. During these population peaks, starved and exhausted lemmings, foraging by day as well as by night, are easily seen by predators amongst the sparse vegetation.

a *How could you investigate the hypothesis that lemming migrations are necessary to regulate the lemming population?*

Sometimes the population size of a predator varies cyclically with that of its prey. The Hudson Bay Company has shot animals for fur for well over a hundred years. Analysis of their records shows a ten-year cycle in abundance in both the snowshoe hare and its predator, the lynx (*figure 349*).

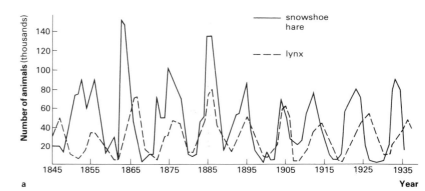

Figure 349
a A graph showing fluctuations in the numbers of snowshoe hare (*Lepus americanus*) and the lynx (*Lynx canadensis*) in the Canadian Arctic. The numbers show the pelts received by the Hudson Bay Company.
b A snowshoe hare (*Lepus americanus*) in winter coat.
c A Canada lynx (*Lynx canadensis*).
The diagram in a is after Odum, E. P. Fundamentals of ecology, 3rd edition, Saunders, Philadelphia, 1972.
Photograph **b** *Irene Vandermolen/Frank Lane Picture Agency;* **c** *Leonard Lee Rue/Frank Lane Picture Agency.*

b *How might hunting have affected these fluctuations?*

It seems likely that the periodicity in lynx numbers is caused by the variation in abundance of its prey, but it is by no means certain that the snowshoe hare undergoes its population cycles because of cyclical predation by lynx. Remove the lynx, and the regular fluctuations in hare numbers would probably continue.

28.5 Human populations

The population size of *Homo sapiens* is known much more precisely than that of any other widely-distributed large animal. It is about 4670 million (1983). It increases at about 1.7 per cent a year, 213 000 a day, 8900 an hour or 150 a minute. This is the excess of births over deaths. Every minute about 253 births and 96 deaths occur. The 150 children to be born whilst you have been reading this paragraph will have to be fed and educated. Potentially, they will form 75 family units which, at present world birth rates, will produce 292 children between them.

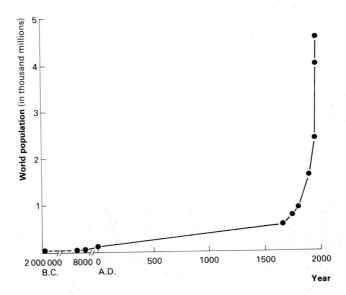

Figure 350
The increase in the human population of the World.
Estimates to 1960 are based on Deevey, E. S. (1960) 'The human population'.
Copyright © 1960 by Scientific American Inc. *All rights reserved. Population sizes and projections since 1980 based on 1982* World Population Data Sheet *of the Population Reference Bureau, Inc., Washington D.C.*

This geometric increase in the human population (*figure 350*) resembles that of many organisms when they reach a new habitat to which they are suited (section 28.2). This crude, overall picture of human population increase, however, obscures the fact that in many parts of the world the population is stable or even declining.

STUDY ITEM

28.51 **The current World distribution of population increase**

Table 69 shows some of the population characteristics of selected countries in 1981.

	Population 1981	2020 (est.) (millions)	Births /1000	Deaths /1000	Infant deaths /1000	Total births/ woman (est.)	% aged 15 years	% aged 65 years	Income /head (£/year)
W. Germany	62	49	10	12	13	1.5	20	15	13590
Sweden	8.3	7.4	12	11	7	1.7	20	16	13520
U.S.A.	232	274	16	9	12	1.9	23	11	11360
U.K.	56	56.5	14	12	12	1.9	22	15	7920
U.S.S.R.	270	346	18	10	36	2.3	24	10	4550
Mexico	72	140	32	6	56	4.8	42	3	2130
Ecuador	8.5	23	42	10	82	6.3	45	4	1220
Kenya	18	59	53	14	87	8.1	50	4	420
China	1000	1400	22	7	45	2.8	32	6	290
Gambia	0.6	1.8	49	21	198	6.4	42	2	250
India	714	1197	35	15	123	5.3	40	3	240
Burma	37	76	39	14	101	5.5	40	4	180

Table 69
Population characteristics of selected countries in 1981.
(*Data from 1982 World Population Data Sheet of the Population Reference Bureau, Inc., Washington D.C.*)

a *Discuss the influence of geographical position and economic development on the population characteristics of the countries listed in table 69.*

How has this situation arisen? It is most easily understood by stating that most countries exhibit a characteristic series of population changes, together known as the 'demographic transition':
1 At first, family size is high and fertility is high, but there is a high death rate and high infant mortality. Because of the prevalence of infectious disease, the expectation of life is low.
2 In the second stage, the death rate falls. Improvements in medical care and nutrition *i* reduce infant mortality, *ii* increase the expectancy of life at birth, and *iii* increase the number of individuals in each family who live to reproductive age. Although birth rate per couple stays the same, there are now more couples in each successive generation and the population increases rapidly.
3 The adoption of methods of birth control lowers the birth rate. Couples have fewer children, on average. Eventually the birth rate declines to match the death rate. The population size levels off.

b *Suggest why couples in economically-developed countries decide to have fewer children.*

The population size in the United Kingdom has been through all these stages. On the other hand, most of the less developed countries (LDCs) have only experienced the first part of the cycle. The life expectancy of those living in the LDCs has increased dramatically under the impact of improved sanitation, medical care and public health measures. In Sri Lanka, for example, campaigns to eradicate the malaria-carrying anophelene mosquitoes with DDT halved the death rate in ten years.

About 40 per cent of the world's population ekes out a subsistence living, poor and undernourished, debilitated by disease, and an easy prey to famine. The governments of mainly pastoral countries often have little money to invest in industrialization.

During famines, and in large concentrations of the poor such as those in the Calcutta slums, the limited food supply may reduce the rate of population increase. There is, however, not much evidence for this. On a world scale, the annual increase in food production (2 per cent) keeps pace with the increase in population size. Most of the increased food output, however, occurs in the United States, Canada, and northern Europe. In the LDCs, food production only increases by 0.4 per cent a year, whilst their populations increase at 1.6 per cent a year. Food shortages are therefore the consequence of inefficient distribution between areas of production and areas of consumption.

As the imbalance between food and population in the LDCs becomes more marked, some governments, often helped by foreign aid, have adopted birth control programmes in the form of contraceptives or advice in an attempt to reduce the birth rate.

STUDY ITEM

28.52 The population–food equation: the Indonesian experience

Indonesia is a developing country with a rapidly-growing population of 151 million, over 75 million of them squeezed onto Java. In 1940 its population was 70 million and by 2020 it is predicted to reach 236 million, at present rates of increase. It has a low income per head, and a slow rate of economic growth. Family sizes are quite high, and 42 per cent of the population is under 15 years old. At the height of the population boom it was calculated that the capital city, Jakarta, would need to build a school a day to allow the children born to be educated.

In 1968–70 Indonesia embarked on a vigorous and large-scale family planning programme, which has been effective particularly on Java and Bali. In 1965 there were 46 children born per 1000 females in the population; in 1981 the birth rate was down to 34 per 1000. The population increased at an average rate of 2.3 per cent a year between the mid 1960s and mid 1980s, and food production increased considerably during this period.

The following passage is slightly modified from an article by Gordon S. Conway and David S. McCauley in *Nature*, **302**, 288, 1983. Read it thoroughly and answer the questions beneath. (Certain words in italics relate to question **a**.)

1 Rice production, currently standing at over 22 million tonnes a year, has risen by an average annual rate of about 4.5 per cent over the last decade – well in excess of the 2.3 per cent population growth rate. Inevitably, however, the speed and scale of this development has brought with it a variety of problems – agronomic, ecological, social, and economic.

2 About a quarter of the increased rice production has come from extending rice cultivation to *marginal land*, in particular the tidal

swamps of Sumatra and Kalimantan. Most, though, has come from the doubling and trebling yields on existing paddy land by the introduction of new short-stalked fertilizer-responsive rice varieties (developed at the International Rice Research Institute (IRRI) in the Phillipines), the rehabilitation of irrigation systems, and the provision of *subsidized inputs* and credit to farmers.

3 Severe setbacks were experienced in the mid 1970s when recurrent outbreaks of the brown plant-hopper (*Niliparvata lugens*) devastated crops. Losses in 1977 were over 2 million tonnes – enough to have fed six million people for a year. The outbreaks were brought under control through the introduction of a new resistant variety, IR 36. Its use was vigorously encouraged and the growing of traditional varieties in the lowlands was prohibited. By 1980, IR 36 had come to occupy well over one-half of Java's paddy area. It matures in around 105 days and is *non-photosensitive*, so that double cropping became feasible even in most rainfed conditions. With year-round irrigation, either continuous rice cropping or a double rice crop followed by sweet potatoes, groundnut, or soybeans was possible. Crop production began to rise dramatically.

4 The brown plant-hoppers were encouraged by the widespread planting of susceptible varieties, continuous and overlapping cultivation, increased application of nitrogenous fertilizers and the destruction of the insect's natural enemies by non-selective insecticides. Massive increases in the use of pesticides has also resulted in many other serious side effects, including, it seems, reductions in the traditional ricefield harvest of fish – normally a major source of animal protein for the rural population.

5 Control of brown plant-hopper requires an *integrated* approach involving an appropriate mixture of resistant cultivars, selective insecticides and cultural measures. Breaks in the planting schedule followed by synchronous planting can help reduce the size and severity of the outbreaks.

6 In the meantime, the perils of putting all one's eggs in one basket have been demonstrated – IR 36 has proved susceptible to tungro virus (carried by the green leafhopper, *Nephotettix cinticeps*). In 1981, serious losses were reported in Bali, Central Java, and East Java. In response, the latest IRRI varieties (IR 50, 52 and 54) were introduced, and following a crash programme of seed increase, several hundred tonnes were distributed for the 1981–2 crop season.

7 The recent pest and disease problems help to emphasize that agricultural development is not solely a question of increased *productivity*. The question now is: how sustainable is this recent growth in agricultural productivity and relative rural prosperity?

a *Explain the italicized terms 'marginal land', 'subsidized inputs', 'non-photosensitive', 'integrated', 'productivity'.*

b *Discuss the possible advantages of short-stalked (paragraph 2) varieties of rice over long-stalked varieties.*

c *Suggest three advantages of growing groundnuts or soybeans (paragraph 3)* in the annual cycle instead of merely another rice crop.

d   Explain how continuous and overlapping cultivation and increased application of nitrogenous fertilizers *(paragraph 4)* can encourage the brown plant-hopper.

e   Suggest three ways in which the application of pesticides might have reduced the harvest of fish *(paragraph 4)*.

f   Discuss methods of diversifying Indonesian agriculture so that the recent increases in crop production can be maintained. Bear in mind that Indonesia's recent oil and gas boom seems to be over.

Reduction in the birth rate is not considered a priority in all countries. France in the past, and Brazil at present, have encouraged population increase. In some countries, artificial methods of birth control are discouraged or illegal. The governments of many nations, however, worried by the rate of population increase, introduced contraceptive measures.

The human population of the globe will eventually stop increasing. It must, for the resources of the planet are finite. Already, the expanding population and the increased food production are altering parts of the planet beyond recognition. Large-scale soil erosion, desertification, pollution of water by fertilizers and sewage, deforestation, and overfishing are changing the relative abundances of other plant and animal species, and the chemical composition of the atmosphere and oceans.

The impacts of these changes are easier to understand and predict by thinking of the biosphere as a series of related compartments. Each compartment has a different range of species, and its own pattern of energy flow and nutrient cycling. Populations do not exist in isolation, but they are inextricably linked with populations of other species and the environment in *ecosystems*, the subject of the next chapter.

Summary

1   Population dynamics is the study of the variation in numbers of individuals, or parts of individuals, with time. **(28.1)**.

2   There is a fundamental difference between unitary organisms and modular ones. In unitary organisms, such as most animals, individuals are discrete and can be counted. In modular organisms, which include higher plants, hydroids, bryozoans, and most colonial ascidians, it is more difficult to assess 'population' size. **(28.1)**.

3   The abundance of plants in a population is usually assessed in small sample areas by estimating the number of individuals, percentage cover, or percentage frequency. **(28.1)**.

4   Animal population sizes can be estimated by counting them in herds and quadrats, beating, sweep-netting, pitfall-trapping, or catching animals in traps. For small mobile animals the capture–mark–recapture method is valuable. **(28.1)**.

5   Fluctuations in population size from year to year are the result of increases through birth and immigration, and decreases through death and emigration. (**28.2**).
6   Populations invading new favourable habitats often exhibit an S-shaped ('logistic') increase in numbers with time. A constant population size can be the result of many different types of stable age distributions and mortality patterns. (**28.2**).
7   The numbers of great-tits (*Parus major*) in British woodlands are influenced by competition for territories, competition for food, predation, and winter death. Factors such as these regulate the population sizes of many species. (**28.3**).
8   Some of the factors which act directly or indirectly to kill individuals in a population are density-dependent, because the intensity with which they act increases with the population size. Other factors are density-independent. Density-dependent factors are the most important in 'regulating' the population sizes of species in nature. (**28.3**).
9   Predators and insect parasitoids can be used to reduce the population sizes of specific pest species. This is known as 'biological control'. (**28.3**).
10 Some animal species, such as lemmings, the snowshoe hare, and the lynx, show regular cycles of abundance and rarity. Lemmings reach population peaks every three or four years. (**28.4**).
11 The human population of the Earth is increasing rapidly. The reasons for the present rapid increase, and its consequences, are discussed. (**28.5**).

Suggestions for further reading

BEGON, M., and MORTIMER, M. *Population ecology: a unified study of animals and plants.* Blackwell Scientific Publications, 1981. (A clear, modern account of some of the research in this rapidly-changing field.)

BONNER, J. Carolina Biology Reader 122. *The world's people and the world's food supply.* Carolina Biological Supply Company, 1980. (Are increases in food production keeping pace with increases in population? This brief discussion provides excellent perspective and also deals with the prospects.)

SAMWAYS, M. J. Studies in Biology 132. *Biological control of pests and weeds.* Edward Arnold, 1981. (A clear, lively, and well-illustrated introduction.)

SOLOMON, M. E. Studies in Biology 18. *Population dynamics.* 2nd edition. Edward Arnold, 1976. (A brief discussion of the main principles.)

VAN EMDEN, H. F. Studies in Biology 50. *Pest control and its ecology.* Edward Arnold, 1974. (Concentrates on insect pests, but also deals with the killing of other organisms.)

CHAPTER 29 COMMUNITIES AND ECOSYSTEMS

29.1 Introduction

Populations of animals and plants live in communities. The species in a community constantly interact with one another and their environments. These interactions are complicated. Besides feeding, parasitizing, or fighting, many organisms remove from their surroundings light, water, or nutrients which others might use (see Chapter 27). The environment constantly fluctuates.

In order to simplify the myriads of interactions into a form which is more comprehensible, ecologists group the organisms into feeding levels, known as trophic levels. Green plants, for example, occupy the first trophic level because they manufacture the high energy organic compounds on which all the other organisms in the community depend. The herbivores constitute the next trophic level, and so on (see the definitions that follow). Ecosystem ecologists study the rates at which each trophic level obtains food, energy, and nutrients from other species and the environment. By concentrating on the transfer of energy and the movements of nutrient ions, models can be built which summarize the interactions between the trophic levels in a particular place.

The main definitions used in ecosystem ecology are given on pages 543–5.

An ecologist may select an area of any size from a puddle to the biosphere and study it as an ecosystem. The first step is to reduce the number of feeding interactions in the data by assigning each species to a particular trophic level. Autotrophs, such as green plants and autotrophic bacteria, are called 'primary producers'. They take up inorganic nutrients and energy and turn them into organic compounds. Other organisms, the heterotrophs, live by eating them. The

Figure 351
A general model of ecosystem structure. The three subsystems (plant, grazing, and decomposer) are shown together with their main components. The major pathways for the transfer of matter and energy within the ecosystem are shown by arrows. Organic matter pools are shown as rectangles, inorganic as 'clouds'.
Modified from Swift, M. J., Heal, O. W., and Anderson, J. M. Decomposition in terrestrial ecosystems, Blackwell Scientific Publications, 1979.

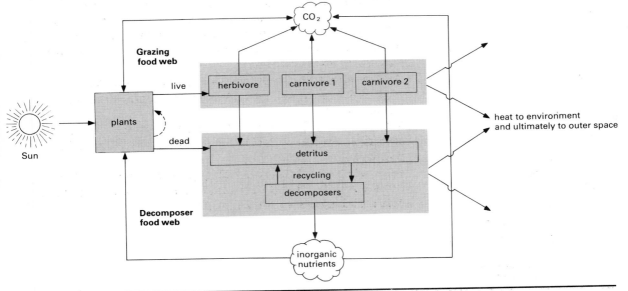

heterotrophs are of different kinds. Some, the herbivores or primary consumers, feed on the primary producers. Secondary consumers feed on primary consumers, and tertiary consumers on secondary consumers. Dead organisms and egesta are acted on by decomposers, which may be animals, fungi, or bacteria. In their metabolic activities, these organisms lose, ultimately to outer space, all the energy originally trapped by the autotrophs in photosynthesis (*figure 351*).

STUDY ITEM
29.11 Trophic levels and food webs

A convenient means of expressing the feeding relationships of the organisms in a community is a food web. A food web is a diagram, such as that in *figure 352*, which shows what eats what in a community. It is convenient to express on the food web only the most important feeding interactions, those which contribute most to energy flow through the

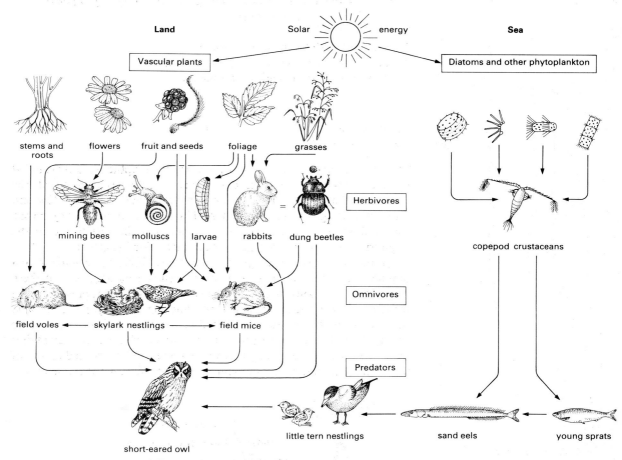

Figure 352
A simplified food web for the short-eared owl (*Aseo flammeus*) inhabiting coastal sand dunes near Blakeney, Norfolk.
From Darlington, A. and Leadley-Brown, A. *One approach to ecology*, Longman, 1975.

community. Apart from the difficulties in quantifying energy flow between individual species in nature, there is the more practical problem of discovering just what an organism does eat.

a   *How could the diet of the following organisms be determined?*
    1 *fish*
    2 *predatory tiger beetle living on the forest floor*
    3 *caterpillar grazing*
    4 *zebra grazing on the Serengeti plains (do not kill the zebra!)*
    5 *vulture*
    6 *earthworm*

The construction of a food web for a community is essential if the community is to be analysed as an ecosystem. This is because studies of energy flow and nutrient cycling in ecosystems require each organism to be assigned to at least one trophic level.

b   *Discuss the trophic levels occupied by the following organisms:*
    1 *bug which sucks phloem sap*
    2 *earthworm*
    3 *blackbird which eats an earthworm*
    4 *fungus associated with an alga in a lichen*
    5 *'carnivorous' plant such as the sundew.*

STUDY ITEM

29.12   Ecological pyramids

The study of an ecosystem can begin when the trophic level(s) occupied by each species have been determined. Ecologists can count the organisms, find their mass, and measure their roles in energy flow and nutrient cycling. Overall estimates of numbers, biomass (dry mass), energy flow, or nutrient cycling are made and the results for all the organisms at the same trophic level are then added together.

In the early days of ecosystem analysis, ecological pyramids were a popular way of expressing the results of such investigations. In these diagrams, the various trophic levels of a community are represented as a series of steps of the same height but differing width (*figure 353*). The width or area of each step is proportional to the numbers, biomass, or energy flow of all the organisms at a particular trophic level. Parasites and decomposers are usually represented by extra bars placed on top of the primary producers, at the same level as the primary consumers.

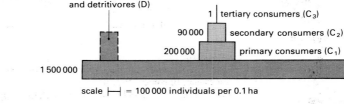

Figure 353
A pyramid of numbers for a derelict grassland in Michigan.
From Odum, E. P. *Fundamentals of Ecology*, 3rd edition, *Saunders, Philadelphia, 1971.*

Figure 354
Two pyramids of biomass in g m^{-2}.
a oak forest at Wytham Wood, England
b English Channel.
Forest from Varley, G. C., The concept of energy flow applied to a woodland community; in Watson, A. (ed.) Animal populations in relation to their food resources, Symp. Brit. Ecol. Soc., **10**, *389, 1970. Sea from Harvey, H. W. J. Mar. Biol. Soc. U.K. n.s.,* **29**, *97, 1950.*

a Why are pyramids of numbers not a valid basis for comparing ecosystems?

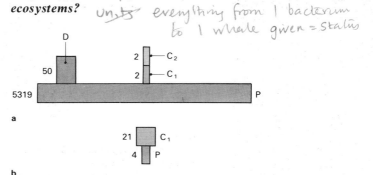

b How might the pyramids of numbers and biomass in figures 353 and 354 differ at different times of year?

c The mass of primary consumers in the English Channel is more than five times that of the primary producers which support them. How can the organisms survive for some time at these biomasses?

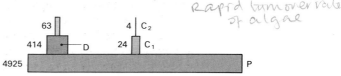

Figure 355
Pyramid of energy flow for arctic tundra on Devon Island, Canada. The figures are in $kJ\,m^{-2}\,year^{-1}$.
From data in Bliss, L. C. 'Devon Island, Canada', in Rosswall, T., Heal, O. W. (editors) Structure and function of tundra ecosystems, Ecol. Bull. (Stockholm) **20**, *17, 1975.*

d Why is the pyramid of energy flow for an ecosystem always pyramidal in shape (figure 355)?

e Should one use the energy eaten by the organisms or the energy assimilated by the organisms to construct the pyramid? What is the difference?

f Suggest how you might estimate a pyramid of energy flow for 1 a forest 2 a meadow and 3 the sea.

> Practical investigation. *Practical guide 7*, investigation 29A, 'A quantitative study of an ecosystem'.

29.2 Energy flow and nutrient cycling in ecosystems

The biosphere can be viewed as a series of related boxes. Their vertical walls delimit types of ecosystems with markedly different patterns of energy flow and nutrient cycling. Each box is an ecosystem. All the boxes exchange water, nitrogen, carbon dioxide, and oxygen with one another. Many have significant inputs of detritus and/or nutrient ions from other ecosystems.

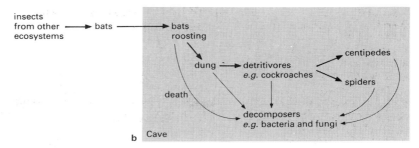

Figure 356
a A tomb bat (*Taphozous melanopogon*).
Photograph, S. C. Bisserôt.
b A greatly simplified food web, based on bat dung, in the bat caves of Borneo. The energy which maintains the system comes from other ecosystems.

The ecosystems which appear to be self-supporting and do not exchange many materials with ecosystems nearby are known as *closed*. Some ecosystems, however, are *open*. They lack autotrophs or depend on detritus from other ecosystems for much of their energy. The bat caves of Borneo are of this type (*figure 356*).

The ecosystem concept focuses attention on the differing patterns of energy flow and nutrient cycling in different communities. We have established that energy flows through an ecosystem. On the other hand, nutrients such as iron, calcium, and magnesium are cycled round and round in it, being used again and again. The autotrophs absorb nutrients, and the heterotrophs may acquire the nutrients by eating the autotrophs. Whatever sorts of organisms die with nutrients inside them, they and their excreta are decomposed and the nutrients are released to the environment. They become available again for the growth of autotrophs (*figure 357*).

Figure 357
Litter traps within a group of beech trees, Wytham Wood, near Oxford. The traps were emptied frequently. Branches, twigs and leaves were dried and weighed separately, and their nutrient contents analyzed. The results contributed to a model of energy flow and nutrient cycling within the ecosystem.
Photograph, J. R. Flowerdew. Department of Applied Biology, University of Cambridge.

The ecosystem concept is valuable because in simplified and easily comprehensible form, the data can be modelled on computers. If experiments are carried out on an ecosystem the results can be used to improve the computer model. Models of nutrient cycling are useful because they allow, for example, the consequences of tree-felling and erosion on tree productivity to be estimated. The effects of the cycling of radioactive isotopes can also be predicted.

In the next two sections, energy flow (29.3) and nutrient cycling (29.4) will be discussed in more detail.

29.3 Energy flow

As energy flows through an ecosystem, the species accumulate some of it in carbon compounds with a high potential energy content. It is important for farmers and fish biologists, for example, to understand the factors which affect the accumulation of this energy if higher sustained yields are to be obtained.

The dry mass or energy content of the organisms at a particular trophic level at any one time is known as the *standing crop*. The standing crop will, of course, change with time, especially in an agricultural crop or a forestry plantation. The rate at which dry mass or energy accumulates in a species, trophic level, or community is known as its productivity.

The *net primary production* (NPP) of the plants in an ecosystem is the rate at which they accumulate dry mass, measured in $kg\,m^{-2}\,year^{-1}$. This represents the mass of organic compounds made in photosynthesis (P, sometimes known as gross primary production GPP), minus the mass of organic compounds broken down in catabolic activities, such as respiration (R). Thus, NPP = P − R. Net primary production can also be expressed in terms of the rate at which energy is fixed by the plants in the ecosystem, in $kJ\,m^{-2}\,year^{-1}$.

This energy, accumulated in net primary production, is available to be eaten by animals. Some of the dry mass of plants becomes the dry mass of animals. The rate at which energy or dry mass is accumulated by the animals in an ecosystem is called the secondary production. In fact for any animal or trophic level:

FE = secondary production + E + R + U,

where FE = energy in food eaten, E = energy in egesta, R = energy lost in respiration and other catabolic activities, and U = energy lost in urine.

STUDY ITEM

29.31 The productivity of plants

Humans depend on photosynthesis not only directly, for food, but also indirectly, for timber, firewood, rope, and fossil fuels. The primary productivity of plants, and the factors which influence it, are therefore of great importance.

The efficiency with which the energy in photosynthetically-active useful radiation is converted into dry matter by a crop differs at different stages in its growth. In an agricultural crop such as wheat or maize, the conversion efficiency is low during the early stages of crop growth because the leaves cover only a small proportion of the area. At a later stage several tiers of leaves may exist. At high light intensities, the upper leaves will be *light-saturated* and the lower leaves may also photosynthesize rapidly. The leaves of the crop are in continual flux. Leaves which are shaded or diseased drop off and are replaced by efficient new leaves higher up. In the fruiting stage, the photosynthetic rate of the leaves declines.

How can the *net primary production* of the crop be estimated? Three sorts of techniques allow the net production to be measured over short periods of time.

1 Measure variation in carbon dioxide concentration in the air at different heights above the plants and within the stand of vegetation throughout twenty-four hours. Calculate the uptake of carbon dioxide.
2 Enclose the area of vegetation in a transparent container and measure the disappearance of carbon dioxide from an air stream passed over it.
3 Use a harvesting method: take samples of plants at intervals to determine changes in dry mass.

Once the net primary production has been measured, the efficiency with which the solar energy is trapped in plant mass can be calculated. The productivities of different species, or the growth of the same species in different environmental conditions, can be compared.

Theoretically, the maximum efficiency for a crop is the fixation of about six per cent of the total solar radiation in plant mass (*figure 358*) at

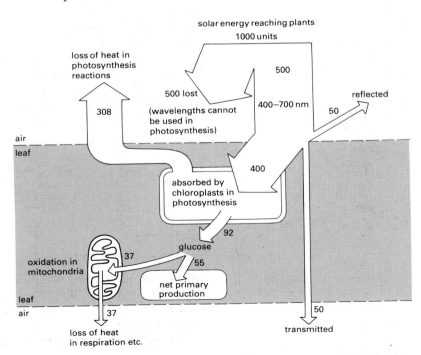

Figure 358
The average efficiency with which total solar radiation is trapped in net primary production in a productive crop plant (5.5 %). Imagine that the plant grows at sea level and receives 1000 units of light energy per unit time. Much of the 'wasted' energy is used to evaporate water.
After Hall, D. O. Biologist, **26**, *17, 1979.*

Chapter 29 Communities and ecosystems 515

any one time. Efficiencies approaching this have been achieved by dense crops of maize growing in moist sub-tropical environments, and in cultures of single-celled algae provided with high light intensities and carbon dioxide concentrations. In one year of maximum cropping, efficiencies approach 2.5 per cent for crops of C_4 plants such as maize and sugar cane, and 1.5 per cent for crops of C_3 plants like wheat.

Environmental conditions during the growth of a crop have a major impact on the rate of net primary production. In particular, crop growth can be limited by low water potentials in the soil or deficiencies of certain nutrient ions. Water stress may cause the stomata to close. This greatly reduces transpiration but it also prevents the uptake of carbon dioxide for photosynthesis (see *Study guide I*, Chapter 7, Study item 7.32 'Limiting factors in the rate of photosynthesis'). Lack of nutrients reduces the rate at which the cells divide.

a Give brief explanations of the following terms:
1 light-saturated *Photo no effect*
2 net primary production *rate*

b Explain the difference between 'photosynthetically-active useful radiation' and 'total solar radiation'. *not trappable*

c 1 Which one of the three methods of measurement of net primary production given above would be best for testing a model of the effect of the environment on production in a natural plant community?
2 Give reasons for your choice.
3 State two disadvantages of using the method.

method reduces wm

d The leaves and plants in a community are continually dying. What effect will this have on the harvesting technique for the direct measurement of net primary production? *Under estimate take frequently*

e Account for the low efficiency of production of agricultural crops compared with algal cultures in the laboratory. *Optimal conditions*

f Suggest two reasons why plants with the C_4 mechanism of photosynthesis are often more efficient at converting solar energy to dry mass than C_3 plants.

C_4 More efficient at using H_2O than C_3 so C_4 can still photosynthesise when C_3 plants have stomata closed

Photorespiration occurs in C_3 plants but hardly at all in C_4 plants
Max photo occurs at higher temp in C_4 plants so warmer climates favour C_4

STUDY ITEM
29.32 Energy flow and agriculture

incl in temp.

The secondary production of beef in modern industrialized agriculture is based on rearing bullocks on fertilized sown leys of ryegrass (*Lolium perenne*) and clover (*Trifolium* spp.). Let us consider first the energy flow through the grass-clover ley.

A typical square metre of pasture in Britain receives about 1 046 700 kJ of visible light energy (wavelengths 400–700 nm) from the Sun each year. This energy can be accounted for as follows:

Energy reflected by the leaves	165 002 kJ
Energy leaving the plant as a result of the evaporation of water	523 350 kJ

Energy transmitted to the ground	334 944 kJ
Energy locked up in new growth of plants	21 436 kJ
Energy transferred during respiration of plants	1 968 kJ

a How much energy is trapped by the plants in photosynthesis?

b What is the efficiency of net primary production over the year, expressed as per cent of total visible light energy?

The fate of a year's energy in such a ley is summarized in *figure 359*. Examine this figure carefully and answer the questions below.

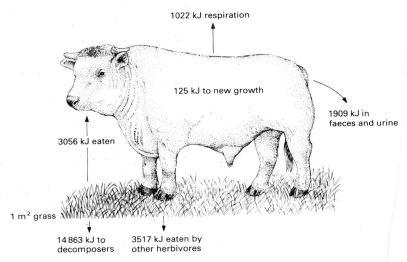

Figure 359
The fate of a year's growth of 1 m² grass when grazed by one bullock.

c What proportion of the net primary production of the ley was eaten by the bullocks? Why is this proportion so low?

d What proportion of the energy in the plants eaten by the bullocks was devoted to increasing the dry mass of bullocks (secondary production)?

e Would you advise that farmers should concentrate more on improving the net primary productivity of grasses than on improving the conversion efficiencies of the bullocks?

f Why does our diet probably contain more meat from herbivores than from carnivores? Which carnivores, if any, do you eat?

g In some tropical regions, native herbivores such as zebra, wildebeest, and antelope yield more energy per unit area than introduced species of cattle. Suggest some reasons why this is so.

Practical investigation. Practical guide 7, investigation 29B, 'The energetics of the stick insect (Carausius morosus)'.

Chapter 29 Communities and ecosystems 517

STUDY ITEM

29.33 The efficiency of energy transfer between trophic levels

A widely-quoted generalization about natural ecosystems is that the efficiency of energy transfer from one trophic level to the next is about 10 per cent. That is, about 10 per cent of the net production of the plants ends up as the net production of herbivores, about 10 per cent of this makes its way into the net production of carnivores, and so on. This generalization was largely based on studies in the 1950s of freshwater lakes and laboratory aquaria. Unfortunately, later research on a wider variety of ecosystems has overthrown this idea.

The efficiency of energy transfer from body mass at one trophic level to body mass at the next depends on three main factors (*figure 360*).

1 What proportion of the dry masses of the organisms at the first trophic level is eaten?

2 What proportion of the food eaten by the organisms at the first trophic level is *assimilated* (taken across the gut wall into the body and absorbed by the cells) by the organisms at the next trophic level, and how much is egested?

3 Of the assimilated energy, how much is used in respiration (maintenance costs) and how much is devoted to net production (growth and reproduction)?

Let us consider each of these three points in turn.

1 Table 70 shows some estimates of the per cent of plant dry mass consumed by animals feeding in different ecosystems.

Table 70
Per cent of net primary production consumed by primary consumers in various ecosystem studies.
Data from Pimental, D., Levin, S. A. and Soans, A. B. On the evolution of energy balance in some exploiter-victim systems, Ecology **56**, *381*, 1975.

Plant community	Consumers	Productivity consumed (%)
Beech trees	invertebrates	8.0
Oak trees	invertebrates	10.6
Grass and herbs (habitat *1*)	invertebrates	4–20
Grass and herbs (habitat *2*)	invertebrates	0.5
Grass	invertebrates	9.6
Marsh grass	invertebrates	7.0
Sedge grass	invertebrates	8.0
Aquatic plants	bivalve molluscs	11.0
Aquatic plants	herbivores	18.9
Algae	zooplankton	25
Phytoplankton	zooplankton	40

a *What patterns emerge?*

2 Examine the assimilation efficiencies of the animals in table 71.

Table 71
Proportion of energy in food eaten which is assimilated and egested in various organisms. Imagine that each organism has ingested 100 units of energy.
Data mainly from Open University. S323: Ecology course, Unit 3, 1973.

Animal	Feeding preference	Assimilated (%) (ingestion minus egestion)	Egested (%) (ingestion minus assimilation)
Grasshopper	herbivore	37	63
Caterpillar	herbivore	41	59
Wolf Spider	carnivore	91.8	8.2
Perch	carnivore	83.5	16.5
Owl	carnivore	85	15
Elephant	herbivore	33	66
Cow	herbivore	40	60

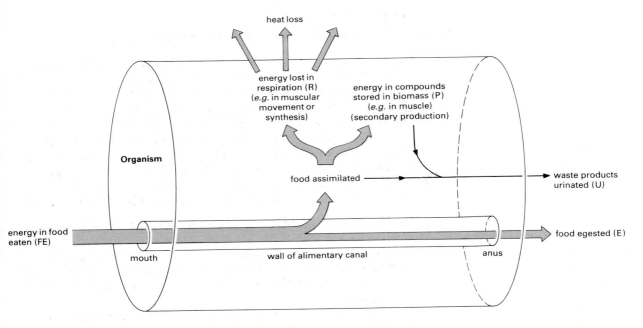

Figure 360
Diagrammatic representation of the fate of the energy in the food eaten by a mammal.

b **State the main factor which seems to affect the proportion of food energy assimilated (absorbed across the gut wall into the bloodstream and absorbed by the cells). Explain in biochemical terms the principal differences which emerge from table 71.**

3 What happens to the assimilated energy, the energy absorbed by the cells of the body? Table 72 summarizes, for 232 different animal species, the proportion of the assimilated energy which is devoted to production (growth and the mass of eggs or offspring at birth). The rest is used in respiration (maintenance costs).

Table 72
The proportion of assimilated energy which is devoted to growth and reproduction (net production) in various groups of animals. The rest of the assimilated energy is used in cellular respiration and escapes from the animal ultimately as heat.
Data from Humphreys, J. anim. Ecol. **48**, 427, 1979. (Discussed in May, R. M. Production and respiration in animal communities, Nature **282**, 443, 1979.)

Group (with, in brackets, number of species in sample)	Per cent assimilated energy devoted to net production
Invertebrates which are not insects:	
herbivores (15)	21
carnivores (11)	28
detritivores (23)	36
Non-social insects:	
herbivores (49)	39
carnivores (5)	56
detritivores (6)	47
Social insects (ants, bees, wasps) (22)	10
Fish (22)	10
Birds (9)	1.3
Mammals:	
insectivores (6)	0.9
small mammals (8)	1.5
large mammals (56)	3.1

c **Discuss, giving reasons where appropriate:**
1 why, in both non-social insects and other invertebrates, herbivores have the lowest production efficiencies; is this what you would expect?

2 the inefficiency of fish and social insects;
3 the difference in production efficiency between poikilotherms and homoiotherms;
4 the possible influence of size on the production efficiencies of mammals.

d Using the data in tables 70 to 72, make some speculative calculations of the per cent of net plant production which ends up in the herbivore production in 1 a pond, with phytoplankton being grazed by invertebrate animal plankton and 2 an oak forest, being grazed by non-social insects. Show your working clearly.

e Do you think that the '10 per cent rule' is likely to be valid for many ecosystems? Is it more likely to be valid for secondary consumers eating herbivores than primary consumers grazing plants?

STUDY ITEM

29.34 Energy flow in a grassland ecosystem

Figure 361 illustrates the energy flow through a small portion of a grassland ecosystem. Examine it closely and answer the questions below.

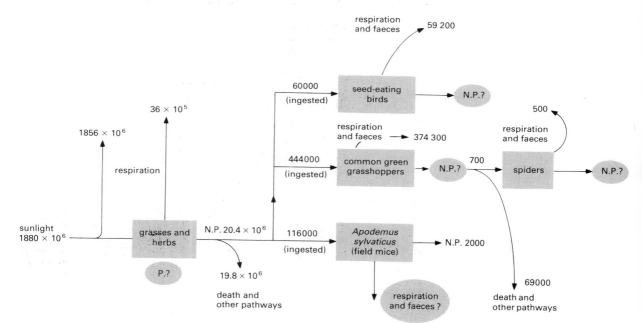

Figure 361
Energy flow through a small portion of a grassland ecosystem. The figures given are in kJ m^{-2} year^{-1}; N.P. = net production.
Data and questions modified from Tribe, M. A. et al. Ecological Principles, Cambridge University Press, **155**, 1975.

a *Which of the labelled boxes represent producers, primary consumers, and secondary consumers?*

b *How much energy is trapped in photosynthesis by the grasses and herbs?*

c *Why are the grasses and herbs so inefficient at trapping the incident light energy?*

d *What is the net production of each of the following groups of animals: 1 seed-eating birds 2 common green grasshoppers 3 spiders?*

e *How much energy is lost through respiration and faeces by field mice?*

f *State in detail what happens to the energy which enters 'death and other pathways'.*

This Study item is based on Tribe, M. A. *et al.*, Ecological principles, Cambridge University Press, 1975.)

STUDY ITEM
29.35 Energy flow in different ecosystems

Figures 362 to *364* show energy flow through three types of ecosystem; a salt-marsh in Georgia, a forest, and a flowing stream.

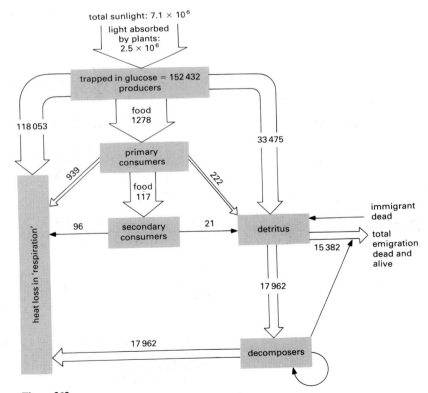

Figure 362
Energy flow through a salt marsh in Georgia, in units of kJ m^{-2} year^{-1}.
Data from Teal, J. M. Energy flow in the salt marsh ecosystem of Georgia, Ecology **43**, 614, 1962.

Figure 363
Energy flow through a forest ecosystem, in units of kJ m^{-2} year^{-1}.
*After Phillipson, J. Studies in Biology 1. Ecological energetics, Edward Arnold, 1966; adapted from Odum, E. P. Jap. J. Ecol., **12**, 108, 1962.*

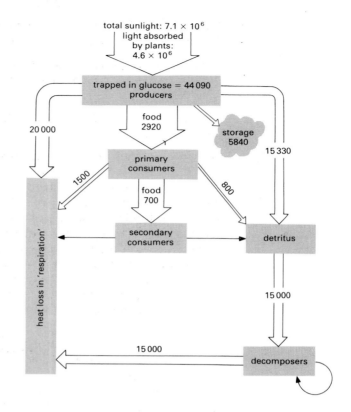

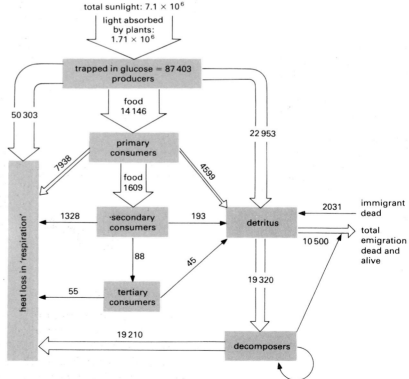

Figure 364
Energy flow through Silver Springs, Florida in 1953–4, in units of kJ m^{-2} year^{-1}.
*Adapted from Odum, H. T. Trophic structure and productivity of Silver Springs, Florida, Ecological Monographs, **27**, 55, 1957.*

a Compare the efficiency with which the incident light energy is incorporated in dry mass by the plants in each ecosystem.

b Compare the proportion of plant production which enters the grazing and the detritus food webs in all three ecosystems. Account for the differences.

c Which of these ecosystems would yield the largest annual crop, and how would you harvest it?

☐ **d** Which of these ecosystems, if any, appear to be in energy balance?

Humans rely on ecosystems for food, and the energetics of ecosystems in which humans live can be measured. It has been estimated that of the 80 million million people who have ever lived on Earth since the emergence of the species *Homo sapiens*, over 90 per cent have lived as hunters and gatherers. About six per cent of humans have been agricultural. Only three per cent have lived in industrialized societies. The energetics of hunter–gatherers and shifting agriculture on the one hand, and that of modern industrialized agriculture on the other, are compared in *figure 365*.

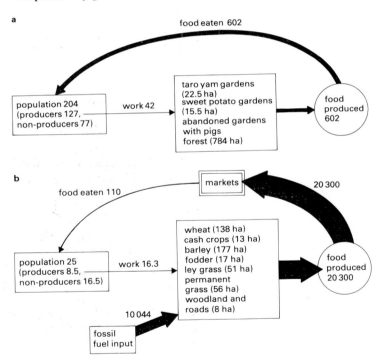

Figure 365
Energy flow in subsistence and industrialized agriculture. The figures represent energy flows in GJ (10^9 J) m^{-2} year^{-1}.
a Flow of energy through the shifting agricultural system of the Tsembaga Maring tribe, in the mountains of New Guinea.
b Energy flow through a mixed farm at West Lavington, Wiltshire, in 1971. Notice the massive energy subsidy from the burning of fossil fuels.
Adapted from Bayliss-Smith, T. P., The ecology of agricultural systems, Cambridge University Press, 1982.

Decay and decomposition

When animals eat plants or other animals, many of the compounds are egested instead of being absorbed. Dead leaves, plants, and animals also fall to the base of the ecosystem, either onto the soil or onto the silt at the bottom of the water. Many different species of decomposers live there. These are animals and micro-organisms that live on detritus. They play an important role in ecosystems, for they break open dead cells and release to the soil or water the nutrient ions which primary producers need for healthy growth. In some ecosystems, such as bogs and temperate forests, the grazing food web is so weak that most of the net primary production enters the decomposer food web.

There are two kinds of decomposer. Animals such as vultures, hyenas, termites, mites, springtails, flies, earthworms, and dung beetles, are usually the first to shred the tissues. They transfer much of the energy in detritus in cellular respiration, and release in their excreta many of the decomposed cells. Dead detritivores may themselves be eaten by detritivores. Of course, living detritivores such as earthworms may themselves be eaten by omnivores or secondary consumers (such as blackbirds). In this case much of their energy may enter the grazing food web.

Figure 366
A detritivore in action. A green scarab beetle (*Khefer aegyptorum*) rolling her ball of buffalo dung, with another scarab beetle riding on it. The ball of dung is rolled into a hole in the ground and used as food for the offspring. Dung beetles like these were successfully imported from Africa into Australia to remove the accumulated dung of introduced cattle.
Photograph, Jane Burton, Bruce Coleman.

The activities of the animals produce finely divided organic matter and egesta, with a considerable surface area. Fungi and bacteria act on this organic sediment. They secrete enzymes through their cell walls and absorb the digestive products. Dead fungi and bacteria are themselves decomposed by enzymes, and the enzymes are themselves broken down by other enzymes. In the end, the energy in the detritus is all released as long-wave radiation (heat) from the detritivores and decomposers in metabolism (*figure 351*). It is not recycled.

The activity of the decomposers is influenced by three main environmental factors.

1 Temperature: almost all detritivores and decomposers are poikilotherms and their metabolic rates increase rapidly with increasing temperature.

2 The ratio of carbon to nitrogen in their diet: bacteria and fungi need nitrogenous compounds to make not only their nucleic acids, phospholipids, and enzymes, but also their cell walls. A poor nitrogen supply may limit the growth of micro-organisms. That is why gardeners often add ammonium sulphate or other nitrogenous compounds to their compost heaps to increase the rate of decomposition.

3 In acid anaerobic conditions, such as those in sphagnum bogs or acidic lakes in upland areas, decomposition of detritus is extremely slow and the detritus accumulates. The 'bog people' of Scandinavia (*figure 367*) were preserved for two or three thousand years under such conditions. Ecosystems in which detritus accumulates are known as 'storage ecosystems' because organic compounds build up in them. Fossil fuels such as oil, gas, and coal probably originated in storage ecosystems millions of years ago.

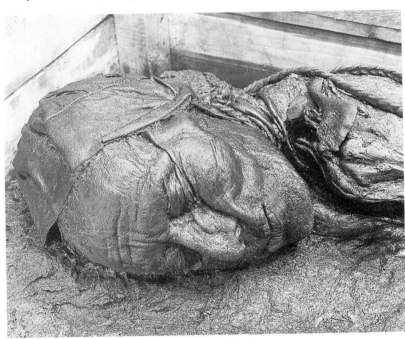

Figure 367
The Tollund man, preserved in a peat bog for about two thousand years under acid anaerobic conditions in Denmark. Even the fingerprints were visible when this man was found on 8 May 1950.
Photograph, L. Larsen, The National Museum, Copenhagen.

Practical investigation. *Practical guide* 7, **investigation 29C, 'A study of decomposer organisms in the soil'.**

29.4 Nutrient cycling

The atoms of which organisms are composed are continually recycled. The cycles of elements in ecosystems can be worked out in terms of the nutrient contents of the soil and the organisms at each trophic level, and the rates of exchange between each compartment.

For healthy growth, plants need carbon dioxide, water, nitrates or ammonium, sulphates, phosphates, potassium, iron, calcium, magnesium, and a host of micronutrients. All these elements cycle in ecosystems at different rates. The cycles of carbon, oxygen, hydrogen, water, nitrogen, and sulphur are said to be 'open', because they contain gaseous stages and the gases can be easily exchanged between ecosystems. On the other hand, the mineral element cycles are said to be 'closed', because the elements are recycled within the system and the exchanges of ions with other systems are limited.

We shall concentrate upon the cycles of carbon and nitrogen. Organisms are composed of carbon compounds, many of which also contain nitrogen.

STUDY ITEM

29.41 The carbon cycle

Organisms consist of carbon compounds. Carbon exists in the biosphere mainly as carbon dioxide gas on the one hand, and organic compounds, made by living organisms, on the other. The cycle of carbon in the biosphere is illustrated in *figure 368*, with estimates of the fluxes from one part of the cycle to another.

a *Describe how you might estimate the annual flux of carbon dioxide from the atmosphere into the primary producers in the sea. Discuss the difficulties which you might encounter and the ways in which they might be overcome.*

b *Many bacteria which live in watery habitats are chemo-autotrophic. Where do they fit into the diagram of the carbon cycle (figure 368)?*

Figure 368
The cycle of carbon in the biosphere.
From King, T, J., *Ecology*, Nelson, 1980; figures from Baes, C. F. Jr. et al., *Carbon dioxide and climate: the uncontrolled experiment*. American Scientist **65**, 310, 1977; and Woodwell, G. M. The carbon dioxide question. Scientific American, **238**, 34, 1978.

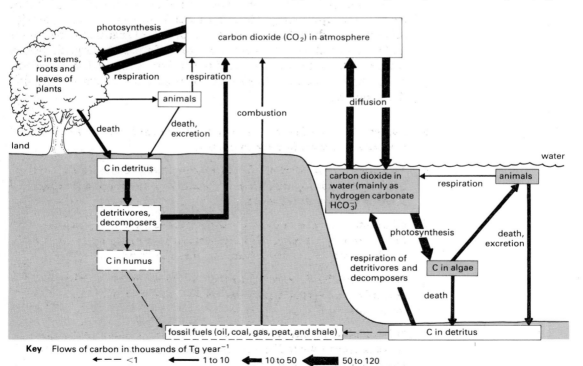

Key   Flows of carbon in thousands of Tg year^{-1}
←--- <1   ← 1 to 10   ← 10 to 50   ← 50 to 120

c *Atmospheric methane levels have risen at 1.7 per cent a year since at least 1975. Suggest some reasons for this increase.*

The carbon cycle is no longer balanced. Bubbles of air in frozen ice 15 000 years old contain about 270 p.p.m. by volume (p.p.m./v) of carbon dioxide. By the year 1800 the carbon dioxide level in the atmosphere had reached 290 p.p.m./v. When detailed measurements began at Mauna Lao, Hawaii (20° N) in 1957, the concentration was 315 p.p.m./v. In 1983 it was about 340 p.p.m./v, increasing at 1.3 p.p.m./v a year (0.4 per cent).

The effect of carbon dioxide and methane on the climate is potentially considerable. This is because they can absorb some of the long wave infra-red radiation which leaves the Earth and would otherwise reach outer space. They radiate it back, like the glass in a greenhouse. Most recent computer models suggest that an increase of 30 per cent in atmospheric carbon dioxide might warm the Earth by 0.9 $\pm$ 0.5 °C. A doubling might warm the planet by 3 $\pm$ 1.5 °C. This is known as the 'greenhouse effect'. It might considerably alter the pattern of climate, rainfall and glaciation across the Earth's surface.

Why are the carbon dioxide levels increasing? One possible contributory factor is the combustion of fossil fuels by humans, at present increasing by 2.3 per cent per annum. In 1980, for example, 19.4×10^{12} kg of carbon dioxide were released into the atmosphere from fossil fuels, joining the 700×10^{12} kg already present. The carbon dioxide concentration in the air at Mauna Lao only increases at half this rate. Presumably much of the extra carbon dioxide becomes dissolved in the oceans (see *Study guide I*, Chapter 7, 'Photosynthesis').

The recent increase in atmospheric carbon dioxide may also be the result of the rapid felling of forests, especially tropical forests. Beneath a mature forest about half the soil mass consists of organic matter. In the arable crop soil which results from forest felling, the organic matter content falls to 25 per cent. Half the carbon in the soil is lost to the atmosphere as carbon dioxide in the respiration of detritivores and decomposers.

d *Suggest two possible reasons why the extra carbon dioxide is not removed by plants in photosynthesis.*

e *Speculate on the effect of phytoplankton in the oceans on the carbon cycle. If the phytoplankton were killed, how might the atmospheric levels of carbon dioxide change?*

f *Climatologists are waiting for a clear signal, above the random fluctuations of climate, of global warming. What types of signal might indicate that the climate was warming up?*

g *What would be the effect of a temperature increase on 1 the global rate of photosynthesis and 2 the global rate of respiration of decomposers?*

h *When a clear signal of climatic warming is received, what steps could nations take to limit the increase in global carbon dioxide levels? Discuss the problems.*

STUDY ITEM

29.42 The nitrogen cycle in a temperate forest ecosystem

Detailed studies, beginning in 1963, have quantified the inputs and outputs of water and chemical elements in several small, undisturbed forested areas in the Hubbard Brook Experimental forest in the White Mountains of New Hampshire in the United States. Thousands of measurements, made over ten years, have established the main fluxes of nitrogen compounds. This is of great value, considering that forests cover 38 per cent of the total continental area of the Earth.

One small forest unit, known as Watershed Six (13.2 ha) was selected for detailed study. The dominant trees are sugar maple (*Acer saccharum*), American beech (*Fagus grandifolia*), and yellow birch (*Betula alleghensis*). They grow on thin acidic soils, well-drained podzols (see section 26.5) overlying impermeable granite. All the water which falls in precipitation runs off in streams in which the concentrations of ions can be measured (*figure 369*).

Figure 369
Watershed Two in the Hubbard Brook Experimental Forest, after felling. The gauge-house and weir are obvious in the foreground.
Photograph by Gene E. Likens.

The trees in this area are no more than 55 years old. They established themselves in about 1920, after all the trees on the site had been cut down. In this secondary forest (net primary production 1040 g m^{-2} year^{-1}) about a third of the net primary production (393 g m^{-2} year^{-1}) accumulates each year in tree trunks, roots, and dead litter on the forest floor. Since nitrogen is present in proteins and lignin, the whole ecosystem is accumulating vast quantities of nitrogen (about 20.7 kg ha^{-1} year^{-1}).

The nitrogen cycle for Watershed Six is shown in *figure 370*. At any one time, 90 per cent of the total nitrogen in the ecosystem is in soil organic matter, 0.5 per cent is in available nitrogen in the soil, and 9.5 per cent is in the vegetation.

The cycle is 'tight', in the sense that the inputs to and outputs from the ecosystem are minor compared with the flux of nitrogen between its various working parts. At any one time, the system contains

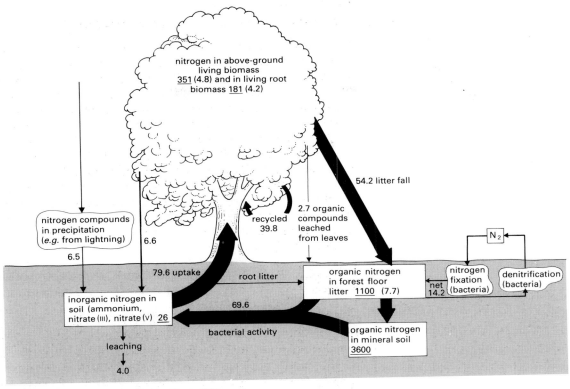

Figure 370
The nitrogen cycle through the Hubbard Brook ecosystem. The figures in circles in each box are the nitrogen contents of the ecosystem components in kg ha^{-1} year^{-1}. The figures in brackets are the rates at which the nitrogen contents are increasing each year. Figures beside the arrows represent fluxes of nitrogen in kg ha^{-1} year^{-1} between the various ecosystem components. The role of animals in the cycling of nitrogen is ignored. Data from Bormann, F. H., Likens, G. E., Melillo, J. M. 'Nitrogen budget for an aggrading northern hardwood forest ecosystem', Science, **196**, 981, 1977. Copyright © 1977 by A.A.A.S.

5000 kg ha^{-1} of nitrogen, but it only gains 20.7 and loses 4.0 kg ha^{-1} year^{-1}.

Of the nitrogen added to the ecosystem each year (20.7 kg ha^{-1} year^{-1}), 60 per cent is from nitrogen gas fixation and 32 per cent from precipitation. The gain from weathering of rocks is negligible. Of the total added, 81 per cent is held in the ecosystem in the accumulating biomass. Of this nitrogen destined for storage, 54 per cent is added to living biomass and 46 per cent to the organic matter of the forest floor.

Each year about 119 kg ha^{-1} of nitrogen is used in plant growth. Two-thirds comes from nitrogen uptake from the soil as nitrate and ammonium ions. The rest is withdrawn from the stores in the twigs, branches, and trunks in spring, and used to make new leaves. In autumn, about 40 kg ha^{-1} of nitrogen is withdrawn from the leaves before they fall, and stored in branches, trunk, and roots. The evidence for this is that at the height of the growing season the leaves contain 70 kg ha^{-1} of nitrogen. Fresh leaf litter, however, only contains 30 kg ha^{-1}!

Dead plant organs, containing nitrogen compounds, form the bulk of the litter which accumulates on the soil, forming a layer 3–15 cm deep. As it decomposes, it gradually becomes incorporated in the mineral soil below, as humus. The nitrogen in humus is slowly released to the soil as inorganic compounds, such as ammonium ions, as a result of the activity of soil organisms. The other source of nitrogen in litter is nitrogen fixation by bacteria (see Study item 27.62), which exceeds denitrification by bacteria by 14.2 kg ha^{-1} year^{-1}.

Ammonium ions are transformed to nitrate(V) ions by soil bacteria in a process known as 'nitrification'. The bacterium *Nitrosomonas*, for example, converts ammonium ions to nitrate(III) ions. Similarly, species like *Nitrobacter* oxidize nitrate(III) to nitrate(V). Both these types of 'nitrifying bacteria' are chemo-autotrophic, in other words, the energy which they gain from the oxidations is used to manufacture glucose and oxygen from carbon dioxide and water.

As a result of the activities of these micro-organisms, there is a large pool in the soil of inorganic nitrogen compounds. This pool is augmented by relatively small amounts of nitrates from rainfall (produced as a result of electrical discharges in the atmosphere), and leaching from the leaves of trees.

Ammonium and nitrate ions are absorbed by plants in large quantities. Two main processes apart from uptake by plants deplete the soil pool of inorganic nitrate. First, denitrifying bacteria such as *Pseudomonas denitrificans*, mainly anaerobes, are heterotrophs which convert compounds such as nitrate(V) ions to atmospheric nitrogen gas. Secondly, some of the ions are washed through the soil to groundwater, and leave the ecosystem in streams.

a *Suggest some of the inputs and outputs of nitrogen which have been omitted from the simplified nitrogen budget for Hubbard Brook shown in* figure 370.

b *If this ecosystem were allowed to mature into a climax system, what differences would you see between its nitrogen budget and the nitrogen budget in* figure 370?

c *Most estimates suggest that nitrogen atoms are recycled much more rapidly in tropical than in temperate ecosystems. Suggest three reasons why this is the case.*

One of the watersheds at Hubbard Brook, Watershed Two, was completely deforested in November and December 1965 (*figure 369*). All trees, saplings and shrubs were felled, but no timber was removed. Herbicides were applied in each of the next three summers to prevent the regrowth of the vegetation.

After deforestation, the annual run-off of water in streams increased by over thirty per cent. The concentrations of all the major nutrients in stream water increased dramatically about five months after cutting, and reached a peak during the second year. In this second year, nitrate concentrations in stream water from the site averaged 52.9 mg dm^{-3}, as compared with 1.8 mg dm^{-3} in streams from the undisturbed Watershed

Six. The pH of the stream water decreased by one unit! In the three years after cutting, the losses of nitrogen (mainly as nitrate) exceeded 340 kg ha^{-1}. The losses of ammonium ions, however, hardly increased at all.

- d *Discuss the reasons why the output of nitrate increased so dramatically when the forest was felled.*
- e *On the basis of the numbers in figure 370, how long might it take before the nitrogen lost was replaced?*
- f *Would the results have been different if the felled vegetation had been removed?*
- g *How might the method of felling and/or after-treatment be modified so as to reduce the loss of nitrate from the ecosystem after felling?*
- h *Discuss the possible effects on rivers and streams of the increase in sediment load and nutrient ion concentration after forest felling.*
- i *Discuss the following statement made by the investigators. 'These facts strongly suggest to us that the forest ... is almost totally dependent for its existence on this thin (2–20 cm), rather fragile, organic layer.'*

The study of nutrient cycles has much practical significance. In particular:

1 A shortage of nutrients may limit plant productivity, and thus limit both primary and secondary production in ecosystems. If the reasons why nutrients are scarce can be identified, the shortages may be alleviated.
2 Radioactive elements may be released into the environment and incorporated into nutrient cycles in ecosystems. Mathematical models of the nutrient cycles may allow the long-term futures of these ions in ecosystems to be predicted.
3 The loss of nitrates from fertilized fields and felled forests such as Hubbard Brook, and the input of phosphates from sewage and detergents, may lower water quality.

STUDY ITEM

29.43 Nutrient cycling in lakes and oceans

In many temperate seas and lakes there is a characteristic annual cycle in the abundance of phytoplankton and zooplankton in the upper layers of the water. An idealized pattern is shown in *figure 371*. This cycle may partly be due to the changes in the thermal stratification of the water throughout the year.

In autumn and winter, the upper and lower layers of water are at approximately the same temperature and considerable mixing occurs. Nutrients are cycled from the decomposers on the sea bed to the upper layers, where algae are photosynthesizing. In spring and summer, however, the upper layers of the water warm up. An illuminated aerobic

Figure 371
Annual fluctuations in the numbers of producers and consumers in a typical temperate sea, such as the North Sea, together with some of the fluctuations in environmental factors which may affect numbers.
From Tribe, M. A., Eraut, M. R., and Snook, R. K. *Basic Biology Course Book 4*, Ecological principles, *Cambridge University Press*, 1975.

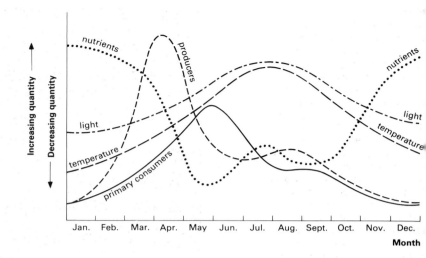

layer of warm water, called the epilimnion, lies over a dark, anaerobic layer of cold water, known as the hypolimnion. Between them is a zone of rapid temperature transition, the thermocline.

The algae and cyanobacteria in the phytoplankton need nutrient ions to grow and reproduce. They need in particular nitrates and phosphates, which they absorb from the water. When they die, their bodies sink and leave the epilimnion, passing through the hypolimnion on their way to the sea bed. This represents a loss of nitrates and phosphates from the epilimnion. These nutrients cannot be replaced from below because in summer there is no exchange between the layers.

a *Why does the warmer water float on top of the cooler water?*

b *Account for the increase in the numbers of producers during March.*

c *Suggest two reasons for the decline in the producers in May.*

d *Account for the shape of the curve for primary consumers.*

e *Amongst the factors which might alter this idealized picture are, first, some species of algae and cyanobacteria are more palatable and digestible to carnivores than others, and second, some cyanobacteria fix nitrogen gas from the atmosphere. Speculate on the influence of these factors on the curves of relative abundance shown in figure 371.*

STUDY ITEM

29.44 Nutrient cycling in the Amazonian tropical rain forest

Tropical rain forests occupy about ten per cent of the land area on Earth. These humid jungles cover 940 million hectares, but they are being destroyed at the rate of over ten million hectares a year. The largest area of remaining forest (484×10^6 ha) is Amazonian rain forest, most of which is in Brazil.

In 1970, the Brazilian government announced plans to integrate the

Figure 372
An experiment on nutrient enrichment in Blelham Tarn, Cumbria, viewed from the shore. The three circles visible on the surface of this small lake (*figure 372a*) are the upper rims of 'Lund tubes'. These are large watertight tubes of butylene rubber 45.7 m in diameter and 11–12 m deep, with their lower rims buried in the bottom mud and their upper rims supported by floating rings. The furthest of these mini-lakes has been enriched with compounds containing nitrates, phosphates, and silica. An algal bloom has resulted from this eutrophication. The two nearer tubes are controls to which no nutrients were added. Amongst the planktonic algae present in the bloom are those belonging to the genera of *Oscillatoria* (*figure 372b*, × 140) and *Rhodomonas* (*figure 372c*, × 1200).
The experiment indicates why algal numbers in the lake began to increase in the 1960s. A small sewage works, which opened in 1962, increased the phosphate concentration in the lake. The use of nitrogenous fertilizers on surrounding farms dramatically increased the levels of nitrates.
Photographs, **a** *Freshwater Biological Association,* **b** *and* **c** *Dr H. M. Canter-Lund, Freshwater Biological Association.*

a

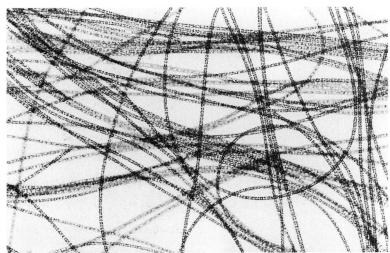
b

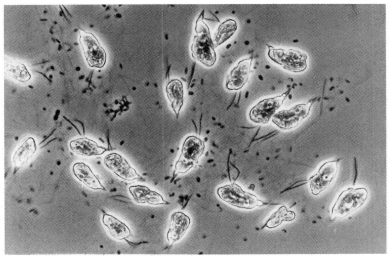
c

Amazonian region with the rest of the country. The 3300 km Transamazon highway was bulldozed through by 1975. By 1980, however, only 1800 families had settled. The aim of unlocking the agricultural, mineral, and timber resources of the area had only been partially successful.

The basic reason for the failure of the Brazilians to transplant agriculture to the Amazonian basin is soil infertility. The rain forest is simply 'a desert covered by trees'. Over 75 per cent of the soils are deep, well-drained, very acid (down to pH 1.6), and very poor in essential plant nutrients. The main constraints which are likely to limit crop growth are phosphorus deficiency (90 per cent of the area), aluminium toxicity (73 per cent), low potassium reserves (56 per cent), and poor drainage and flooding (24 per cent). Only 6 per cent of the Amazonian basin has soils, mainly of volcanic origin, which can support productive agriculture. Nevertheless, they cover over 32 million hectares!

Why are the Amazonian basin soils so poor in nutrients? A clue comes from an examination of the nutrient contents of trees and soil from temperate zone and tropical forest ecosystems.

	Nutrient elements (kg ha^{-1}):		
	Nitrogen	Phosphorus	Potassium
Temperate deciduous ash and oak (Belgium) (Biomass 380 tonnes ha^{-1})			
Soil:	14 000	2 200	767
Vegetation:	1 260	95	624
Tropical deciduous (Ghana) (Biomass 333 tonnes ha^{-1})			
Soil:	4 587	13	649
Vegetation:	1 794	124	808

Table 73
The mass of certain elements in the soil and vegetation in two representative deciduous forests, one temperate and the other tropical.
Data from Ovington, J. D. Organic production, turnover and mineral cycling in woodlands, Biological reviews, **40**, 295, 1965 and Duvigneaud, P. and Denaeyer-de-Smet, S. Biological cycling of minerals in temperate deciduous forests. In Reichle, D. E. (ed.) Analysis of temperate forest ecosystems, Springer-Verlag, N.Y., 1970.

a *Calculate for each element at each site the ratio of its mass in the soil to its mass in the vegetation.*

b *Is the felling of the temperate or the tropical forest more likely to produce soil more suitable for the growth of plants? Which of the three elements in the tropical soil is most likely to be deficient?*

Why do the plants in a tropical forest contain such a high proportion of certain nutrient elements in the ecosystem? Some experimental evidence comes from plots in the Amazonian rain forest, near San Carlos in the Rio Negro in Venezuela. Radioactive ^{45}Ca and ^{32}P were added both in leaves and twigs placed on the forest floor and as a direct chemical spray, to simulate 'decomposed' litter, on the plots. In nine of the ten study plots, less than 0.1 per cent of the labelled nutrient leached past the root mat in six months. All leaching ceased after one or two years. Analysis of root samples showed that the isotopes had indeed been taken up and translocated by living roots.

The complex root mat therefore enables the forest to survive even on a poor soil. Many species also produce 'canopy roots' from their branches and twigs. These adventitious roots take up nutrient ions from the decomposing remains of epiphytic plants on their trunks and branches, and may also trap nutrients from precipitation. Add this to the fact that the litter decomposition is probably five times faster in the tropics than in temperate regions, and it does not seem surprising that the residence time for nutrient ions in the soil is much shorter in tropical than in temperate forests.

c *If leaf litter decomposes five times more rapidly in tropical than in temperate forests, when leaves fall one might expect five times the nutrient ions to be suddenly released in tropical soils than in temperate soils. Why, then, is the nutrient concentration in tropical forest soils often* **lower** *than in temperate soils?*

29.5 Pollution

Human activities have had a dramatic effect on many ecosystems. Many of the compounds which man has released into the environment have never previously come into contact with living organisms (dioxin, DDT (dichlorodiphenyltrichloroethane), polychlorinated biphenyls). Some pollutants are more serious than others. Many pollutants are broken down by decomposers and do not accumulate in ecosystems (carbon monoxide, oil, sewage). The most dangerous long-term pollutants are likely to be atmospheric carbon dioxide (Study item 29.41), compounds which cannot be broken down by organisms and build up in food webs (lead, radioactive isotopes, mercury, dioxin, DDT and its derivatives), and 'acid rain'.

> Practical investigation. *Practical guide 7*, investigation 29D, 'A comparison of the growth of tolerant and non-tolerant seedlings when exposed to metal ions'.

STUDY ITEM
29.51 The influence of pollution on river communities

Waste materials from industry and domestic sewage are sometimes discharged directly into a river. The waste from laundries, the food and drink industries, and paper mills contains many unstable organic compounds. They are readily oxidized and decomposed.

The effect on the river is similar to that produced by the decay of large masses of dead vegetation. The influence of such an effluent on the levels of oxygen and suspended solids in the water downstream is shown in *figure 373*. The graphs are results averaged from a large number of investigations. *Figure 374* shows corresponding changes in the

Figure 373
Changes in the concentration of dissolved oxygen in river water with distance from a point of effluent discharge. The oxygen used by decomposing organic waste, and the quantities of suspended solids in the water, are also illustrated.
Figures 373–5 are based on Hynes, H. B. M. The biological effects of water pollution; in Yapp, W. B. (ed.) The effects of pollution on living material, *Symposium of the Institute of Biology*, **8**, 1958.

Figure 374
Changes in the concentrations of various ions with distance from the point of effluent discharge in a polluted river.

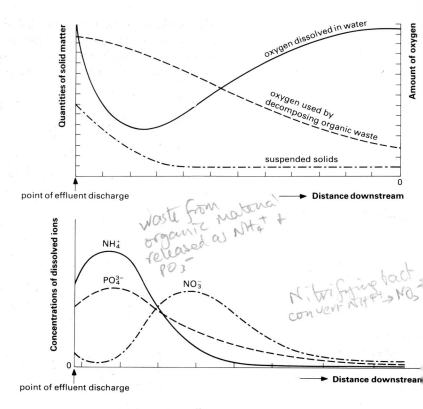

concentrations of ammonium, nitrate(V) and phosphate ions in water downstream from an effluent outlet.

a *Explain in detail the data in* **figure 374** *on the basis of the rates of decomposition of organic matter documented in* **figure 373**.

b *What are the ecologically important physical and chemical characteristics of water polluted by organic residues?*

The distributions of different species of river organisms downstream from the effluent outflow are shown in *figure 375*. A clear water fauna consists, for instance, of fish, freshwater pulmonate snails, water shrimps (*Gammarus* spp.), caddis fly larvae, and so on. Use keys or reference books to find out what the various organisms look like.

c *What evidence is there for three fairly well-defined communities downstream from the point where the effluent was discharged? List the chief members of each community.*

d *Name two relevant physiological characteristics which the members of the polluted water community might possess.*

e *What is the practical value of the information obtained from these studies of river pollution?*

f *The pollution of a river with organic matter is sometimes measured in terms of the biochemical oxygen demand (B.O.D.) of its waters.*

This is shown in the curve labelled 'oxygen used by decomposing organic waste' in figure 373. From your knowledge of respiration and the rate of diffusion of oxygen from air to water, suggest the reasons underlying the use and value of this term.

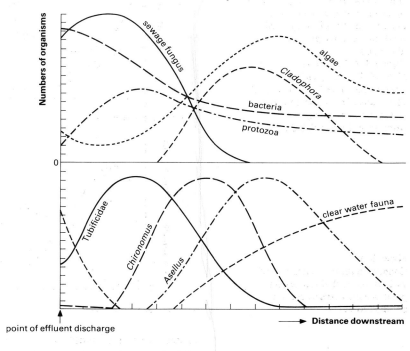

Figure 375
The distributions of organisms downstream from the point of effluent discharge in a polluted river.

STUDY ITEM

29.52 The influence of sulphur dioxide on lichen patterns in Britain

Lichens are slow-growing, often brightly-coloured organisms growing on stones, walls, roofs, tree-trunks, and branches. Each lichen consists by

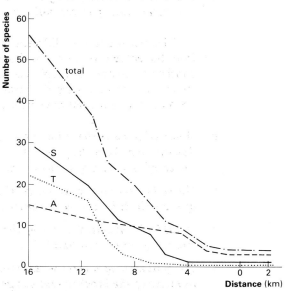

Figure 376
The numbers of lichen species occurring on asbestos and cement (A), sandstone (S) and trees (T) along a transect through the centre of Newcastle-upon-Tyne. The horizontal axis represents the distance from the centre of the city, where O = centre.
From Gilbert, O. L. Symp. Brit. Ecol. Soc., **5**, 35, 1965.

Chapter 29 Communities and ecosystems 537

mass of about 95 per cent fungus and 5 per cent alga, intimately bound together in an association which may be mutualistic (section 27.6). The fungus and the alga, grown separately, look quite unlike the lichen they might form when grown together. There are nearly 1400 species in Britain.

Any transect along a line which stretches from the countryside to the middle of a major industrial city shows that the number of lichen species declines towards the centre. It is usual to find tens of species in unpolluted graveyards outside a city, but only one or two in the very centre (*figure 376*). The order in which the species become extinct along such a transect is roughly the same for all cities. Some species are more sensitive to pollution than others.

Researchers were able to define ten groups of lichens on the basis of their different susceptibilities to pollution. These 'lichen zones' have been mapped across the British Isles. The general picture for England and Wales is shown in *figure 377*. As one might expect, even the countryside around the major conurbations lacks lichens. The western uplands of Britain, however, generally lack pollution and that is where the most dramatic lichen floras are still to be found (*figure 378*).

Lichens absorb substances of all kinds. There are several possible components of industrial smoke or car exhausts which might affect lichen patterns, including carbon monoxide, lead, hydrocarbons, sulphur dioxide, carbon particles, and oxides of nitrogen. Correlations between the lichen zones and the winter levels of sulphur dioxide near the ground in the area have proved striking (see table 74, page 540). In fact the most sensitive indication of the mean winter sulphur dioxide level in

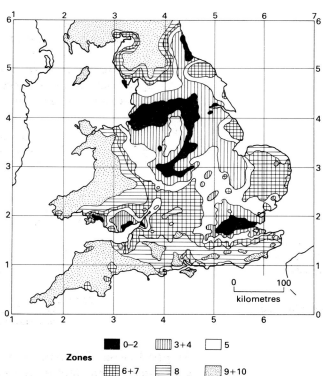

Figure 377
Approximate limits of lichen zones in England and Wales. Lichen zone 0 is the most polluted, zone 10 is the least polluted.
From Hawksworth D. L. and Rose F. *Studies in Biology 66*. Lichens as pollution monitors, *Edward Arnold, 1966*.

Figure 378
Lichens on trees in Devonshire. The long trailing lichens are *Usnea* species, which indicate unpolluted areas. *Photographs, Dr M. C. F. Proctor, Department of Biological Sciences, University of Exeter.*

the atmosphere at a site is to census the lichen species present, to find the lichen zone, and from that to estimate the sulphur dioxide level (table 75, page 540).

These correlations have been investigated experimentally by exposing lichens to various concentrations of sulphate(IV), the ion formed by sulphur dioxide when it dissolves in water (*figure 379*).

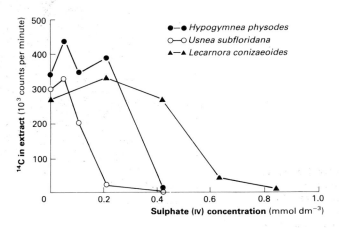

Figure 379
The effects of sulphate(IV) on the incorporation of ^{14}C in *Usnea subfloridana*, *Hypogymnea physodes*, and *Lecanora conizaeoides*.
From Hill, D. J. New Phytologist, 70, 831, 1971.

a   *Suggest how the ^{14}C isotope was incorporated into the fungus in the experiment which yielded the results illustrated in* figure 379.

Table 74
Comparison of mean winter SO$_2$ 1967–70 smoke and sulphur dioxide levels and the lichen vegetation at selected sites.
Data from Hawksworth, D. L. and Rose, F. Studies in biology 66, Lichens as pollution monitors, Edward Arnold, 1976.

Monitoring station	Smoke (µg m^{-3})	Sulphur dioxide (µg m^{-3})	Lichen zone(s)
Leicester	89	175	0–1
Kew	38	150	2–3
Buxton	17	126	3
Sheffield	43	88	3
Dursley	22	87	3–4
Hayfields	102	84	3–4
Plymouth	97	82	3–4
Didcot	22	39	7
Torquay	33	32	8
Llanberis	29	27	9

b *Relate the results in tables 74 and 75 to the data in* **figure 379.**

Table 75
A very much simplified zone scale for the estimation of mean winter sulphur dioxide levels in England and Wales.
Adapted from Hawksworth, D. L. and Rose, F., Nature, 227, 155, 1970.

Zone	Characteristic species	Mean winter SO$_2$ (µg m^{-3})
1	*Pleurococcus* (an alga) grows on tree trunks	above 170
2	*Lecanora conizaeoides* present on trees and acid stone	150–160
4	*Parmelia saxatilis* on trees and acid stone; *Grimmia pulvinata* (a moss) on mortar	100
4–6	*Hypogymnia physodes* on trees	70–90
7–8	*Evernia prunastri* and other shrubby lichens appear	40–60
9–10	*Usnea* species become abundant	about 35

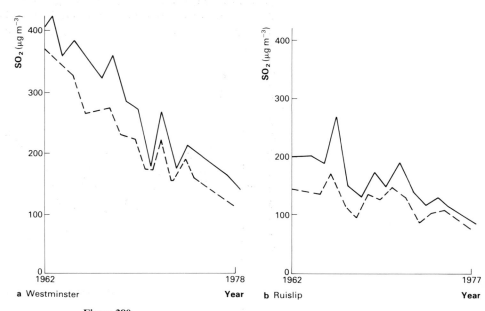

Figure 380
Changes in the mean annual (black) and winter (unshaded) SO$_2$ values at Westminster (centre of London) and Ruislip (16 km from centre) from 1962 to the late 1970s.
From Rose, C. I. and Hawksworth, D. L. Lichen recolonization in London's cleaner air. Nature **289**, 289, 1981. Reprinted by permission from Nature. Copyright © 1981 by Macmillan Journals Ltd.

In London, 129 lichen species were lost from an area within a 16 km radius of Trafalgar Square between 1800 and 1970. Since the mid 1950s, however, the sulphur dioxide concentrations at ground level have been falling as a result of the Clean Air Act (1956). (See *figure 380*.) A census in 1979 of 29 sites in north-west London showed that *Hypogymnia physodes* had returned to sixteen, and *Evernia prunastri* to eight. *Usnea subfloridana*, a member of a genus last seen in London before 1800, now occurs at Ruislip, 14 km from the centre.

STUDY ITEM

29.53 The effect of DDT on the bald eagle

The bald eagle (*Haliaeetus leucocephalus*) is an American national symbol, but although it is one of the most legally protected organisms in the country, it is on the danger list. One reason for its decline might have been the widespread use of organochlorine insecticides, such as DDT. Several studies in Britain and America have shown for other birds of prey that **1** the birds, being at the tops of food webs, accumulate DDT, **2** DDE, a metabolite of DDT, accumulates in the body fat and **3** it lowers reproductive success, in some cases by reducing eggshell thickness so that the eggs break when incubated.

Of course, DDT has undoubtedly saved millions of human lives as an effective insecticide. It has been particularly valuable in the control of mosquitoes which carry malaria. Although the environmental problems associated with DDT and its metabolites led the Environmental Protection Agency to ban the use of DDT in the United States at the end of 1972, the chemical is still used on a large scale in the less developed countries.

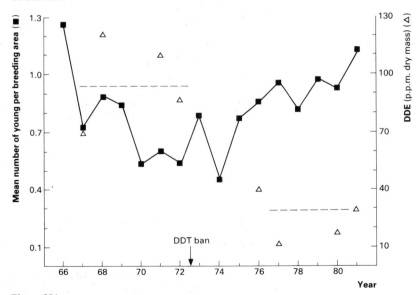

Figure 381
Summary of average annual bald eagle reproduction and DDE residues in addled eggs in north-western Ontario, 1966 to 1981. The dashed lines indicate weighted mean concentrations of DDE residues in clutches before (94 p.p.m.) and after (29 p.p.m.) the ban of DDT.
From Grier, J. W. 'Ban of DDT and subsequent recovery of reproduction in bald eagles', Science, **218**, *1232, 1982. Copyright © 1982 by A.A.A.S.*

A population of bald eagles in northwestern Ontario was studied from 1966 to 1981. The birds nest in a part of Canada which has suffered little exposure to DDT. However, they spend the winter in the United States, where they consume prey contaminated with DDT.

During the 16-year study, the average number of young produced was 0.82 per breeding area per year. The mean concentration of DDE in addled eggs was 57 parts per million. The measurements are summarized in *figure 381*.

a *Summarize what the graph in* figure 381 *shows.*

b *Construct a hypothesis to explain the decline and recovery of the bald eagle population in the area shown.*

c *How could you test your hypothesis experimentally?*

d *Suggest three other explanations for the fluctuations in bald eagle numbers.*

e *Why do birds at the tops of food webs have higher levels of DDE in their body fat than organisms lower down?*

f *DDE in food webs does not break down. Suggest two reasons why the average DDE concentration in bald eagle eggs declined from 1974 to 1981.*

29.6 Nature conservation

At the same time that biologists have learnt to recognize and to classify the vast numbers of species on Earth, they have realized that many species are becoming extinct or are teetering on the brink of extinction. Species in nature may be dying out at an accelerating rate because of habitat destruction for farming, mining, and forestry, urbanization, pollution, overfishing, and the killing of animals for natural products. In various ways and for various reasons, local, national, and international action has been taken by conservation organizations to preserve endangered species and habitats, and to change political opinions so that wildlife resources are used for the long-term benefit of humankind.

Large areas of natural habitat are valuable for various reasons. They act as watersheds, soaking up rainfall and releasing it slowly in rivers and streams, thus preventing soil erosion. Recreation areas where people can view and study species under natural conditions are often also valuable because tourism, fishing, and shooting often bring money into economically depressed areas. In places such as the uplands of Wales and Scotland, the most appropriate use for the land may often be to manage it for wildlife.

One reaction of governments and voluntary organizations in both rich and poor countries has been to set up national parks and nature reserves in which examples of natural habitats can be preserved. In Britain, for example, the Nature Conservancy Council, a government organization, has the brief of conserving examples of the main types of natural communities, and manages nearly 200 National Nature

Reserves. This preservation is an active process, for many communities would undergo succession, and the range of animal and plant species might change, if the vegetation was not grazed, cut, or protected from visitors.

Voluntary organizations have also mounted vigorous campaigns to make the public and politicians aware of the rate at which wildlife habitats and aesthetically-pleasing species are disappearing. There have been protests over whaling, the use of crocodile skin and fur pelts to make garments, and farming practices like the destruction of hedges and stubble burning. The increasing rate of disappearance of tropical rain forests, with their fascinating diversity of species, is currently in the public eye. Zoos may be used to breed rare species with a view to releasing their descendants back to the wild, and particular species, such as the giant panda, may be used as a symbol to raise a great deal of money for the conservationist cause.

Conservation is a political and social matter. Conservation of a natural habitat is only one of several possible uses for valuable land in an intensively-cultivated natural landscape. In the real world, are not efforts to conserve species merely an exercise in sentimentality and nostalgia?

After all, about ninety-nine per cent of all the species which have ever lived are thought to have become extinct, and a few more will not make much difference. Although the disappearance of species has altered the balance of some communities, ecologists have been unable to show that there are critical species whose disappearance will radically alter life on the planet or threaten human survival. Is it morally justifiable for ecologists from intensively-cultivated European countries, in which for centuries the landscape has been managed for human gain and species have been systematically exterminated, to press governments in other continents to leave large areas of their natural vegetation intact? The remaining semi-natural vegetation of the world is a natural resource ripe for exploitation by peoples whose average income per head is minute. They may have no other way of making a living and raising their expectations.

STUDY ITEM

29.61 The value of nature conservation (Essay)

a *Discuss whether the conservation of wild species of animals or plants is either necessary or desirable, using named species of animals or plants to illustrate your arguments.*

Main terms used in ecosystem ecology

Autotrophs Green plants and chemo-autotrophic bacteria, organisms which 'feed themselves' from inorganic compounds and an energy source such as the Sun.

Biomass Dry mass. It may refer to the dry mass of an organism, a population of organisms or, more usually in ecosystem ecology, a whole trophic level, as in 'pyramids of biomass'.

Carnivores Animals which catch living animals, kill and eat them.

Chemo-autotrophs Autotrophic bacteria which obtain energy not from the sun, but by the oxidation of inorganic ions or molecules such as $Fe(II)$, ammonia (NH_3), methane (CH_4) or hydrogen sulphide (H_2S). The energy is used to make glucose from carbon dioxide. Chemo-autotrophs, for example nitrifying bacteria in the nitrogen cycle, can live in the dark.

Decomposers Mainly bacteria and fungi, which secrete enzymes onto dead organic matter and absorb through their cell walls the breakdown products of digestion.

Detritus Dead organic matter produced by the plants and animals in an ecosystem. Food for detritivores and decomposers.

Detritivores Animals which feed on dead organic matter.

Ecosystem Community of organisms, and their environments, in which ecologists study energy flow and nutrient cycling through trophic levels.

Energy flow The passage of energy through the trophic levels in an ecosystem. The energy in sunlight is absorbed by the leaves of green plants. It is then transferred from one trophic level to the next. At each stage energy is lost as heat by respiration and egestion. Ultimately the rest of the energy is lost by decomposer respiration.

Herbivores Animals which feed on green plants. Primary consumers.

Nutrient cycling The movements of essential elements between the trophic levels and between the organisms and the soil, air and water in an ecosystem. The nutrients are absorbed from the environment by green plants and pass from one trophic level to another. Ultimately all the nutrients will pass into the environment again. Most nutrient atoms cycle round and round in the same ecosystem. Many nutrient atoms are, however, exchanged between ecosystems.

Omnivores Animals which eat both green plants and other animals.

Parasites Organisms which feed in or on another living organism (the host) from which they obtain organic compounds. The host is harmed.

Primary consumers Herbivores which eat green plants (producers).

Primary producers Green plants which absorb inorganic compounds and use them, and sunlight energy, to make in photosynthesis the organic compounds on which the other organisms in the ecosystems depend.

Primary production The production of plant material. This provides the food for primary consumers, decomposers, and detritivores.

Production Dry mass of plants or animals harvested or produced per unit area per unit time. Measured in units of $kg\ m^{-2}\ year^{-1}$.

Productivity Rate at which dry mass is produced. Sometimes used to mean rate of dry matter production per unit dry mass.

Secondary consumers Carnivores which eat primary consumers.

Secondary production The production of the animals in an ecosystem, or the animals at a particular trophic level.

Standing crop The mass of a particular species, a trophic level, or the plants or animals present at any one place at one time. Available to be harvested. Often expressed in terms of dry mass (to compensate for differences in water content).

Trophic level A group of organisms which all obtain their energy in basically the same way, *e.g.* primary producers, primary consumers, decomposers.

Summary

1. Ecosystems consist of communities of organisms and the environments in which they live. When ecologists investigate ecosystems they assign the organisms to trophic levels on the basis of food webs, and study energy flow or nutrient cycling in the whole system or part of it.
2. Food webs, trophic levels, and ecological pyramids are briefly discussed (**29.1**).
3. The factors affecting the net primary production of the plants in an ecosystem, that is, the rate at which they accumulate dry mass, are described with reference to crop plants (**29.3**).
4. The animals in an ecosystem convert food to their own dry mass. The rate at which they do so is known as secondary production. The efficiency of secondary production by the primary and secondary consumers in an ecosystem depends on the proportion of primary production eaten, whether the animals are homoiotherms or poikilotherms, and whether they eat plants or animals (**29.3**).
5. The high levels of productivity in modern intensive crop production are maintained by a massive input of energy from the burning of fossil fuels (**29.3**).
6. Energy from the producers and consumers, in dead parts and egesta, becomes accessible to the detritivores and decomposers in the system. They ultimately release the incoming energy as heat, and make the incoming elements available for uptake by producers (**29.3**).
7. The carbon cycle of the Earth, and the nitrogen cycle in the Hubbard Brook temperate forest ecosystem in New Hampshire, U.S.A., are described in detail. Forest clearance disrupts nutrient cycles and this is illustrated for both temperate and tropical forests (**29.4**).
8. The species composition of many communities is affected by pollutants (**29.5**).
9. This chapter ends with a discussion of the value of nature conservation (**29.6**).

Suggestions for further reading

ANDERSON, J. M. *Ecology for environmental sciences: biosphere, ecosystems and man.* Edward Arnold, 1981. (A concise, modern account of energy flow and nutrient cycling in ecosystems, and applications in agriculture, forestry and waste disposal.)

ANDERSON, J. M. 'Life in the soil is a ferment of little rotters.' *New Scientist* **100** (1378), 1983, 29. (A lengthy and beautifully-illustrated summary of decomposition in the soil.)

COLLINVAUX, P. *Why big fierce animals are rare.* Penguin, 1980. (Entertaining series of brief, well-written essays which include much interesting common-sense about ecosystems, avoiding jargon.)

DELWISCHE, C. C. 'The nitrogen cycle'. *Scientific American* **223**(3), 1970, 137. (A short clear account of this complex cycle and the underlying chemical changes.)

DUFFEY, E. *Nature reserves and wildlife.* Heinemann, 1974. (Discusses principles and practice of nature conservation, with emphasis on Britain.)

HAWKSWORTH, D. L. and ROSE, F. Studies in Biology 66. *Lichens as pollution monitors*, Edward Arnold, 1976. (Useful for project work.)

KING, T. J. *Ecology.* Nelson, 1980. (Elementary introduction to the principles of ecosystems, pollution, and nature conservation.)

MASON, C. F. Studies in Biology 74. *Decomposition.* Edward Arnold, 1976. (Deals in simple fashion with the activities of detritivores and decomposers, especially in land habitats.)

MELLANBY, K. Studies in Biology 38. *The biology of pollution.* Edward Arnold, 1972. (A readable introductory account of the effects of the major pollutants on humans and wildlife.)

PHILLIPSON, J. Studies in Biology 1. *Ecological energetics.* Edward Arnold, 1966. (A modern classic. Difficult to read from cover to cover, but worth dipping into for its excellent examples.)

WOODWELL, G. M. 'The carbon dioxide question.' *Scientific American* 238, 34, 1978. (Deals with the carbon cycle and its possible climatic impact.)

CHAPTER 30 EVOLUTION

30.1 Two centuries of evolution theory

The ladder of nature

For several centuries before 1800, people's views of the natural world were dominated by one powerful idea: that of the *scala naturae* or 'ladder of nature'. All animals, plants, and inanimate substances were thought to form a single series. At one end were the simplest and most elemental materials, while at the other was the highest form of life.

In its fullest version, the *scala* included everything, and there were no gaps between the steps of the ladder. But the highest form of life was God, and man was less than halfway up: so how was the enormous gap between man and God to be filled? To solve this problem, the medieval theologians introduced the doctrine of angels, nine orders of spiritual beings who ranged from the man-like to the God-like and who were thought to govern the motions of the planets.

Clearly this vision of nature bears little resemblance to modern evolution theory. The structure was static; each substance or being had its appointed place in the series. It was not possible for anything to climb up or down the ladder and so transform itself into an entity of a different kind. In earlier days, the *scala* was intimately associated with Christian doctrine: it was a theory of the relationship of man and nature to God. Since the 17th century, however, naturalists were not so interested in the spiritual beings above man: following the work of Kepler and Newton they were no longer thought necessary to explain the motion of the planets. The top part of the ladder was put aside, and attention was directed instead to the detailed arrangement of minerals and living creatures. Many versions were suggested, and *figure 382* gives an example. The main groups of organisms are shown in bold type, and you can see how the gaps were thought to be filled by intermediate kinds.

a   *How does the sequence of major groups in* **figure 382** *differ from a modern evolutionary series?*

b   *What other intermediate kinds could have been regarded as evidence for the existence of the* **scala naturae**?

Lamarck and progressive improvement

The *scala naturae* made a profound impression on the French biologist Lamarck. He came from the semi-impoverished lesser nobility of northern France; following in his father's footsteps he spent several of his early years in military service. Spells of duty in eastern France and near the Mediterranean encouraged his interest in botany. In 1779 his *Flore françoise* was published, describing the plants found in France. It was a notable book if only because for the first time it used dichotomous keys, like the ones we still use for identifying plants and animals. In the

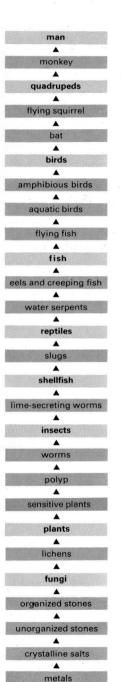

Figure 382
An example of the *scala naturae*, adapted from the eighteenth-century *Contemplation de la Nature* of Charles Bonnet.

Figure 383
The duck-billed platypus, *Ornithorhynchus anatinus*, showing its rubbery bill and webbed feet. This mammal lays eggs and suckles its young through pores in the mammary area.
Photograph, Australian Information Service, London.

Figure 384
Jean-Baptiste de Lamarck (1744–1829).
Photograph, Mary Evans Picture Library.

upheavals following the French Revolution, the Musée National d'Histoire Naturelle was created in Paris, and Lamarck was appointed to it as a professor of zoology.

Although Lamarck originally believed that species could not be transformed into other species, in 1799 or 1800 he changed his opinion. In the course of his teaching at the museum, he developed a theory of evolution. At the time, the question whether species ever became extinct was a subject of lively debate, and Lamarck found he could not accept the possibility of extinctions. A much more acceptable explanation for the disappearance of species from the fossil record was that they simply changed into other species.

Having asserted that species can change into others, Lamarck described the approximate course that evolution had followed and also attempted to show how it happened. Central to both parts of the theory was the *scala naturae*. Life originated as small and simple forms at the bottom of the ladder of organisms. New species then came into being, gradually progressing higher up the ladder, until man eventually appeared. The *scala* was the over-riding pattern in nature, and the reason was that living things had an innate tendency to become 'higher', more complex kinds of organisms. This spontaneous progressive improvement was the central feature of his theory.

Lamarck was well aware that the *scala* was imperfect, however. There certainly was no smooth gradation from the simplest organism to man. To explain the all-too-obvious deviations from the ideal pattern, Lamarck suggested that there were other mechanisms of change. These interfered with progressive improvement. In particular an animal may experience certain needs arising from its environment, such as the need of an antelope to escape from lions. This would stimulate the antelope to develop strong muscles, long legs, and acute senses during its lifetime. Then, Lamarck supposed, some of these changes could be handed on to

Figure 385
Sir Charles Lyell (1797–1875). *Photograph, The Mansell Collection.*

the next generation. In other words, the changes acquired by the parent through its own experience could be inherited. This theory of the inheritance of acquired characters was one way of explaining the *scala*'s imperfections.

The early nineteenth century

Lamarck's theory gained few adherents in France. In 1818 he became blind, and when eventually he died in 1829 his family were unable to pay for his funeral. It was a sad end to the career of a highly talented man.

Nor were there many supporters of Lamarck's ideas in Britain, although on this side of the Channel they did at least reach a wide audience – through the writings of a man who wholeheartedly rejected them. This man was the great geologist Charles Lyell.

Lyell was born near Kirriemuir in Scotland, but before he was a year old his family took him to England. During his boyhood in the New Forest and Sussex Lyell's favourite hobby was collecting insects, but while a student at Oxford he turned to geology. He became convinced that the planet Earth was immensely older than the 6000 years allowed by orthodox theologians. In addition he came to believe that the extraordinary changes that must have occurred in geological history, such as the creation of mountain ranges, were not the result of occasional gargantuan upheavals. Quite the opposite: given enough time, all geological phenomena could be explained as the result of the slow, barely perceivable changes going on around us today. For instance, if an area of land is pushed upwards by only 1 cm each year, after a million years it will be a mountain range 10 000 m high.

This theory was heretical to the majority of geologists in the 1820s. Nevertheless, when Lyell published these ideas in his *Principles of geology* in 1830–33, the book attracted much attention and had a far-reaching influence. It is important to our story for two reasons. First, Lyell showed how nature, given immense periods of time, could be seen to make profound changes merely by the slow action of familiar processes. Once accepted in geology, the idea could the more easily be accepted in biology (as we shall see shortly). Secondly, Lyell disagreed profoundly with Lamarck's theory of evolution. So, in order to say so, he first gave a full and fair account of Lamarck's views. The *Principles of geology* was widely read not only at its first appearance, but also for several decades as new and improved editions appeared, and as a result Lamarck's theory became well known in Britain between 1830 and 1860.

Part of Lyell's success was due to his talent as a writer. His books were a pleasure to read, even for the layman. This was also true of the writings of another Scot, Robert Chambers. Born in Peebles and later living in Edinburgh and St Andrews, Chambers came from humbler origins than those of Lyell and Lamarck. After leaving school he took jobs as a teacher and a junior clerk, but eventually found his feet when he started a bookshop. From small beginnings, prosperity quickly followed. With his brother he started the famous publishing firm of W. & R. Chambers, but while William ran the business Robert wrote books, many about Edinburgh and Scottish history.

Figure 386
Robert Chambers (1802–1871).
Photograph, The Mansell Collection.

So Robert Chambers knew the publishing business inside out, and was also a fluent and successful author. He decided to write a book advocating a theory of evolution. In order to safeguard the firm's good name and his own literary reputation, the book was to be published anonymously by a firm in London. In 1844 the *Vestiges of the natural history of Creation* appeared, and it was an immediate success. It went through edition after edition over the next four decades, and it was only in the twelfth edition of 1884 that the secret of its authorship was disclosed, long after Chambers' death. Meanwhile rumours abounded; some people accused Lyell of having written it, and others thought it was by Prince Albert.

The theory advocated in the book was quite similar to Lamarck's. Species were thought to have climbed the ladder of nature, with man finally emerging from a modified ape. Compared with the orthodox theological account of man's origin – by a supernatural act of God – Chambers' account of man rising from the lower animals seemed scandalous. The book excited a great deal of public interest, although few if any scientists expressed their support for the theory. Well-written and persuasive though the *Vestiges* was, the anonymous author was clearly not a biologist with a first-hand knowledge of the things he was writing about.

By the 1850s, then, evolution theory was a common topic of discussion, even though the leaders of science refused to accept it. The theory moreover was still closely tied to the *scala naturae*; species were portrayed as climbing the ladder of nature from humble beginnings to the eventual emergence of glorious man. But this situation changed in the late 1850s.

Wallace and Darwin

Like Chambers, Alfred Russel Wallace grew up without the comfortable and privileged upbringing of Lyell and Darwin. His first five years were spent in Usk, a small town in Gwent, and then the family moved to Hertford. After an education in the one-roomed Hertford Grammar School, which did not impress him very much, he was apprenticed as a surveyor to his brother before becoming a school teacher for a while. Just as Chambers had made his breakthrough when he tried his hand at bookselling, so Wallace decided on yet another occupation – collecting natural history specimens in the Amazon basin. He was there from 1848 to 1852, but unfortunately lost most of his collections when the ship sank on the return voyage. Undaunted, he was again exploring and collecting from 1854 to 1862, this time in Malaysia and Indonesia.

Before setting out for the Amazon, he read widely in the fields of travel and science, including Chambers' *Vestiges*, a book he admired. For ten years from the start of his explorations in the tropics, he kept trying to solve the problem of whether one species could change into another. While on a collecting trip in the Moluccas in 1858, he was in the throes of a bout of malaria when a solution sprang into his mind. He recalled the argument of another book he had read back in the 1840s, a book on human populations by Thomas Malthus. It demonstrated that

Figure 387
Alfred Russel Wallace (1823–1913).
Photograph, The Mansell Collection.

Figure 388
Charles Darwin (1809–1882) as a young man.
Photograph reproduced by kind permission of the President and Council of the Royal College of Surgeons of England.

human populations could increase dramatically over the course of a decade or two. But normally they remained in fact at a fairly steady level. Why was this? Malthus argued that unless population growth was held in check by disease, war, or voluntary controls, there was bound to be the misery of massive poverty and starvation.

In his fever, Wallace saw that the same was true of all animals and plants in nature. They could all increase in numbers so fast that food supplies would quickly be outstripped. So, many individuals must be dying before they could produce any offspring. This meant there was competition for inadequate resources. If some individuals were better able than others to survive the competition, these would be the parents of the next generation. To go back to Lamarck's example, some antelopes would have more acute senses and greater speed than others. So these would have a better chance of contributing offspring to the next generation; and if their superior senses and speed were heritable, the next generation would be slightly improved in these respects. It was as if nature were selecting the best adapted to be the parents of the next generation.

Wallace soon wrote out his theory of natural selection. Knowing Charles Darwin would be interested, he posted it to England where Darwin received it in June 1858. It came to him as a bombshell, for, unknown to Wallace, Darwin had been thinking along the same lines, and had written a summary of his ideas some twenty years earlier, also as a result of reading Malthus' book. Unfortunately he had not yet published it, even though he had written it out as an essay in 1844, the year the *Vestiges* appeared. Obviously Darwin did not want to lose the credit for having thought of the theory first, and so advice was sought from Darwin's friends Lyell and Joseph Hooker, the botanist. They decided to present both Darwin's and Wallace's versions of the theory at a meeting of the Linnean Society in London, which took place on 1 July 1858. Darwin thereupon set to work writing a book on his theory. He had been accumulating evidence ever since the 1830s and had already started on a series of volumes describing the natural selection theory. But speed was now the top priority, and as a result a much shorter book was published at the end of 1859. It was called *On the origin of species by means of natural selection*.

c *Do you think there could have been any connection between the theory of natural selection and the opinions prevailing in politics at the time of its formulation?*

The *Origin of Species*

In the *Origin*, Darwin showed how domesticated animals and plants vary. The seedlings from one plant, or the litter mates from one mother, are not identical. By selecting the desirable variants, man has brought into existence the wheats and pears, the dogs and pigeons, in all their great variety. Darwin then showed how all animals and plants produce far more offspring than survive to reproduce their kind. He calculated that even the slow-breeding elephant could produce nineteen million elephants in seven hundred and fifty years from one ancestral pair. But this does not happen: there are checks to such increases, and vast numbers of seeds and seedlings, eggs and young, are destroyed. These checks can be lack of space or food, unfavourable climate, epidemics, or predation.

The size of a particular population is often the result of complex causes. For example, Darwin suggested that the number of cats in a field affects the frequency of red clover plants. The cats determine the number of mice; the mice determine the number of bees by destroying nests; and the bees determine the number of clover plants because they pollinate them. Thus: no cats, no clover. The checks on natural increase lead to a struggle for existence and the survival of those adapted to the environment. The predatory hawk selects the grouse it sees on the moor, so the best camouflaged grouse survive. The hawk keeps the camouflage pattern constant. But if the climate and vegetation change, the hawk may establish a new camouflage pattern for the grouse.

There is divergence, too. As Darwin wrote: 'isolation, also, is an important element in the process of natural selection'. If a population is divided by a physical barrier, the separated groups are likely to diverge as they become adapted to slightly different environments. Isolated, they are prevented from interbreeding and have time to evolve independently. But, Darwin cautioned, if the isolated populations are small there may not be enough variety in the population for selection to act on; so a new species may not be produced.

Darwin then searched for a cause of the variation that was the essential raw material for selection. But he was not successful and confessed his failure. The effects of habit, use and disuse, developmental changes, reversions to an earlier state – all were discussed and found unsatisfactory. He had to admit that variation between individuals of a species existed all around him, but he could not find the cause. Further difficulties lay in the frequent discontinuities between living species and in the fossil record. There were no intermediates between bats and their possible ancestors, for example, but no one could be certain that there never had been.

So, in the first part of the *Origin of species*, Darwin put forward a theory of how evolution has been brought about. He identified a slow and unobtrusive process, *natural selection*, which over immense periods of time can produce extraordinary changes in organisms. In this sense his theory had something in common with Lyell's vision of geological history.

Next he argued that the theory explained many geological observations; two chapters of the book looked at the geological record. Darwin agreed with Lyell's account of how the earth had been formed, and then considered the fossils. He had to admit that there were gaps in the fossil record but attributed such gaps to inadequate collecting or the hazards of fossilization. He believed that missing links would be found to fill the gaps. In conclusion, Darwin thought that the fossil record provides evidence for gradual evolution; it does not support a theory of catastrophism or the story of the Flood. Even though it shows that certain animals, say, have become extinct, such extinctions are not

Figure 389
a The orange-rumped agouti, *Dasyprocta agouti*.
b The vizcacha, *Lagidium viscacia*.
c The hare, *Lepus* sp.
d The rabbit, *Oryctolagus cuniculus*.
Photographs: **a** *Rod Williams/Bruce Coleman;* **b** *Udo Hirsch/Bruce Coleman;* **c** *Hermann Schünemann/Frank Lane Picture Agency;* **d** *François Merlet/Frank Lane Picture Agency.*

evidence of world-wide catastrophe. Animals have become extinct because the environment changes and they do not, or because their populations are small, or because of the effects of competition. In short, they have become extinct through the operation in the past of the same forces that operate today.

This led to a comparison of events in different parts of the world and to an evaluation of how the geographical distribution of animals and plants supports evolution. Why, he asked are there agoutis and vizcachas on the plains of South America but hares and rabbits on the plains of Europe (*figure 389*)? Independent evolution from dissimilar ancestors to fit similar environments was the only sensible explanation.

Migration, colonization, and isolation at different stages of the earth's history and at different stages of animal evolution would account for the regional faunas of today.

But why did some animals and plants get from one place to another, and others not? Darwin thought that some required a land bridge (across the Bering Straits, for instance) while others were able to cross the sea. He tested the survival of seeds and found that dried seeds of wild celery survived to germinate after floating for ninety days in salt water. Such accidental dispersal by water and wind was the means by which islands like the Galápagos had acquired their daisy trees, iguana lizards, tortoises, and finches.

Finally, Darwin supported his theory by arguing from classification. Naturalists, however unconsciously, use the idea of descent and evolutionary relationship in systematics. Turning the argument round, classification points to ancestry and descent. Thus the theory of evolution accounts for similarities of structure, embryology, and rudimentary organs. But the systematist must beware of superficial similarities such as the shape of whales and fishes. Like Lamarck, Darwin realized the important differences between analogy and homology:

'When the views advanced by me in this volume, and by Mr Wallace in the *Linnean Journal*, or when analogous views on the origin of species are generally admitted, we can dimly foresee that there will be a considerable revolution in natural history.'

Why did biologists not accept Darwin?

The *Origin of species* was a huge success with the general public – less so, however, with the professional biologists. But whether biologists approved of the theory or not, they had to take notice of it. A revolution in natural history had occurred and the subject would never be the same again.

Like Lamarck and Chambers before him, Darwin was suggesting that man arose by natural processes from the apes. This in itself forced many Christians to oppose the theory vehemently. Perhaps surprisingly, however, scientists also found fault with Darwin's book. Indeed, not until another eighty years had passed were most biologists prepared to recognize natural selection as the principal mechanism of evolutionary change. By the early 1870s it is likely that the majority of biologists had come to believe in evolution, but it was not quite the same as the concept Darwin had in mind.

Darwin made it clear that he saw natural selection as a process causing adaptation. Under its influence an animal or plant species became better able to survive its environment. But this meant that many sorts of change could occur. For instance, a predator species might become more complex, enabling it to catch prey more easily; on the other hand, an internal parasite such as a tapeworm might benefit by becoming simpler in its organization. Stags might become larger, because they needed to fight against others of the same species; but mice might become smaller if that allowed them to escape from predators by running down holes.

In short, as Darwin made clear, natural selection could not only account for increasing complexity or advancement in organisms; it could equally well cause degeneration – or any other change. It was a theory of adaptation, particularly good at explaining adaptive radiation. But the concept of evolution that came to be widely accepted in the late nineteenth century was really rather like Lamarck's and Chambers'. The fossil record was perceived as demonstrating that life had started as very simple primitive forms, and that through geological time progressively more advanced forms had appeared. Life had climbed up the ladder of nature, which was now visualized as resembling the trunk of a tree, with numerous though rather unimportant branches. How this had happened was not certain, but natural selection did not seem to be the sole cause. The evolution theory now accepted was not quite what Darwin had hoped for.

d *Why is it misleading to think of invertebrates → fish → amphibia → reptiles → mammals as an evolutionary series?*

Many scientific criticisms were aimed at the *Origin of species*, and it is worth mentioning a few of the more important ones. First, just as Lyell's theory of geology demanded an almost unimaginable amount of time, so did the theory of evolution by natural selection. With all the authority of contemporary physics, Lord Kelvin had calculated a figure of 20 million years for the earth's age, and many people believed that natural selection could not possibly have turned a blob of protoplasm into man in that time.

Second, Darwin's insistence on the slow and gradual pace of evolution was thought to be unrealistic. Species were known to appear suddenly in the fossil record, and then to remain unchanged over long periods of time. Other sudden changes could be seen to be taking place at the present time, such as a normal sheep giving birth to a short-legged offspring. (Incidentally this particular offspring gave rise to a new breed of short-legged sheep which were useful because they could not jump over fences. See *figure 390*.)

Third, natural selection theory assumed that all the stages leading to a complex organ gave an advantage to the organism. For instance, the electric organ of a fish might be effective in stunning prey or an enemy only if it develops a certain minimum voltage. So how could the fish's ancestors have gained any advantage by having an organ producing only a very weak voltage, which would have been a necessary stage in the gradual evolution from a fish with no electric organ at all?

Fourth, palaeontologists were describing some remarkable series of fossils which were said to demonstrate *orthogenesis* or straight-line evolution. This meant that, over millions of years, a line of fossils showed a very consistent direction of modification (see *figure 391*). Sometimes this seemed to have an unstoppable momentum, so that eventually quite bizarre forms were produced, so badly adapted that the line became

Figure 390
An Ancon ram (*left*) and Ancon ewe (*right*) either side of a normal ewe.
Photograph from Landauer, W. and Chang, T. 'The Ancon or otter sheep: history and genetics', Journal of Heredity, **40**, *105, 1949.*

extinct. One example concerned oysters: the fossils became progressively more tightly coiled, with the result that the latest forms apparently could not have opened their shells. It seemed that natural selection could not account for orthogenesis.

A fifth difficulty was particularly troubling to Darwin. This concerned the origin of the varieties from which nature selected the best-adapted forms. Before 1900 there was no generally accepted theory of heredity, but most biologists favoured some form of 'blending inheritance'. According to this view, offspring tend to be intermediate between the parental forms. So as the members of a species interbred, the range of variation would continually be reduced, making them all more and more like the average. Moreover, any new, better-adapted forms would tend to be lost by interbreeding with less well-adapted members of the species.

Darwin tried to surmount this problem by inventing his own theory of heredity. He called it 'pangenesis', and it was equivalent in effect to the inheritance of acquired characters. The action of the environment was supposed to make changes in tiny particles in the body cells, and these particles were then transferred to the germ cells from which the gametes were formed. This theory never gained much support.

e *How many of these objections to Darwin's theory might still hold water today?*

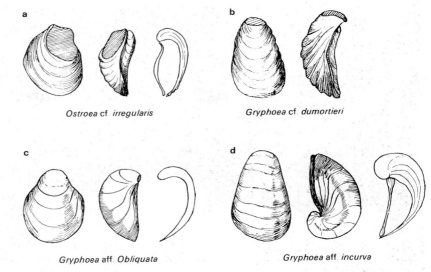

Figure 391
Four members of the lineage of *Gryphoea incurva*, showing the left valve, the shell as viewed from the posterior side, and the left valve in section. The area of attachment is shaded.
From Trueman, A. E., 'The use of Gryphoea in correlation of the Lower Lias', The Geological Magazine, **59**, *1922*.

Neo-Darwinism – the 'modern synthesis'

Considering the state of biological knowledge in the late nineteenth century, and the evidence available, it is easy to see why Darwin's theory could not be accepted whole-heartedly. But today the theory is highly venerated; indeed, it has become a part of everyday life. This is because the currently accepted theory is a slight modification of Darwin's and to emphasize this it is called 'neo-Darwinism'.

Biologists' understanding of variation and its inheritance changed dramatically in 1900. This was when Mendel's description of the fundamental laws of genetics came to light, after having been forgotten for more than thirty years. It is an interesting though perhaps fruitless speculation to wonder what Darwin would have made of Mendel's work, had he known of it – very little, in all probability, if only because Darwin was firmly convinced of the validity of the 'blending inheritance' hypothesis.

The geneticists of the early twentieth century were impressed by the striking differences between normal and mutant forms of the same species, such as the short-legged sheep. In addition a Dutch botanist, while studying the evening primrose *Oenothera* in the 1890s, found that sudden and marked changes occurred occasionally in the progeny. As the new science of genetics rapidly developed, many biologists were persuaded that new varieties arose by 'jumps' or large mutations from the parental forms. There seemed to be no room for a theory postulating the gradual and continuous change that Darwin imagined.

So Mendelian genetics was not immediately helpful to Darwinists. In fact the two groups of biologists formed opposing camps. Neo-Darwinism was not to be born until the two camps reconciled their differences, and when this came about the reconciliation and the theory it gave rise to were called the 'modern synthesis'.

As their science grew more sophisticated, geneticists realized that not all heritable variation was like the discontinuous mutations they were used to. It was discovered, for instance, that human height was inherited, even though it is a continuous kind of variation. A character such as height could be determined by genes if in fact many genes were concerned, rather than just one. Each of these genes would have only a slight effect on its own. Thus if all the genes involved coded for greater height, a very tall person would result; if only three-quarters of them did so, the person would be not quite so tall but still above average, and so on.

This expanded view of inheritance allowed the 'modern synthesis'. Genes are agreed to be the basis of inheritance, and the slow and gradual effects of natural selection postulated by Darwin can indeed occur where inheritance of a character involves many genes, each with only a slight effect. New variations arise by small or large mutations, and Darwin's worries over blending inheritance can be forgotten since the alleles of a gene do not generally blend. Neo-Darwinism arose from the marriage of Darwin's natural selection theory and Mendel's genetics, and this happened as recently as the late 1930s.

f *Would the neo-Darwinist theory of evolution have been established sooner if Darwin had never been born?*

How do things stand now?

Despite the general acceptance of neo-Darwinism, some questions are still puzzling modern biologists. They are not very different from those of the past hundred years. Is evolution directed from within the organism? Is evolution a series of jumps? Are there neutral genes not subject to selection? Can the environment cause adaptive mutations? Are the gaps in the fossil record real?

Few doubt that evolution has occurred. The arguments are about how it has occurred. And yet how it occurred was the main hypothesis of the Darwin–Wallace papers of 1858. The two men had found, they thought, a mechanism in selection – the differential check to population growth.

Only in the past fifty years have serious attempts been made to test the validity of the selection hypothesis. Selection has now been seen in action. It has changed pale populations of moth in industrial areas into dark populations, and metal tolerance has been selected for in grasses on old slag heaps. These changes have occurred fast – faster than anyone anticipated. But the new moths and grasses are not new species. There is, however, circumstantial evidence for fast speciation: for example, five species of banana moth have appeared on the Hawaiian Islands since the banana was introduced 1000 years ago. But how do species diverge from a common ancestor? Is geographical isolation a prerequisite?

Observations suggest that species have been formed after isolation, like the Galápagos finches. But a sudden heritable change – a jump – might lead to speciation without geographical isolation. Polyploidy is one example well known in plants. Many believe that other less dramatic chromosome changes are also important in animal speciation.

More than a hundred years after the *Origin of species*, the theory of evolution by natural selection is still provocative, still stimulating.

30.2 Punctuated equilibria

One of the most characteristic features of Darwin's theory of evolution was his emphasis on *gradual* change. The continued operation of natural selection on even minute variations could produce significant transformations provided there was plenty of time. This belief that species change only gradually was probably derived, at least in part, from Lyell's theory of geology. Geological events could be explained, Lyell believed, without invoking any processes not seen at the present day. Thus a huge mountain range could be raised as a result of numerous minor uplifts, such as those we know take place during earthquakes. Unlike Lyell, however, Darwin adopted a similar idea to explain evolutionary change in species. He insisted so strongly on gradualism that one of his staunchest supporters, T. H. Huxley, berated him for ruling out the possibility of species changing by means of 'saltations' or jumps.

Unfortunately for Darwin, he could find little support for gradualism in the fossil record. It revealed instead numerous breaks between species found in successive rock strata, and few series of fossils showed a smooth transition from one species to another. He accounted for this by suggesting there were breaks in the sequence of rocks, produced by intervals of erosion, so that here and there the fossil record had been destroyed. There might, for instance, be a period of several million years from which no rocks appear to have survived.

In the decades following publication of the *Origin of species* many palaeontologists were prepared to accept Darwin's view. They assumed that slow, gradual change in organisms had been the norm. This came to be known as *phyletic gradualism* ('phyletic' just means 'evolutionary'). But scepticism increased as research continued across the world; gaps in the sequence of rocks were being filled without the corresponding gaps in the fossil sequence being filled at the same time – even though occasionally 'missing links' were discovered such as the earliest bird *Archaeopteryx* (*figure 392*).

Eventually, in 1972, phyletic gradualism was challenged by two American palaeontologists, Niles Eldredge and Stephen Jay Gould. Eldredge and Gould put forward an alternative theory. They suggested that species persist unchanged in form for perhaps millions of years; then a comparatively abrupt change may take place to a new and clearly distinct form, which remains stable until a further disturbance causes another relatively sudden change in the species. This new interpretation was called the *punctuated equilibrium* model. 'Equilibrium' refers to the concept of a species as a population of organisms preserved

Figure 392
A fossil *Archaeopteryx* – a missing link? (Artist's impression)
Photograph, by permission of the Trustees of the British Museum (Natural History).

for most of the time in a condition of equilibrium by stabilizing selection. This is natural selection that merely weeds out badly adapted variants as they arise in a species already well adapted to its environment.

At infrequent intervals, however, a change in the physical environment might upset the equilibrium. A rise in sea level, for example, could create islands, and so split a species into geographically isolated populations. These might then give rise to new species in a geologically very brief interval of time, perhaps only thousands of years. This would lead to new species appearing suddenly in the fossil record. Abrupt changes of this kind are called 'punctuations'.

The difference between the conventional Darwinian view of

Figure 393
Hypothetical phylogenies representing extreme views.
a All evolution is concentrated in speciation.
b All evolution is phyletic.
From Stanley, F. M., Macroevolution: pattern and process, W. H. Freeman, 1979.

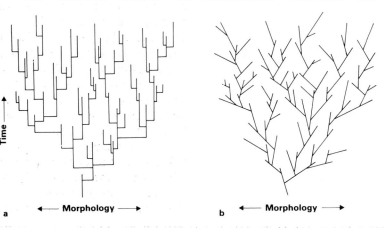

560   Ecology and evolution

evolution (phyletic gradualism) and the new punctuated equilibrium theory is illustrated in *figure 393*. Darwin's 'tree of life' (**b**) is replaced by what has been called the 'rectangular' model of speciation (**a**). The two evolutionary trees are graphs, with time on the vertical axis, and morphology or any other kind of variation on the horizontal axis. In the punctuated equilibrium model species are stable for most of the time, so each species is represented by a vertical line. But when change does occur, it is rapid and takes very little time. So the evolution of a new species, known as a *speciation event*, appears as a horizontal line. This is why *figure 393a* has so many right angles in it.

One of the major problems arising from the rectangular model is how to account for long-term morphological trends in the fossil record. For many years the existence of such trends had been widely recognized by palaeontologists. They had found examples in such familiar groups as trilobites, graptolites, and ammonites, and between the reptiles and mammals. Another example is shown in *figure 394*.

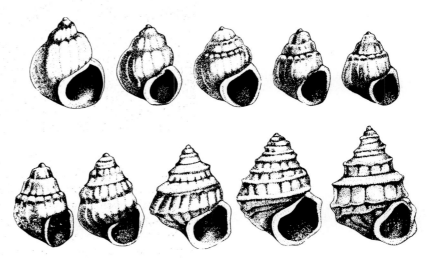

Figure 394
An example of gradual and continuous evolutionary change in the shape of the shell of the snail Paludina, from successive geological strata of Pliocene age.
From Dobzhansky, T., Evolution, genetics and Man, *John Wiley, 1955.*

The answer put forward by supporters of punctuated equilibrium theory involves a process called *species selection*. In *figure 395* a morphological trend is apparent as the general movement of the evolutionary tree towards the right. Suppose that the horizontal axis represents an increasing size of antlers in deer; at the start of the period all the deer have small antlers, but after 50 or 60 million years their descendants all have large antlers. Now the speciation events are depicted as random in direction compared with the long-term trend; half the speciations are to the left, meaning the antlers are getting a bit smaller, and the other half of the speciations result in slightly larger antlers. So the trend is not caused by most of the speciations producing larger antlers. According to the species selection concept, the trend is caused instead by a process analogous to natural selection: species are being selected to survive rather than individuals. From the point of view of antler size, the species are being produced at random; but more of the new larger-antlered species go on to produce further species in the future than of the new smaller-antlered species.

STUDY ITEM

30.21 Species selection

Use *figure 395* to answer the following questions.

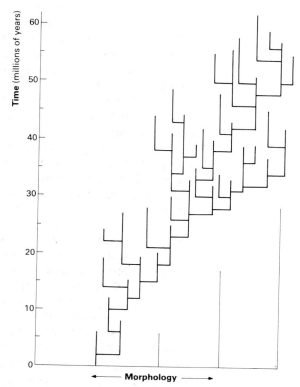

Figure 395
Hypothetical phylogenetic trend produced by species selection.
From Stanley, F. M., Macroevolution: pattern and process, *W. H. Freeman, 1979.*

a How many speciation events occur in the 60 million years?

b How many of these are to the left (*producing smaller antler size*) and how many to the right (*increasing antler size*)?

c How could you test whether the trend was caused by the average increase in antler size (in speciations increasing antler size) being greater than the average decrease in antler size in the other speciations?

d How many speciations resulting in smaller antlers produce a species which becomes extinct before it in its turn gives rise to a new species? Make the same count for speciations resulting in increased antler size.

If the species selection theory is true, it has important implications. For several decades, population geneticists have been studying natural selection in the field, and noting how it produces distinct but small-scale effects. These include the increasing frequency of dark moths in industrial areas, compared with pale moths of the same species: pigmentation provides camouflage from predators when moths settle on soot-blackened tree-trunks. Such changes are called *microevolution*. On the other hand *macroevolution* comprises the origin of new species and long-term morphological trends, best studied in the fossil record. The species selection theory suggests that microevolution and macroevolution might be quite separate processes. So perhaps natural selection has nothing to do with the origin of species and visible changes in the fossil record; it may produce only the small-scale effects studied by population geneticists. Quite different processes may be producing species randomly, and determining which survive.

As you might expect, these new ideas have aroused much discussion and debate. Many palaeontologists support the notion of punctuated equilibria, others (probably a minority) prefer the older gradualist model, while some accept that a mixture of the two may operate. It is becoming hard to deny, however, that many species have persisted more or less unchanged for extended periods of time. Even groups as rich in species as the beetles (implying numerous speciation events in their history) are not excluded. Study of their wing-cases shows no change in many species right through the Quaternary period – that is, the last two million years, including the Ice Ages. The species have responded to the strong climatic fluctuations of this period by migrating to other favourable environments rather than by going extinct and being replaced by newly evolved forms. This suggests strongly that not just the hard parts (wing-cases) but also the beetles' climatic tolerance and physiology have remained unchanged. So it appears that some species can remain stable for several million years.

The idea of species selection is more controversial; so far no supporting evidence has been brought forward based on detailed study of fossil sequences. But the challenge to orthodox neo-Darwinism has been vigorously countered by population geneticists. The punctuated equilibrium model may assume instantaneous speciation events, appearing to take no time at all in terms of the geological time scale. But population geneticists point out that a geologist's 'instant' may in fact be several tens of thousands of years. This is because even a small cliff face can represent many millions of years from the oldest rocks at the bottom to the youngest at the top. If a metre of rock represents, say, a hundred thousand years, one cannot expect to find in it a detailed record of a speciation event, with all the intermediate forms linking ancestor and descendant species. So speciations that, to the palaeontologist, are apparently instantaneous could well have been the result of ordinary natural selection.

On the other hand, geneticists have been less successful in explaining away the phenomenon of morphological stability for periods as long as several million years. It can hardly be assumed that the environment has remained unaltered for such lengthy time periods, so the organisms must have been effectively buffered against change. This is something that is not adequately explained by orthodox neo-Darwinian theory. Much more research is needed if we are to solve these major problems concerning the fossil record.

30.3 Cladistics and evolution theory

In section 30.1 it was pointed out that Darwin used classification as support for his theory of evolution. It might be expected therefore that biological classification and evolution theory exist in harmony. But this is not the case, and in the next few pages we shall explore the reasons, with particular reference to one type of classification – *cladistics*.

Cladistics is a recently developed method of classification which has been enthusiastically adopted by some biologists, but vehemently criticized by others. The critics believe that cladistics is incompatible with evolution theory. They have four main objections (two of which we shall discuss later).

1 Cladistics has a narrow, simplistic view of speciation.
2 Cladistic classifications take no account of evolutionary divergence.
3 Cladistics refuses to identify ancestors.
4 Cladistics implies that evolution always takes the simplest route.

These charges are serious indeed, but to understand the arguments fully we need to remind ourselves of the primary aim of biological classification (see *Systematics and classification*).

Biological classification, or systematics, reflects our ideas of order in the diversity of life. Species can be grouped by similarity into genera, genera into families, and so on, with the result that animals and plants can be classified into a series of successively more inclusive sets (for example, see *figure 396*). This is the system set up by Linnaeus; it was used by Darwin and is still in use today. Before 1860, many people believed that the order in the classification was a plan of God's creation.

Darwin's theory of evolution by natural selection provided a causal explanation for such order. Organisms could be classified in this way because they had descended through time, with varying degrees of structural modification, from common ancestors. It was Darwin's hope that future classifications would incorporate such evolutionary relationships. These evolutionary relationships are not immediately

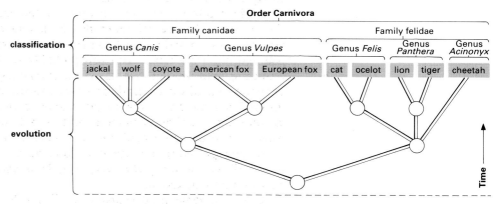

Figure 396
Classification and evolution. Species can be grouped into genera, genera into families, and so on because they have descended with modification from ancestors (open circles) more or less remote in time.
*After Patterson, C., 'Cladistics and classification', New Scientist, **94** (1303), 1982, 303.*

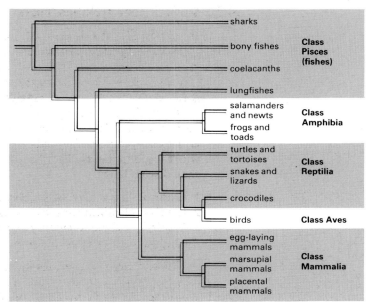

Figure 397
Classification and evolution. The traditional 'textbook' classification of jawed vertebrates (shown to the right) recognizes five classes. The presumed evolutionary relationships are shown by the branching lines. Only in three instances (amphibians, birds, and mammals) do the phylogeny and classification correspond.
After Patterson, C., 'Cladistics and classification', New Scientist, **94** *(1303), 1982, 303.*

obvious, however; they have to be inferred from the data around us. Furthermore, evolutionary relationships are more complicated than the relationships of similarity and difference used by earlier systematists; this is because they contain the additional idea of ancestry and descent (genealogy). These evolutionary and similarity relationships are not mutually exclusive, but they often conflict with one another when we try to express them in classifications.

In *figure 397*, for instance, only three of the classes correspond with the genealogy. These are the amphibia, birds, and mammals. The reason why they correspond is that these groups are monophyletic, while the fish and reptiles are not. You can probably see from the pattern of branching lines what this means. The Mammalia group has a common ancestor, and *all* the descendants of that ancestor are included in the group. The same is true of Aves and Amphibia. But Reptilia and Pisces are different: each has a common ancestor, but not all the descendants from the ancestor are included in the named group. To make Reptilia into a monophyletic group, the birds would have to be included too; and to make Pisces monophyletic, everything else in *figure 397* would have to be drawn into the group.

It is a fundamental principle of cladistics that only monophyletic groups are allowed in classifications. This is one of the major contributions to systematics made by the founder of cladistics, Willi Hennig. It is a simple idea, and it may seem obvious, but merely from the widespread acceptance of groups like Reptilia and Pisces it is clear that earlier systematists did not recognize the principle. Cladistically speaking, to recognize a group like the Reptilia is not very useful because it is not a genealogical entity like Aves or Mammalia, and therefore not directly comparable.

Making cladograms

The branching patterns in *figures 396* and *397* are genealogies, phylogenetic trees, or evolutionary trees. These terms all mean the same thing: they show how present-day species are thought to have diverged through the course of evolution from their common ancestors.

At the bottom of the facing page, however, are shown fifteen diagrams which are very different from evolutionary trees, even though they look superficially similar. They are called *cladograms*, and are the way cladists present their classifications. They have been drawn here in single rather than double lines to distinguish them from evolutionary trees.

Cladistic method

This example concerns the classification of four taxa: frog, duck, bat, and anteater. With four taxa there are 15 possible hypotheses of relationship in which all taxa are placed at the tips of branches and in which we recognize only dichotomous possibilities. The problem is to choose the hypothesis that best fits the characters of the four organisms.

From our knowledge of these animals we can recognize the following character matrix, in which a shaded box opposite the name of a character means that the character is present in the organisms concerned. 14 characters are shown here but we could apply more.

	Character	A (frog)	B (duck)	C (bat)	D (anteater)
1	Hair			■	■
2	Mammary glands			■	■
3	Wings		■	■	
4	Webbed feet	■	■		
5	Fingers and toes	■	■	■	■
6	Sticky tongue	■			■
7	Ovipary	■	■		
8	3 ear ossicles			■	■
9	Spiral cochlea			■	■
10	Endothermy		■	■	■
11	Amnion		■	■	■
12	Diaphragm			■	■
13	Placenta			■	■
14	Internal fertilization		■	■	■

Next, in order to group the taxa, we look for similarities in presence characters, and we notice that the characters fall into several categories which specify different groupings.

1 Some characters are seen in all four taxa and allow us to say only that we have a group ABCD. In this example we have one such character (5–possession of fingers and toes).
2 Some characters are shared by three of the four taxa. (There are four groups of three: ABC, ABD, ACD, and BCD.) From the matrix:

Characters 10, 11, 12, 14 specify a group BCD
No characters suggest a group A, B, D
No characters suggest a group A, C, D
No characters suggest a group A, B, C

3 Some characters are shared by only two of the four taxa. (There are six possible groups of two: AB, AC, AD, BC, BD, and CD.) From the matrix:

Characters 4 and 7 suggest a group A and B
No characters suggests a group A and C
Character 6 suggests a group A and D
Character 3 suggests a group B and C
No characters suggests a group B and D
Characters 1, 2, 8, 9, and 13 suggest a group C and D.

There might be characters seen in only one of the four taxa (such as the possession of feathers in this example). These are clearly of no use for grouping and have therefore not been included in the matrix.

In the diagram that follows we show the fifteen alternatives, to which we have applied the characters against the groups we have identified.

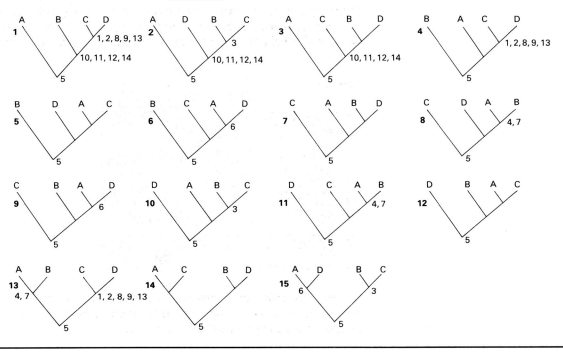

Diagram (1) is evidently the most successful hypothesis in explaining the character distribution: of the fifteen alternatives, this one takes into account the greatest number of characters and becomes the cladogram. It tells us, firstly, that organisms C and D form a group recognized by the common possession of characters 1, 2, 8, 9, and 13, and also that B, C, and D form a larger group recognized by the characters 10, 11, 12, and 14. We can reasonably expect that other characters, not yet investigated, of the four organisms will also fit this classification.

We can interpret the cladogram in evolutionary terms:

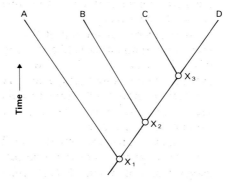

This shows that C and D are more closely related to each other than either is to A or B, because they share a common ancestor (X_3), which is unique to them. We can think of characters 1, 2, 8, 9, and 13 as evolutionary novelties which arose in the ancestor X_3 and which were inherited by C and D. Similarly, characters 10, 11, 12, and 14 are evolutionary novelties which arose in ancestor X_2 and which were inherited by B, C, and D.

In the example, hypothesis (1) was chosen as the cladogram because that was the alternative which explained the data in the most economical manner. But it was not totally successful. For instance, some characters (4–webbed feet, 7–ovipary) suggest that frog and duck should be grouped together and they consequently imply a different hypothesis of relationship. We must therefore explain why these do not fit the chosen cladogram. To do this, we might call on other knowledge and suggest that ovipary is a primitive condition seen, for instance, in most fish-like vertebrates and lost in the bat and anteater. Additionally, we might suggest that character 4–webbed feet–is best interpreted as an evolutionary convergence; that is, a character which arose twice. But we must remember that these explanations are additional hypotheses, introduced to support our case. Clearly, the fewer additional hypotheses we are forced to assume the stronger our case must be. This is why we choose the most economical solution.

Can you identify any other characters which suggest a grouping other than that in the chosen cladogram? And can you think of any explanations for these non-congruent characters?

The method of producing a cladogram for four organisms is illustrated in some detail in the panel. It is worth following it through carefully, and as you go one thing will become clear: absolutely no knowledge of a particular evolutionary history is needed for its construction.

First a table of characters is drawn up. Fourteen characters have been used in this example, from possession of hair to internal fertilization. For each organism, the state of the character is recorded in the table, so that each character is represented by a pattern of white and black boxes. One could equally well use plus and minus signs, or '0' and '1'. Normally, of course, many more organisms would be involved, and at this stage the information in the table could be fed into a computer for automatic processing.

It is obvious at the outset that the bat and anteater are more like each other than either is like the frog or duck. This is expressed in the table by characters 1, 2, 8, 9, and 13 all having the same pattern – white/white/black/black. We also know that the duck, bat, and anteater are more similar than any of them is to the frog. Again the table shows this in characters 10, 11, 12, and 14; the pattern is white/black/black/black.

Because the bat and anteater are like each other in so many ways, it is sensible to group them into one set. This can be depicted as a Venn diagram, shown in *figure 398a*. Similarly that group and the duck can be included in a larger set, because of the similarity shown in characters 10,

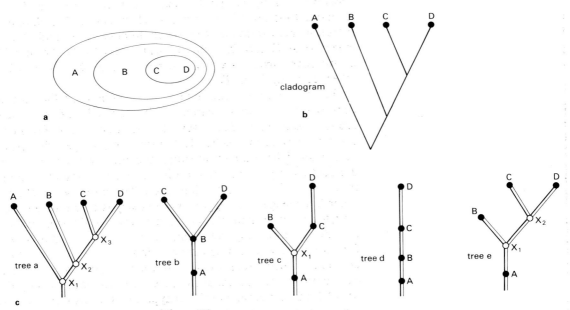

Figure 398
The cladogram **b** is equivalent to the Venn diagram **a**. It can be translated into a set of evolutionary trees **c** by the addition of a vertical time axis and notions of ancestry, speciation, and descent with modification. Each tree is a precise statement of an evolutionary history which includes known and unknown ancestors (the unknown ancestors are shown as open circles). The diagram shows only a few of the twelve possible trees equivalent to the four-taxon cladogram.

Chapter 30 Evolution

11, 12, and 14. A third set will include all four animals, because they all have character 5 in common. The cladogram in *figure 398b* is simply another way of showing the same information. Each branching point corresponds to one set: the members of the set are all the organisms reached by going up the lines from the branching point.

The first cladogram in the panel is the same as *figure 398b*, but in addition it shows which characters were used to make each of the three sets. Altogether ten out of the fourteen characters were used. Although we might agree that this is the best way to classify these four organisms, it is not the only possible arrangement. The panel shows fourteen other possible ones. Have a look at one of them, the second one for instance. This shows how we could have grouped the duck and bat because of their wings, then put that set together with the anteater in a larger set because of characters 10, 11, 12, and 14, and finally made the largest set on the basis of character 5. Why is this not as good as the first cladogram? The reason is that only six of the characters have been used from the table, leaving eight that do not fit with the classification. You will see that one of the other cladograms uses eight characters, but the first cladogram is the best one because it uses more of the information than any of the others.

Cladograms and evolutionary trees

The process of classification described in the last section made no use of any particular theory of evolution. Each branching point in *figure 398b* indicates merely which organisms have been grouped into a set. The whole process could have been done on ethnic cultures or postage stamps instead of the four organisms.

Conflicts between cladistics and evolution theory arise because many biologists interpret cladograms as literal evolutionary trees. An evolutionary tree, like the one in *figure 396*, differs from a cladogram in three ways;

1 it has a time axis running vertically up the diagram;
2 each branch point represents an ancestor species splitting into two or more new species;
3 the lines connect ancestors with descendants.

The difference between the two kinds of diagram is clear from *figure 398*. This shows how several evolutionary trees can be constructed to explain one cladogram. There are, in fact, twelve trees equivalent to this four taxon cladogram but only five are shown here. Each tree is a separate hypothesis about the evolutionary origins of A, B, C, and D. For instance, tree **c** tells us that an ancestor (A) gave rise to an unknown descendant (X_1); this subsequently speciated to give rise to two species, one of which (C) underwent a gradual transformation to give rise to D. The other trees suggest different evolutionary histories which you can work out for yourself.

All of the trees shown in *figure 398* are compatible with the cladogram. Cladograms are therefore more general than trees. But it also follows that if we can draw twelve trees from one four-taxon cladogram the information used to construct the cladogram is not sufficient to choose one tree in preference to another. The cladogram does not tell us which evolutionary history is correct.

STUDY ITEM

30.31 An example of cladistic analysis

Suppose you need to do a cladistic analysis on a daisy, fern, tulip, and moss. You have collected the following information:

1 only the fern and moss have a photosynthesizing gametophyte;
2 only the daisy and tulip produce seeds;
3 all four have chlorophyll;
4 only the moss and fern disperse by means of spores;
5 all except the moss have vascular tissue;
6 only the tulip and daisy have flowers;
7 the daisy, fern and tulip show the sporophyte as the dominant generation;
8 the daisy, fern and tulip have true roots.

a *Construct a table to show the state of each character in each plant.*

b *Find the cladogram using most characters.*

c *How many characters are not used?*

d *Consider the various evolutionary trees that fit the cladogram, and give reasons for the one you think most likely.*

Criticisms of cladistics

A major objection levelled at cladistics is that it takes no account of evolutionary divergence.

To explain what this is all about, let us go back to *figure 397*. You will remember that cladists dislike the idea of a group comprising only Reptilia. The reason is that the Reptilia is not a monophyletic group; to be monophyletic the Aves would have to be included. The Reptilia is paraphyletic, which means that it is a monophyletic group with part missing. In other words, not all the descendants from the most recent common ancestor are included.

Although cladistic classification can proceed a long way without any consideration of particular theories of evolution, Hennig himself wished to see classifications representing phylogeny correctly. Now let us suppose everyone agrees that the evolutionary tree in *figure 397* is correct. What Hennig recommended, quite logically, was that any two monophyletic groups originating from a common ancestor should be given equal rank in the classification. Birds and crocodiles are two such groups; so if birds are made into a class, so should the crocodiles be. Similarly sharks should have an importance equal to that of the group that includes everything else from bony fish to placental mammals.

Orthodox systematists cannot agree with this approach. Certainly some of the cladistic arrangements represent a massive departure from the classification we are used to. In the opinion of orthodox systematists the birds should be given an important rank equal to that of the Reptilia. Moreover they would say the paraphyletic Reptilia is a valid grouping. Their argument is that the birds have diverged very far from the reptiles because they have acquired endothermy (like mammals), feathers, and flight (like some mammals). Classification is, after all, based on similarities and differences; so birds and reptiles, because they are so different in these very obvious features, should be placed in different high-rank groups such as Classes.

There is really no easy solution to this conflict of opinion. It depends on what purpose the classification is supposed to serve. If the main objective is to reconstruct the phylogenetic history, probably cladistics offers the best approach. If on the other hand you think grouping organisms by similarity and difference, to take account of adaptational differences, is most important, then the conventional method is appropriate, such as the one described in Systematics and classification in 1.3 Numerical Taxonomy. If you want to achieve both purposes at once then you have to accept there will be areas of the classification where a presumed phylogeny does not match ideas of adaptation.

A second objection levelled at cladistics concerns the fact that many believe cladistics assumes evolution to have taken the simplest course. For example, consider the evolution of flight from flightless ancestors. It might be suggested that some birds acquired the power to fly as they evolved from one group of reptiles, while other birds evolved from other reptiles and acquired flight independently. But unless we had evidence to suggest a different phylogeny it would be much simpler to suppose that flight evolved in birds only once.

For another example, look again at the cladograms in the panel. The first cladogram is the most informative. If we translate this into an evolutionary tree (*figure 398*, tree a), there are only four characters left over not explained. So four extra theories are needed to explain the distribution of characters which do not fit. Now all the other cladograms in the panel and their associated trees show more characters in the table that are left unexplained. So more extra theories would be needed to account for all the information. This is what is meant by saying that cladistics assumes evolution has taken the simplest route.

Cladists do not deny that they make this assumption. However, some orthodox systematists point out that neo-Darwinism does not predict that evolution takes the simplest course. Characters may evolve more than once, either in the same lineage or independently in different groups of organisms. The possession of wings is one of the characters left unexplained by the first cladogram, and it is generally agreed that flight has evolved on at least four occasions.

In reply the cladists argue that, for constructing classifications choosing the simplest arrangement is a convenient procedure which can be universally adopted for the sake of consistency. When it comes to the evolutionary histories associated with a cladogram, cladists are merely following a rule adopted by all scientists, that of Occam's Razor: this requires that where two theories explain the facts equally well, the simpler is to be preferred.

Cladistics provides a method for suggesting evolutionary trees by giving equal importance to all the characters that have been measured. Sometimes, however, other evidence suggests that some characters are better than others as indicators of evolutionary relationship, because they are more stable. For sorting out the major groups of plants, for example, possession of an embryo sac is important, because it is likely to have evolved only once. So all plants with an embryo sac probably belong to a single monophyletic group. But hairiness of leaves is pretty useless by comparison: it could have evolved many times, either in different groups or even along one lineage. So this character is really only of much value in separating species or subspecies.

It should be clear from this account that there are two aspects to cladistics. It is primarily a method of classification. This method can be quite mechanical, once the table of characters has been drawn up. The classification is presented in the form of a cladogram. After that, the cladist can take off the systematist's hat and put on the evolutionary theorist's hat. The various evolutionary trees compatible with the cladogram are considered, and one is chosen on ancillary evidence as the one most likely to document a particular evolutionary history.

30.4 Looking for laws in biology

Genetics can tell us a great deal about the inherited differences between organisms, but it cannot tell us about their similarities. This is simply because a gene can only be recognized by the differences caused by its alleles. Consider the experiment described on page 5 involving the exchange of rhizoids and nuclei between different species of *Acetabularia*. We can conclude that genetic factors are involved in determining the differences in structure of the cap; but nothing can be concluded about what determines the basic rhizoid–stem–cap organization common to both species.

Our attention is drawn to inherited morphological differences rather than to inherited similarity, and this is an emphasis running through the whole of modern biology. It can be traced back to Darwin's interest in the inheritance of adaptive differences between organisms. For instance, why is a male chaffinch more brightly coloured than a female? or why is the bill of one Galápagos finch larger than another's? The major problem in the *Origin of species* was to understand how inherited diversity arises in populations. Similarity is merely what is left after the differences have been explained, rather than something to be studied in its own right.

Scientifically speaking, this is a very peculiar emphasis. What gave rise to the great achievements of the exact sciences, such as Newton's laws of motion and the Periodic Table of the elements, was the attempt to understand basic similarities of process and structure. Once these regularities were recognized, it became possible to understand both the similarities and the differences in natural phenomena. For example, why do chlorine and iodine resemble one another in certain ways and differ in others? Answer: they belong to the same column of the Periodic Table, but to different rows.

Figure 399
Frog's egg before and after fertilization, showing the formation of the grey crescent.

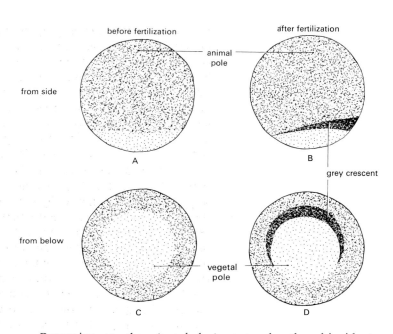

Returning to the *Acetabularia* example, the rhizoid–stem–cap structure reveals a basic property of cells and organisms. This is called polarity: one end or pole of the cell or organism differs from the other. It is evident even in the unfertilized frog egg, where the animal pole differs from the vegetal pole (*figure 399*). The same kind of polarity occurs in most cell types and nearly all organisms. Regularities of this kind suggest basic principles of spatial organization. Another example is the pattern of early cell divisions in all species which undergo what is called holoblastic cleavage (*figure 400*). Yet another pattern is seen in the limbs of all four-legged vertebrates; they have the same basic arrangement of bones. Such underlying similarities of form extending across a great range of species suggest laws of order in biology. Moreover, these laws could be as fundamental to our understanding of biological principles as those which have been discovered in the exact sciences.

Studies in cell and developmental biology are beginning to show what these principles may be based upon. They concern several

Figure 400
Early cell divisions in the salamander *Ambystoma maculatum*. From Hamburger, V., A manual of experimental embryology, University of Chicago Press, 1960.

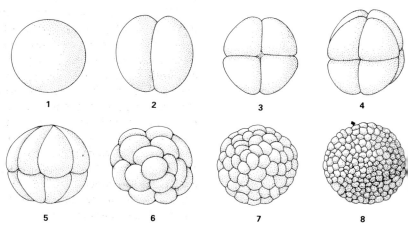

574   Ecology and evolution

phenomena: fundamental properties of the living state, such as electric potentials across membranes and ion flows which cause electric currents; the viscous and contractile properties of cytoplasm; biochemical reactions, and the diffusion of chemicals within and between cells. Then particular genes acting within this organized, living state can result in inherited differences of form, such as the details of cap morphology in *Acetabularia*, or the size and shape of wings in *Drosophila*. So it could become possible to understand both the similarities and differences in biology, just as in the exact sciences.

A simple analogy will clarify the argument. When you get out of the bath and pull out the plug, the water runs down the plughole in a certain way: it takes on either a right- or a left-handed spiral motion. This doesn't depend on which hemisphere you're living in, as you can easily verify by keeping track of right- and left-handed spirals after a number of baths. It depends upon the water movements you make on stepping out of the bath and pulling the plug. Suppose on one occasion you get a right-handed spiral. Then you can swirl the water around with your hand in the opposite direction and get a left-handed spiral. Now ask yourself this question: did your hand make the left-handed spiral? Strictly speaking, it didn't. The motion of your hand merely selected a left-handed spiral flow down the hole. This is one of just two possible stable motions for a liquid such as water under these conditions, the other being a right-handed flow. It is the properties of water, expressed in the laws of liquid flow, which make such forms of motion possible. All your hand did was to choose one particular form from the possibilities permitted by the laws.

Returning to organisms and genes, it seems that genes affecting a particular character, such as cap shape in *Acetabularia*, are doing what your hand did in the bath. They aren't 'making' the shape; they are producing conditions which ensure that one of the possible shapes, determined actually by the laws of organization of the living state, is selected. Until we understand these laws better, we cannot understand development, and so we cannot understand how species are produced. According to this view, they are not made by genes, though genes may be said to choose the conditions which favour the appearance of a particular form.

This view has some interesting consequences when we look at evolution. Biology is usually described as an historical science: by this we mean that, to understand the origins of particular types of organism, we need to understand the sequence of their ancestors. This is why evolutionary trees are thought to be central to biological explanations of origins. The trouble is that historical descriptions, which now mean changes in gene frequencies in evolving populations, simply do not explain how the forms of organisms are generated in the first place, nor why basic features are conserved over large taxonomic groups, making a hierarchical classification possible. What does this tell us about the transformation of one type of organism to another? Genes and the external environment undergo random changes, but organisms do not. Why not? What are the laws which cause a degree of order through the whole of the biological realm?

STUDY ITEM

30.41 Differences and similarities

In this section we have contrasted the 'similarities' and 'differences' aspects of organisms. For instance, the genetic differences between organisms are contrasted with the features they share, such as polarity, cleavage patterns in embryos, and bone patterns in limbs. For each of the following 'differences' aspects of a set of objects suggest what the corresponding 'similarities' aspects might be.

a *differences in colour, shape, density, hardness, and chemical composition of the mineral crystals found in rocks;*

b *differences between languages in vocabulary and sentence structure;*

c *differences in the structure of human societies in various parts of the world, ranging from tribal societies lacking any kind of visible leadership to modern bureaucratic dictatorships.*

Genetics and evolution theory have answered many important questions about hereditary differences between organisms. But they leave unanswered many other problems about the similarities of organisms, and these are just as fundamental to an understanding of classification and evolution. These questions, which were raised before Darwin, are being asked again with renewed vigour today.

Summary

1 The 'ladder of nature' or *scala naturae* was the concept which had the greatest influence on early evolution theory (**30.1**).
2 The interest in evolution which existed in the first part of the nineteenth century is confirmed by the writings of a number of scientists at that time (**30.1**).
3 Many scientists did not accept Darwin's theory of evolution by natural selection when it was first published (**30.1**).
4 A rediscovery of Mendelian genetics resulted in a general acceptance by biologists of Darwin's theory of natural selection, which became known as the modern synthesis or neo-Darwinism (**30.1**).
5 Punctuated equilibrium is a new interpretation of the mechanism for the evolution of a new species (**30.2**).
6 Cladistics is a recently developed method of classification which has proved to be controversial (**30.3**).
7 The morphological similarities of organisms are as fundamental to an understanding of classification and evolution as are their differences (**30.4**).
8 Many questions still remain to be answered.

Suggestions for further reading

ATTENBOROUGH, D. *Life on Earth*. Collins/BBC, 1979.
BERRY, R. J. Studies in Biology No. 144. *Neo-Darwinism*. Edward Arnold 1982.

DARWIN, C. *The origin of species.* 1st edn. John Murray, 1859. (Worth dipping into although the style makes it difficult reading.)

DAWKINS, R. *The selfish gene.* Oxford, 1976. (An account of how natural selection influences behaviour.)

DOWDESWELL, W. H. *Evolution,* Heinemann, 1984.

GEORGE, W. *Darwin.* Fontana, Collins, 1982.

GOULD, S. J. *Ever since Darwin.* W. W. Norton, 1977. (A collection of essays.)

HUTCHINSON, P. *Evolution explained.* David and Charles, 1974. (Easy to understand, clear account of evolution.)

LEAKEY, R. E. *The illustrated origin of species.* Faber, 1979 (An easier to read, abridged version of Darwin's classic.)

MILLER, J. *Darwin for beginners.* Writers and Readers, 1982. (An unusual, amusing, and highly informative book.)

PATTERSON, C. *Evolution.* Routledge and Kegan Paul, 1978. (Clear account of modern evolutionary thought: recommended.)

INDEX

A

abscisins, 343, 356–8, 361, 378
abscission, 356, 360
absorption spectrum, chlorophyll, 381–2
Acacia spp., mutualistic associations, 465
Accipiter nisus, see sparrowhawk
Acetabularia, 5–9, 573, 574
acidophilous plants, 440–43
acquired characters, inheritance, 549, 556
acquired immune deficiency syndrome, *see* AIDS
acquired immunity, 360, 363–72
    cell-mediated, 363, 364–70
    humoral, 363–4, 370, 371
acriflavine, mutagen, 109
acrosome, 250, 251
actin, 23, 290
action spectrum, photosynthesis, 381–2
adaptation, 308, 554–5
adaptive immunity, *see* acquired immunity
adder, embryo protection, 226
adenine, 102–105, 354
adenosine phosphates
    ATP, 117, 324, 345
    cAMP, 320–21, 324–5
adenylate cyclase, 321, 324
adrenaline
    mechanism of action, 324, 325
    synthesis and release, 317–19
aestivation, 168
aflatoxin, 89
age structure of population, 486–8
agonist, 324
agoutis, 553
agriculture
    biotechnology in, 212–13
    energy flow in beef production, 516–17
    energy flow in subsistence and industrialized, 523
    productivity of crop plants, 514–16
AIDS, 211, 370
air
    factors influencing distribution patterns, 444–5
    pollution, 422–3, 537–41
albinos, 140, 152
    plants, 253–4
    theoretical frequency, 158
alcohol, *see* ethanol
alcohol dehydrogenase, polymorphism in *Drosophila,* 173–4
alder, 476
Alexander, Professor R. D., 187, 190
algae, 304
    distribution in polluted river, 537
    fertilization, 226, 254
    in lichens, 465, 467, 538
    population cycling, 531–2, 533
    specialization in, 2, 3
alginates, 196
allatectomy, 336
alleles, 57, 59
    origin of new, 85, 87; *see also* mutation
Allium cepa, see onion
Allolobophora spp., *see* earthworms
Alnus glutinosa, see alder
alternation of generations, 232
amino acids
    genetic coding, 91–2
    production by fermentation, 196
ammonium ion, pollution by, 536
amniocentesis, 38, 52, 174
amniotic fluid, 270, 271
amnion, 270
amphibious bistort, 62
amylase, 206
    insertion of gene for, 203
    production in germinating barley, 350–51
anaphase
    meiosis, 43, 44
    mitosis, 16, 19, 36, 37
Ancon (short-legged) sheep, 555, 556
Andalusian fowl, 71–2
aneuploidy, 37–8, 39, 48, 51
angiosperms
    fertilization, 252–3
    gametes, 233
    gametogenesis, 237–40
    life cycle, 240, 241
    reproductive structures in, 262
    seed production, 267
anisogamy, 226
Anopheles spp., *see* mosquitoes
anther, 237–9
antibiotics, 194–6, 374
    resistance to, 172, 201
antibodies, 210, 360, 361–2, 363–5
    monoclonal, 210–12
antigens, 210, 360
    response, 364, 365;
        hypersensitive, 366, 367
    tumour, 372–3
Antirrhinum heterozygote, 144
ants, 161, 163
    ecological study of desert, 423–4
    inter-specific competition, 457–8
    mutualistic association, 465
aphids, life cycle, 222–3
apical ectodermal ridge, 293–4, 296
Apis mellifera, see honey bee
apogamy, 229, 232
apolysis, 329
apospory, 232
apple
    breaking of dormancy, 378
    flowering, 385
    ripening, 358–9, 360
aquatic factors influencing distribution patterns, 426–30
Arabidopsis thaliana, meristem cell, 17
Archaeopteryx, 559, 560
archegonium, 266
arginine, biochemical synthesis, 128, 130
armadillos, asexual reproduction, 220, 221
artificial insemination, 178–9
artificial organs, 212
artificial selection, 175–83
Arum maculatum, see cuckoo pint
Asellus, distribution in polluted river, 537
Aseo flammeus, see short-eared owl
aseptic fermentation, 193, 194, 195
asexual reproduction, 218–25, 261
Aspergillus flavus, aflatoxin production, 89
Aspergillus niger, commercial use, 194
Aspergillus spp., cross-feeding complementation, 127
Asperula cynanchica, see squinancywort
aspirin, 319
assimilation efficiencies, 518–19
assortive mating, 159
asthma, 366
ATP, *see under* adenosine phosphates
Atropa belladonna, see belladonna
audiospectrographs
    bird song, 401–402
    cricket song, 188–9
autecology, 422
autosomal linkage, 79–82
autosomes, 52
autotrophic organisms, 308, 509, 513, 544
auxins, 41, 342, 343, 348, 361
Avery, O. T., 100
Azotobacter, 196, 469

B

baboon
    coloration changes at ovulation, 245
    social grouping, 265
Bacillus subtilis, 206
Bacillus thuringiensis, use in biological pest control, 497
bacteria
    amino acid coding in, 117
    chemo-autotrophic, 543, 544
    distribution in polluted river, 537
    DNA replication in, 106
    genetic maps, 81
    immunization against, 371–2
    immune response, 362
    in deoxygenated water, 428, 429
    in nitrogen cycle, 530
    mutualistic associations, 467
    nitrogen-fixing, 468–9, 530
    'overproducing', 193
    population growth, 484–5
    protein synthesis in, 110–12
    soil, 431
    transformation, 99–101, 201, 202–204

bacteriophages, 96, 97–9
  RNA production in, 115
  use in genetic engineering, 198–9
badger, delayed implantation, 242–3
Balanus balanoides, see under barnacles
bald eagle, effect of DDT, 541–2
banana moth, 558
bananas, ripening, 358–9
bar eye (*Drosophila*), 36
barley
  effect of earthworms on growth, 437
  germination: following irradiation, 90; role of gibberellin in, 350–51
barnacles, distribution, 447–8
Barr body, 51, 135
basidiospores (*Coprinus lagopus*), 56
basophilous plants, 440–43
Bateson, W., 147
bats
  delayed fertilization, 242
  food web, 513
  pollination by, 255
Bazalgette, Sir Joseph, 429
BCG vaccine, 371, 373
Beadle, G. W., 57, 125–7
beans, see broad bean; French bean
beating, sampling by, 481
beech
  mycorrhiza, 466
  'sun' and 'shade' leaves, 459
beef production, 516–17
bees, 161, 163
  see also honey bee
beetles, species stability, 563
behavioural isolation, 186–90
belladonna, tissue culture, 4
benzyladenine, 354, 361
bindweed, 224–5
biochips, 214
biological clock, 385, 388, 500
biological pest control, 497–500
biomass, 543
  pyramid of, 512
biomes, 418, 419, 420–22
biometric studies of variation, 83–5
biosensors, 208–209, 212, 214
biosphere, 418, 507
biotechnology, 192–215
  enzymes in, 205–209
  fermentation, 192–8
  future trends in, 212–14
  genetic engineering, 198–204
  immunology in, 210–12
biotic factors, 451–79
birds
  breeding season, 384
  breeding systems, 265
  dispersal agents, 471
  egg, 268; cleavage divisions in, 278; parental care of, 269
  migration, 382
  pollination by, 255
  song development, 401–404
  tissue interactions in development, 287

birth control, 505, 507
  'morning after' contraceptive pill, 243
Biston betularia, see peppered moth
bistort, amphibious, 62
Black, J. N., 164–6
black bean aphid, life cycle, 222–3
blastocyst, 243, 247, 270
blastomeres, 279–81
blastula, 274–5
  fate maps, 285, 286, 289–90
bleomycin, 89
blood
  electrophoretic studies, 29
  see also lymphocytes; polymorphonuclear cells; red blood cells
blood groups, 147
  frequency in population, 158–9
  inheritance of antigens, 174
'bog people', 525
bog-rush, distribution, 443
bones
  extracellular matrix in tissue, 24
  growth, 287–8
  marrow, 363, 364, 365
boreal coniferous forest, 421
bot-flies, 416–18
brachiopods, 1, 20
bracken, 224–5, 232, 266
Brassica spp., self-incompatibility in, 263–4
breeding systems, 261–6
bristlecone pine, 217
broad bean
  DNA replication in, 107
  nutrient reserve of seed, 268
  root growth, 302–303
Broadbalk wilderness, 437–8
bromodeoxyuridine, 107, 108
brown plant-hopper, 506
Brussels sprouts, effect of cytokinin, 356
Bryophyllum spp., vegetative reproduction, 219
bryophytes, 219, 233, 254, 266–7
Buchner, E., 205
budgerigar, plumage colour, 85–6
bud-selfing, 264
bullhorn acacias, mutualistic associations, 465
bulrush, 475
bursa of Fabricius, 363–4
bursicon, 332
butanol, production by fermentation, 194
trans-butenedioic acid, production by fermentation, 194
buttercups
  distribution patterns, 425
  pollen formation, 239

C
Cactoblastis cactorum, use in biological pest control, 497
caddis flies, distribution patterns of nymphs, 430, 431

calcium ion, role in hormone release, 320–21
Callitroga spp., see screw-worm fly
Calluna vulgaris, see ling heather
callus, plantlet regeneration from, 4, 5, 306
Calmette, A., 371
cambium, 304
campion, speciation in, 259–60
Canadian pondweed
  dispersal, 470
  vegetative reproduction, 219
cancer, 91, 96, 211–12, 372–3
Candida utilis, foodstuff, 197
canopy roots, 535
capacitation, 250
Carausius morosus, see stick insect
carbon cycle, 526–7
carbon dioxide
  enrichment of glasshouse atmospheres, 393–4
  ethene inhibitor, 359, 361
  rising levels in atmosphere, 527
capture–mark–recapture technique, 481–3
carcinogens, 91, 372
cardiac muscle cells, myofibrillar system, 23
Carex spp., see sedges
carnivores, 544
carrot, tissue culture, 5, 307
cartilage
  cultured cells from, 138–9
  in growth plate, 287–8
castration, 175
cat
  coat colour inheritance, 52, 62, 73–4, 135
  polydactylism, 149
catarrhines, oestrous cycle, 245–6
catecholamines
  mechanism of action, 324, 325
  synthesis and release, 317–19
cattle
  beef production, 516–17
  effect on nettle population, 426
  selective breeding, 85, 178
  traditional varieties, 182, 183
causal relationship, determination, 422–3
cell-mediated immunity, 363, 364–70
cell wall, 24
  effect of IAA, 344–6
  formation, 17–18
cells
  activities during development, 277–8
  development and differentiation, 1–30
  differentiation, 276–7; as population phenomenon, 138–40; gene control of, 135; in plants, 301; surface changes in, 23–4
  division, see meiosis, mitosis
  fixation, 18–19
  see also cytoplasm; nucleus
centrifugation, density gradient, 93
centriole, 15–16
centromere, 15, 16, 20, 34, 42–3, 44, 123
Cepaea nemoralis, see snail
cephalosporins, 195

cereal proteins, improvement, 213
Cervus elaphus, see red deer
chaffinch, song, 402, 403, 404
Chain, Ernst, 194
Chambers, Robert, 549–50
chameleon, 461
Chargaff, E., 102–103
chemo-autotrophic organisms, 543, 544
chestnut-headed oropendola bird, 416–18
chiasmata, 41, 43, 44, 81, 161
chicks, *see under* fowl
chimerae, 253, 284
chimpanzee, chromosome banding, 35
Chironomus spp.
   distribution in polluted river, 537
   polytene chromosomes, 135
chloramphenicol, 195
chlorophyll, 381–2
chloroplasts, 22, 32, 253–4
chloroquine, 464
cholesterol, 317
chorion, 270
chorionic gonadotrophin, 247
Chorthippus parallelus, karyotype, 34
chromagranin, 317
chromatids, 15, 16, 21, 33, 41, 43, 44, 81
chromatin, 12–13, 14
   composition, 93, 121–3
chromomeres, 35
chromosomes, 31, 32–4
   abnormalities in maize, 81–2
   aneuploidy, 37–8, 39, 48, 51
   banding, 34–6, 57, 121, 136
   genetic maps, 81
   in mitosis, 13, 14–17, 20–21
   in meiosis, 40–48
   isolation, 93
   labelled, for replication study, 106–107
   polyploidy, 36–7, 39, 47–9, 175–7, 225, 559
   puffed, 136–7, 325, 333
   rearrangement, 86–8, 89
   sex, 49–53
   ultrastructure, 122
chrysanthemum, flowering, 389, 292
Chrysolina quadrigemina, control of Klamath weed, 461–2, 497
Chthalamus stellata, see under barnacles
cicadas, population cycles, 500
Cirsium acaule, see stemless thistle
cisternae, 23
citric acid, production by fermentation, 194
Citrus spp.
   asexual reproduction, 220–21
   hybrids, 225
   seedless fruit, 352
cladistics, 564–73
cladograms, 566–71
Cladophora
   distribution in polluted river, 537
   life cycle, 230–31
   classification, 554
   see also cladistics
clawed toad
   fate map for embryo, 286

nuclear transplantation, 10–11
production of clone, 282
clay, 433
cleavage, 274, 275, 278–9, 574
Clematis vitalba, see old man's beard
climacteric (fruit ripening), 359
climax
   metamorphic, 336, 337
   succession, 452, 475, 476, 477–8
clonal propagation, 224–5, 282, 283
Clonorchis sinensis, see liver fluke
Clostridium acetobutylicum, commercial use, 194
clover, interaction between strains, 164–6
Clupea harengus, see herring
codon, 115
colchicine, 33, 36, 52
coleoptiles
   cell extension in, 346
   hormone activity in, 342
collagen, 24, 340
collared dove, population growth, 484
commensalism, 452, 460, 461, 464
communities, 418, 422, 509–46
compensation point, 460
competition, 423, 448, 451–60
complement, 360–62, 364, 372
congenital malformations (defects), 278, 291
coniferous forest, 421
Connell, J., 448
conservation, nature, 542–3
constitutive mutants, 131–2
consumers, 510, 518, 544
continuous variation, 60–61, 64
   frequency distribution, 163–4
   inheritance, 82–5, 163, 558
continuity of life, 1, 20
contraception, *see* birth control
Convolvulus spp., *see* bindweed
Conway, G. S., 505
Coprinus lagopus, life cycle, 55–6
corpora allata, 329, 331, 332, 336
corpora cardica, 313, 329, 332
corpus luteum, 246, 247
cottony cushion scale insect, biological control, 497
couch grass, 224–5
cover, per cent, 481
cowbird, 416–18
Crick, F., 105, 109, 113
crickets, isolated populations, 187–90
cristae, 22
cross-fertilization, 226–7, 254, 262, 263
crossing over, *see* recombination of characters
Crocus candidus, phases of mitosis, 19–21
cuckoo, 269
cuckoo pint, pollination, 255–6
cyanide, effect on ion uptake by roots, 435
cyclic-3,5-adenosine monophosphate (cAMP), *see under* adenosine phosphates
cystic fibrosis, 145, 174
cytochrome a_3 (cytochrome oxidase), 435
cytokinins, 343, 354–6, 357, 361

cytoplasm
   changes with differentiation, 22–3
   division, 17–18
   link with nucleus, 9, 11, 112–14
cytosine, 102–105
cytosol, 320
cytotoxic lymphocytes, 368, 370, 372

D
2,4-D, 360, 361
daffodil, flowering, 385
dandelion
   fruit production, 40
   triploidy in, 39
Darwin, Charles, 64, 163, 185, 551–7, 559, 564, 573
Darwin, Erasmus, 64
Dasyprocta agouti, see orange-rumped agouti
Davidson, D. W., 423–4
day length
   related to flowering, 349
   see also photoperiodism
DDE, 541, 542
DDT, effect on bald eagle, 541–2
decay, 523–5
decomposers, 510, 523–5, 544
deer, *see* Muntjac deer; red deer
deflected climax, 477
deforestation, 527, 530–31
delayed fertilization, 242
delayed hypersensitivity, 366
delayed implantation, 242–3
deletion mutation, 86, 144
demographic transition, 504
denitrifying bacteria, 530
density dependent/independent factors in population fluctuations, 494–6
density gradient centrifugation, 93
deoxyribonucleic acid, *see* DNA
deoxyribose, 112
Deschampsia flexuosa, see wavy hair grass
desert locust, egg protection, 268
detergents
   enzymes in, 206
   synthetic, in river water, 429
determination (developmental), 289–90
detritivores, 524, 544
detritus, 544
development, 274–311, 327
   behavioural, *see* ontogeny
   hormones in control of, 327–41
   plant and animal patterns compared, 300–310
Dicrostonyx groenlandicus, see lemmings
differences, inherited, 573–6
'diffuse haemoglobin' gene, 185–6
Digitalis purpurea, see foxglove
dihybrid crosses, 68, 75–9
diphtheria immunization, 371
Diplectrona felix, see caddis flies
diploid cell, 32–3
diplontic life cycle, 229–30, 234
discontinuous variation, 60–61, 64
disease resistance, 179–81, 182, 373–4
dispersal, 233, 267, 452, 470–72

Index   581

dispersal barrier, 449, 470
distribution patterns, 425–6
  acidophiles and basophiles, 440–43
  buttercups, 425
  caddis fly nymphs, 430, 431
  determined by food plant, 462
  effect of human policies, 472
  evidence for evolution, 553–4
  nettles, 426
  stemless thistle, 444–5
Dittmer, H. C., 433–4
diversity, 1–3, 22–5, 552
division of cells, see meiosis; mitosis
DNA, 93, 94
  chemical structure, 101–104
  chromatin content, 121
  generation of fragments, 199–200
  genetic code in, 109–14
  identified as 'hereditary material', 100–101
  labelling of viral, 97–9
  mass per nucleus, 94–5
  molecular model, 104–105
  recombinant, 202, 212
  replication, 105–108
  RNA synthesis from, 115–17
  splicing, 200–202
DNA ligase, 202
DNA polymerase I, 200
dog
  changing behaviour patterns, 399
  variation between breeds, 177–8
dominant character, 70, 72–3, 144, 150
dopa, 141, 318
dopamine, 318
dormancy, 306, 350, 357, 377–9, 384
dormin, 357
  see also abscisins
Down syndrome, 37–8, 237, 271
Drosophila melanogaster
  bar eye, 36
  body colour inheritance, 73
  body colour/wing shape inheritance, 79–80
  chiasma frequency variation, 161
  competing larvae, 453
  enzyme polymorphism, 172–3
  eye colour inheritance, 74
  eye colour/wing form inheritance, 75–7
  linkage studies, 79–80, 81
  polytene chromosomes, 135–7
  population changes, 154–5
  vestigial-wing mutation, 147
  X-ray related mutation, 90
drug resistance, 464
Dryopteris affinis, life cycle, 228, 229
duck-billed platypus, 269, 548
duckweed, separation of meristems, 306
dung beetles, 524
Dunn, J., 410–11

E
earthworms, 436–7
ecdysis, 329
ecdysones (ecdysis-inducing hormones), 137, 317, 325, 329, 330, 331, 332, 333
echinoderms, 15–16, 220

ecological isolation, 186–90
ecological pyramids, 511–12
ecology, 415–16
  air factors, 444–5
  aquatic factors, 426–30
  distribution patterns, 425–6
  making and testing hypotheses, 422–5
  soil factors, 430–43
  terminology used in, 418, 543–5
  zonations, 446–9
ecosystems, 418, 422, 509–46
ectoderm, 275, 286–7, 293–4
egg
  development of fertilized, 274–5, 278–9, 574
  human, 237, 249–52
  parental care, 269
  polarity in frog's, 574
  protection, 268
ejaculation, 235
Elaeis guineensis, see oil palm
Eldredge, Niles, 559
electrophoresis, 27–9, 173
  in gravity-free environment, 214
Elodea canadensis, see Canadian pondweed
Elton, C., 500
Elymus repens, see couch grass
embryo
  development from single blastomere, 279
  development from transplanted nucleus, 282–3
  fate maps, 285, 286, 289–90
  protection and nutrition, 266–71
  similarity in early stages, 275–6
embryo sac, 235, 239, 240, 268
embryogenesis
  in animals, 309
  in plants, 301
Encarsia formosa, use in biological pest control, 498
endangered species, 542–3
endocrine cells, ultrastructure, 314–16
endocrine system, 312–16
  insect, 329, 332
  see also hormones, and specific hormones
endoderm, 275
endogenous rhythms, 388
endoplasmic reticulum, 23, 110, 119
endosperm, 268
energy, biotechnological production, 213–14
energy flow, 513, 514–25, 544
  between trophic levels, efficiency, 518
  in agriculture, 516–17, 523
  forest, 522
  grassland, 520–21
  pyramid of, 512
  salt-marsh, 521
  stream, 522
environment, role compared with inheritance, 61–3
enzyme electrode, 208–209
enzymes, 113, 115–16, 117, 118, 205–209
  analytical applications, 207–209

defects in mutants, 127, 130
immobilized, 206, 207, 208
inactive form, 119
induction, 131
industrial applications, 206–207, 214
phosphorylation, 325
purification, 205–206, 214
see also specific enzymes
Eotetranychus sexmaculatus, see six-spotted mite
epidermis (tadpole), effect of thyroid hormone, 339–40
epididymis, 235
epilimnion, 532
epiphytes, 464
epistasis, 149–52
erythromycin, 195
Escherichia coli
  DNA replication in, 106
  genetic engineering applications, 201, 204–205
  lactose-mobilizing enzymes, 131–3
  protein synthesis in, 110–12
  ribosomes, 117
ethanedioic acid, production by fermentation, 194
ethanol, production by fermentation, 193, 213, 350–51; for use as fuel, 198
ethene, 342, 343, 358–60, 361
Ethephon, 360, 361
euchromatin, 13, 14, 22, 135
European starling, dispersal, 470
Eutheria, embryo protection and nutrition, 269–70
eutrophication, 468, 533
evergreen oak, 'sun' and 'shade' leaves, 459
Evernia prunastri, 540, 541
evolution, 547–77
evolutionary trees, 564–5, 570–71, 572, 573
exocytosis, 316, 321
exotoxins, 371
extracellular matrix, changes with differentiation, 24–5

F
facilitation (succession), 477
factor complex, 425
facultative anaerobes, 469
Fallopian tube, 249, 250
Faraday, Michael, 428
fate maps, 285, 286, 289–90
feedback control of hormone release, 319–20, 338
fermentation, 192–8
  of galactose, 133–4
ferns
  apogamy and apospory, 232
  inbreeding in, 261
  life cycle, 229
  spore production, 267
fertilization, 31
  delayed, in bats, 242
  human, 237, 248–52; *in vitro*, 243, 251–2
  in flowering plants, 252–3

species recognition at, 260
types, 226
fertilizers, 468
α-fetoprotein, assay, 211
fish, effect of river pollution, 428–9
fixation of cells, 18–19
flax, competition between plants, 454–5
Fleming, Alexander, 194
Flore française, 547
Florey, Howard, 194
'florigen', 390–92
flounder, in Thames water, 429
flour beetles, inter-specific competition, 456–7
flower initiation, 385
flowering
    hormone control, 348–50
    photoperiodism, 384–92
follicle cells, 236, 237, 246
follicle stimulating hormone, 246–7
food
    preferences, 400–401
    production related to population, 505–506
food webs, 510–11, 513
Ford, E. B., 169
forests
    boreal coniferous, 421
    climax, 476, 477–8
    effect of felling, 527, 530–31
    energy flow in, 522
    secondary succession in, 473
    temperate, nitrogen cycle in, 528–31
    tropical rain, 420, 478; nutrient cycling in, 532, 534–5
Forsythia, flowering, 385
fossil record, 1, 552, 555–6, 561, 563
fowl
    Andalusian character inheritance, 71–2
    comb shape inheritance, 147–9
    egg, 268
    intensively farmed, 181
    limb development in chicks, 292–300, 301
    pecking in chicks, 399–400
    selection for various purposes, 184–5
foxglove, seed production, 267
Fragaria, see strawberry
French bean, seed size inheritance, 84
frequency of occurrence, per cent, 481
frogs
    egg protection in common, 268
    genes in specialized cells from, 282–3
    larval development in gastric brooding, 268
    metamorphosis in *R. pipiens*, 337
    polarity in egg, 574
fructose, enzymic production, 206
fruit
    ripening, 358–9, 360
    seedless, 352
fruit fly, *see Drosophila melanogaster*
FSH, *see* follicle stimulating hormone
Fucus, fertilization, 226, 252
fuels
    from renewable resources, 213
    microbially derived, 198

fungi
    in lichens, 465, 467, 538
    in mycorrhiza, 465, 466
    inheritance in, 54–9
    reproduction in, 218, 219, 260
    self-incompatibility in, 160
furfuryladenine, *see* kinetin
Fusarium moniliforme, 347

G

galactose, fermentation by yeast, 133–5
β-galactosidase, 131
Galápagos Islands, 185
Galápagos vent-worm, mutualistic association, 467–8
Galium spp., distribution, 441, 443
Galleria spp., *see* locust
Galton, Francis, 83
gametes, 31, 32
    transfer of male to female, 254–8
gametogenesis, 234–5
    in flowering plants, 237–40
    in humans, 235–7
gametophyte, 230
Gammarus, commensalism with *Vorticella*, 464
Garrod, A. E., 140
Gärtner, 65, 70
gastric brooding frog, larval development, 268
gastrulation, 276, 285–6
gel electrophoresis, 27–9, 173
gene pools, 162
    conservation, 181–3
genes, 3, 22, 93–4, 122
    action, 575
    common DNA sequence, 116
    enzymatic synthesis, 200, 201
    in specialized cells, 281–3
    linkage, 79–81, 87, 161; *see also* sex-linked inheritance
    manipulation, *see* genetic engineering
    'one gene–one polypeptide' hypothesis, 25–6, 29
    pleiotropy, 146–7
    role in development, 278
genetic code, 109–14, 117
genetic drift, 162, 186
genetic engineering, 101, 198–205, 212
genetic maps, 81
genetic marking, 99–100
genetic material, 3, 13, 15, 22
    nature, 93–124
genetical transformation, 99–101
genome, 123
genotype, 70
    conservation, 217–18, 261, 262
    generation of variations, 217–18, 227, 261
geographical races, 183
geological change, 549, 552, 559
geotropism, 342
gerbil, oestrus in Mongolian, 242
germination
    effect of light, 378
    role of gibberellin in, 350–51

giant chromosomes, *see* polytene chromosomes
giant cowbird, 416–18
giant tortoise, 217
gibberellins, 343, 347–54, 357, 361, 378, 391
ginkgo trees, dispersal, 449
glasshouse cultivation
    economics, 392–8
    pest control, 497–500
glucagon
    mechanism of action, 324
    role in insulin release, 320–21, in glycogen breakdown, 324
gluconic acid, production by fermentation, 194
glucose
    access to muscle and adipose tissue, 322
    enzymic detection, 208
    enzymic production, 206
    role in insulin release, 320–21
glucose electrodes, 208–209
glucose oxidase, 208
glucose isomerase, 205, 206
L-glutamic acid, production by fermentation, 196
glycogen
    breakdown in muscle, 325
    synthesis, 322
glycogen phosphorylase, 324
Golgi apparatus, 23, 314
Gonium, 24, 25
Gould, Stephen Jay, 559
Graafian follicle, 246, 247
grapes, seedless, 352
grasses
    asexual reproduction, 219
    energy cost of sexual reproduction, 227
grasshoppers
    capture–mark–recapture study, 482
    karyotype (*Chorthippus parallelus*), 34
grassland
    energy flow in, 520–21
    pyramid of numbers, 511
gravitropism, 342
great tit, population dynamics in Marley Wood, 488–95
green leafhopper, 506
green scarab beetle, 524
green tortrix, 490
greenhouse effect, 527
grey squirrel, dispersal, 470
Griffith, 99
Grohmann, 398
gross primary production, 514
growth
    animal, 275, 287–9; hormone control of, 327–41
    human, measurement of, 288–9
    plants: effect of light, 380–98; modular pattern, 305; roots, 302–304
growth hormone
    bacterial production of human, 101
    effect of thyroid hormone, 326
    genetically engineered, 213
    see also somatostatin

growth plate, 287–8
growth regulators, synthetic, 361
growth velocity curves, 288
Gryllus spp., *see* crickets
Gryphoea spp., 557
GTP, 118, 120
guanine, 102–105
guanosine triphosphate, *see* GTP
Gudernatsch, 337
Guérin, C., 371
guppy, embryo protection, 226
gymnosperms, gametes, 233

H

habitat, 418
   conservation of endangered, 542–3
Haeckel, E., 275, 276
haemoglobin, 25–9, 119, 276
   in isolated mouse population, 186
   in sickle-cell anaemia, 147, 168–9
   of plant origin, 469
haemophilia, 145, 271
Haliaeetus leucocephalus, *see* bald eagle
halophytes, 433
Hammerling, Joachim, 5
Haplochromus burtoni, *see* mouthbrooder fish
haplo-diplontic life cycle, 229, 230–31, 232, 234
haploid cell, 32–3
   formed by meiosis, 43
haplontic life cycle, 229, 230, 234
Hardy–Weinberg model of population composition, 155–9, 160
hares, 553
   see also snowshoe hare
Harlow, H. F., 405–409
Hartsoeker, Niklaas, 31
hawkweeds, Mendelian studies, 66–7
hay fever, 366
Helianthemum chamaecistus, *see* rock-rose
hemimetabolous insects, 332, 333
henbane
   flowering, 349
   grafted to tobacco, 391
Hennig, Willi, 565, 571
herbivores, 544
herring, in Thames water, 429
Hess, Ekhardt, 399–400
heterochromatin, 13, 14, 17, 22, 122, 135
heterogametic sex, 50, 159
heteromorphic life cycle, 231
heterosis, 145–6, 182
heterosporous plants, 232
heterotrophic organisms, 308, 309, 509–10, 513
heterozygote, 72
   gene expression in, 144–6
   proportion in mixed population, 156–7
Hieracium spp., *see* hawkweeds
Hinde, R. A., 409–10
Hippocampus spp., *see* sea horse
histones, 121, 122–3
HLA system, *see* human leucocyte antigen system
Hogan, 400

holoblastic cleavage, 574
holometabolous insects, 333, 334
homeostasis, 320
homogentisic acid, 140, 141
homozygote, 72
'homunculus', 31
honey bee, 32, 66, 221, 228
Honkenya peploides, *see* sea sandwort
Hooker, Joseph, 551
hormones, 312–14
   activation/inactivation, 323
   defined, 312
   dose–response relationships, 321–3
   flowering control by, 389–92
   mechanisms of action, 323–7
   menstrual cycle control, 246–8
   plant, 342–60
   protein binding of, 323
   synthesis and release, 316–21
Hubbard Brook Experimental Forest, 528–31
human leucocyte antigen system, 367
humans
   allele frequencies in populations, 157–9
   chromosome banding, 35
   effect of mutagens, 91
   embryo implantation and nutrition, 270–71
   fertilization, 248–52
   gametogenesis, 235–7
   genes influencing metabolic reactions, 140–42
   haemoglobin polymorphism, 168–9
   height, 63, 558
   life expectations, 487
   measurement of growth, 288–9
   menstrual cycle, 245–8
   personality, 63
   polydactylism, 149
   polymorphism, 168–9
   population dynamics, 503–507
   races, 183
   red blood cell types, 26–7
   semen storage, 178–9
   sex inheritance, 49–50
   sex ratio, 53–4
   synchronization of oestrus, 245
   'test-tube' baby, 243
humoral immunity, 363–4, 370, 371
humus, 431, 432, 530
Huxley, T. H., 559
hybrid vigour, *see* heterosis
hybridomas, 210–11
hybrids, 65
   asexual reproduction by, 225
   genome, 123
   mechanisms of avoidance, 258–9
   reduced fertility, 48–9, 87, 258
   see also heterosis
Hydra, asexual reproduction, 220
hydrogen, bacterial generation, 213
hydrogen sulphide, energy source, 467–8
Hydropsyche spp., *see* caddis flies
hydrosere succession, 475
Hyoscyamus niger, *see* henbane
Hypericum perforatum, *see* Klamath weed

hypersensitivity reaction, 366, 367
Hypogymnia physodes, 539, 540, 541
hypolimnion, 532
hypothalamus, 246, 313, 320, 338
hypotheses, making and testing (in ecology), 422–5

I

IAA, 342, 343
   interactions: with gibberellic acid, 348, 349, with kinetin, 354
   mechanism of action, 344–6
   metabolism, 343–4
IAA oxidase, 344
Icerya purchasi, *see* cottony cushion scale insect
imaginal discs, 333
immobilized enzymes, 206, 207, 208
immune complex, 364
immune response, 360–73
immunity
   acquired (adaptive), 360, 363–72; cell-mediated, 363, 364–70; humoral, 363–4, 370, 371
   natural (non-specific), 360–63
immunization, 363, 371–2
immunology, biotechnological applications, 210–12
implantation of embryo, 243, 270
   delayed, in badgers, 242–3
inbreeding, 159–62, 261–2
indigo bunting, song, 403
indirect flowering, 385
indolebutanoic acid, 361
indole-3-ethanoic acid, *see* IAA
Indonesia, food production, 505–506
induction (tissue interaction), 286
industrial melanism, 169–72
infertility, 250, 252
inheritance, 31–59
   acquired characters, 549
   coat colour in mice, 152
   comb shape in poultry, 147–9
   continuous variation, 82–5, 163, 558
   human haemoglobin variations, 27–9
   in fungi, 54–9
   leaf colour in *Pelargonium*, 253–4
   matri- and patri-lineal, 31
   pigmentation in sweet pea, 151–2
   protein variations, 29
   role compared with environment, 61–3
   sex, 49–54
   sex-linked, 32, 53, 73–4
inhibition (succession), 477
insect pollination, 255–6
insecticides, 499, 506
insects
   control of metamorphosis, 327–36
   ecdysone production, 317
   temperature effects on life cycle and growth, 380
insulin
   bacterial production of human, 101, 204
   dose–response curves, 321–2
   half-life, 323
   mechanism of release, 320–21

synthesis and release, 316
integument, 267
interferon
 assay of α-, 211
 microbiologically produced, 204–205
 response to viral infection, 362
interphase
 meiosis, 40, 44
 mitosis, 13, 14, 15, 19
intron, 200
inversion of chromosome material, 86, 87
islets of Langerhans, 316
isolation
 effects on infant monkeys, 408
 reproductive, 183–90
isomorphic life cycle, 231
isotope labelling
 DNA, 97–9, 106–107
 study of protein synthesis, 110–111
itaconic acid, production by fermentation, 194

Jacob, F., 131
Janzen, 245
Japweed, dispersal, 470
Jenner, Edward, 371
JH, see juvenile hormone
Johannsen, W. L., 83, 84
juvenile hormone (JH), 331–3

K cells, see killer cells
karyotype, 34
Kelvin, Lord, 555
ketoconazole, 195
Kettlewell, H. B. D., 169, 170–71
Khefer aegyptorum, see green scarab beetle
killer (K) cells, 372
kin selection, 163
kinetin, 4, 343, 354, 355, 356, 357
kinins, 361
Klamath weed, control by beetle, 461–2, 497
Klebsiella spp., 469
Klinefelter's syndrome, 51
Koelreuter, 64, 65
Krebs, J. R., 490

Lacerta vivipara, see lizard, common
Lack, D., 488
Lactobacillus bulgaricus, population growth, 485–6
lactose permease, 131
ladder of nature (*scala naturae*), 547–9, 550, 555
Lagidium viscacia, see vizcacha
lakes
 nutrient cycling in, 531–2
 succession round, 474–6
Lamarck, J.-B. de, 547–9
lapwing, life expectation, 487
Lark, river, 427
Lasius spp., see under ants
Lathyrus odoratus, see sweet pea

laws of biology, 573–6
learning, behaviour changes due to, 399
leaves, 'sun' and 'shade', 459–60
Lebistes reticulatus, see guppy
Lecanora conizaeoides, 539, 540
Leeuwenhoek, A. van, 193
leghaemoglobin, 469
leguminous plants, nitrogen fixation by, 469
Lejeune, 38
lemmings (*Lemmus* spp.), population cycles, 500–502
Lepus spp., see hares
lethal mutants, 89
lettuce
 effect of light on germination, 378
 varieties, 176
LH, see luteinizing hormone
Libellula depressa, moulting, 328
lichens, 169
 effect of pollutants, 422–3, 537–41
 mutualism in, 465, 467
life cycles
 evolution, 233–4
 types of, 229–31
life expectancy
 human, in LDCs, 504
 of different species, 487
light
 effect on germination, 378
 effect on plant growth, 379, 380–98
lignin, 25
Ligustrum ovalifolium, see privet
lily
 embryo sac in ovule, 239
 pollen formation, 238, 239
limbs
 development, 292–300, 301
 similarities in vertebrates, 574
lime chlorosis, 443
Limnodrilus hoffmeisteri, see tubificid worms
Linckia, asexual reproduction, 220
ling heather, acidophilous nature, 440
Lingula, 1, 20
linkage, see under genes
Linnaeus, C., 64
Linnean classification, 564
Linum usitatissimum, see flax
lipolysis, insulin-dependent, 322
litter traps, 513
Littlefield, J. W., 110
liver, treatment of failure, 212
liver fluke, asexual reproduction, 220
lizard, common, embryo protection in, 226
loam, 433
locus of gene, 57
 related to chromosome puffs, 136–7
locust
 changes in wing pads, 331
 β-ecdysone titre in, 330
 egg protection, 268
 JH titre in, 331
Longworth traps, 482–3
Lumbricus terrestris, see earthworms
luteinizing hormone (LH), 246–7

Lyell, Charles, 549, 551, 552, 559
lymph nodes, 369
lymphocytes, 363, 368, 369–70
 B, 364, 368–9, 370
 K, 372
 NK, 371, 372, 373
 T, 365–6, 367, 368, 370
lymphoid organs
 primary, 364, 369
 secondary, 369
lymphokines, 366, 372
lynx (*Lynx canadensis*), population cycles, 502–503
lysine, bacterial production, 193
lysogenic cycle, 95, 96
lysozyme, 360

M
Macaca mulatta, see rhesus monkey
Macleod, C. M., 100, 101
macroevolution, 563
macronutrients, 433
macrophages, 362, 371, 372, 373
Magicicada spp., see cicadas
maize
 chromosome 9 abnormalities, 81–2
 effect of density on growth of monoculture, 453–4
 linkage studies, 80
 root growth, 433
major histocompatibility complex (MHC), 367, 368
malaria parasite, 462–4
Malthus, Thomas, 550–51
malting, 350–51
mammals
 cleavage pattern, 275
 embryo protection and nutrition, 226, 269–71
 estimating population size of small, 482–3
 sexual cycles, 240–48
Man, see humans
mangrove, vivipary in, 267
Marler, P., 402–403
Marley Wood study (great tit population), 488–95
marrowstem kale, self-incompatibility in, 263–4
marsh warbler, song, 404
Marsilea vestita, see shamrock fern
marsupials
 distribution, 470
 embryo protection and nutrition, 269–70
mast cells, 366
Mathei, 113
mating-type alleles, 160
matrilineal society, 31
maturation, behaviour changes due to, 399
mayfly, 216
McCarty, M., 100, 101
McCauley, D. S., 505
McQuillen, K., 110, 112
megasporangium, 233
megaspore, 233, 239

Index 585

meiocytes, 234
meiosis, 32–3, 36, 40–47, 79, 81, 121, 123
 abnormal, 36, 37
 in hybrids and polyploids, 47–9
 in sexual reproduction, 226, 234–5, 237
 in subsexual reproduction, 228–9
melanin, 140, 141
melanism, industrial, 169–72
membranes, see plasma membrane, and under nucleus
Mendel, Gregor, 66–71, 79, 557
menstrual cycle, 245–8
Meriones unguiculatus, see gerbil
meristems, 300, 301, 302, 304–308
mermaid's wineglass, see *Acetabularia*
mesoderm, 275, 286–7
messenger RNA, 113–14, 115, 117, 118–19, 137, 325, 333
metamorphic climax, 336, 337
metamorphosis, 327, 340–41
 amphibian, 336–40
 insect, 327–36
metaphase
 meiosis, 41–3, 44, 79
 mitosis, 16, 19
metaphase plate, 16
methane
 effect on climate of atmospheric, 527
 production by fermentation, 198
methionine, 117
Methylophilus methylotrophus, 197
MHC, see major histocompatibility complex
microevolution, 563
micronutrients, 433
micropyle, 239
microspore, 233
Miller, 364
Milstein, Cesar, 210, 211
minimal growth medium, 125
Mirsky, A. E., 94
mites
 six-spotted, population fluctuations, 496
 soil, 432
 two-spotted spider, biological control, 497, 498–9
mitochondria, 22, 32
 changes during life cycle, 232
 effect of thyroid hormone, 327
mitosis, 13–21, 33, 36, 37, 121, 223–4
 abnormal, 36, 48
mixed lymphocyte reaction (MLR), 367–8
modular organisms, 480, 481, 495–6
'mongolism', see Down syndrome
monoclonal antibodies, 210–12
monoculture, 392
 effect of density on growth, 453–6
monocytes, 363, 366
Monod, J., 131
monohybrid crosses, 68, 71–4
monosodium glutamate, 196
morph, 166
 see also polymorphism
morphogen, 298–9
morphogenesis, 276, 277, 290–92, 326

morphological stability, 563
mosaic individual, 88
mosquitoes, 183
 eradication campaign, 504
 role in malaria transmission, 462–4
mosses, 32
 on Surtsey, 471, 472
mother–infant relationships, 404–11
moulting (insect), 328, 329
 induced in adult, 330–31
 stationary, 330
mouse
 coat colour inheritance, 152
 development from single blastomere, 279
 effect of males on oestrus, 243–5
 isolated populations, 185–6
 nuclear volumes, 22
 potential of embryo cells, 283–4
mouse stock box, 244
mouthbrooder fish, 269
mule, 258
Müller, K. O., 374
Muntjac deer, dispersal, 449, 470
muscle cells, myofibrillar system, 23
muscles
 development, 299–300
 growth, 288
muscular dystrophy, 271
mutagens, 89, 91, 109
mutant, 86
 complementation, 125–31
 declining frequency, 155
mutation, 85–91, 557
 biotechnological applications, 193
mutualism, 452, 461, 465–70
myco-protein, 198
mycorrhiza, 465, 466
myelomas, hybrid, see hybridomas
Myobacterium tuberculosis, 371
myofibrils, 23
myosin, 23

N
Nägeli, K. W. von, 66, 67
naphthalene-ethanoic acid, 4, 361
natural (non-specific) immunity, 360–63
natural killer (NK) cells, 371, 372, 373
natural selection, 64, 551–9, 564
nature conservation, 542–3
neo-Darwinism, 557–8
Nephotettix cinticeps, see green leaf-hopper
nerve supply to limb, 300, 301
net primary production, 514–16
nettle, distribution related to cattle population, 426
neural crest cells, 291–2
neural plate, 286
neural tube, 276, 286
 defects, 271, 291
 morphogenesis, 290–91
neurocrine system, 313
neurohaemal organs, 313
neurohormones, 312–13
neurones, 312
neurosecretory cells, 312, 313, 314–16

Neurospora crassa
 mutant complementation, 125–7, 130
 spore formation, 57–9
niche, 418
Nicotiana rustica, see tobacco
night heron, life expectation, 487
Nile blue sulphate staining, 285
Nilaparvata lugens, see brown plant-hopper
Nilsson-Ehle, 83, 85
Nirenberg, M. W., 113
nitrate(v) ion
 losses following deforestation, 530–31
 pollution by, 533, 536
nitrifying bacteria, 530
nitrogen
 fixation, 468–9; by non-leguminous crops, 213, 469; in forest litter, 530
 importance in decomposition, 525
nitrogen cycle, 528–31
nitrogen mustards, 89
nitrogenase, gene for, 213
nitrous acid, mutagen, 109
NK cells, see natural killer cells
non-specific immunity, see natural immunity
noradrenaline, synthesis and release, 317–19
Norway spruce, 476–7
nucellus, 239, 267, 268
Nucifraga caryocatactes, see nutcracker
nucleic acids, see DNA; RNA
nucleolus, 12, 14, 17, 20, 22
nucleosomes, 123
nucleus, 31–59
 changes with differentiation, 22
 division, see meiosis; mitosis
 DNA content, 94–5
 link with cytoplasm, 9, 11, 112–14
 membrane, 13, 14, 15, 17, 43, 44
 transplantation, 7–11
nutcracker, 451
nutrient cycling, 512–14, 525–35, 544
 in lakes and oceans, 531–2, 533
 in tropical rain forest, 532, 534–5
nutrient film cultivation technique, 394–5, 396
nutrient ion uptake (plants), 433–5
Nycticorax nycticorax, see night heron

O
obligate mutualism, 469
obturacular plume (tube-worm), 467, 468
Oenothera, see evening primrose
oestrogen, 247
oestrous cycle, 246
 see also menstrual cycle
oestrus, 242
 in mice, effect of males, 243–5
oil palm, cloning, 225, 256
old man's beard, basophilous nature, 440
omnivores, 544
onion, nutrient reserve of seed, 268
ontogeny, 398–411
oocytes, 236–7, 249
oogamy, 226
oogenesis, 236–7

operator locus, 131, 132
operon, 132
Operophtera brumata, see winter moth
Ophrys spp., see under orchids
Opuntia spp., see prickly pear cactus
orange-rumped agouti, 553
orcein staining, 38
orchids
   basophilous nature, 440
   insect mimicry in *Ophrys* spp., 255, 256
organ transplantation, 368
organic compounds, 544
organogenesis, 275, 286–7
Origin of species, 551–5, 573
Ornithogalum zeiheri, chromosomes, 34, 35
Ornithorhynchus anatinus, see duck-billed platypus
oropendola bird, 416–18
orthogenesis, 555–6
Oryctolagus cuniculus, see rabbit
Oscillatoria spp., 533
outbreeding, 159–62, 262, 265
ovary (mammal), 246
ovulation, 242, 245, 246, 247
ovule, 258, 267
owls, see short-eared owl; tawny owl
Oxford ragwort, dispersal, 470
oxygen, in river water, 427, 428–9, 536

P
pancreas
   artificial, 212
   exocrine cell, 23
   insulin synthesis and release in, 316, 320–21
Pandorina, 2
'pangenesis', 64, 556
Paramecium aurelia, fission, 216
parasitism, 452, 460, 461, 462–4, 544
Park, T., 456
Parmelia saxatilis, 540
parthenogenesis, 221, 222, 223, 229
Parus major, see great tit
Pasteur, Louis, 192, 193, 371
Pasteurella pestis, see plague
patrilineal society, 31
pattern formation in development, 277, 292–300, 301
pea
   effect of gibberellic acid on plants, 347, 348
   height of plants, 61, 68
   Mendelian studies, 66, 67–70, 79
   nutrient reserve of seed, 268
   self-pollination, 160
peaty-gley soil, 441
pecking by chicks, 399–400
Pelargonium
   inheritance in, 253–4
   polyploidy in, 39
penicillin, 194, 195
peppered moth, polymorphism in, 169–72
peptide hormones
   mechanism of action, 324–5, 340
   synthesis and release, 316

peptidyl transferase, 118
pertussis vaccine, 372
pest control, 497–500
pesticides, 499, 506
phages, see bacteriophages
phagocytosis, 362, 364
Pharbitis nil, effect of day length on flowering, 386
phase contrast illumination, 19
Phaseolus vulgaris, see French bean
pheasant, population growth, 484
phenotype, 70
phenylalanine metabolism, 140, 141–2
phenylketonuria, 140–42, 174
pheromone, 245
phosphate ion, pollution by, 536
phosphorylases, 324, 325
photomorphogenesis, 382, 383
photoperiodism, 382–92
photorespiration, 213
photosynthesis, 380–82
   energy from, 213–14
   in 'sun' and 'shade' leaves, 459–60
phototropism, 342
Phragmites australis, see reed
phyletic gradualism, 559, 560, 563
phytoalexins, 374
phytochrome, 378, 379, 388–9
Phytophthora infestans, 374
phytoplankton, population cycles, 531–2, 533
Phytoseiulus persimilis, use in biological pest control, 498–9
Picea abies, see Norway spruce
pig
   extinct breeds, 182
   varieties, 176
pigeons
   flight, 398–9
   plumage colour inheritance, 74
pigments, plant, 381–2, 388
   see also chlorophyll; phytochrome
Pigott, D., 445
Pinus aristata, see bristlecone pine
Pinus cembra, pattern on mountainside, 451
Pinus sylvestris, see Scots pine
'pioneering' plants, 224–5
Pipa pipa, see Surinam toad
Pisum sativum, see pea
pitfall-trapping, sampling by, 481
pituitary gland, 313
   control of growth hormone secretion, 326
   FSH and LH secretion, 246, 247
   PRL and TSH secretion, 337, 338, 339, 340
placenta, 270–71
plague, dispersal, 470
plaice, age structure of population, 486–7
plankton, population cycles, 531–2, 533
plantains (*Plantago* spp.), effect of density on growth, 455
plasma membrane
   changes with differentiation, 23–4
   gametes, 250–51

role in hormone release, 320–21
plasmids, 198–9, 200–202, 203
plasmodesmata, 17
Plasmodium spp.
   life cycle, 462–4
   replicating DNA in, 108
plastic development, 308
plastids, cleavage during life cycle, 232
Platichthys flesus, see flounder
pleiotropy, 146–7
Pleodorina, 2
Pleuronectes platessa, see plaice
Pneumococcus, genetical transformation in, 99–101
podzol, 438, 439
point-mutation, 87–8, 147
point quadrat, sampling by, 481
polar body, 237, 249
polarity of cells, 574
polarizing region (limb bud), 297–300
poliomyelitis, 371
pollen, 233, 237–9, 255
   rejection of foreign, 259, 263–4
pollination, 252, 255–8
pollution, 535–42
polychaetes, asexual reproduction, 220
polydactylism, 149
Polygonum amphibium, 62
polymorphism, 166–75
polymorphonuclear cells, 362, 364
polyoxin D, 195
polypeptides, 25
   'one gene–one polypeptide' hypothesis, 25–6, 29
   synthesis, 117–21
   synthesis of artificial, 114
polyploidy, 36–7, 39, 47–9, 175–7, 225, 559
polysaccharides, production by fermentation, 196
polysome, 119
polyspermy, 251
polytene (giant) chromosomes, 35
   gene activity in 'puffs', 135–7, 325, 333
Pontin, J., 458
population cage, 154
population dynamics, 480–508
population explosion, 470
population genetics, 154–91
populations
   assessment of size, 480–83
   cycles, 500–503
   growth, 483–8
   human, 503–507
   limitation, 488–500
potassium, temperature effect on root uptake, 435
potatoes
   blight resistance, 374
   effect of gibberellic acid on stored, 352–4
poultry, see fowl
predation, 452, 460, 461–2
pregnancy
   diagnosis, 211
   termination, 39, 175
premetamorphosis, 336

presumptive regions, 285, 289–90
prickly pear cactus
  biological control, 497
  dispersal, 470
primary producers, 544
primary production, 544
primrose (*Primula vulgaris*), breeding systems, 264
Principles of geology, 549
privet, modular growth pattern, 305
PRL, *see* prolactin
PRL-IH, *see* prolactin release-inhibiting hormone
producers, 509, 544
production, 544
  see also net primary production; secondary production
productivity, 514–16, 544
progesterone, 247
proinsulin, 316
prolactin (PRL), 338, 339, 340
prolactin release-inhibiting hormone (PRL-IH), 338, 339
prometamorphosis, 336, 337, 338
promoter region of DNA, 132
propanone, production by fermentation, 194
prophase
  meiosis, 40–41, 42–3, 44, 81, 123
  mitosis, 15, 19
prostaglandins, 319
proteases, 205
protein hormones
  mechanism of action, 324–5
  synthesis and release, 316
protein kinase, 325
proteins, 25–9
  binding, 323
  chromatin content of, 121–3
  electrophoretic studies, 27–9, 173; in gravity-free environment, 214
  improvement of cereal, 213
  inherited variations, 29
  plasma membrane, 24
  repressor substances, 135
  single cell, 196–8
  synthesis, 109, 117–21
protein-secreting endocrine cell, 314, 315
proteolysis, insulin-dependent, 322
proton pumps, 345, 346
protozoa
  distribution in polluted river, 537
  mutualistic associations, 467
Pruteen, 197
Psarocolium wagleri, *see* chestnut-headed oropendola bird
Pseudomonas denitrificans, 530
Pteridium aquilinium, *see* bracken
pteridophytes, 219, 233, 254, 266–7
punctuated equilibria, 559–63
Punnett, R. C., 147
pure strain, *see* true breeding variety
Puschkinia libanotica, karyotype, 34
pyramid of biomass, 512
pyramid of energy flow, 512
pyramid of numbers, 511
pyrimethamine, 464

Q

quadrat, sampling by, 481
Quercus ilex, *see* evergreen oak
quinacrine staining, 35
quinine, 464

R

rabbit, 553
  dispersal, 470
  ovulation, 242
  population growth, 484
rabies vaccine, 372
races, geographical, 183
rad (defined), 90
radiation-induced mutation, 89–91
radishes, effect of light and nitrate limitation, 341
rain forest, 420, 478
  nutrient cycling in, 532, 534–5
Rana spp., *see under* frogs
Ranunculus spp., *see* buttercups
Rare Breeds Trust, 183
rat
  coat colour inheritance, 77–8
  food preferences, 401
  Warfarin resistance in, 172–3
recessive character, 70
reciprocal cross, 32, 52, 68–9, 73
reciprocal replacement (climax species), 478
recombination of characters, 41, 79–82
red blood cells
  nuclei of chick, 22
  types of human, 26–7
red deer, social grouping, 265–6
redwing, song, 401
reed, 475
reed-mace, 475
regulatory genes, 132
relatives, mating of, 160
releasing hormone (RH), 246
rendzina soil, 440
rennin, 205
replica plating technique, 128–30
replication origins, 108
repressed mutants, 131–2
repressor proteins, 135
reproduction, 216–73
  asexual, 218–25, 261
  breeding systems, 261–6
  embryo protection and nutrition, 266–71
  gametogenesis and fertilization, 234–60
  requirement for, 216–18
  sexual, 218, 226–7, 261
  subsexual, 218, 228–9
  variations in life cycles, 229–34
  vegetative, 219, 225, 355–6
reproductive isolation, 183–90
RER, *see* rough endoplasmic reticulum
resistance
  antibiotics, 172, 201
  disease, 179–81, 182, 373–4
  drugs, 464
  pesticides, 499
  Warfarin, 172–3
respirometer, Warburg, 133

restriction enzymes, 199–200, 201, 202
reverse transcriptase, 200
RH, *see* releasing hormone
Rhacomitrium moss, on Surtsey, 471
Rheobatrachus silus, *see* gastric brooding frog
rhesus monkey
  mother–infant relationships, 404–10
  pheromone production, 245
Rhizobium, 469
Rhizophora, vivipary in, 267
Rhodnium prolixus, 333–6
Rhodomonas spp., 533
riboflavin, production by fermentation, 194
ribonucleic acid, *see* RNA
ribose, 112
ribosomal RNA, 113
ribosomes, 110, 113
  arrangement within cell, 22–3
  protein synthesis in, 117–21
  sedimentation, 112
  variations in numbers, 22, 232
ribulose bisphosphate carboxylase, 213, 468
ribulose-5-phosphate kinase, 468
rice, production in Indonesia, 505–506
rifampicin, 195
Riftia pachyptila, *see* Galápagos vent-worm
Ris, H., 94–5
rivers and streams
  effect of changes in, 426–30
  effect of pollution on communities in, 535–7
  energy flow in, 522
RNA, 22, 110, 112
  constituent of chromatin, 121
  synthesis, 115–17
  types of, 112–13
  see also messenger RNA; ribosomal RNA; transfer RNA
RNA polymerase, 115–16, 132–3
robin, 159–60
rock-rose, basophilous nature, 440
rocky shores, zonation on, 446–8
rodents, competition with ants, 423–4
Rodiola cardinalis, *see* vedalia beetle
roots
  gravitropic response, 342
  growth, 302–304, 433–4; effect of temperature, 379
  meristem, 302, 304
rough endoplasmic reticulum, 23, 314, 315
rubisco, *see* ribulose bisphosphate carboxylase
ruminants, gut mutualists, 467
rye, root growth, 433–4

S

St Kilda mice, 186
salmon (*Salmo salar*), in Thames, 428, 429
saltations, 559
salting out, 205
salt-marsh, energy flow in, 521

ampling, 480–82
and, 432–3
sardine, life expectation, 487
Sargassum muticum, see japweed
satellite (chromosome region), 33
savannah, 421
Scabiosa columbaria, see small scabious
scaffolding proteins, 123
scala naturae, see ladder of nature
Scaphidura oryzivora, see giant cowbird
scent, use by women, 245
Schistocerca gregaria, see desert locust
Schoenus nigricans, see bog-rush
Sciurus caroliniensis, see grey squirrel
scorpions, embryo protection, 226
scots pine, acidophilous nature, 440
SCP, see single cell protein
screw-worm fly, biological control, 497
sea arrow grass, distribution, 442
sea horse, brood pouch, 269
sea lion, 159
sea sandwort, on Surtsey, 472
sea urchins, 15–16
    cleavage pattern, 275
    fate map of embryo, 285
    gastrulation, 285–6
    partitioning of egg cytoplasm, 283
    polyspermy prevention, 251
seas, nutrient cycling in, 531–2
Secale cereale, see rye
secondary consumers, 544
secondary production, 514, 516–17, 544
sedges, 475
sedimentation analysis of bacterial cell sap, 111–12
seedling, quality of growth, 379
seeds, 233, 267, 268, 301
    dormancy, 306, 350, 357, 377–9
selection, 162–6
    artificial, 175–83
    natural, 64, 551–9, 564
    species, 561–3
self-fertilization, 159–60, 161, 226, 262
self-incompatibility, 160, 263–4, 265
selfing, see self-fertilization
semen, 235
    storage, 85, 178–9
Senecio squalidus, see Oxford ragwort
senescence, effect of cytokinin, 356
Sertoli cells, 235
sewage pollution in Thames, 428–9
sex
    evolution, 159
    inheritance, 49–54
sex-linked inheritance, 32, 53, 73–4
sex ratio (human), 53–4
sexual reproduction, 218, 226–7, 261
shamrock fern, fertilization, 252
sharks, embryo protection, 226
sheep
    life expectation of dall mountain, 487
    short-legged, 555, 556, 557
shoot meristem, 304, 305
shrimp, see *Gammarus*
'Siamese' twins, 281
Siboglinum fiordicum, 468

sickle-cell anaemia, 27, 29, 147, 168–9, 174
silkworms, 334–5
similarities, inherited, 573–6
single cell protein, 196–8
six-spotted mite, population fluctuations, 496
skin grafts, 368, 369
Skokholm mice, 185–6
Slater, P. J. B., 404
small scabious, distribution, 442, 443
smallpox vaccination, 371
Smith, N., 416
snails
    evolution, 561
    polymorphism, 166–8
snakes, food preferences, 400
snowshoe hare, population cycles, 502–503
soil mite, 432
soil profiles, 438, 439–41
soils, 430–43
    effect of felling on forest, 527
    in rain forest, 534–5
somatostatin, 200, 204
somites, 293
sonograms, bird song, 401–402
sound spectographs, see audiospectrographs
sparrowhawks, great tit predation by, 494
sparrows, songs, 402, 403
specialization
    cells, 2–3
    leading to vulnerability, 306, 309
speciation event, 561
species, 418
    conservation of endangered, 542–3
    difficulty of definition, 183
    extinction, 552–3
    origination of distinct, 64, 186, 551–63
    recognition, 258–9, 260, 263
species selection, 561–3
spermatids, 235
spermatocytes, 235, 236
spermatogenesis, 235, 236
spermatogonia, 235
spermatozoa (sperm), 235, 249–51
Sphagnum bog, 475, 476
spina bifida, 278, 291
spinach, flowering, 349
spindle (nucleus), 15–16, 41–2, 43
spiny ant-eater, 269
Spirogyra, fertilization, 226
Spirulina, protein source, 197
spleen, 370–71
sponges, 24
spontaneous mutation, 86
sporangia, 232
spores, 219, 232–3, 234–5
sporophyte, 230, 266–7
'sports', 85
squinancywort, distribution, 443
squirrel, dispersal of grey, 470
staining procedures
    Nile blue sulphate, 285
    orcein, 38

    quinacrine, 35
    toluidine blue, 18
standing crop, 514, 544–5
star of Bethlehem, see *Ornithogalum zeiheri*
starfish, 15–16
    asexual reproduction, 220
starling, dispersal of European, 470
stationary moults, 330
statistical studies, inheritance, 83, 85
stemless thistle
    distribution, 444–5
    susceptible to aluminium, 443
steroid hormones
    mechanism of action, 325–6
    protein binding of, 323
    synthesis and release, 317
steroid-secreting endocrine cells, 314, 315
stick insect, growth in, 327, 329
stigma, 240, 258, 259
stomata, effect of abscisins, 357–8
storage ecosystems, 525
stratification, 378
strawberry, vegetative reproduction, 219
streptomycin, 195
Strix aluco, see tawny owl
structural genes, 132
'struggle for existence', 163
Sturnus vulgaris, see European starling
style, 240, 258, 259
subsexual reproduction, 218, 228–9
subtilisin, 205, 206
succession, 418, 452, 472–7
sulphonamide, 464
sulphur dioxide, effect on lichen patterns, 422–3, 537–41
suppressed weaklings, 453, 455
Surinam toad, egg development, 269
Surtsey, 471–2
'survival of the fittest', 163
suspensor, 267–8
Svedberg units (sedimentation rate), 112
Sweat Mere, 474
sweep-netting, sampling by, 481
sweet pea, inheritance of pigmentation, 151–2
synecology, 422
synergism (hormones), 314, 333, 340
syngamy, 226
systematics, 554

T
2,4,5-T, 360
T_3, see tri-iodothyronine
T_4, see tetraiodothyronine
Tachyglossus spp., see spiny ant-eater
tanning (insect cuticle), 329, 332
Tansley, A., 443
Taphozous melanopogon, see tomb bat
Taraxacum officinale, see dandelion
Tatum, E. L., 57, 125–7
tawny owl, habitat and niche, 418
teeth, continuity of structure, 1
Tegeticula yuccasella, 256
telophase
    meiosis, 43, 44
    mitosis, 16, 19

Index   589

temperature
  effect on decomposer activity, 524
  effect on metabolism and development, 379–80
  effect on potassium uptake by root, 435
teratomas, 284–5
termites
  gut mutualists, 467
  stationary moults, 330
test cross, 77, 79
testis, 235
  photoperiodic control of development, 384
'test-tube' baby, 243
Testudo gigantea, see giant tortoise
tetanus immunization, 371
tetracyclines, 195
tetraiodothyronine, 317
Tetranychus spp., see two-spotted spider mite
tetraploidy, 36, 48
TH, see thyroid hormones
thalassaemia, 27, 29
Thalidomide effects, 296, 297
Thames, river, restoration of tidal, 427–9
thermocline, 532
thiamine, influence on mutant tomatoes, 142–4
Thorpe, W. H., 401
thymine, 102–105, 112
thymus gland, role in immune response, 364–6
thyroglobulin, 317, 318
thyroid hormones (TH), 317
  mechanism of action, 326–7
  protein binding of, 323
  role in amphibian metamorphosis, 337–40
thyroid releasing hormone (TRH), 338
thyroid stimulating hormone (TSH), 337–8
thyroxine, 317
  half-life, 323
tissue culture, 3–5, 356
tissues, 138
  interactions, 286–7
  typing, 368
TMV, see tobacco mosaic virus
TNF, see tumour necrosis factor
toads, see clawed toad; Surinam toad
tobacco
  henbane grafted to, 391
  heterosis in, 145
  pith callus, 354, 355
tobacco mosaic virus, resistance, 179–81
toluidine blue staining, 18
tomato
  demand for uniform product, 181
  economics of glasshouse cultivation 392–8
  influence of thiamine on mutant, 142–4
  tobacco mosaic virus resistance, 179–81
tomb bat, 513
tongue rolling character, 157–8

tortoise, giant, 217
Tortrix viridiana, see green tortrix
totipotency, 218
toxins, bacterial, 371
toxoids, 371
Tradescantia virginiana, chromosomes, 15
transcription, 115–17, 132–3
  control in higher organisms, 135–7
transfer RNA, 113, 117–18
translation, 115, 117, 118–21
translocation of chromosome material, 86, 87
traps
  litter, 513
  Longworth, 482–3
  pitfall, 481
traveller's joy, basophilous nature, 440
TRH, see thyroid releasing hormone
Trialeuroides vaporariorum, see whitefly
Tribolium spp., see flour beetles
Trichostema spp., pollination syndromes in, 257
Trifolium spp., see clover
Triglochin maritima, see sea arrow grass
tri-iodothyronine, 317
triploidy, 36, 39, 47
trisomy, 38–9
Triticum aestivum, see under wheat
trophic levels, 509, 510–11, 545
  energy transfer between, 518–20
trophoblast, 247, 270, 271
tropical rain forest, 420, 478
  nutrient cycling in, 532, 534–5
tropical savannah, 421
trout, effect of temperature on hatching, 380
true breeding variety, 31–2
trypsinogen, 119
TSH, see thyroid stimulating hormone
tuberculosis (BCG) vaccine, 371, 373
tubificid worms
  distribution in polluted river, 537
  in Thames water, 429
Tubifex tubifex, see under tubificid worms
tumour immunology, 372–3
tumour necrosis factor (TNF), 373
tundra, 420
Turdus iliacus, see redwing
Turner's syndrome, 51
twins
  identical, 279–81, 368
  non-identical, 249
two-spotted spider mite, biological control, 497, 498–9
Typha latifolia, see bulrush
Typhlodromus occidentalis, population fluctuations, 496
tyrosine metabolism, 140, 141

U
ultra-violet light, mutagenic activity, 89, 109
umbilical cord, 271
unicellular organisms, reproduction, 218
unitary organisms, 480, 481, 495
uracil, 102, 112, 114
urea, determination, 207–208

urea cycle, 128
urease, 207
Urtica dioica, see nettle
Usk, river, caddis fly in, 430, 431
Usnea spp., 539, 540, 541
uterus, 270–71

V
vaccination, see immunization
vaccines, 210, 212, 213, 363, 371–2
variation, 60–92, 552
  continuous, 60–61, 64; frequency distribution, 163–4; inheritance, 82–5, 163, 558
  discontinuous, 60–61
variegation in plants, 32
  inheritance in *Pelargonium*, 253–4
vectors
  genetic engineering, 199
  pollination, 255–6
vedalia beetle, use in biological pest control, 497
vegetal pole (embryo), 285–6
vegetative reproduction, 219
  exploitation, 225, 355–6
vernalization, 349
Vestiges of the natural history of Creation, 550
Vicia faba, see broad bean
Vipera berus, see adder
viruses, 95–6, 123
  AIDS, 370
  bacterial, see bacteriophages
  DNA in, 97–9
  elimination by antibodies, 368
  immunization against, 363, 371–2
  tobacco mosaic, resistance, 179–81
  tumours induced by, 372
'vitalism', 205
vitamin B_1, see thiamine
vitamin B_2, see riboflavin
vivipary
  mammals, 269
  plants, 267
vizcacha, 553
Vorticella, commensalism with *Gammarus*, 464

W
Wallace, Alfred Russel, 64, 550–51
Warburg respirometer, 133
Warfarin resistance, 172–3
wasps, 161, 163
water, see lakes; rivers and streams; seas
Watson, J., 105, 109, 113
wavy hair grass, distribution, 442
weasels, great tit predation by, 493 494–5
wheat
  effect of earthworms on growth, 436–7
  polyploidy and hybridization, 49, 175–7
  seed colour inheritance, 83
  varieties, 177, 181
White, Gilbert, 436
white mustard seedlings, effect of light on growth, 379